Algae in Diabetes Management

Rathinam Raja
Shanmugam Hemaiswarya
Muthusamy Sanjivkumar
Tamil Selvan Silambarasan • Grasian Immanuel
Editors

Algae in Diabetes Management

Therapeutic Properties and Applications

Springer

Editors
Rathinam Raja
Centre for Integrative Medical Research (CIMR)
Sree Balaji Medical College and Hospital (SBMCH)
Chennai, Tamil Nadu, India

Muthusamy Sanjivkumar
K.R. College of Arts & Science (KRCAS)
Kovilpatti, Tamil Nadu, India

Grasian Immanuel
Centre for Marine Science & Technology
Manonmaniam Sundaranar University
Rajakkamangalam, Tamil Nadu, India

Shanmugam Hemaiswarya
Centre for Global Health Research,
Saveetha Medical College and Hospital
Saveetha Institute of Medical and Technical Sciences, Saveetha University
Chennai, Tamil Nadu, India

Tamil Selvan Silambarasan
Center for Global Health Research
Saveetha Medical College and Hospitals
Chennai, Tamil Nadu, India

ISBN 978-981-95-0570-8 ISBN 978-981-95-0571-5 (eBook)
https://doi.org/10.1007/978-981-95-0571-5

This Springer imprint is published by the registered company Springer Nature Singapore Pte Ltd.
The registered company address is: 152 Beach Road, #21-01/04 Gateway East, Singapore 189721, Singapore

To my father A. Rathinam

Preface

A chronic metabolic disorder, diabetes mellitus causes a major global health challenge, necessitating innovative and sustainable approaches to its management. *Algae in Diabetes Management: Therapeutic Properties and Applications* explores the untapped potential of algae bioactive-rich resource in combating Diabetes. Drawing on recent advances in phycology, biotechnology, and nutritional science, this book offers a comprehensive analysis of the antidiabetic compounds found in various macro- and microalgal species, their mechanisms of action, and their clinical relevance. Bridging the gap between traditional knowledge and modern research aims to provide researchers, healthcare professionals, and students with important insights into algae-based therapeutic strategies that are effective for mankind and also environmentally sustainable.

Chennai, India	Rathinam Raja
Chennai, India	Shanmugam Hemaiswarya
Chennai, India	Muthusamy Sanjivkumar
Kovilpatti, India	Tamil Selvan Silambarasan
Rajakkamangalam, India	Grasian Immanuel

Acknowledgements

This book, *Algae in Diabetes Management: Therapeutic Properties and Applications*, is the result of the collective efforts of contributors to whom we are deeply grateful. We extend our heartfelt thanks to the research community and institutions dedicated to the study of phycology and metabolic disorders, whose groundbreaking work laid the foundation for this compilation. We express our sincere gratitude to the publishers for their professional guidance throughout the publication process.

Rathinam Raja
Shanmugam Hemaiswarya
Muthusamy Sanjivkumar
Tamil Selvan Silambarasan
Grasian Immanuel

Contents

Editors and Contributors

About the Editors

Rathinam Raja pursued his Ph.D. in Microalgae at the University of Madras in 2004. He did his Post-doctoral Fellowships in Taiwan, South Korea, and IIT Madras, and worked for couple of years as a Lead Scientist at Aquatic Energy LLC, Louisiana. For more than a decade, he worked as an Auxiliary Investigator at the University of the Algarve, Portugal. Currently, he is a Professor and Director of the Centre for Integrative Medical Research (CIMR) at Sree Balaji Medical College and Hospital (SBMCH), Chennai, and is supervising eight Ph.D. scholars. Dr. Raja has 47 international and 2 national publications, 11 patents, 16 books, and 18 book chapters. He has organized three training programs and two national conferences. Being a scientist, he has presented his research articles at several international conferences/symposiums in Belgium, Denmark, Germany, Hungary, Morocco, New Zealand, Portugal, Spain, South Korea, Taiwan, the Netherlands, and the USA.

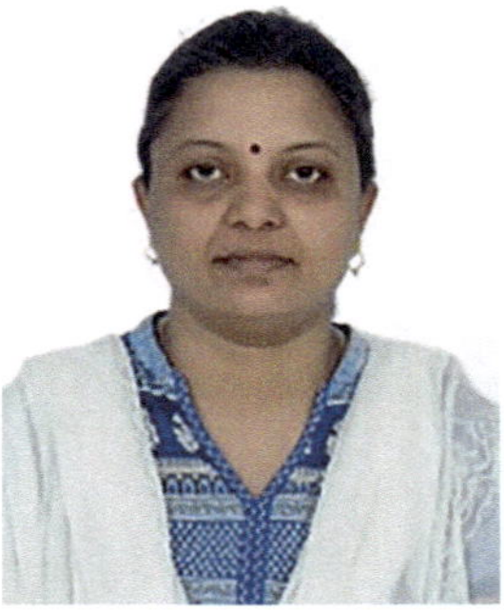

Shanmugam Hemaiswarya received her Ph.D. from the Department of Biotechnology, IIT Madras (India), followed by research experience at the University of Algarve, Portugal, for couple of years. She received a three-year Women Scientist (WOS-A) fellowship from the Department of Science and Technology, India, and completed the project at AUKBC Research Centre, Anna University, Chennai. She also worked as a Project Officer at the Department of Biotechnology, IIT Madras and as a Scientist at IIT M Research Park, Taramani, Chennai. Currently she is working as a Senior Scientist at Fib-Sol Life Technologies Pvt. Ltd., Kottivakkam, Chennai. Dr. Hemaiswarya has published 35 international articles and 13 book chapters, co-authored 6 books, and holds 4 patents. She has delivered her research findings at Germany, Spain, the UK, and the USA.

Muthusamy Sanjivkumar is currently working as an Assistant Professor of Microbiology at K.R. College of Arts and Science, K.R. Nagar, Kovilpatti, Tamil Nadu, India. He received his Ph.D. in Microbiology-Marine Biotechnology (Interdisciplinary) from the Centre for Marine Science and Technology, Manonmaniam Sundaranar University, Tirunelveli, in 2018, and he qualified for the ICAR-NET exam in 2018 as well as 2019. He has expertise in Microbial-active compounds, mangrove microbiome, and actinobacterial research. He has published 38 research articles in peer-reviewed international journals with a cumulative impact factor of 77 and an H index of 13 and has also published 21 book chapters. He has edited a book, *Mangrove Microbiome: Diversity and Bioprospecting* published by Springer Publishers. Dr. Sanjivkumar has teaching experience of 10 years and 11 months and has supervised 20 M.Sc. students for their projects. He has served as an Editorial Board Member and Reviewer of 15 international journals.

Tamil Selvan Silambarasan is currently serving as an Assistant Professor at the Helix Research Center, Department of General Medicine, Saveetha Medical College and Hospital, part of the Saveetha Institute of Medical and Technical Sciences (SIMATS), Saveetha University, Chennai, Tamil Nadu, India. He received his Ph.D. in Microbiology-Biotechnology (Interdisciplinary) in 2017. A dedicated researcher and academician, Dr. Silambarasan has published over 65 research articles in peer-reviewed international journals and contributed 15 book chapters. He has successfully completed 10 intramural research projects and has more than 5 years of teaching experience. Dr. Silambarasan's research interests are in Algal Biotechnology, Enzyme Technology, and Drug Discovery. His contributions to the scientific community extend beyond research, as he serves as a Guest Editor for the Industrial Biotechnology section of *Frontiers in Bioengineering and Biotechnology*. Dr. Silambarasan is also a member of the Editorial Board for several reputed journals, including *Discovery Applied Sciences*, *Discovery Biotechnology*, *Discovery Sustainability*, and *Scientific Reports*. Additionally, he actively contributes as a reviewer for numerous international journals.

Grasian Immanuel is working as an Associate Professor and Director of the Centre for Marine Science and Technology (CMST) and Post Graduate Extension Centre, Manonmaniam Sundaranar University, Kanyakumari District, Tamil Nadu, India. He pursued his Ph.D. (Marine Biotechnology) at CMST, Manonmaniam Sundaranar University in 1997 and Post-doctoral Research at the Department of Biology, SPAS, University of South Pacific, Suva, Republic of Fiji in 2002. He has served as Principal Investigator and completed 14 funded research projects with >3.50 crores Indian rupees received from the Federal Government of India and International agencies. He has sound expertise in Marine Food Processing and Marine Natural Products Production and Applications, and has 32 years of research and 27 years of teaching experience. He has published 215 national and international articles with a cumulative impact factor of 276.42 (Scopus), 29 proceedings, 26 book chapters, 5 books

(Co-editor), 1 monograph, and 68 NCBI gene bank microbial accessions. He received the DST Young Scientist Fellowship 2005; the Academic Excellence Award from Manonmaniam Sundaranar University (2009 and 2024); Shri. P.K. Das Memorial Best Faculty Award (2011-12); Eminent Scientist award-2023, presented by the Society for Advancement of Biological Sciences, ICER, Bangalore; Top 2% position of World Ranking Scientist List in Fisheries/Biotechnology/Agriculture (2020-24) by Stanford University; and World Scientist and University Ranking 2024 in Fisheries: #1 in Manonmaniam Sundaranar University, #5 in India, # 95 in Asia and in World (AD Scientific Index). Dr. Immanuel has organized and co-organized 10 training programs, 3 conferences, 3 workshops as organizer/co-organizer, and one international special symposium as convener. He has guided 102 M.Phil. and 28 Ph.D. scholars.

Contributors

Priya Josson Akkara Department of Life Sciences, School of Biological and Forensic Sciences, Kristu Jayanti (Deemed to be University), Bangalore, India

M. Sonia Angeline Department of Life Sciences, School of Biological and Forensic Sciences, Kristu Jayanti (Deemed to be University), Bengaluru, India

J. R. Anusha Centre for Marine Science & Technology, Manonmaniam Sundaranar University, Kanyakumari District, Tamil Nadu, India

M. V. Arularasu Sustainable Energy and Environment Research Unit, Department of Physiology, Saveetha Medical College, Saveetha Institute of Medical and Technical Sciences, Saveetha University, Chennai, India

Neethu Asokan Department of Life Sciences, Sri Satya Sai University for Human Excellence, Kalaburagi, Karnataka, India

R. Babujanarthanam Department of Biotechnology, Thiruvalluvar University, Vellore, Tamil Nadu, India

Kulvinder Bajwa University School for Graduates Studies, Chaudhary Devi Lal University, Sirsa, India

S. Bharathi Centre for Modern Organic Agriculture Research (CMOAR), Sathyabama Institute of Science and Technology, Chennai, Tamil Nadu, India

Prasannakumar Chinnamani Department of Life Sciences, School of Biological and Forensic Sciences, Kristu Jayanti (Deemed-to-be University), Bengaluru, Karnataka, India

T. Citarasu Centre for Marine Science & Technology, Manonmaniam Sundaranar University, Kanyakumari District, Tamil Nadu, India

Deepak Paramasivam Department of Life Sciences, School of Biological and Forensic Sciences, Kristu Jayanti (Deemed-to-be University), Bengaluru, Karnataka, India

Mariany Costa Deprá Rural Sciences Center, Federal University of Santa Maria, Santa Maria, RS, Brazil

Rosangela Rodrigues Dias Rural Sciences Center, Federal University of Santa Maria, Santa Maria, RS, Brazil

Darissa Alves Dutra Rural Sciences Center, Federal University of Santa Maria, Santa Maria, RS, Brazil

Kranti Kiran Reddy Ealla Department of Oral and Maxillofacial Pathology, Malla Reddy Institute of Dental Sciences, Malla Reddy Vishwavidyapeeth, Hyderabad, India

Jayada Gallege School of Health Sciences, Queen Margaret University, Edinburgh, UK

Govindaraj A. Shivaranjani Molecular Evolution Laboratory, Department of Microbiology, School of Life Sciences, Bharathidasan University, Tiruchirappalli, Tamil Nadu, India

Priya Gupta Amity Institute of Biotechnology, Amity University Chhattisgarh, Raipur, Chhattisgarh, India

Grasian Immanuel Centre for Marine Science and Technology (CMST), Manonmaniam Sundaranar University, Rajakkamangalam, Nagercoil, Tamil Nadu, India

S. Janani Iswarya Department of Biotechnology, Thiruvalluvar University, Vellore, India

Eduardo Jacob-Lopes Rural Sciences Center, Federal University of Santa Maria, Santa Maria, RS, Brazil

Ankita Jain Tilak College of Science and Commerce Affiliated to University of Mumbai, Mumbai, India

Vairavanathan Jayavarshini Department of Biotechnology, Karpagam Academy of Higher Education, Coimbatore, India

Department of Biotechnology, Dr. N.G.P. Arts and Science College, Coimbatore, India

Sangeetha P. Kademani Department of Zoology, Maris Stella College, Benz Circle, Vijayawada, Andhra Pradesh, India

K. M. G. M. M. Kariyawasam Department of Biosystems Technology, Faculty of Technological Studies, Uva Wellassa University of Sri Lanka, Badulla, Sri Lanka

Department of Food Science and Biotechnology of Animal Resources, Konkuk University, Seoul, South Korea

K. M. G. R. M. Kariyawasam Department of Food Science, Obihiro University of Agriculture and Veterinary Medicine, Hokkaidao, Japan

Manikandan Kathirvel Department of Life Sciences, School of Biological and Forensic Sciences, Kristu Jayanti (Deemed-to-be University), Bengaluru, Karnataka, India

P. Madhan Kumar Department of Biotechnology, Thiruvalluvar University, Vellore, Tamil Nadu, India

J. Lesitha Jeeva Kumari Department of Biotechnology, Marudhar Kesari Jain College for Women, Thirupathur, Tamil Nadu, India

Neema Kumari Department of Microbiology, Malla Reddy Institute of Medical Sciences, Malla Reddy Vishwavidyapeeth, Hyderabad, India

Richard Luan Silva Machado Rural Sciences Center, Federal University of Santa Maria, Santa Maria, RS, Brazil

Manon Mani Vellingiri Department of Life Sciences, School of Biological and Forensic Sciences, Kristu Jayanti (Deemed-to-be University), Bengaluru, Karnataka, India

Udari Menaka Department of Food Science and Technology, Faculty of Applied Sciences, University of Sri Jayewardenepura, Nugegoda, Sri Lanka

Faculty of Graduate Studies, University of Sri Jayewardenepura, Nugegoda, Sri Lanka

Gangatharan Muralitharan Molecular Evolution Laboratory, Department of Microbiology, School of Life Sciences, Bharathidasan University, Tiruchirappalli, Tamil Nadu, India

National Repository for Microalgae and Cyanobacteria (NRMC – F & M), Bharathidasan University, Tiruchirappalli, Tamil Nadu, India

Nathiya Thiyagarajulu Department of Life Sciences, School of Biological and Forensic Sciences, Kristu Jayanti (Deemed-to-be University), Bengaluru, Karnataka, India

Prabhudas Nelaturi Department of Medical Biotechnology, Aarupadai Veedu Medical College and Hospital, Vinayaka Mission's Research Foundation (Deemed to Be University), Kirumampakkam, Puducherry, India

S. Jeraldin Nisha Centre for Marine Science & Technology, Manonmaniam Sundaranar University, Kanyakumari District, Tamil Nadu, India

R. Parameswari Department of Biotechnology, Thiruvalluvar University, Vellore, Tamil Nadu, India

S. Azhagu Pavithra Department of Biotechnology, Thiruvalluvar University, Vellore, India

Samiya Pawne Tilak College of Science and Commerce Affiliated to University of Mumbai, Mumbai, India

Dhamodharan Prabhu Department of Biotechnology, Karpagam Academy of Higher Education, Coimbatore, India

Centre for Bioinformatics, Karpagam Academy of Higher Education, Coimbatore, India

S. Prakash Marine Biotechnology Research Laboratory, Department of Basic Sciences, Institute of Fisheries Post Graduate Studies (IFPGS), Tamil Nadu Dr. J. Jayalalithaa Fisheries University, Vaniyanchavadi, Chennai, Tamil Nadu, India

Prasannakumar Chinnamani Department of Life Sciences, School of Biological and Forensic Sciences, Kristu Jayanti (Deemed-to-be University), Bengaluru, Karnataka, India

Sundararaj Rajamanikandan School of Pharmaceutical Science & Technology, Tianjin University, Tianjin, People's Republic of China

Center for Bioinformatics, Karpagam Academy of Higher Education, Coimbatore, India

Deepa Ramachandran Department of Life Sciences, School of Biological and Forensic Sciences, Kristu Jayanti (Deemed-to-be University), Bengaluru, India

R. Ramasubburayan Centre for Marine and Aquatic Research, Department of Prosthodontics, Saveetha Dental College and Hospitals, SIMAS, Saveetha University, Chennai, Tamil Nadu, India

D. Rameshbabu Centre for Marine Science & Technology, Manonmaniam Sundaranar University, Kanyakumari District, Tamil Nadu, India

Shiva Ranjani Molecular Evolution Laboratory, Department of Microbiology, School of Life Sciences, Bharathidasan University, Tiruchirappalli, Tamil Nadu, India

National Repository for Microalgae and Cyanobacteria (NRMC – F & M), Bharathidasan University, Tiruchirappalli, Tamil Nadu, India

G. Rathnayake Department of Biochemistry, Faculty of Medicine, Uva Wellassa University of Sri Lanka, Badulla, Sri Lanka

Subaranjani Ravichandiran Center for Global Health Research, Saveetha Medical College & Hospital, Saveetha Institute of Medical and Technical Sciences, Chennai, Tamil Nadu, India

Baby Anusha Satravada Computational Oncology, Memorial Sloan Kettering Cancer Centre, New York, NY, USA

Adriane Terezinha Schneider Rural Sciences Center, Federal University of Santa Maria, Santa Maria, RS, Brazil

T. M. Shashanka Department of Biochemistry, Chikkamagaluru Institute of Medical Sciences, Tegur Village, Chikkamagaluru, Karnataka, India

K. N. Shashidhar Department of Biochemistry, Sri DevarajUrs Medical College, Sri DevarajUrs Academy of Higher Education and Research, Tamaka, Kolar, Karnataka, India

R. Esther Shoba Department of Life Sciences, School of Biological and Forensic Sciences, Kristu Jayanti (Deemed to be University), Bangalore, India

Tamil Selvan Silambarasan Helix Research Center, Department of General Medicine, Saveetha Medical College, Saveetha Institute of Medical and Technical Science, Saveetha University, Chennai, India

Nalini Soni Amity Institute of Biotechnology, Amity University Chhattisgarh, Raipur, Chhattisgarh, India

Yolin Angel P. A. Soundara Rajan Centre for Nanoscience and Nanotechnology, Sathyabama Institute of Science and Technology, Chennai, Tamil Nadu, India

Saummi Uduwana Department of Food Science and Technology, Faculty of Applied Sciences, University of Sri Jayewardenepura, Nugegoda, Sri Lanka

G. Uma Centre for Marine Science & Technology, Manonmaniam Sundaranar University, Kanyakumari District, Tamil Nadu, India

Balasubramanian Velramar Amity Institute of Biotechnology, Amity University Chhattisgarh, Raipur, Chhattisgarh, India

Jayachandran Venkatesan School of Dental Sciences and Technology, Department of Biomaterials and Transplant Sciences, Malla Reddy Vishwavidyapeeth, Hyderabad, Telangana, India

P. Vinitha Advanced Materials, Energy and Environmental Research Lab, Department of Community Medicine, Saveetha Medical College, Saveetha Institute of Medical and Technical Sciences, Chennai, Tamil Nadu, India

Yajushi Yashitha Vinnakota Department of Life Sciences, School of Biological and Forensic Sciences, Kristu Jayanti (Deemed to be University), Bangalore, India

Indira Wickramasinghe Department of Food Science and Technology, Faculty of Applied Sciences, University of Sri Jayewardenepura, Nugegoda, Sri Lanka

T. Yogesh Department of Biotechnology, Thiruvalluvar University, Vellore, India

Leila Queiroz Zepka Rural Sciences Center, Federal University of Santa Maria, Santa Maria, RS, Brazil

Chapter 1
Mechanism of Action and Signaling Pathways of Diabetes

Prabhudas Nelaturi, Sangeetha P. Kademani, Tamil Selvam Silambarasan, T. M. Shashanka, M. V. Arularasu, and K. N. Shashidhar

Abstract Diabetes mellitus (DM) is a chronic disorder considered as sustained hyperglycemia, affected by insufficient insulin production and/or action, leading to progressive organ damage. It primarily presents as type 1 (T1DM) or type 2 (T2DM). T1DM is an autoimmune-related disease resulting in beta-cell destruction as well as complete insulin deficiency, while T2DM involves insulin resistance and deficiency in insulin associated with lifestyle changes. Emerging research suggests that bioactive compounds derived from *algae*, including phlorotannins, polyphenols, and carotenoids, possess antioxidant, anti-inflammatory, and insulin-sensitizing effects. These bioactives appear to enhance key insulin signaling pathways, notably PI3K/Akt and AMPK, and reduce inflammation, offering protective benefits to β-cells. Species such as *Spirulina platensis* and *Ulva lactuca* contain molecules that support glucose homeostasis, mitigate oxidative stress, and improve lipid profiles, position-

P. Nelaturi (✉)
Department of Medical Biotechnology, Aarupadai Veedu Medical College and Hospital, Vinayaka Mission's Research Foundation (Deemed to Be University), Puducherry, India
e-mail: prabhudasmahe2@gmail.com

S. P. Kademani
Department of Zoology, Maris Stella College, Benz Circle, Vijayawada, Andhra Pradesh, India

T. S. Silambarasan
Helix Research Center, Department of General Medicine, Saveetha Medical College, Saveetha Institute of Medical and Technical Sciences, Saveetha University, Chennai, India

T. M. Shashanka
Department of Biochemistry, Chikkamagaluru Institute of Medical Sciences, Tegur Village, Chikkamagaluru, Karnataka, India

M. V. Arularasu
Sustainable Energy and Environment Research Unit, Department of Physiology, Saveetha Medical College, Saveetha Institute of Medical and Technical Sciences, Saveetha University, Chennai, India

K. N. Shashidhar
Department of Biochemistry, Sri DevarajUrs Medical College, Sri DevarajUrs Academy of Higher Education and Research, Tamaka, Kolar, Karnataka, India

R. Raja et al. (eds.), *Algae in Diabetes Management*,
https://doi.org/10.1007/978-981-95-0571-5_1

ing *algae*-based bioactives as promising adjuncts in diabetes management. These natural compounds could help alleviate hyperglycemia-induced tissue damage and preserve pancreatic function. This chapter highlights the mechanisms through which *algae*-derived bioactives combat oxidative stress, inflammation, and disrupted insulin signaling in diabetes. Advances in precision nutrition further underscore the potential of these compounds to improve metabolic health as well as quality of life in patients with DM.

Keywords Diabetes mellitus · Algae · Insulin therapy · Molecular pathway · Diabetes management

1 Introduction

Diabetes mellitus (DM) is a metabolic disorder indicated by persistent hyperglycemia due to reduced insulin production, action, or both are leading to progressive damage in organs such as the nerves, eyes, kidneys, heart, and blood vessels. Symptoms comprise of polydipsia (excessive thirst), polyphagia (increased hunger), polyuria (frequent urination), weight loss, and blurred vision. The World Health Organization classifies DM into two main categories: type 1(T1DM) and type 2 (T2DM). Without treatment, diabetes can escalate to severe complications, including diabetic ketoacidosis, particularly in type 1 diabetes (T1DM). T1DM, an autoimmune disorder, results from the immune-mediated destruction of pancreatic β-cells, leading to an absolute insulin deficiency and necessitating lifelong insulin therapy (Cheng et al. 2023). Type 2 diabetes (T2DM), the most common form of diabetes, develops due to insulin resistance and a relative deficiency of insulin. These conditions are associated with obesity, physical inactivity, and genetic factors. Management of T2DM typically begins with lifestyle modifications and oral hypoglycemic agents but may progress to insulin therapy over time. Other rare forms of diabetes include genetic variations affecting β-cell function, such as maturity-onset diabetes of the young and insulin receptor mutations. The pathogenesis of DM is complex, involving mechanisms such as oxidative stress, chronic inflammation, and metabolic dysfunction.

Increased glucose levels and fatty acids trigger pathways like glycolysis, oxidative phosphorylation, and β-oxidation, leading to the production of byproducts, including diacylglycerol and methylglyoxal (Yameny 2024). These byproducts activate protein kinase C (PKC) and/or pro-inflammatory nuclear factor-kappa B (NF-κB), disrupting cellular signaling and promoting inflammation, especially in endothelial cells and podocytes. Persistent metabolic stress damages cellular structures, causing microvascular complications including retinopathy, neuropathy, nephropathy, and atherosclerosis. Protective mechanisms against diabetes-related oxidative damage include cellular antioxidant defense and insulin-related signaling pathways (Deshpande et al. 2008). The phosphoinositide 3 kinase (PI3K)/Akt pathway is crucial for glucose uptake and is often impaired in DM leading to reduced cellular glucose intake. Natural antioxidants—especially those derived from algae, such asphlorotannins, polyphenols, and carotenoids—can counter oxidative stress, protect β-cell function, and improve insulin sensitivity (Maffei et al. 2018). These *algae*-derived bioactives target key molecular pathways, supporting glucose homeostasis and lipid metabolism while reducing inflammation and cellular stress. Compounds from algae such as *Spirulina platensis, Ulva lactuca*, and *Sargassum* have shown promise in diabetes management. They exert anti-inflammatory, antioxidant, and insulin-mimetic actions, for the improved glucose uptake by activating AMP-activated protein kinase(AMPK) and facilitating glucose transporter-4 (GLUT4) translocation to cell membranes. Furthermore, these bioactives may reduce tissue damage associated with hyperglycemia, preserve pancreatic β-cells, and thereby sustain endogenous insulin production (Pereira and Valado 2023). This chapter highlights the potential of algae-based bioactives in diabetes care, focusing on their mechanisms to counter hyperglycemia-induced oxidative stress, inflammation, and impaired insulin-related signaling pathways, such as PI3K/Akt and NF-κB. By supporting glycemic control and reducing inflammation, *algae*-derived bioactives offer promising natural therapeutic options. Advances in precision nutrition further enhance these strategies, paving the way for improved metabolic health and quality of life in individuals with diabetes.

2 Algae

Algae, specifically marine macroalgae, are recognized for their potential health benefits, particularly in the management of chronic diseases like diabetes. These seaweeds, commonly found in coastal areas, are rich in bioactive compounds that offer unique properties due to their adaptation to saltwater environment. Unlike terrestrial plants, *algae* have no true roots, stems, or leaves, yet they produce a range of beneficial compounds as secondary metabolites, which are highly effective in health applications. *Algae* are classified as three main groups as per pigmentation: red *algae* (*Rhodophyta*), green *algae* (*Chlorophyta*), and brown *algae* (*Phaeophyceae*). Among these, brown *algae* contain the highest levels of phenolic compounds and polysaccharides, which exhibit strong antioxidant, anti-inflammatory, and antidiabetic properties (Abo-Shady et al. 2023). This pigment-based diversity in *algae* enhances their therapeutic efficacy, making them valuable in pharmaceutical and nutraceutical fields. The bioactive compounds found in *algae* include phlorotannins, fucoxanthin, and polysaccharides like fucoidan, each of which has been studied for its health benefits, particularly in glycemic control. Phlorotannins, exclusive to brown *algae*, are marine polyphenols synthesized through the acetate-malonate pathway and are known for their potent antioxidant and anti-diabetic effects (Abo-Shady et al. 2023; Ahmed et al. 2024). They are especially effective in inhibiting enzymes such as alpha-glucosidase and alpha-amylase that are involved in carbohydrate metabolism and glucose release. By inhibiting these enzymes, phlorotannins help reduce postprandial hyperglycemia, which is crucial for managing blood sugar levels in diabetes. This makes brown *algae* a valuable resource for developing alternative therapies for type 2 diabetes, where enzyme inhibition can play a significant role in controlling blood glucose. Additionally, brown *algae* contain fucosterol, a bioactive sterol that supports lipid metabolism and has shown potential in reducing blood glucose levels.

Fucoxanthin, another pigment found in brown algae, has demonstrated antioxidant and antidiabetic effects. Studies have shown that fucoxanthin can promote the release of adiponectin, a hormone involved in glucose regulation and lipid metabolism, thus helping to prevent metabolic disorder (Sharifuddin et al. 2015; Lee et al. 2009). Fucoidan, a sulfated polysaccharide, is also abundant in brown algae and exhibits anti-inflammatory, antiviral, and antidiabetic activities. Research suggests that fucoidan helps protect pancreatic beta-cells that are responsible for insulin production, and may improve insulin-sensitivity, offering a potential avenue for diabetes management. Red and green algae also provide therapeutic benefits, though their bioactive compounds differ slightly. Red algae contain bromophenols, which have demonstrated alpha-glucosidase inhibition, and are known for their high levels of dietary fibers that support blood sugar regulation. These fibers slow down carbohydrate absorption, reducing postprandial glucose spikes. Green algae, particularly species like *Ulva lactuca* and *Ulva reticulata*, contain soluble fibers and have been studied for their hypoglycemic activity (Lee et al. 2009). Extracts from these species have shown enzyme-inhibiting properties, suggesting they could aid in lowering blood glucose levels following meals. The potential of *algae*-derived bioactive compounds has driven interest in their use for diabetes management, particularly as an alternative to synthetic drugs that often carry adverse side effects. Incorporating edible *algae* into the diet or isolating their active compounds for pharmaceutical use could provide a natural, sustainable option for managing diabetes and reducing reliance on synthetic medications. As research findings advance, algae's role in health and medicine is likely to expand, further establishing them as valuable resources for combating oxidative stress and regulating metabolic processes essential for managing DM (Pereira and Valado 2023).

3 Types of Diabetes

Type 1 diabetes mellitus (T1DM) is defined as the targeted death of pancreatic beta cells, resulting in a severe insulin lack. This defect causes chronic hyperglycemia, requiring lifetime insulin treatment to maintain glucose homeostasis. Bioactive substances produced from *algae* are now gaining emphasis as prospective supplementary therapy in T1DM management due to their unique anti-inflammatory, antioxidant, and insulin-mimetic capabilities. Marine algae, in particular, are rich in compounds such as polyphenols, carotenoids, phlorotannins, and sulfated polysaccharides, which can modulate cellular pathways implicated in T1DM pathogenesis (Rahman et al. 2022; Animish and Jayasri 2023). The autoimmune aspect of T1DM is associated with heightened oxidative stress and chronic inflammation, largely driven by pathways like NF-κB that intensify the immune response against beta cells. Algal compounds like phlorotannins and fucoidans, especially from brown algae, inhibit NF-κB activity, thereby reducing inflammation by downregulated expression of pro-inflammatory cytokines namely IL-6 and TNF-α. Decreased levels of pro-inflammatory cytokines may modulate the autoimmune response, reducing cellular damage in beta cells. Furthermore, *algae*-derived antioxidants effectively neutralize reactive oxygen species(ROS), which are concerned in beta-cell apoptosis and/or exacerbate the inflammatory cascade in T1DM. By attenuating ROS levels, these antioxidants may help preserve remaining beta-cell function and slow the immune-mediated destruction (Toren et al. 2021). Beyond anti-inflammatory and antioxidant actions, certain *algae*-derived polysaccharides exhibit insulin-mimetic effects, particularly through activation of the AMP-activated protein kinase (AMPK) pathway, a key controller of cellular homeostasis. AMPK activation facilitates glucose uptake and improves lipid metabolism, supporting better glycemic control. Carotenoids like fucoxanthin further assist by enhancing lipid metabolism, which may reduce insulin resistance and improve cellular energy utilization, indirectly supporting glucose management. Additionally, some algal bioactives, such as phlorotannins and polyphenols, have demonstrated protective effects on beta cells by activating survival pathways like PI3K/Akt (Fig. 1.1). This pathway promotes cellular resilience against apoptosis, offering a crucial benefit in T1DM, where beta-cell regeneration is limited, and preserving existing cells can mitigate further autoimmune damage (Pereira and Valado 2023). While insulin therapy remains essential for T1DM, *algae*-derived bioactives can be a complementary approach by addressing underlying oxidative stress, inflammation, and impaired cellular metabolism. Although current evidence largely stems from in vitro and animal studies, it suggests that *algae* could play a valuable role in comprehensive diabetes management. Clinical studies are necessary to confirm these benefits in human T1DM

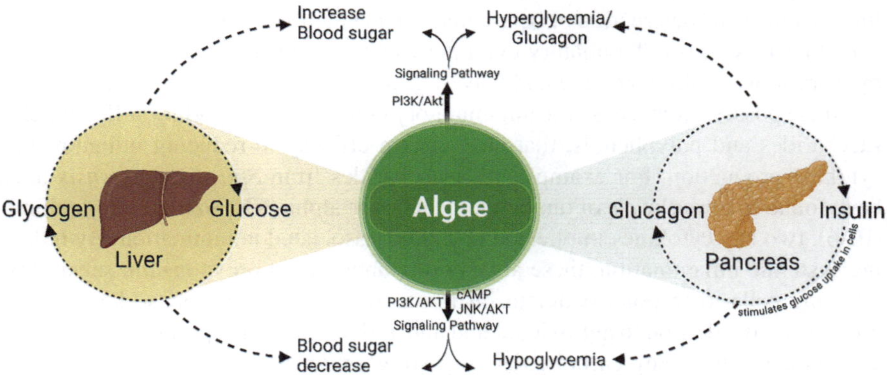

Fig. 1.1 Schematic representation of role of algae and impaired insulin receptor signaling in diabetes, highlighting the PI3K-Akt and MAPK pathways that regulate glucose uptake and cellular health. Chronic inflammation, lipolysis in adipose tissue, and β-cell dysfunction disrupt metabolic balance, leading to insulin resistance and hyperglycemia

patients, potentially establishing *algae*-based bioactives as a natural adjunct to conventional treatment strategies (Behl et al. 2022).

Type 2 diabetes mellitus (T2DM) is a progressive dysfunction of insulin-producing pancreatic beta cells. This impairment hinders the body's ability to regulate blood glucose, resulting in prolonged hyperglycemia. The management of T2DM focuses on maintaining blood glucose levels and preventing complications associated with oxidative stress, lipid dysregulation, and inflammation. Recent studies highlight *algae* as promising natural sources of bioactive compounds with anti-diabetic properties, offering alternative or complementary support to traditional treatments. Algal bioactive molecules, including polysaccharides, phenolic compounds, carotenoids, and peptides, exhibit various therapeutic effects by addressing key T2DM pathophysiological pathways such as oxidative stress, inflammation, and enzymatic inhibition of glucose digestion (Bhatti et al. 2022; Truong et al. 2023). Oxidative stress is a condition characterized by excessive ROS relative to the body's antioxidant defenses is a critical factor in T2DM progression. ROS molecules are highly reactive which can damage cellular components including lipids, proteins, and DNA, contributing to insulin resistance and beta-cell dysfunction. This oxidative burden is further intensified by hyperglycemia, as glucose metabolism generates additional ROS.

Algae-derived compounds, particularly phenolics, carotenoids, and phlorotannins, are potent antioxidants that help neutralize ROS and strengthen the body's antioxidant defense systems. For instance, carotenoids in *algae* such as *Ulva lactuca, Dictyopteris polypodioides,* and *Spirulina* have demonstrated scavenging activity of free radicals, reducing oxidative damage to pancreatic cells (Truong et al. 2023). Additionally, phlorotannins from brown algae are particularly effective in mitigating oxidative stress by inhibiting the generation of ROS, preserving cellular integrity, and potentially protecting beta cells from damage. By lowering oxidative stress, these algal antioxidants may slow T2DM progression and reduce the risk of complications. Inflammation is another hallmark of T2DM, closely related to insulin resistance and destruction of beta-cells. Persistent inflammation triggers immune cells to release pro-inflammatory cytokines, which contribute to insulin resistance by disrupting insulin signaling pathways(Duan et al. n.d.; Besednova et al. 2022).

Algae possess numerous anti-inflammatory compounds, including sulfated polysaccharides and polyphenols, that have shown efficacy in reducing inflammatory cytokine production. For example, polysaccharides from *Spirulina platensis* have been found to lower levels of tumor necrosis factor-alpha (TNF-α) and interleukin-6 (IL-6), two key cytokines implicated in T2DM-associated inflammation. By reducing systemic inflammation, these algal compounds can improve insulin sensitivity, allowing cells to respond better to insulin to regulate blood glucose levels more effectively. In addition to antioxidant and anti-inflammatory properties, certain algal compounds inhibit key carbohydrate-digesting enzymes, which can reduce post-prandial (after-meal) blood glucose spikes, a significant concern for T2DM patients (Matin et al. 2024). The enzymes α-amylase and α-glucosidase are accountable for breaking down complex carbohydrates into simple energy molecules, making them more readily absorbable. By inhibiting these enzymes, algal polysaccharides slow carbohydrate digestion, thereby controlling the release of glucose into the bloodstream.

Studies have demonstrated that sulfated polysaccharides, commonly found in brown algae like *Fucus vesiculosus*, exhibit potent α-glucosidase and α-amylase inhibitory activities. This enzymatic inhibition helps to lower blood glucose levels, improving postprandial glucose control, which is crucial for preventing glucose-induced oxidative stress and reducing the demand on pancreatic beta cells. Furthermore, T2DM is often accompanied by dyslipidemia, characterized by elevated triglycerides and cholesterol levels, which increase the risk of cardiovascular diseases (Siddiqui et al. 2024). Algae-derived compounds contribute to lipid regulation, another pathway beneficial for T2DM management. Certain brown and green algae contain lipophilic compounds that modulate lipid metabolism, including the carotenoid fucoxanthin found in *Undaria pinnatifida* and *Laminaria japonica.* Fucoxanthin is specified to reduce levels of triglyceride and improve cholesterol profiles, aiding in the management of dyslipidemia. Additionally, *algae* like *Spirulina* contain gamma-linolenic acid (GLA), an essential fatty acid that positively influences lipid metabolism and may improve insulin sensitivity (Mohibbullah et al. 2022). By regulating lipid levels, algal compounds not only support glucose control but also reduce cardiovascular risks associated with T2DM. Moreover, the potential of algal compounds exhibits extended insulin-signaling and increasing glucose uptake in cells, mechanisms that are often impaired in T2DM. Algal bioactives can activate key pathways like the AMP-activated protein kinase (AMPK) pathway, a central regulator of cellular energy homeostasis. Activated AMPK enhances glucose uptake and lipid metabolism, supporting energy utilization within cells and thereby improving insulin sensitivity.

Studies on green algae, such as *Chlorella vulgaris*, have shown that its extracts can activate AMPK in muscle and liver tissues, facilitate glucose transport across cell membranes, and support glucose homeostasis. This effect can be particularly beneficial for T2DM patients, as enhanced glucose uptake can lower blood glucose levels and reduce the need for insulin, alleviating strain on pancreatic beta cells. The broad spectrum of bioactive compounds in *algae*, including antioxidants, anti-inflammatory agents, enzyme inhibitors, and lipid modulators, illustrates the potential for *algae* as adjuncts in T2DM management (Tong et al. 2024). These compounds collectively address multiple aspects of T2DM pathology, contributing protection of antioxidants, reducing inflammation, inhibiting glucose absorption, and improving good lipid levels. Notably, *algae* have shown efficacy in both in vitro and animal studies, suggesting promising therapeutic potential. Although clinical trials in humans are limited, ongoing research indicates that algal extracts could serve as a complementary approach alongside traditional therapies for T2DM. By improving glycemic control, enhancing insulin sensitivity, and potentially lowering cardiovascular risks, *algae*-derived compounds represent a promising natural solution that targets the complex mechanisms underlying T2DM. Further investigation into dosage, long-term effects, and the integration of *algae*-based supplements into current T2DM treatment regimens will be critical in confirming their therapeutic role (Arabshomali et al. 2023).

4 Signaling Pathway

Protective factors against diabetes complications act by either neutralizing harmful byproducts of metabolic processes or activating alternative mechanisms to safeguard cells. They can directly counteract toxins, like antioxidants neutralizing oxidative byproducts or reactions between high glucose levels and reactive molecules. Excess glucose and fatty acids are metabolized through β-oxidation, oxidative phosphorylation, and glycolysis, often creating toxic byproducts such as diacylglycerol and methylglyoxal, which impair cellular signaling and can activate pathways that lead to cellular stress. These byproducts can activate enzymes like protein kinase C, which influences stress responses, or stimulate nuclear factor kappa-B (NF-KB), promoting inflammation (Yu et al. 2023). Such pathways can contribute to cellular damage and dysfunction, but natural protective mechanisms help to maintain cellular homeostasis and prevent damage to critical cell types, such as podocytes and endothelial cells which play vital role in maintaining cellular integrity. In patients with diabetes, levels of various growth factors and cytokines fluctuate in response to the functional needs of targeted cells and tissues, often as a result of the chronic stress induced by the disease. These factors, such as TGF-β, VEGF, connective tissue growth factor, and PDGF, are commonly elevated during the progression of complications like diabetic retinopathy, diabetic nephropathy, atherosclerosis, and wound healing. However, the changes in their expression are typically a physiological response to hyperglycemia-induced tissue damage. Inflammatory cytokines like TNF-α and IL-6 associated with vascular complications in diabetes, also play protective roles in acute stress situations, such as during exercise or after hypoglycemia. While chronic elevations of these cytokines contribute to detrimental effects on organs, their transient increases can be beneficial in neutralizing the adverse effects of hyperglycemia and mitigating injury to tissues (Zhang et al. 2023).

5 Anti-oxidative Signaling

Endogenous networks of antioxidant and inflammatory response play crucial roles in protecting cells and organs from the harmful effects of oxidative stress, which is particularly relevant in diabetes. Hyperglycemia leads to increased production of ROS through multiple cellular pathways, especially in mitochondria, where the electron transport chain is disrupted, generating superoxide. Elevated glucose levels also promote the formation of advanced glycation end-products (AGEs), which can trigger inflammation and further cellular damage. The body counters this oxidative stress through several antioxidant systems, including small molecules like glutathione, enzymes like catalase and superoxide dismutase, and dietary antioxidants such as vitamins C and E. One of the key systems is the Keap1/Nrf2 pathway that regulates the expression of various antioxidant enzymes in response to cellular stress. Activating Nrf2 has shown promise in experimental treatments for diabetic complications, such as diabetic retinopathy and diabetic nephropathy, through compounds like bardoxolone methyl, though clinical trials have had mixed results (Mallik et al. 2024). Despite the challenges in demonstrating consistent therapeutic efficacy, these endogenous protective networks are essential in managing the oxidative and inflammatory damage caused by diabetes. Antioxidants are essential for protecting the body from oxidative stress, as they neutralize free radicals—unstable molecules that can harm cells and accelerate aging by damaging DNA, proteins, and cell membranes. Although free radicals are produced naturally through metabolism, environmental factors like pollution, smoking, and radiation can increase their levels, amplifying the risk of cellular damage. Antioxidants counteract this damage by donating electrons to stabilize free radicals or breaking the chain of oxidation reactions. While many antioxidants come from fruits and vegetables, certain *algae* are also valuable sources, producing compounds like tocopherols, ascorbic acid, carotenoids, polyphenols, α-lipoic acid, glutathione, phlorotannins, and alkaloids. These natural antioxidants from *algae* act swiftly with free radicals, preventing cellular damage and even extending the shelf life of foods. Unlike synthetic antioxidants—such as butylated hydroxytoluene (BHT), propyl gallate (PG), and tert-butylhydroquinone (TBHQ)—which are widely used in food preservation but raise safety concerns due to potential toxicity and instability, *algae*-based antioxidants are seen as promising, safe alternatives with a lower risk of harmful side effects (Tangvarasittichai 2015). Moreover, *algae* produce these antioxidants as secondary metabolites to protect against environmental stressors, which, when consumed, also provide similar health benefits in humans. Studies suggested that *algae*-derived antioxidants can reduce the risk of chronic diseases such as cardiovascular disease and cancer and may help guard against genetic mutations. Additionally, these natural compounds are being investigated as economical and safe treatments for managing diabetes and protecting organs from oxidative stress-related damage. Beyond health, algae-based antioxidants support food preservation by maintaining nutritional quality and extending shelf life. As research continues, the potential of *algae*-derived antioxidants to enhance long-term health and prevent disease is becoming increasingly apparent, indicating these natural compounds may play a central role in the future of health and food sciences (Pradhan et al. 2021).

6 Peroxisome Proliferator-Activated Receptors

Peroxisome proliferator-activated receptors (PPARs) are essential ligand-activated nuclear receptors that regulate gene expression involved in lipid metabolism, glucose homeostasis, and inflammatory responses. There are three primary PPAR isoforms: PPARα, PPARγ, and PPARδ (or PPARβ). Each isoform plays a distinct role in cellular metabolism, with PPARα largely responsible for fatty acid oxidation in the liver, PPARγ involved in adipogenesis and insulin sensitivity in adipose tissue, and PPARδ broadly affecting lipid metabolism and energy balance in various tissues. PPAR activation has therapeutic potential for managing metabolic disorders, particularly diabetes mellitus, through mechanisms that enhance insulin sensitivity and control lipid levels. In recent years, algae have emerged as a significant source of bioactive compounds that modulate PPAR activity, particularly through oxo-fatty acids (oFAs) derived from algal species like *Chaetoceros karianus* (D'Aniello et al. 2023; Grygiel-Górniak 2014). Oxo-fatty acids such as [(7E)-9-OHE] and [(10E)-9-OHE] exhibit dual agonist activity toward PPARα and PPARγ, stimulating both lipid catabolism and glucose uptake. Research demonstrates that these compounds activate PPAR-related pathways in hepatic and adipose cells, promoting fatty acid oxidation and improving insulin sensitivity. For instance, when human hepatocyte (Huh-7) and adipocyte (SGBS) cells were treated with oFAs, there was a marked upregulation of genes associated with lipid metabolism, including *CPT1A*, *ACSL3*, and *PLIN1*, which are crucial for fatty acid transport and breakdown. Additionally, upregulation of the *ANGPTL4* gene, known for its role in energy regulation, suggests that oFAs from algae could help manage hyperglycemia and insulin resistance associated with diabetes of PPARγ in adipose tissue and its impact on insulin sensitivity has led to its use as a target for managing diabetes (Sæther et al. 2018).

Activation of PPARγ by algal oFAs not only increases the expression of genes involved in glucose uptake, such as *GLUT4*, but also elevates levels of insulin-sensitizing adipokines like *adiponectin* and *leptin*. These effects were observed alongside reduced expression of pro-inflammatory cytokines, including TNF-α, IL-6, and IL-1B, which are typically elevated in metabolic disorders. By decreasing inflammation and enhancing insulin response in adipose tissue, algal-derived PPARγ agonists show promise as therapeutic agents with less side effects compared to synthetic drugs. Furthermore agonists may offer therapeutic benefits beyond diabetes management, potentially impacting cardiovascular health and obesity. Activation of PPARα, particularly by compounds derived from marine algae, improves lipid profiles by promoting fatty acid oxidation in liver, reducing blood triglyceride levels, and mitigating the risk of fatty liver disease. Additionally, by engaging PPARδ pathways, these compounds may enhance energy expenditure, making them potential agents for obesity treatment. The unique bioactivity of algal oFAs in stimulating dual PPAR activation (α and γ) provides a synergistic approach that could address multiple facets of metabolic syndrome, including dyslipidemia, insulin resistance, and inflammation (Dubuisson et al. 2011; Choi et al. 2014). Bicularly oxo-fatty acids represent a promising area of research for PPAR-targeted therapies in metabolic disorders. These compounds demonstrate potent dual agonist activities that stimulate both lipid metabolism and glucose regulation, with beneficial effects on insulin sensitivity and inflammatory response. The potential of *algae*-based PPAR modulators in clinical applications suggests an effective, natural alternative for managing diabetes, obesity, and related metabolic diseases, paving the way for safer, sustainable therapeutic options in metabolic health management.

7 Mitogen-Activated Protein Kinase Pathway

Mitogen-activated protein kinase (MAPK) pathway is a critical cellular signaling mechanism involved in various biological/cellular processes, including inflammation, insulin signaling, and cellular stress responses—all of which are important in the context of diabetes. Algae, with their diverse bioactive compounds, have shown potential in modulating the MAPK pathway, particularly in managing oxidative stress and inflammation associated with diabetes. Microalgae like *Arthrospira platensis* (commonly known as *Spirulina*) have been studied for their effect on the MAPK pathway. This species contains antioxidants, phenolic compounds, and pigments that exert anti-inflammatory and antioxidant effects, which can help modulate MAPK signaling. For instance, studies showed that *A. platensis* inhibits the activation of p38, ERK, and JNK proteins, which are essential components of the MAPK pathway linked to inflammation and insulin resistance (Vaibhav and Anjaneyulu 2016). Downregulation of these proteins, *Spirulina's* bioactive compounds can help protect pancreatic β-cells from oxidative damage, reduce insulin resistance, and enhance glucose uptake. In diabetic models, supplementation with *A. platensis* demonstrated a reduction in markers of oxidative stress and inflammation. Treated diabetic rats showed decreased plasma glucose levels, reduced lipid peroxidation (indicated by lower malondialdehyde levels), and improved antioxidant enzyme activities, including superoxide dismutase (SOD) and catalase. Furthermore, *A. platensis* downregulated pro-apoptotic markers in the MAPK pathway, thereby promoting β-cell survival. This modulation of the MAPK pathway in pancreatic tissues suggests that algal compounds could effectively intervene in diabetes by reducing cellular stress and preserving insulin-secreting cells. Overall, *algae*-derived compounds offer a promising avenue for natural diabetes management by modulating the MAPK pathway, highlighting the therapeutic potential of microalgae in both antioxidant support and insulin sensitivity improvement in diabetic care (Hannan et al. 2020).

8 Insulin Signaling Pathway

The insulin signaling pathway is essential for glucose uptake and regulation, as well as overall metabolic homeostasis. The pathway begins when insulin binds to the insulin receptor on cell surfaces, activating intracellular cascades that facilitate glucose uptake, primarily through the recruitment of GLUT4 to the cell membrane. In T2DM, signaling pathway is disrupted by insulin resistance, characterized by cells' reduced responsiveness to insulin. This impairment in the insulin pathway leads to diminished glucose uptake and chronic hyperglycemia, which can escalate to complications like cardiovascular disease, neuropathy, and kidney damage. *Algae*, with their diverse bioactive compounds, have shown potential to support insulin signaling, offering a natural means of modulating glucose levels and combating insulin resistance in T2DM. Insulin signaling is mediated by two main pathways: the PI3K/

Akt pathway and the MAPK pathway. The PI3K/Akt pathway is particularly critical for glucose metabolism as it triggers *GLUT4* translocation to cell membranes, allowing glucose entry into cells, especially in skeletal muscle and adipose tissues (Petersen and Shulman 2018). However, insulin resistance impedes the activation of PI3K/Akt, reducing *GLUT4* translocation and resulting in lower glucose uptake by cells.

Chronic inflammation and oxidative stress further exacerbate this by disrupting cellular signaling, leading to persistent hyperglycemia. Compounds derived from algae—such as polysaccharides, phlorotannins, carotenoids, and polyphenols—exhibit antioxidant and anti-inflammatory properties that mitigate oxidative stress and inflammation, thereby supporting the insulin signaling pathway and potentially enhancing insulin sensitivity. One of the primary roles *algae* play in modulating insulin signaling is through the regulation of oxidative stress, a major disruptor of insulin activity. Elevated levels of reactive oxygen species (ROS) due to chronic hyperglycemia impair the PI3K/Akt pathway and damage insulin receptors, leading to insulin resistance. *Algae* contain powerful antioxidants like carotenoids, phlorotannins, and polyphenols, which neutralize ROS and help restore cellular redox balance. For example, phlorotannins from brown algae (*Ecklonia cava*) have shown significant free-radical-scavenging activity, protecting insulin receptors from oxidative damage and thereby preserving the integrity of insulin signaling (van Gerwen et al. 2023).

Furthermore, carotenoids like fucoxanthin in brown algae are known to reduce oxidative damage within insulin-responsive tissues, which supports sustained insulin signaling and improves glucose homeostasis. In addition to antioxidant properties, algae-derived compounds exhibit potent anti-inflammatory effects, which are vital for improving insulin signaling in T2DM. Inflammation, especially from elevated levels of pro-inflammatory cytokines like TNF-α and IL-6, interferes with insulin signaling by blocking the phosphorylation of insulin receptor substrates, a critical step in the insulin pathway. Algal polysaccharides, such as those from *Spirulina platensis* and *Fucus vesiculosus,* are indicated to inhibit the release of inflammatory cytokines, reducing inflammation-induced disruptions in the insulin pathway. Lowering inflammation, these compounds support the downstream signaling required for effective *GLUT4* translocation and glucose uptake, enhancing insulin sensitivity in cells. Algal polysaccharides also directly influence the insulin signaling pathway by activating AMP-activated protein kinase (AMPK), which plays a significant role in cellular energy balance and insulin sensitivity (Din et al. 2022).

AMPK activation has been linked to improved glucose uptake and fatty acid oxidation, both of which are beneficial in T2DM management. Sulfated polysaccharides from brown algae, for instance, can activate AMPK, which subsequently supports the PI3K/Akt pathway, promoting *GLUT4* translocation and facilitating glucose entry into cells. This dual pathway activation not only helps in managing glucose levels but also improves lipid metabolism, reducing the lipid accumulation often seen in insulin-resistant tissues. Moreover, *algae*-derived bioactives impact insulin signaling by modulating lipid profiles, which is another factor that influences insulin resistance. High levels of free fatty acids and lipid intermediates in muscle and liver tissues disrupt insulin signaling, particularly by inhibiting the PI3K/Akt pathway. Compounds like fucoxanthin from brown algae can enhance lipid metabolism, reducing triglyceride levels and decreasing the lipid interference in insulin signaling. By improving lipid profiles, algae help restore the sensitivity of insulin receptors and support the effective functioning of the insulin pathway, thus enhancing overall metabolic health. Another significant contribution of algae to insulin signaling is through the maintenance and protection of pancreatic beta cells, which produce insulin (Chung et al. 2021). In T2DM, prolonged exposure to high glucose levels and oxidative stress damages these beta cells, diminishing insulin production. *Algae*-derived polyphenols, particularly those from green algae, have been observed to protect beta cells from apoptosis by reducing oxidative damage and stabilizing cellular structures. This protection not only maintains insulin production but also enables a more consistent insulin response, which is critical for activating the insulin pathway in target tissues. *Algae*-derived compounds offer a multifaceted approach to enhancing the insulin signaling pathway in T2DM. Through antioxidant, anti-inflammatory, and lipid-modulating effects, these bioactives address key disruptors of insulin signaling, restoring cellular responsiveness to insulin (Golovinskaia and Wang 2023). By supporting AMPK and PI3K/Akt activation, *algae* enhance *GLUT4* translocation and glucose uptake, aiding in glycemic control. Additionally, the protective effects on pancreatic beta cells contribute to sustained insulin production. Collectively, these mechanisms position algae as promising natural adjuncts in managing T2DM, providing support for insulin signaling and potentially mitigating the progression of insulin resistance. Further research into the specific bioactive components in *algae* will deepen our understanding of their therapeutic potential, paving the way for new interventions in diabetes management.

9 Phosphoinositide 3-Kinase/Protein Kinase B (PI3K/Akt) Pathway

The phosphoinositide 3-kinase/protein kinase B (PI3K/Akt) pathway is essential in managing cellular glucose homeostasis, insulin signaling, and cell survival, making it highly relevant in T2DM. This pathway is often impaired due to chronic inflammation, oxidative stress, and insulin resistance, contributing to decreased glucose uptake and impaired metabolic regulation. This disruption is further intensified by the production of inflammatory cytokines and oxidative stress, which inhibit the activation of the PI3K/Akt pathway, leading to reduced insulin sensitivity in muscle, liver, and adipose tissues. Bioactive compounds derived from *algae* have shown significant potential in modulating the PI3K/Akt pathway, thereby offering a natural approach to improve insulin signaling and glycemic control in T2DM (Savova et al. 2023; Cao et al. 2023). *Algae*, such as brown algae (*Fucus vesiculosus*), green algae

(*Ulva lactuca*), and cyanobacteria (*Spirulina platensis*), produce a range of bioactives, including phlorotannins, polysaccharides, carotenoids, and polyphenols. These compounds exhibit strong antioxidant and anti-inflammatory properties, which counteract the inhibitory effects of ROS and cytokines on insulin signaling pathways. Reducing oxidative stress and inflammation, algal compounds support the activation of the PI3K/Akt pathway, allowing for enhanced glucose uptake and metabolic function in insulin-target tissues (Pradhan et al. 2021). Phlorotannins, for instance, have been observed to inhibit pro-inflammatory cytokines like TNF-α and IL-6, which are known to interfere with insulin signaling, thus preserving PI3K/Akt activity.

The action supports improved insulin sensitivity, promoting cellular glucose uptake and utilization. Additionally, polysaccharides from algae have been shown to directly enhance the PI3K/Akt pathway. For example, sulfated polysaccharides from brown algae can activate AMPK while concurrently promoting PI3K/Akt signaling, which plays a crucial role in facilitating glucose transport into cells. Through this dual pathway activation, algal polysaccharides not only reduce blood glucose levels but also improve lipid metabolism and energy balance, both of which are compromised in T2DM. Moreover, certain carotenoids, such as fucoxanthin from brown algae, have demonstrated insulin-sensitizing effects by promoting the translocation of GLUT4 to the cell membrane in muscle and adipose tissues (Tong et al. 2024; Kaushik et al. 2023).

The GLUT4 translocation is a key outcome of PI3K/Akt activation, directly supporting glucose uptake and enhancing cellular response to insulin. The modulation of the PI3K/Akt pathway by algae also extends to beta-cell protection, which is vital in T2DM to maintain insulin production. *Algae*-derived antioxidants, such as polyphenols from green algae, can prevent beta-cell apoptosis by protecting against oxidative damage and maintaining cellular integrity. By safeguarding beta-cell function and survival, these antioxidants contribute indirectly to sustained insulin production, which is crucial for activating the PI3K/Akt pathway in target tissues. Furthermore, algae bioactive compounds demonstrate pleiotropic effects, meaning they can simultaneously target multiple cellular pathways (Kaushik et al. 2023). This multi-targeted approach is particularly beneficial in T2DM, where cellular dysfunctions exist on multiple fronts, including insulin resistance, inflammation, and beta-cell degradation. Algae-derived bioactive compounds represent a promising complementary approach to T2DM management by enhancing PI3K/Akt signaling. Through antioxidant, anti-inflammatory, and insulin-sensitizing actions, these compounds improve insulin sensitivity, glucose uptake, and cellular resilience. This modulation of the PI3K/Akt pathway by algae offers a natural means to counteract insulin resistance and support metabolic health, indicating the potential of algae-based treatments as adjuncts in T2DM therapy (Shan et al. 2024). Further research into the specific bioactive molecules in *algae* that influence PI3K/Akt signaling will be valuable for optimizing their use in clinical diabetes management.

10 Conclusion

In conclusion, algae-derived bioactive compounds hold great promise as natural agents for diabetes management due to their antioxidant, antidiabetic, and cardioprotective properties. With diabetes, especially type 2 diabetes, on the rise, safe and effective strategies are essential to control hyperglycemia, oxidative stress, and related metabolic issues. Species like *Ulvalactuca*, *Sargassum*, *Spirulina platensis*, and *Chlorella vulgaris* are rich in polyphenols, carotenoids, and flavonoids, which help neutralize ROS, reducing oxidative damage to pancreatic beta cells and supporting insulin secretion. Additionally, *algae* inhibit carbohydrate-hydrolyzing enzymes such as α-amylase and α-glucosidase, slowing glucose absorption and moderating post-meal blood glucose spikes. Algae also benefit lipid metabolism, as their polyunsaturated fatty acids and sterols improve cholesterol and triglyceride levels, bolstering cardiovascular health—a critical factor for diabetic patients. Certain *algae* enhance insulin sensitivity, improving glucose uptake in tissues and potentially reducing insulin resistance. Animal studies have shown that algal extracts can lower blood glucose, improve lipid profiles, reduce inflammation, and protect pancreatic health. Collectively, these findings suggest *algae* as valuable adjuncts in diabetes management, supporting glycemic control and metabolic health. Further clinical research is needed to validate these promising effects in humans.

References

A.Y. Cheng, M.B. Gomes, S. Kalra, A.-P. Kengne, C. Mathieu, J.E. Shaw, Applying the WHO global targets for diabetes mellitus, Nature Reviews Endocrinology 19(4) (2023) 194–200.

A.A. Yameny, Diabetes Mellitus Overview 2024, Journal of Bioscience and Applied Research 10(3) (2024) 641–645.

A.D. Deshpande, M. Harris-Hayes, M. Schootman, Epidemiology of Diabetes and Diabetes-Related Complications, Physical Therapy 88(11) (2008) 1254–1264.

A. Maffei, G. Lembo, D. Carnevale, PI3Kinases in Diabetes Mellitus and Its Related Complications, International Journal of Molecular Sciences 19(12) (2018) 4098.

L. Pereira, A. Valado, Algae-Derived Natural Products in Diabetes and Its Complications—Current Advances and Future Prospects, Life 13(9) (2023) 1831.

A.M. Abo-Shady, S.F. Gheda, G.A. Ismail, J. Cotas, L. Pereira, O.H. Abdel-Karim, Antioxidant and Antidiabetic Activity of Algae, Life 13(2) (2023) 460.

N. Ahmed, M.A. Sheikh, M. Ubaid, P. Chauhan, K. Kumar, S. Choudhary, Comprehensive exploration of marine algae diversity, bioactive compounds, health benefits, regulatory issues, and food and drug applications, Measurement: Food 14 (2024) 100163.

Y. Sharifuddin, Y.-X. Chin, P.-E. Lim, S.-M. Phang, Potential Bioactive Compounds from Seaweed for Diabetes Management, Marine Drugs 13(8) (2015) 5447–5491.

S.-H. Lee, Yong-Li, F. Karadeniz, M.-M. Kim, S.-K. Kim, α-Glucosidase and α-amylase inhibitory activities of phloroglucinal derivatives from edible marine brown alga, Ecklonia cava, Journal of the Science of Food and Agriculture 89(9) (2009) 1552–1558.

M.M. Rahman, P.S. Dhar, Sumaia, F. Anika, L. Ahmed, M.R. Islam, N.A. Sultana, S. Cavalu, O. Pop, A. Rauf, Exploring the plant-derived bioactive substances as antidiabetic agent: An extensive review, Biomedicine & Pharmacotherapy 152 (2022) 113217.

A. Animish, M.A. Jayasri, A retrospective review of marine algae and the strategies employed for prospective diabetes management, Algal Research 74 (2023) 103209.

E. Toren, K.S. Burnette, R.R. Banerjee, C.S. Hunter, H.M. Tse, Partners in Crime: Beta-Cells and Autoimmune Responses Complicit in Type 1 Diabetes Pathogenesis, Frontiers in immunology 12 (2021) 756548.

T. Behl, A. Gupta, M. Albratty, A. Najmi, A.M. Meraya, H.A. Alhazmi, M.K. Anwer, S. Bhatia, S.G. Bungau, Alkaloidal Phytoconstituents for Diabetes Management: Exploring the Unrevealed Potential, Molecules (Basel, Switzerland) 27(18) (2022) 5851.

J.S. Bhatti, A. Sehrawat, J. Mishra, I.S. Sidhu, U. Navik, N. Khullar, S. Kumar, G.K. Bhatti, P.H. Reddy, Oxidative stress in the pathophysiology of type 2 diabetes and related complications: Current therapeutics strategies and future perspectives, Free Radical Biology and Medicine 184 (2022) 114–134.

T.P.T. Truong, T.M. Tran, T.X.T. Dai, C.L. Tran, Antihyperglycemic and anti-type 2 diabetic activity of marine hydroquinone isolated from brown algae (Dictyopteris polypodioides), Journal of Traditional and Complementary Medicine 13(4) (2023) 408–416.

X. Duan, O.T. Agar, C.J. Barrow, F.R. Dunshea, H.A.R. Suleria, n.d. Improving potential strategies for biological activities of phlorotannins derived from seaweeds, Critical Reviews in Food Science and Nutrition 1–23.

N.N. Besednova, B.G. Andryukov, T.S. Zaporozhets, T.A. Kuznetsova, S.P. Kryzhanovsky, S.P. Ermakova, I.V. Galkina, M.Y. Shchelkanov, Molecular Targets of Brown Algae Phlorotannins for the Therapy of Inflammatory Processes of Various Origins, Marine Drugs 20(4) (2022) 243.

M. Matin, M. Koszarska, A.G. Atanasov, K. Król-Szmajda, A. Jóźwik, A. Stelmasiak, M. Hejna, Bioactive Potential of Algae and Algae-Derived Compounds: Focus on Anti-Inflammatory, Antimicrobial, and Antioxidant Effects, Molecules (Basel, Switzerland) 29(19) (2024) 4695.

S.A. Siddiqui, P. Singh, D.T. Utama, M.Y. Samatra, A. Ahmad, S.A. Wani, Encapsulation of bioactive compounds in foods for diabetics - sources, encapsulation technologies, market trends and future perspectives—A systematic review, Food and Bioproducts Processing 147 (2024) 277–303.

M. Mohibbullah, M.N. Haque, A.A.M. Sohag, M.T. Hossain, M.S. Zahan, M.J. Uddin, M.A. Hannan, I.S. Moon, J.-S. Choi, A Systematic Review on Marine Algae-Derived Fucoxanthin: An Update of Pharmacological Insights, Marine Drugs 20(5) (2022) 279.

A. Tong, D. Wang, N. Jia, Y. Zheng, Y. Qiu, W. Chen, H.R. El-Seed, C. Zhao, Algal Active Ingredients and Their Involvement in Managing Diabetic Mellitus, Biology 13(11) (2024) 904.

A. Arabshomali, S. Bazzazzadehgan, F. Mahdi, Z. Shariat-Madar, Potential Benefits of Antioxidant Phytochemicals in Type 2 Diabetes, Molecules (Basel, Switzerland) 28(20) (2023) 7209.

M.G. Yu, D. Gordin, J. Fu, K. Park, Q. Li, G.L. King, Protective Factors and the Pathogenesis of Complications in Diabetes, Endocrine reviews 45(2) (2023) 227–252.

K. Zhang, Z. Fu, Y. Zhang, X. Chen, G. Cai, Q. Hong, The role of cellular crosstalk in the progression of diabetic nephropathy, Frontiers in endocrinology 14 (2023) 1173933.

S. Mallik, B. Paria, S.M. Firdous, H.S. Ghazzawy, N.K. Alqahtani, Y. He, X. Li, M.M. Gouda, The positive implication of natural antioxidants on oxidative stress-mediated diabetes mellitus complications, Journal of Genetic Engineering and Biotechnology 22(4) (2024) 100424.

S. Tangvarasittichai, Oxidative stress, insulin resistance, dyslipidemia and type 2 diabetes mellitus, World journal of diabetes 6(3) (2015) 456–80.

B. Pradhan, R. Nayak, S. Patra, B.P. Jit, A. Ragusa, M. Jena, Bioactive Metabolites from Marine Algae as Potent Pharmacophores against Oxidative Stress-Associated Human Diseases: A Comprehensive Review, Molecules (Basel, Switzerland) 26(1) (2021) 37.

E. D'Aniello, P. Amodeo, R.M. Vitale, Marine Natural and Nature-Inspired Compounds Targeting Peroxisome Proliferator Activated Receptors (PPARs), Marine Drugs 21(2) (2023) 89.

B. Grygiel-Górniak, Peroxisome proliferator-activated receptors and their ligands: nutritional and clinical implications - a review, Nutrition Journal 13(1) (2014) 17.

T. Sæther, S.M. Paulsen, J.E. Tungen, A. Vik, M. Aursnes, T. Holen, T.V. Hansen, H.I. Nebb, Synthesis and biological evaluations of marine oxohexadecenoic acids: PPARα/γ dual agonism and anti-diabetic target gene effects, European Journal of Medicinal Chemistry 155 (2018) 736–753.

O. Dubuisson, E.J. Dhurandhar, R. Krishnapuram, H. Kirk-Ballard, A.K. Gupta, V. Hegde, E. Floyd, J.M. Gimble, N.V. Dhurandhar, PPARγ-Independent Increase in Glucose Uptake and Adiponectin Abundance in Fat Cells, Endocrinology 152(10) (2011) 3648–3660.

S.S. Choi, J. Park, J.H. Choi, Revisiting PPARγ as a target for the treatment of metabolic disorders, BMB reports 47(11) (2014) 599–608.

S. Vaibhav, K. Anjaneyulu, The Regulatory Roles of Mitogen-Activated Protein Kinase (MAPK) Pathways in Health and Diabetes: Lessons Learned from the Pancreatic β-Cell, Recent Patents on Endocrine, Metabolic & Immune Drug Discovery (Discontinued) 10(2) (2016) 76–84.

J.M.A. Hannan, P. Ansari, S. Azam, P.R. Flatt, Y.H.A. Abdel Wahab, Effects of Spirulina platensis on insulin secretion, dipeptidyl peptidase IV activity and both carbohydrate digestion and absorption indicate potential as an adjunctive therapy for diabetes, British Journal of Nutrition 124(10) (2020) 1021–1034.

M.C. Petersen, G.I. Shulman, Mechanisms of Insulin Action and Insulin Resistance, Physiological Reviews 98(4) (2018) 2133–2223.

J. van Gerwen, A.S. Shun-Shion, D.J. Fazakerley, Insulin signalling and GLUT4 trafficking in insulin resistance, 51(3) (2023) 1057–1069.

N.A.S. Din, A.S. Mohd Alayudin, N.-S. Sofian-Seng, H.A. Rahman, N.S. Mohd Razali, S.J. Lim, W.A. Wan Mustapha, Brown Algae as Functional Food Source of Fucoxanthin: A Review, Foods 11(15) (2022) 2235.

M.-Y. Chung, H.-K. Choi, J.-T. Hwang, AMPK Activity: A Primary Target for Diabetes Prevention with Therapeutic Phytochemicals, Nutrients 13(11) (2021) 4050.

O. Golovinskaia, C.-K. Wang, The hypoglycemic potential of phenolics from functional foods and their mechanisms, Food Science and Human Wellness 12(4) (2023) 986–1007.

M.S. Savova, L.V. Mihaylova, D. Tews, M. Wabitsch, M.I. Georgiev, Targeting PI3K/AKT signaling pathway in obesity, Biomedicine & Pharmacotherapy 159 (2023) 114244.

R. Cao, H. Tian, Y. Zhang, G. Liu, H. Xu, G. Rao, Y. Tian, X. Fu, Signaling pathways and intervention for therapy of type 2 diabetes mellitus, 4(3) (2023) e283.

A. Kaushik, R. Sangtani, H.S. Parmar, K. Bala, Algal metabolites: Paving the way towards new generation antidiabetic therapeutics, Algal Research 69 (2023) 102904.

S. Shan, Z. Zhang, J. Nie, Y. Wen, W. Wu, Y. Liu, C. Zhao, Marine algae-derived oligosaccharide via protein crotonylation of key targeting for management of type 2 diabetes mellitus in the elderly, Pharmacological Research 205 (2024) 107257.

Chapter 2
Diabetes: Current Treatment and Its Management

M. V. Arularasu, P. Vinitha, and Prabhudas Nelaturi

Abstract The prevalence of diabetes has been consistently rising, and it has become a substantial global public health issue. It is a metabolic condition that disrupts carbohydrate management, marked by the body's inability to produce or properly respond to insulin, leading to challenges in regulating normal blood sugar levels. All forms of diabetes may be managed with medication and lifestyle changes; however, most are chronic conditions. Increased appetite, thirst, and frequent urination result from this elevated blood sugar. Numerous issues can arise from diabetes if left untreated. Nonketotic hyperosmolar coma and diabetic ketoacidosis are examples of acute complications. Long-term significances include heart disease, neuro-problems, foot ulcers, liver damage, and vision loss. The two main causes of diabetes are insufficient insulin production by the pancreas and improper insulin cellular response. Many forms of diabetes are covered in this chapter, along with the current treatments and strategies for managing people with diabetes before and after treatment. The guidelines for diagnosing diabetes mellitus provided by the American Diabetes Association are also addressed.

Keywords Diabetes · Insulin · Type of diabetes · Treatments

M. V. Arularasu (✉)
Sustainable Energy and Environment Research Unit, Department of Physiology, Saveetha Medical College, Saveetha Institute of Medical and Technical Science, Chennai, Tamil Nadu, India

P. Vinitha
Advanced Materials, Energy and Environmental Research Lab, Department of Community Medicine, Saveetha Medical College, Saveetha Institute of Medical and Technical Sciences, Chennai, Tamil Nadu, India

P. Nelaturi
Department of Medical Biotechnology, Aarupadai Veedu Medical College and Hospital, Vinayaka Mission's Research Foundation (Deemed to Be University), Kirumampakkam, Puducherry, India

R. Raja et al. (eds.), *Algae in Diabetes Management*,
https://doi.org/10.1007/978-981-95-0571-5_2

1 Introduction

1.1 Diabetes Mellitus

Chronic high blood sugar is a metabolic condition resulting from either reduced insulin effectiveness, irregular insulin secretion, or a combination of both. Insulin is a crucial anabolic hormone that influences how proteins, fats, and carbs are metabolized (Poznyak et al. 2020). Skeletal muscles, liver, and adipose tissue are the primary tissues affected by metabolic abnormalities linked to diabetes due to insulin resistance. Symptom intensity varies with both the kind and duration of diabetes. These symptoms include increased hunger, polydipsia, dysuria, weight loss, and eyesight issues in people with high blood sugar, especially in youngsters who have no insulin at all. Some diabetics may not exhibit symptoms, especially in the early stages (Rossi et al. 2019). Uncontrolled diabetes can cause several problems if left untreated, including unconsciousness, disorientation, and in rare instances, death from untreated nonketotic hyperosmolar syndrome or ketoacidosis (Poznyak et al. 2020).

1.2 Classification of Diabetes

1.2.1 Type 1

This form of diabetes is primarily caused by the autoimmune destruction of pancreatic β-cells. Individuals with type 1 diabetes are insulin-dependent and at an increased risk for ketoacidosis (Jensen et al. 2021; Rewers et al. 2015). Anti-β-cell antibodies can target various antigens, primarily insulin, glutamic acid decarboxylase (GAD), and proteins associated with insulin storage granule membranes, such as islet antigen-2 and zinc transporter 8 (ElSayed et al. 2023a). In high-risk individuals, autoantibodies, such as those targeting GAD, can predict the onset of diabetes mellitus, as antibodies emerge in the plasma during the preclinical stage. Type 1 diabetes mellitus is a polygenic condition influenced by environmental factors, rather than following a straightforward Mendelian inheritance pattern. It is thought that autoimmune destruction of β-cells may be triggered by specific infectious agents, like the coxsackie virus, through a process known as molecular mimicry (Hill-Briggs et al. 2021). Of all the diabetes cases, type 1 diabetes makes up around 5–10%. While most cases develop during childhood or adolescence, about 20% of new diagnoses occur in adults.

1.2.2 Type 2

The problems with insulin resistance and insulin production deficiencies are the hallmarks of this most common type of diabetes (Galicia-Garcia et al. 2020). Type 2 diabetes mellitus can also refer to diabetes mellitus that develops in young people due to mutations in the glucose kinase gene, which was previously explored. Obesity is the major cause that leads to insulin resistance develop type 2 diabetes mellitus (Tonstad 2009). Type 2 diabetes is the most prevalent form of diabetes, accounting for roughly 90% of all cases. Its prevalence is increasing globally, with significant variations both between and within countries. While it primarily affects adults, especially older individuals, it can also occur in children and adolescents (Polsky and Akturk 2017).

1.2.3 Gestational Diabetes

Gestational diabetes is a transient disease linked to pregnancy. This condition manifests in certain individuals during gestation. Gestational diabetes often resolves following pregnancy. Individuals with gestational diabetes face an elevated risk of subsequently acquiring type 2 diabetes (Gucciardi et al. 2019; Zhang et al. 2015).

2 Causes, Symptoms, and Diagnosis of Type 1 Diabetes

2.1 Type 1 Diabetes Causes

The specific origin of type 1 diabetes is unclear, however, it is believed to be a mix of hereditary and environmental factors. Insulin resistance develops when the immune system negligently targets and kills the pancreatic cells responsible for generating insulin. Insulin plays a crucial role in enabling glucose to enter cells and serve as an energy source. Without insulin, glucose leftovers in the blood flow, resulting in elevated blood sugar levels while depriving cells of the energy they need (Durlach et al. 2022; Volkova et al. 2021). People with type 1 diabetes must treat their illness with daily insulin injections or a continuous insulin pump, as well as constantly monitoring their blood sugar levels. The FDA has approved a new treatment, Tzield (teplizumab), that helps delay the progression of type 1 diabetes. This innovative therapy targets the immune system to prevent it from attacking insulin-producing cells in the pancreas, potentially delaying the onset of advanced symptoms by several years. This medication may be an option for those aged 8 and older who are at high risk or have early indications of the illness (Plows et al. 2018).

2.2 Symptoms of Type 1 Diabetes

Type 1 diabetes frequently manifests abruptly. Symptoms may encompass:

Excessive thirst, Frequent and excessive urination, Severe appetite accompanied by weight loss, Visual impairment, Nausea and vomiting, Profound debilitation and diminished vitality (asthenia), Irritability and fluctuations in mood, Fungal (yeast) dermal infections in the groin region. In children, symptoms may resemble those of influenza. Bedwetting may signify a significant concern. Other illnesses or health issues, including some medications, may be the source of these symptoms. Always consult healthcare physician for a diagnosis (American Diabetes Association Professional Practice Committee 2022).

2.3 Diagnosis of Type 1 Diabetes

Diagnostic tests include:

Glycated hemoglobin (A1C) assay: The glycated hemoglobin (A1C) test is a key method for assessing long-term blood sugar levels, reflecting an average over the past 2–3 months. It measures the hemoglobin percentage in red blood cells that has bonded with glucose. Higher glucose levels result in a greater proportion of glycated hemoglobin. An A1C level of 6.5% or higher on two separate tests confirms a diabetes diagnosis. Levels between 5.7% and 6.4% indicate prediabetes, while levels below 5.7% are considered normal. This test is not only useful for diagnosis but also for monitoring the effectiveness of diabetes management strategies, with a common goal of keeping A1C levels below 7% to reduce complications (Elsayed et al. 2023b).

If the A1C test cannot be performed due to specific factors, such as pregnancy or the presence of certain hemoglobin variants that might affect accuracy, alternative tests may be used. These include the fasting plasma glucose (FPG) test, which measures blood sugar after an overnight fast, and the oral glucose tolerance test (OGTT), which evaluates the body's glucose handling over time following a glucose-rich drink. A random plasma glucose test may also be used to assess blood sugar levels without regard to fasting. These methods help ensure accurate diagnosis and management in situations where A1C results might not be reliable. Random blood sugar test: The random blood sugar test involves taking a blood sample at any point during the day, regardless of food intake. It measures glucose levels in milligrams per deciliter (mg/dL) or millimoles per liter (mmol/L). A reading of 200 mg/dL (11.1 mmol/L) or higher is typically an indication of diabetes. However, this result is usually confirmed with further tests, such as the fasting plasma glucose test or the A1C test, to provide a definitive diagnosis (Rewers et al. 2015; ElSayed et al. 2023a).

Fasting blood sugar test: A fasting blood glucose test is conducted after an overnight fast. A normal fasting blood glucose level is below 100 mg/dL (5.6 mmol/L). A level between 100 and 125 mg/dL (5.6 to 6.9 mmol/L) is considered prediabetes. If the fasting blood glucose level is 126 mg/dL (7 mmol/L) or higher on two separate occasions, it indicates the presence of diabetes. When diagnosing diabetes, clinicians may perform blood tests to detect specific autoantibodies that are commonly present in type 1 diabetes. These tests can be particularly helpful when distinguishing between type 1 and type 2 diabetes if the diagnosis is unclear. Additionally, the presence of ketones in the urine—byproducts of fat breakdown—suggests type 1 diabetes rather than type 2, as this is more typical of type 1 diabetes due to insulin deficiency. These diagnostic tools aid doctors in determining the most appropriate treatment and management for patient care (Tonstad 2009; Eliasson 2003).

3 Causes, Symptoms, and Diagnosis of Type 2 Diabetes

3.1 Type 2 Diabetes Causes

When cells in the body, particularly those in the liver, muscles, and fat, no longer react normally to insulin, a condition known as insulin resistance emerges. This leads to type 2 diabetes. Pancreatic insulin is a hormone that aids in glucose regulation in the blood. In people with insulin resistance, the body needs more insulin to manage blood glucose, which eventually leads to higher blood sugar levels and leads to diabetes (Heerman et al. 2016). In hyperinsulinemia, the pancreas tries to counteract the body's insulin resistance by manufacturing more insulin to bring blood glucose levels down. However, as insulin resistance worsens over time, the pancreas struggles to produce enough insulin to overcome this resistance. The emergence of type 2 diabetes is caused by this failure to manage glucose levels adequately (Rossi et al. 2019).

Multiple factors may lead to insulin resistance, including: Genetics, Excess adipose tissue, particularly in the abdominal region and around internal organs (visceral fat), Lack of physical exercise, Consuming highly processed, carbohydrate-rich meals and saturated fats regularly, Specific drugs, such as prolonged corticosteroid use, Hormonal problems such as hypothyroidism and Cushing's syndrome, Prolonged stress and insufficient quality sleep.

3.2 Symptoms of Type 2 Diabetes

Symptoms of type 2 diabetes include: frequent urination, especially at night, extreme thirst, profound fatigue, unexplained weight reduction, Itching in the vaginal region or recurrent bouts of candidiasis, Lesions or abrasions that exhibit delayed healing, Visual impairment due to desiccation of the ocular lens. The manifestations of type 1 diabetes are often evident and progress rapidly, generally within a few weeks. The signs and symptoms are not always evident and are frequently identified during a regular examination. This occurs due to their often moderate nature and slow development over several years. This indicates that one may develop type 2 diabetes for an extended period without awareness (Williams et al. 2022).

3.3 Diagnosis of Type 2 Diabetes

The subsequent blood tests assist in diagnosing type 2 diabetes:

- Fasting plasma glucose test: This laboratory examination assesses blood glucose levels. This test is often required to be conducted in the morning following an eight-hour fast, during which only water is permitted. A result of 126 mg/dL or greater is classified as diabetes.
- Random plasma glucose test: This laboratory test measures blood sugar levels as well, but it may be performed at any time without requiring fasting. A value of 200 mg/dL or greater indicates a diagnosis of diabetes.
- A1C test: This laboratory test measures the average blood glucose levels over the past two to three months. If the result is 6.5% or higher, it is indicative of diabetes (Williams et al. 2022).

4 Treatments for All Types of Diabetes

The persistence of diabetes treatment is to control blood glucose levels. Achieving this aim is critical for enhancing overall health and reducing the initiation and progression of long-term problems linked with diabetes. HbA1c values can be used to determine the efficiency of a person's diabetic therapy, can bring to normal levels (Williams et al. 2022).

4.1 Diet and Exercise

4.1.1 Diet

People with diabetes are encouraged to follow personalized diets aimed at achieving and maintaining a healthy body weight. Regular physical activity is also recommended, as it helps muscles utilize glucose more effectively and prevents sharp spikes in blood sugar levels after consuming carbohydrates. A diet that is moderate in fat and high in protein is generally advised to support blood sugar control and overall health.. A balanced diet for people with diabetes typically suggests that about 30% of total daily calories come from fats, 20% from proteins, and the remaining calories from carbohydrates. It is recommended to prioritize complex carbohydrates, such as whole grains and vegetables, over simple sugars to help better regulate blood sugar levels (Fig. 2.1). Consistent energy levels and better health management are both aided by this method (Heerman et al. 2016; Georgoulis et al. 2014; Sleiman et al. 2015).

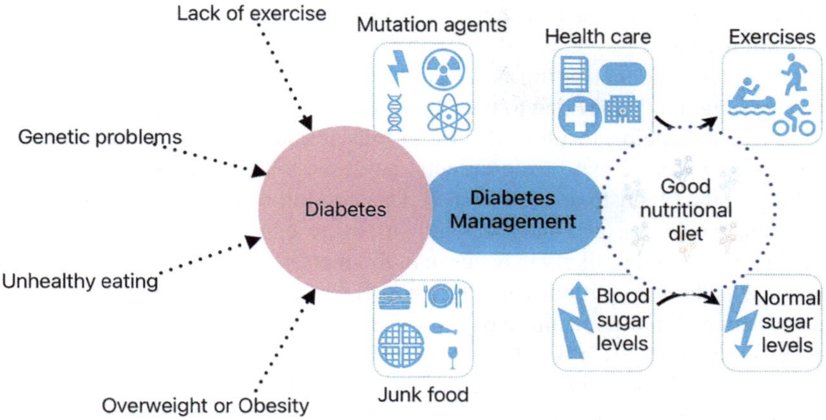

Fig. 2.1 The causes of diabetes, including genetic factors, unhealthy eating, obesity, lack of exercise, and environmental agents. Effective management through healthcare, exercise, and a good diet helps maintain normal blood sugar levels

4.1.2 Physical Activity

All diabetics are recommended to follow diets that promote and maintain a healthy body weight, as well as frequent exercise, which aids in glucose absorption by muscles and reduces blood glucose increases following carbohydrate intake. They are recommended to have a diet that is low in fat but high in protein. Approximately 30% of total calorie consumption should be from fat, 20% from protein, and the rest from carbs, with a preference for complex carbohydrates over simple sugars (Ermeley 2020; Palermi et al. 2022).

Diabetes-friendly exercises include: walking, swimming, dancing, and cycling.

4.2 Treatments

4.2.1 Oral Diabetes Medications

Individuals with diabetes who still make insulin, especially those with type 2 diabetes or prediabetes, can use oral diabetic medicines to control their blood sugar levels. Oral medicine may also be necessary for those with gestational diabetes. Various varieties exist. Various oral medications, such as thiazolidinediones, biguanides, and sulfonylureas, are commonly used to help manage blood glucose levels in individuals with diabetes. Medications of the sulfonylurea family, such as glimepiride and glipizide, reduce blood sugar by increasing insulin secretion by the pancreas. Hypoglycemia, or dangerously low blood glucose levels, is a typical adverse effect of sulfonylureas. This is particularly prevalent in older adults or those with liver or kidney impairments, as these conditions can affect how the body processes the medication during cellular damage (Heerman et al. 2016; American Diabetes Association 2020).

Biguanides, with metformin as the principal representative, are classified as antihyperglycemic medicines due to their mechanism of reducing hepatic glucose generation and enhancing insulin efficacy in muscle and adipose tissues. A rare but serious side effect of metformin is lactic acidosis, which occurs when lactic acid builds up in the bloodstream and tissues. This condition can lead to symptoms such as nausea, weakness, and fatigue, and, in severe cases, can be life-threatening. Lactic acidosis is more likely to occur in individuals with preexisting kidney or liver problems, as these conditions can affect how the body clears metformin. It's important for patients on metformin to be monitored regularly for signs of this complication (Sleiman et al. 2015; Bomberg et al. 2019).

Thiazolidinediones (TZDs), such as rosiglitazone and pioglitazone, work by reducing insulin resistance in muscle and fat cells. This helps the body utilize insulin more effectively, improving glucose uptake into these tissues. As a result, they help lower blood sugar levels in individuals with type 2 diabetes by enhancing the action of insulin, making these medications particularly useful for managing the condition. However, they can come with side effects such as weight gain and fluid retention, and they require careful monitoring, especially in patients with heart conditions. However, these drugs can induce swelling, liver damage, and severe cardiovascular events in certain people. Furthermore, oral hypoglycemic medicines often reduce average blood glucose levels by 50–80 mg per 100 mL (2.8–4.4 mmol per liter), albeit their efficacy may diminish with time (Kaiser et al. 2019; Sharma et al. 2022).

One synthetic hormone that acts similarly to the human hormone amylin is pramlintide, which is administered intravenously. By reducing stomach food absorption and glucagon secretion, it aids diabetics with managing their blood sugar levels. Glucagon is a hormone that signals the liver to produce glucose, so by reducing its release, pramlintide helps lower glucose production. This dual action helps maintain more stable blood sugar levels, particularly after meals, making it an effective treatment for diabetes when used alongside other medications. Incretins, like gastric inhibitory polypeptides, are gastrointestinal hormones that stimulate insulin release from the pancreas and are related to the injectable antihyperglycemic medication exenatide. Exenatide has a longer half-life than natural incretins because it is less susceptible to breakdown by the enzyme dipeptidyl peptidase-4 (DPP-4). This allows it to maintain its effects for a longer period, helping to regulate blood sugar more effectively. On the other hand, sitagliptin works by selectively inhibiting DPP-4, which increases the levels of naturally occurring incretins that promote insulin secretion and reduce glucagon release. Both of these medications typically have minimal side effects. However, pramlintide, which mimics the hormone amylin, can pose a risk of severe hypoglycemia, especially in individuals with type 1 diabetes. This is due to its effects on insulin secretion and its potential to increase the risk of low blood sugar, particularly when used in combination with insulin therapy (Williams et al. 2022; Bomberg et al. 2019).

4.2.2 Insulin

The administration of insulin is essential for the management and maintenance of life for people with type 1 diabetes and, in certain cases, type 2 diabetes as well. Synthetic insulin comes in a variety of forms, each with its own specific timing and duration of action. Injectable insulin, insulin pens, insulin pumps, and fast-acting inhaled insulin are the main ways that insulin is administered. Modern insulin therapies are produced using recombinant DNA technology, allowing the creation of insulin that closely resembles the natural version produced by the pancreas. Insulin is available in different forms, such as short-acting insulin, which works quickly but for a short period, and long-acting insulin, which is designed to last up to 24 hours. These formulations provide flexibility in managing blood sugar levels, depending on the individual's needs and the timing of meals or other factors affecting glucose control. Another kind, rapid-acting insulin, begins to decrease blood glucose levels within 10 to 30 minutes of treatment; this form became available as an inhalable alternative in 2014. For some people, a mix of intermediate-acting and short-acting insulin given twice a day provides a balanced strategy. Patients on insulin therapy may need to adapt their food intake based on their activity level; as exercise frequency and intensity rise, insulin dosages may need to be reduced while food intake increases.

4.2.3 Herbal Remedies

Capsaicin cream: Some people have claimed that capsaicin cream, a topical therapy derived from cayenne, can help reduce diabetic neuropathy-related pain in the hands and feet. Individuals with reduced sensitivity in their hands or feet, such as those with neuropathy, should use caution when applying capsaicin, as they may not be able to feel any burning sensations or discomfort. This can lead to skin irritation or burns if the product is applied improperly. It is advisable to consult a healthcare professional before using capsaicin-based products to ensure they are safe and appropriate condition (Kumar et al. 2020). Evening primrose oil is sometimes suggested as a treatment for diabetic neuropathy due to its potential to reduce inflammation and improve nerve function. However, while some studies have shown promise, conclusive evidence supporting its effectiveness is still lacking. Other herbal remedies often recommended for alleviating symptoms of diabetes include ginkgo, garlic, holy basil leaves, fenugreek seeds, ginseng, and hawthorn. These plants are believed to have various benefits, such as improving circulation, lowering blood sugar levels, and reducing inflammation. However, research on their effectiveness for managing diabetes is still ongoing, and their use should be discussed with a healthcare provider to ensure safety and proper integration into a treatment. It's important to note that while herbal treatments may offer some benefits.. Additional inquiry is necessary to ascertain the possible function of these herbal compounds (Upendra Rao et al. 2010).

Natural substances and their derivatives have generated significant interest as potential agents for managing diabetes. These compounds, derived from plants, fungi, marine life, and algae, encompass a variety of chemical components with diverse pharmacological properties. For example, several plant-based compounds have shown promise in improving insulin sensitivity, regulating blood sugar, and offering anti-inflammatory effects. Additionally, marine organisms and algae may contribute bioactive molecules that aid in glucose metabolism and offer protective benefits against complications associated with diabetes. Research into these natural materials continues to explore their mechanisms of action, bioavailability, and long-term safety, but they often require further clinical evidence before being widely recommended for diabetes management. Nevertheless, their potential as adjunctive treatments alongside conventional therapies remains an area of active investigation. If considering natural remedies, it is essential to consult with a healthcare provider to ensure their safety and compatibility with any ongoing treatments (Abo-Shady et al. 2023; Quitério et al. 2021). Natural compounds and their derivatives have received a lot of interest as prospective diabetes treatment options. These chemicals, derived from plants, fungus, marine creatures, and algae, have a varied chemical composition and pharmacological effects. Exploring the therapeutic potential of natural products in diabetes research holds promise for developing new medications and treatments that effectively manage blood glucose levels while also addressing the complications associated with the disease. Natural substances, including plant-derived compounds, marine organisms, and algae, offer a range of bioactive properties that could support glucose regulation, improve insulin sensitivity, and provide additional benefits like reducing inflammation. Continued research into these natural resources may lead to innovative solutions for diabetes management, complementing traditional therapies and enhancing overall patient care. However, more clinical evidence is needed to confirm their effectiveness and safety before widespread adoption.

4.2.4 Less Common Treatments

Recent experimental therapies have shown a beneficial impact on blood glucose levels and diabetes management.

4.2.4.1 Bariatric Surgery

Bariatric surgery, often considered for individuals with obesity, can also help those with type 2 diabetes achieve better blood glucose control. Studies have shown that the procedure can significantly improve glucose metabolism, often leading to remission of type 2 diabetes. Interestingly, research suggests that bariatric surgery may also benefit individuals with type 1 diabetes by helping them better regulate blood sugar levels, though the mechanisms and long-term outcomes for this group are still under investigation. This surgical intervention could offer a promising complementary approach for diabetes management, especially for those struggling to control their condition through lifestyle changes and medications alone (Zhou and Zeng 2023; Affinati et al. 2019).

4.2.4.2 Artificial Pancreas

A hybrid closed-loop device, referred to as an artificial pancreas, substitutes glucose monitoring, and insulin injections by checking blood glucose levels every five minutes and automatically delivering suitable quantities of insulin and glucagon. Remote oversight by healthcare experts or guardians is feasible to guarantee the system's functionality. Mealtimes need a manual modification of insulin dosage, although they enable individuals with diabetes to sleep uninterrupted throughout the night without the need to monitor blood glucose or provide medicine to lower levels (Nwokolo and Hovorka 2023; Dimou et al. 2020).

4.2.4.3 Pancreatic Islet Transplantation

Islets are aggregates of cells that synthesize insulin. The immune system of an individual with type 1 diabetes assaults these cells. Transplantation involves extracting islets from a functioning, donated pancreas and substituting the damaged islets in an individual with type 1 diabetes. This is an experimental therapy accessible exclusively via participating in research trials. Numerous insurance policies exclude coverage for weight loss therapies or experimental procedures. Consequently, individuals have to consult their physicians prior to undertaking these operations. Outcomes fluctuate and frequently hinge on the individual's insulin usage, the duration of their diabetes, and the degree of weight reduction (Takaki and Shimoda 2020; Cayabyab et al. 2021).

5 Management

5.1 Glucose Monitoring

Monitoring blood glucose levels is essential for assessing the effectiveness treatment plan and managing diabetes on a daily or even hourly basis. Regular testing, whether through glucose meter readings with fingerstick tests or using a continuous glucose monitor (CGM), provides real-time insights into blood sugar levels. Working closely with healthcare provider, this will identify the target blood glucose range that best suits individual needs. This ongoing monitoring helps make informed decisions about diet, exercise, and medication adjustments, ultimately supporting better diabetes control and reducing the risk of complications (Williams et al. 2022).

5.2 *Wear a Medical ID Tag*

Wearing a MedicAlert bracelet or tag indicating your diabetes status is advisable. This will inform others of your situation in the event of a severe hypoglycemia crisis, should you be unable to communicate, or if you experience an accident requiring immediate medical assistance. Self-identification as a diabetic is crucial, as hypoglycemia episodes may be misinterpreted as intoxication, and affected individuals frequently lack the capacity for self-care. In the absence of immediate intervention, hypoglycemia may lead to coma or seizures. Due to heightened stress on your body during illness or injury, medical staff providing emergency treatment will need to monitor your blood sugar levels (Wahowiak 2018).

5.3 *Take Care of your Teeth*

Always remember to floss daily and take excellent care of your teeth. Having diabetes makes gum disease worse.

5.4 *Reduce Stress*

Stress can elevate blood glucose levels and diminish insulin sensitivity. Under excessive stress, our body engages in a "fight or flight" reaction. It ensures the availability of sufficient sugar and fat for energy production. Research on individuals with type 1 diabetes indicates that blood glucose levels increase for the majority under mental stress, but they decrease for others. Experiencing pressure when having type 2 diabetes will elevate glucose levels (Williams et al. 2022).

If anything disturbs, endeavor to implement modifications that facilitate relaxation. You may engage in physical activity, socialize with friends, do meditation, or substitute negative ideas with positive affirmations. Finding strategies that work for you is key to managing diabetes effectively. This may include a combination of medical treatments, lifestyle changes, and emotional support. For some individuals, joining support groups, seeking counseling, or participating in therapy can be very beneficial in managing both the psychological and physical aspects of the condition.

5.5 *Quit Smoking*

Smokers are more prone to severe health issues and have an increased risk of complications from diabetes (Tonstad 2009; Molla et al. 2020). These may encompass:

- Cardiovascular and renal disorders: Diabetes significantly increases the risk of heart disease, stroke, and kidney damage due to its impact on blood vessels and circulation.
- Poor circulation in the legs and feet: This can result in infections, ulcers, and in severe cases, the amputation of toes or feet due to reduced blood flow.
- Retinopathy: A diabetic complication affecting the eyes, which can lead to blindness if not treated.

Peripheral neuropathy: This type of nerve damage primarily affects the limbs, causing symptoms like weakness, numbness, pain, and impaired coordination.

5.6 Alcohol Abstainer

Individuals taking insulin or oral antidiabetic medications such as sulfonylureas or meglitinides may experience dangerously low blood sugar levels while consuming alcohol. While consuming alcohol, the liver prioritizes the elimination of alcohol from the bloodstream above the regulation of blood glucose levels. Intoxication and hypoglycemia may induce dizziness, confusion, and somnolence. The symptoms of excessive alcohol consumption and hypoglycemia may be indistinguishable. A lady should consume no more than one alcoholic beverage each day. The maximum allowance for males is two alcoholic beverages daily (Zhang et al. 2015; Johnson and Melton 2016).

To avoid diabetes or manage existing diabetes, consider opting for light beer or wine spritzers, and consume beverages more slowly or substitute with water or another calorie-free drink. Reducing beverage consumption may enhance dietary choices. Alcohol might diminish your self-control to avoid overeating.

6 Conclusion

In conclusion, effective diabetes management relies on a multifaceted approach, combining lifestyle modifications, routine monitoring, and pharmacological therapies. Diabetes is a global health issue caused by lifestyle factors and carbohydrate metabolism disruptions. Common symptoms of the condition include frequent urination, excessive thirst, and increased hunger. If diabetes is not managed properly, it can lead to serious complications, such as diabetic ketoacidosis, cardiovascular diseases, kidney failure, and vision impairment. Early detection, proper management, and lifestyle adjustments are crucial to preventing or minimizing these risks. Regular monitoring of blood sugar levels, along with medical supervision, plays a key role in reducing complications associated with diabetes.. Management involves lifestyle modifications, pharmacological interventions, and lifestyle adjustments. Treatment approaches vary by type, with type 1 diabetes requiring lifelong insulin therapy, while type 2 diabetes often benefits from lifestyle adjustments and oral medications. Nutraceuticals, including herbal remedies, are recognized for their anti-diabetic effects, stimulating insulin secretion, enhancing glucose uptake, and reducing glucose absorption. Early diagnosis and structured care are crucial to reduce complications and improve quality of life.

References

Poznyak A, Grechko A V., Poggio P, et al (2020) The diabetes mellitus–atherosclerosis connection: The role of lipid and glucose metabolism and chronic inflammation. Int. J. Mol. Sci. 21

Rossi MC, Nicolucci A, Ozzello A, et al (2019) Impact of severe and symptomatic hypoglycemia on quality of life and fear of hypoglycemia in type 1 and type 2 diabetes. Results of the Hypos-1 observational study. Nutr Metab Cardiovasc Dis 29:. https://doi.org/10.1016/j.numecd.2019.04.009

Jensen ET, Stafford JM, Saydah S, et al (2021) Increase in prevalence of diabetic ketoacidosis at diagnosis among youth with type 1 diabetes: The SEARCH for diabetes in youth study. Diabetes Care 44:. https://doi.org/10.2337/dc20-0389

Rewers A, Dong F, Slover RH, et al (2015) Incidence of diabetic ketoacidosis at diagnosis of type 1 diabetes in colorado youth, 1998-2012. JAMA - J. Am. Med. Assoc. 313

ElSayed NA, Aleppo G, Aroda VR, et al (2023a) Addendum. 2. Classification and Diagnosis of Diabetes: Standards of Care in Diabetes-2023. Diabetes Care 2023;46(Suppl. 1):S19-S40. Diabetes Care 46:. https://doi.org/10.2337/dc23-ad08

Hill-Briggs F, Adler NE, Berkowitz SA, et al (2021) Social determinants of health and diabetes: A scientific review. Diabetes Care 44

Galicia-Garcia U, Benito-Vicente A, Jebari S, et al (2020) Pathophysiology of type 2 diabetes mellitus. Int. J. Mol. Sci. 21

Tonstad S (2009) Cigarette smoking, smoking cessation, and diabetes. Diabetes Res. Clin. Pract. 85

Polsky S, Akturk HK (2017) Alcohol Consumption, Diabetes Risk, and Cardiovascular Disease Within Diabetes. Curr. Diab. Rep. 17

Gucciardi E, Yang A, Cohen-Olivenstein K, et al (2019) Emerging practices supporting diabetes self-management among food insecure adults and families: A scoping review. PLoS One 14:. https://doi.org/10.1371/journal.pone.0223998

Zhang L, Wang XH, Zheng XM, et al (2015) Maternal gestational smoking, diabetes, alcohol drinking, pre-pregnancy obesity and the risk of cryptorchidism: A systematic review and meta-analysis of observational studies. PLoS One 10:. https://doi.org/10.1371/journal.pone.0119006

Durlach V, Vergès B, Al-Salameh A, et al (2022) Smoking and diabetes interplay: A comprehensive review and joint statement. Diabetes Metab 48:. https://doi.org/10.1016/j.diabet.2022.101370

Volkova NI, Davidenko IY, Degtyareva YS (2021) Gestational diabetes mellitus. Akusherstvo i Ginekol (Russian Fed 2021:. https://doi.org/10.54393/df.v2i2.18

Plows JF, Stanley JL, Baker PN, et al (2018) The pathophysiology of gestational diabetes mellitus. Int. J. Mol. Sci. 19

American Diabetes Association Professional Practice Committee (2022) Diabetes Care in the Hospital: Standards of Medical Care in Diabetes-2022. Diabetes Care 45:. https://doi.org/10.2337/dc22-S016

Elsayed NA, Aleppo G, Aroda VR, et al (2023b) 2. Classification and Diagnosis of Diabetes: Standards of Care in Diabetes-2023. Diabetes Care 46:. https://doi.org/10.2337/dc23-S002

Eliasson B (2003) Cigarette smoking and diabetes. Prog. Cardiovasc. Dis. 45

Heerman WJ, Wallston KA, Osborn CY, et al (2016) Food insecurity is associated with diabetes self-care behaviours and glycaemic control. Diabet Med 33:. https://doi.org/10.1111/dme.12896

Williams DM, Jones H, Stephens JW (2022) Personalized Type 2 Diabetes Management: An Update on Recent Advances and Recommendations. Diabetes, Metab. Syndr. Obes. 15
Georgoulis M, Kontogianni MD, Yiannakouris N (2014) Mediterranean diet and diabetes: Prevention and treatment. Nutrients 6:. https://doi.org/10.3390/nu6041406
Sleiman D, Al-Badri MR, Azar ST (2015) Effect of Mediterranean Diet in Diabetes Control and Cardiovascular Risk Modification: A Systematic Review. Front. Public Heal. 3
Ermeley Z (2020) Physical fitness and lifestyle behaviors among type2 diabetes Persons over 50 years old. DRASSA J Dev Res Sport Sci Act 6:. https://doi.org/10.31377/jdrssa.v6i1.671
Palermi S, Iacono O, Sirico F, et al (2022) The complex relationship between physical activity and diabetes: An overview. J. Basic Clin. Physiol. Pharmacol. 33
American Diabetes Association (2020) Classification and diagnosis of diabetes: Standards of Medical Care in Diabetes-2020. Diabetes Care 43:. https://doi.org/10.2337/dc20-S002
Bomberg EM, Rosenmoss S, Smith M, et al (2019) Diabetes-Related Health Care Utilization and Dietary Intake among Food Pantry Clients. Heal Equity 3:. https://doi.org/10.1089/heq.2019.0102
Kaiser AH, Hehman L, Forsberg BC, et al (2019) Availability, prices and affordability of essential medicines for treatment of diabetes and hypertension in private pharmacies in Zambia. PLoS One 14:. https://doi.org/10.1371/journal.pone.0226169
Sharma P, Hajam YA, Kumar R, Rai S (2022) Complementary and alternative medicine for the treatment of diabetes and associated complications: A review on therapeutic role of polyphenols. Phytomedicine Plus 2
Kumar S, Mittal A, Babu D, Mittal A (2020) Herbal Medicines for Diabetes Management and its Secondary Complications. Curr Diabetes Rev 17:. https://doi.org/10.2174/1573399816666201103143225
Upendra Rao M, Sreenivasulu M, Chengaiah B, et al (2010) Herbal medicines for diabetes mellitus: A review. Int. J. PharmTech Res. 2
Abo-Shady AM, Gheda SF, Ismail GA, et al (2023) Antioxidant and Antidiabetic Activity of Algae. Life 13
Quitério E, Soares C, Ferraz R, et al (2021) Marine health-promoting compounds: Recent trends for their characterization and human applications. Foods 10:. https://doi.org/10.3390/foods10123100
Zhou X, Zeng C (2023) Diabetes remission of bariatric surgery and nonsurgical treatments in type 2 diabetes patients who failure to meet the criteria for surgery: a systematic review and meta-analysis. BMC Endocr Disord 23:. https://doi.org/10.1186/s12902-023-01283-9
Affinati AH, Esfandiari NH, Oral EA, Kraftson AT (2019) Bariatric Surgery in the Treatment of Type 2 Diabetes. Curr. Diab. Rep. 19
Nwokolo M, Hovorka R (2023) The Artificial Pancreas and Type 1 Diabetes. J. Clin. Endocrinol. Metab. 108
Dimou M, Laimer M, Stettler C, Bally L (2020) What is the current state of the artificial pancreas in diabetes care? Internist 61:. https://doi.org/10.1007/s00108-019-00713-y
Takaki T, Shimoda M (2020) Pancreatic islet transplantation: toward definitive treatment for diabetes mellitus. Glob Heal Med 2:. https://doi.org/10.35772/ghm.2020.01057
Cayabyab F, Nih LR, Yoshihara E (2021) Advances in Pancreatic Islet Transplantation Sites for the Treatment of Diabetes. Front. Endocrinol. (Lausanne). 12
Wahowiak LK (2018) National Standards for Diabetes Self-Management Education and Support: What's Changed. AADE Pract 6:. https://doi.org/10.1177/2325160317742909
Molla GJ, Ismail-Beigi F, Larijani B, et al (2020) Smoking and Diabetes Control in Adults With Type 1 and Type 2 Diabetes: A Nationwide Study From the 2018 National Program for Prevention and Control of Diabetes of Iran. Can J Diabetes 44:. https://doi.org/10.1016/j.jcjd.2019.07.002
Johnson N, Melton ST (2016) Using a social marketing approach to understand barriers to diabetes self-management education utilization. Diabetes 65:

Chapter 3
Diabetes: Secondary Complications

R. Parameswari, P. Madhan Kumar, S. Azhagu Pavithra, S. Janani Iswarya, T. Yogesh, and R. Babujanarthanam

Abstract Diabetes mellitus (DM) is an endocrine-related chronic disease affecting 637 million people worldwide, with the highest mortality rate. It is a metabolic ailment caused by decreased or absent insulin release from the pancreatic islets of Langerhans, resulting in type 1 and type II disorders. Type II diabetes accounts for 75% of all instances worldwide, as are concomitant adaptive illnesses such as obesity, an unhealthy diet combined with a sedentary lifestyle, smoking, and drinking. Poor infrastructure and late diagnosis are the most common causes of high morbidity and death rates in the world. Sedentary and lousy eating habits have become more prevalent in recent years, making middle-aged adults a prime target for this campaign. Developing nations are the most vulnerable to diabetes, and in rural regions of the developing world, people still lack understanding of precautions, and staying healthy is still a possibility. It's crucial to stress the need for early diagnosis, as it can significantly improve outcomes. Secondary complications, such as hypertension, diabetic foot ulcer, glycosuria, stroke, diabetic retinopathy, and nephropathy, are more commonly documented as death-causing causes in people with diabetes mellitus. Furthermore, each year, an additional 3.3% of obese people develop and become diabetics. Nonetheless, effective therapy is necessary for every indication of diabetes, even if no symptoms are present. The current chapter focuses on diabetes and its secondary consequences at each stage, including inadequate insulin secretion and insulin-associated cellular damage in humans.

Keywords Diabetic foot ulcer · Diabetes mellitus · Insulin · Retinopathy · Nephropathy

---

R. Parameswari · P. M. Kumar · S. A. Pavithra · S. J. Iswarya · T. Yogesh ·
R. Babujanarthanam (✉)
Department of Biotechnology, Thiruvalluvar University, Vellore, India

R. Raja et al. (eds.), *Algae in Diabetes Management*,
https://doi.org/10.1007/978-981-95-0571-5_3

1 Introduction

The International Diabetes Federation stated that 417 million individuals were at risk of diabetes mellitus, and it is expected that 486 million will be by 2045, with a potential increase of 111 million between 2019 and 2045 as severe cases range from 35 to 66 years. Diabetes kills 1.5 million people and accounts for around 58% of all deaths in adults under the age of 70. According to data, around 46,000 victims report renal illness caused by increased diabetes, as well as 20% of CVD-related abnormalities exacerbated by diabetes. Furthermore, from 2000 to 2019, 3% of death rates in lower and middle-income countries were recorded as noncommunicable illnesses in the top four noncommunicable diseases (CVD, cancer, pulmonary, and diabetes) among people aged 30 to 69 years old (Tomic et al. 2022). Diabetes mellitus (DM), a form of chronic metabolic condition, is caused by endocrine pancreatic islets of Langerhans secreting insulin, which can lead to hyperglycaemia, or excessive glucose levels in the blood, which is connected with micro- and macrovascular complications. Diabetes mellitus is also characterized by a variety of aetiologic variables regulating major metabolisms, such as carbs, proteins, and lipids, resulting in low insulin or no insulin from beta cells of the Langerhans of the pancreas, resulting in a chronic hyperglycaemic syndrome (Abel et al. 2024). Figure 3.1 depicts the conditions that, according to the Diabetes Federation, are diagnostic hallmarks for confirming diabetes. Most health concerns, including cancer and neurological disorders, are caused by increased reactive oxygen species (ROS) produced by unpaired free electrons, which have also been linked to the promotion of diabetes and are responsible for the highest mortality and morbidity globally (Tattersall and Matthews 2024). According to reports, oxidative stress has a substantial influence on the development and persistence of consequences from endocrine issues in diabetes mellitus. It was agreed upon previously (Dilmurodovna 2024; Pedersen et al. 2024; Yusuf et al. 2024). According to Browning et al., oxidative stress of ROS and free radical electrons are unavoidable participants in the molecular pathophysiology of insulin deficiency/absence that leads to diabetic complications (Brownee et al., 2018). Redox equilibrium and its disruption of the cell damage cellular membranes, followed by important biomolecular involved in translation, transcription, and replication errors that cause physiological changes (Chen et al. 2024). A compromise between insulin production and insulin-disbursed action causes increased oxidative stress, complicating diabetes. It is controversial if oxidative stress is like two sides of the same coin, promoting diabetes when it discharges but also causing the patient's condition to deteriorate (Alam et al. 2014; Lebovitz and Banerji 2004).

Numerous in vitro and in vivo studies demonstrated how cytokines, TNF-α, and AID (autoimmune disease reactions)-induced ROS hinder the development of diabetes. Furthermore, blood hyperglycaemia stimulates the production of ROS by de

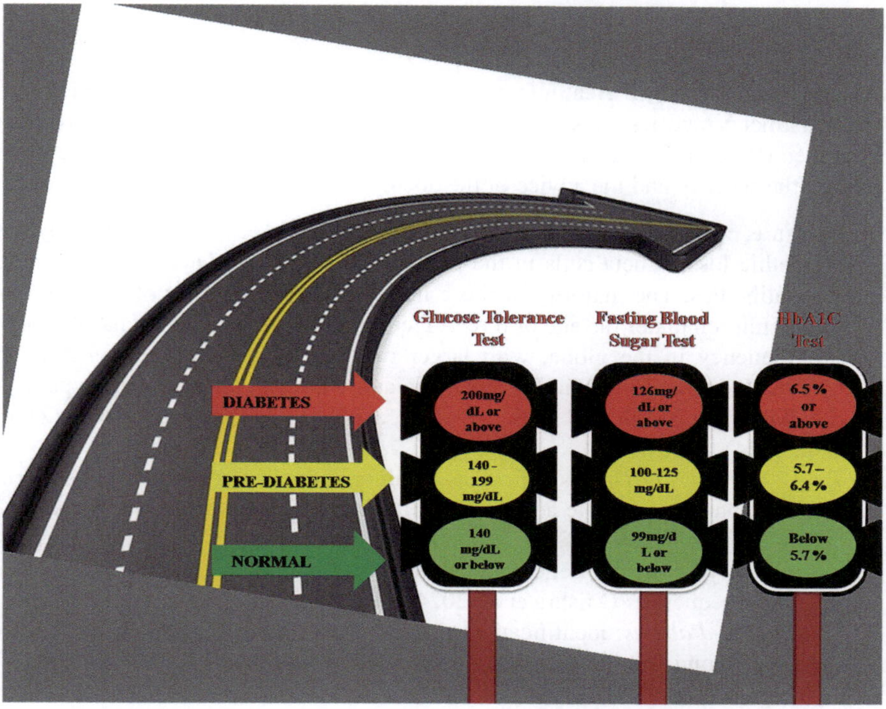

Fig. 3.1 Hall mark value of diabetes mellitus to judge the criteria's

novo synthesis, which directly outweighs the antioxidant scavengers in metabolism (De Deus et al. 2023; Morze et al. 2018). Chronic hyperglycaemic patients showed persistent ROS and permanent enzymatic and non-enzymatic antioxidant decreases in the majority of organs and tissues, which further contributes to the worsening of ROS. Additionally, this explains why diabetic folks have more oxidative cells in their organs than healthy people (Fakhruddin et al. 2017). Through in vitro and in vivo experiments, a variety of molecular and cellular mechanisms address how free radicals exacerbate diabetic problems (Busik et al. 2008; Wu et al. 2022a, b). Serious consequences include hyperosmolar syndrome, ketoacidosis, non-ketotic syndrome, coma, disorientation, and hypertrophy vacuoles could result from poor management of diabetes when it is uncontrolled (Al Shehabi et al. 2024). The current chapter concentrates on several variables that affect glucose oxidation, including glycolysis and lipid metabolism damage brought on by hyperglycaemic situations linked to secondary problems.

1.1 Types

According to the World Health Organization's trademark, diabetes is divided into three distinct categories, plus one undetermined kind; thus far, four types have been identified (Riddell et al. 2024). Future classifications may be possible, but they will rely on the severity and the advice of the doctor.

1. Insulin is completely absent in type 1 diabetes (TIDM), which is caused by the idiopathic loss of beta cells in the pancreas and the total lack of autoimmune autoantibodies. The majority of those impacted are individuals of all ages and small children under the age of three. According to reports, Europe has the largest frequency in the globe, with larger rates seen in some Saharan African nations, India, and the Middle East. Clinically, there is uncontrolled hyperglycaemia, which is followed by dehydration, acidosis, glaucoma, and coma (Zhou et al. 2024).
2. *Type 2 diabetes (TIIDM)* is defined as a failure of the pancreatic beta cells' islets of Langerhans to generate adequate insulin, which is associated with hyperglycaemia aggravated by ROS and secondary complications. This group of persons has a history of insulin deficiency and control, and when exposed, they suffer severe consequences (Mishra et al. 2024; Zhang and Luo 2024).
3. *Gestational diabetes*: identification of hyperglycaemic condition at the present status of second and third trimester of pregnancy will be normal after the delivery (Elgazzaz et al. 2024).
4. *Unidentified specific types of diabetes*: Glucocorticoid-induced drugs cause allergic reactions that can lead to monogenic diabetes syndromes, such as diabetes onset in young people and neonatal diabetes, followed by organ transplantation, which causes poor pancreas function in both exocrine and endocrine systems, resulting in pancreatitis, among other things (Ahmed et al. 2024; Sakini et al. 2024).

2 Diabetic Retinopathy

Diabetic retinopathy (DR) is becoming more and more common around the world at a startling rate. According to WHO predictions, India will have the largest proportion of adults suffering from diabetes worldwide by 2030, rising between 19 million in 1995 reaching 80 million. Numerous undiagnosed and uncontrolled diabetic individuals develop blindness as a result of inadequate screening and treatment facilities, mostly during the primary and secondary care levels. Diabetes prevalence and hyperglycaemia severity are the primary risk factors associated with DR. DR has a lengthy latent phase and is a chronic condition (Chowdhury 2014). Type 1 diabetes affects 10–15% of diabetics, while type 2 diabetes affects the remaining individuals. Between 81 and 90% of people with type 1 diabetes acquire DR within 10 years, and in 20 to 30 years, the incidence increases to 95%. Proliferative DR is present in

30–50% of these patients. Ten years after being diagnosed with type 2 diabetes, 67% of people acquire DR, with 10% exhibiting proliferative DR characteristics.

The prevalence of retinopathy in newly diagnosed diabetics might reach up to 5%. The most common cause of the development of legal eyesight among Americans of working age is complications of (DR). Loss of vision is more likely in cases of advanced DR (Alam 2021). Additionally, those with DR are more likely to die, and have limb amputations, diabetic nephropathy, coronary heart disease, and stroke. Hypertension, inadequate glycaemic management, and the length of diabetes mellitus are the primary causes for the onset or advancement of DR. Vascular problems are the primary cause of diabetes's high death and morbidity rates. Diabetic retinopathy (DR) affects over 45% of the 400 million persons with type 2 diabetes globally. DR is an age-dependent illness that progresses in phases, as diabetes progresses to retinopathy gradually. At 10 years following a diabetes diagnosis, the incidence rises to 50%, and at 25 years, it reaches 90%. The risk of blindness is greatly increased by diabetic macular oedema (DME), the most chronic form of DR. There are two types of diabetic maculopathy: ischemic maculopathy and diabetic mellitus (DME) (Browning et al. 2018). The earlier kind is triggered by complicated and multifactorial changes in the retinal capillaries. Retinopathy typically manifests in DM2 patients seven years after the disease is diagnosed. Reactive oxygen metabolites are elevated due to DR-induced reactive oxygen species (ROS). Nitric oxide (NO), the enzyme catalase, glutathione peroxide, and lipo peroxide are among the macromolecules whose catabolites rise as diabetes retinopathy improves in type 2 diabetes patients. High blood sugar and diabetic-induced retinal vascularization activation are associated with increased arginase enzymatic activity and decreased NO bioavailability. Anti-VEGF medications have been used to treat DME patients, improving their vision and slowing the progression of their DR (Gong et al. 2022).

2.1 Diabetic Retinopathy Mechanism

O-linked β-N-acetylglucosamine (O-GlcNAc) increases the activity of dumping modification after translation of Ser/Thr residue in the hexosamine biosynthesis pathway. Diabetes is caused by the proteins' decreased ability to absorb blood glucose as a result of the increasing input of residues. Numerous intrinsic and cellular proteins are known to be regulated by O-GlcN acylation, just like in protein phosphorylation. Insulin resistance can result from O-GlcNAc alteration, which can interfere with the regular signalling pathways that control insulin sensitivity. Hyperglycaemia is made worse by this alteration, which impacts the activity of important proteins that regulate absorbing glucose and metabolism. Therefore, a contributing component to the onset and progression of diabetes is the deregulation of O-GlcNAc cycling (Williams 2009). The primary cause of myopia in working individuals is caused by DR.

The mechanism behind this is an inflammation within the retinal blood vessel, which is brought on by the accumulation of arteriosclerotic plates from elevated blood glucose. This leaves the plates susceptible to microrupture, which allows fluid to leak into the retina. Retinal microvasculature is distorted by the formation of new vessels, which leads to retinal detachment if left uncontrolled. This type of occlusion causes microaneurysms to develop, which in turn causes greater leucocytosis and thickening of the basal vascular membrane. When blood glucose levels are high, PKC-δ signalling is stimulated, which leads to the death of pericyte cells (Agarwal 2011). Pericytic cells undergo apoptosis as a result of this signalling cascade, which also dephosphorylates PDGF receptors, reduces their downstream signalling, and increases the expression of p38 mitogen-activated protein kinase and protein kinases C-δ (PKC-δ, expressed via Prkcd). Pericytic cell death progressively alters the structure of the retina's micro vessels, which leads to extensive haemorrhage and disruption within the blood-retinal barrier, microaneurysm exudates, and neovascularization on the retinal surface. Cell-cell junctions, which comprise the blood-retinal barrier, are decreased because of the pericytes' cell death (Rudraraju et al. 2020).

The development of microaneurysms, exudates, and neovascularization is caused by the fluid and leakage of proteins made available by the breakdown of cell-cell junctions. Biomolecules such as lipoprotein-PLA2, anti-inflammatory cytokines (the expression of TNF and IL-1), with secretory phospholipase A2 IIA are being implicated as being associated with vascular inflammation. High blood glucose levels are the root cause of this illness, therefore aggressive insulin therapy can either slow or stop the ongoing damage (Khunti and Millar-Jones 2017). Additionally, thioredoxin, an innovative gene and medication, has been discovered to be a powerful therapeutic agent to minimize pericytic cell division and arrest the progression of DR. An antioxidant protein called thioredoxin aids in lowering oxidative stress in cells. Blood-retinal barrier integrity and micro vessels can be preserved by thioredoxin's ability to prevent oxidative damage and prevent pericytes from dying. Preventing the advancement of diabetic retinopathy (DR) is largely dependent on maintaining the retinal microvasculature (Lenin et al. 2018).

2.2 *Inflammation and Immune Cell Activity in DR*

Immune cell regulation and inflammation control are essential for the retina's health and proper operation. This reaction encourages homeostasis and healthy tissue repair, but over time, persistent para-inflammation aids in the development and advancement of certain disease processes. The retina shows several signs of persistent, subclinical inflammation as diabetes increases, such as the creation of inflammatory chemicals and the activation of immune cells. Increased leukocyte-endothelial interactions in the vessels of the retina are one of the immune cell types that become activated during the initial phases of DR. High blood sugar, or hyperglycaemia can cause retinal cells to experience more oxidative stress. Pro-inflammatory cytokines are then released by activated circulating microglial cells along with additional immune cells as a consequence of this stress (Forrester et al. 2020). Diabetic retinopathy advances as a result of these inflammatory chemicals, which worsen retinal damage. Myeloid cells in circulation, such as neutrophils and monocytes, attach to active vascular endothelium, a characteristic of leucocytosis. Leucocytosis can cause hypoxia in the tissues & higher vascular permeability by obstructing circulation in the retinal microvasculature. Oedema, fluid leakage, and further retinal injury are caused by these alterations, which also aid in the collapse of the blood-retinal barrier. These consequences may worsen diabetic retinopathy's severity and visual loss over time.

Leukocyte-endothelial interaction may damage the retinal vascular endothelial cells, and the tissue which sustains them by blocking capillaries and generating inflammatory cytokines and superoxide (Antonetti et al. 2021). Besides the role of intravenous immune cell types, the development of inflammatory conditions carried on by diabetes is probably mediated by the activation of microglia, which are the retina's resident immune cells, as well as by invading monocytes. VEGF may contribute to the development of DR in addition to its well-established effects on vascular permeability, as seen by the slower progress to capillary dropping out seen in DME individuals treated with anti-VEGF inhibitors. Vascular endothelial growth factor, or VEGF, is an amino acid that is essential for encouraging the development of new blood vessels. Elevated VEGF levels are linked to increased vascular permeability in diabetic retinopathy, which may occur in oedema, leakage, and the breakdown of the blood-retinal barrier. In addition, VEGF increases retinal damage and raises the potential risk of serious myopia by promoting the neovascularization process, which results in the formation of fragile and aberrant blood vessels. Diabetes raises retinal concentrations of cytokines that cause inflammation including IL-1β & TNF-α in addition to VEGF (Neelam et al. 2012). These inflammatory cytokines have been associated with contributing to important pathological ends in DR, including vascular permeability and capillary abandonment. By stimulating inflammation and endothelial dysfunction, Cytokines that promote inflammation such as IL-1β & TNF-α contribute significantly to the pathophysiology of diabetic retinopathy. By promoting leukocyte adherence and transmigration, these cytokines can cause endothelial cells to produce adhesion molecules, exacerbating damage to tissues and inflammation. The blood-retinal barrier may also be compromised, which could result in a higher vascular permeability and accelerate the development of retinal inflammation & microvascular occlusion (Yang et al. 2020).

2.3 *Pathology and Its Classification of Diabetes Retinopathy*

Nevertheless, diabetes also affects vascular cells that are crucial to the onset and growth of DR, albeit in tandem with the vasculature. Non-proliferative diabetic retinopathy (NPDR) and the more severe phase of PDR are the two primary stages of DR. Diabetes affects nonvascular cells, including glial and retinal neuron cells, which leads to retinal deterioration and malfunction. The stress and inflammation that these cells experience can worsen vascular damage and hasten the development of diabetic retinopathy. Because it directly impacts the macula, the core region of the retina that is in charge of sharp and detailed vision, macular oedema caused by diabetes is important in DR (Shruti 2019). Swelling and thickness of the macula caused by fluid buildup from leaking blood vessels impairs vision. Among those suffering from diabetic retinopathy, this condition emphasizes the importance of early detection and care to avoid permanent vision loss. Diabetes-induced blood-retinal barrier (BRB) disruption results in vascular fluid and circulating protein leaking within the neuronal retina, which leads to DME. Atypical retina thickening and frequent cystoid oedema within the macula are caused by fluid extravasations into the neural retina. Understanding that vascular and nonvascular elements interact is essential to creating all-encompassing DR treatments. Diabetic macular oedema (DME) is a significant DR manifestation occurring within each DR phase of severity including NPDR and PDR. For those suffering from DR, it's a very prevalent manifestation of loss of vision and is a major new diagnosis in DR. While peripheral lipids, lipoproteins, and DR are not strongly correlated when assessing retinopathy risk in the context of specific individuals, dyslipidaemia and hypertension may also have an impact on DR (Khuu 2015). Antihypertensive medications may also help people with hypertension, which might be correlated to an increased risk for DR. Current research, however, has shown that, in comparison to routine control, more stringent blood pressure management had no additional positive effects on the course of retinopathy. Blurred vision, floaters, black or empty spots in the eyesight, and trouble seeing colours are some of the symptoms of diabetic retinopathy. If treatment is not received, the condition might worsen over time and result in blindness. For the early identification and treatment of these signs to avoid long-term harm, routine eye exams are essential (Abel 2014). While it is readily apparent optimizing systemic risk factors is essential, hyperglycaemia assessed via HbA1c can only contribute about 10% of the overall DR vulnerability and in certain cohorts, the combined risk of hypertension and dyslipidaemia may be less than 10%. Based on these facts, it is highly probable that additional unidentified variables are also important during the progression and development of DR (Kannel 1987).

3 Retinal Microvasculopathy and Hyperglycaemia

It has been widely designated that DR is a microvascular illness. The pathophysiology of retinal microvascular injury is thought to be significantly influenced by hyperglycaemia. Hyperglycaemia-induced vascular injury was previously associated with several metabolic processes, such as the polyol pathway, protein-kinase C (PKC) pathway, accumulation of progressed hydrolysis final products (AGEs), and the route for hexosamine. Because the PKC pathway alters blood flow and vascular permeability, it plays a role in diabetic retinopathy. PKC activation exacerbates retinal damage by altering the expression of genes related to inflammation and angiogenesis (Babel and Dandekar 2021). This triggers aberrant new blood vessel creation and increased vascular leakage, both of which are hallmarks of diabetic retinopathy. These alterations continue to exist as the result of metabolic autoregulation to boost the metabolism of the retina in individuals with diabetes.

Pericytes are in charge of supplying structural support for capillaries; their absence causes the capillary walls to out pouch locally. This procedure is linked to the development of microaneurysms, the first clinical indication of DR. Capillaries that aren't supported grow weak and are more likely to develop tiny outpouchings called microaneurysms (Caplan 2013). Fluorescein angiography and other imaging methods can detect these microaneurysms, which are among the first indications of diabetic retinopathy. Microaneurysms may burst over time, causing haemorrhages and also compromising the retina's blood supply. VEGF or vascular endothelial growth factor is released when retinal tissue experiences hypoxia or ischaemia as the condition increases, encouraging the development of new, delicate blood vessels. Proliferative diabetic retinopathy can develop as a result of this neovascularization, which can lead to more serious side effects such as haemorrhage and retinal detachment. Furthermore, endothelial cell death and basement membrane thickening are also linked to pericyte loss found during the DR pathogenesis, which together lead to the impairment of the BRB (Dorrell et al. 2007). Additionally, a significant loss of endothelial cells and pericytes causes capillary occlusion and ischemia. VEGF is upregulated in response to retinal ischemia/hypoxia because it activates HIF-1, or hypoxia-inducible factor. According to additional data, phospholipase (PLA2) VEGF upregulation is also triggered by an increase in the diabetic state. Diabetes-related increases in phospholipase A2 (PLA2) can accelerate the development of diabetic retinopathy by encouraging the excessive production of a protein known as VEGF or vascular endothelial growth factor.

The activity of this enzyme fuels the retina's inflammatory processes, which in turn cause vascular permeability to rise and new, delicate blood vessels to grow (Mandal et al. 2022). Thus, PLA2 increases the likelihood of serious consequences such as retinal haemorrhages and detachment in addition to speeding up neovascularization. VEGF, a crucial component implicated in the development of DME and PDR, is anticipated to increase vascular permeability by phosphorylating encapsulating proteins, including zonula occludens-1 (ZO-1) and occludin. Additionally, VEGF is an angiogenic agent that activates endothelial cells to enhance their proliferation of MAP (mitogen-activated protein). Angiogenic factors other than VEGF are expected to be contributing to this microvasculature's change during DR, which could lead to new therapeutic targets (Caldwell et al. 2005). Angiopoietins, platelet-derived growth factors (PDGFs), and fibroblast growth factors (FGFs) are additional angiogenic factors that can contribute to diabetic retinopathy. Angiopoietins, especially angiopoietin-2, can destabilize vessels and collaborate with VEGF to stimulate neovascularization; FGFs support the survival and proliferation of endothelial cells; and PDGFs aid in the infiltration of pericytes. These factors are potentially significant in the emergence of diabetes-related retinopathy and might serve as targets for novel therapeutic interventions (Gacche and Meshram 2014).

4 Retinal Neurodegeneration

During the early stages of DR's progression, retinal neurodegeneration occurs. Rats with diabetes can exhibit retinal neuron apoptosis as soon as a month after the disease is induced. Programmed cell death, or apoptosis, plays a role in the degeneration of retinal neurons, which impairs vision and makes diabetic retinopathy worse. The delicate balance necessary for regular visual function is further upset by this process, which intensifies the retinal structure's deterioration. Therefore, apoptosis targeting in treatment plans may help slow the evolution of diabetic patients' retinal degeneration. In diabetic people and animals, retinal neurons have been found to exhibit an increase in pro-apoptotic indicators including cleaved caspase-3, Fas & Bax (Galvão 2013). Degeneration of the retina in DR has been associated with mitochondrial malfunction. The retinal gene expression for cytochrome c and apoptosis-induced factor (AIF), two pro-apoptotic mitochondrial proteins, has been demonstrated to significantly increase the diabetic participants' donor eyes. Due to its disruption of cellular energy generation and promotion of oxidative stress, mitochondrial dysfunction is a major factor in the development of diabetic retinopathy. Apoptotic pathways are triggered by this stress, which exacerbates retinal neurodegeneration.

Novel treatment strategies to preserve retinal neurons and reduce the evolution of diabetic retinopathy may be created by comprehending and addressing mitochondrial malfunction (Kowluru and Chan 2007). High glucose administration was correlated to increased mitochondrial disintegration and cell apoptosis, according to in vitro studies. The role of oxidative stress as diabetes-induced retinal deterioration is being extensively examined along with mitochondrial damage. The formation of ROS or reactive oxygen species is markedly elevated in the retina of diabetic mice. Suppression has great potential for diabetic retinopathy treatment. Reducing ROS levels minimize oxidative stress, which helps maintain retinal neurons' structural integrity and inhibits to prevent deterioration. This method may be able to prevent or even reverse the disease's associated vision impairment, providing diabetes patients with a preventative measure to preserve their visual function (Jenkins et al. 2004). Anti-inflammatory treatments that target the cytokines and chemokines that contribute to inflammation may also be a potential option. Inhibitors of angiogenesis, which stop the growth of blood vessels, can also help treat diabetic retinopathy. Neuroprotective drugs that improve retinal neuron survival and function are also being investigated to offer complete protection against retinal injury (Osborne et al. 1999).

5 Factors Associated with DR in Diabetes Type 2

Diabetic retinopathy is perhaps the most extensive and extensive study of DR, its incidence, prevalence, and associated risk factors that has been performed to date. According to the first cross-sectional study, which had 2366 patients and was conducted between 1980 and 1982, the incidence of DR was 39% among participants with the older start of diabetes, or type II diabetes mellitus, and 71% among those with diabetes beginning before the age of 30, or type I diabetes mellitus. Ocular perfusion pressure and microalbuminuria are additional variables that are independently linked to the occurrence and severity of DR. Micro aneurysm count, as well as characteristics of insulin-resistant syndrome including serum triglycerides and the ratio of the waist to the hips, are indicators of future development. Small levels of protein in the urine, or microalbuminuria, are a sign of kidney impairment early on and are strongly associated with the development of diabetic retinopathy (Nair 2018). It indicates damage to the blood vessels, which may also impact the retinal vessels of the eye. Assessing microalbuminuria levels can therefore aid in anticipating when DR will worsen and help direct prompt actions. Several factors, many of which are connected to prolonged treatment and evolution of type 2 diabetes, affect the chance of developing diabetic retinopathy.

One of the primary indicators of risk for diabetes is its duration. People who have had type 2 diabetes for a long time are more vulnerable because long-term exposure to high blood sugar (hyperglycaemia) damages the retina's tiny blood capillaries (Ramasarma and Rafi 2016). Damage to the retina can worsen over time, resulting in microvascular alterations such as fluid leakage, haemorrhage, or the development of new, aberrant blood vessels that can cause blindness. Diabetes patients should concentrate on maintaining ideal blood sugar levels by routine monitoring and medication adherence to lower their chance of developing diabetic retinopathy. A balanced diet full of vegetables, fruits, and whole grains, as well as frequent exercise, can help control weight and enhance insulin sensitivity. Hypertension, or high blood pressure, exacerbates diabetic retinopathy, increases the chance of severe visual impairment, and speeds up retinal blood vessel damage. Diabetic retinopathy can result from thicker blood vessel walls brought on by high cholesterol, which also impairs circulation. Controlling cholesterol levels is essential for people with type 2 diabetes to maintain eye health (Zhang et al. 2010). High cholesterol can restrict blood vessels and reduce blood circulation to the retina by forming fatty deposits in the arteries. Because poor circulation can aggravate the progression of diabetic retinopathy, patients must manage their cholesterol levels through a heart-healthy diet and medication regimen. For diabetic retinopathy to be identified and treated in its initial phases, before vision loss begins, routine eye exams are essential. Laser therapy is one possible treatment for diabetic retinopathy; it helps to reduce oedema and seal bleeding blood vessels.

Anti-VEGF injections, which function by preventing the development of aberrant blood vessels in the retina, are an additional choice. In increasingly severe cases, vitrectomy surgery might be required to remove the vitreous gel and scar tissue that could pull on the retina and damage vision (Smith and Steel 2015; Savastano et al. 2018). Hypertension, or high blood pressure, makes this risk much worse because it can hasten the deterioration of retinal blood vessels, exacerbating diabetic retinopathy and raising the possibility of serious visual disabilities. The increase in thickness of the walls of blood vessels caused by elevated cholesterol levels can also affect circulation and lead to the development of diabetic retinopathy. Controlling cholesterol is essential for type 2 diabetes to keep their eyes healthy. Excessive cholesterol can cause fatty deposits to form within the blood vessels, narrowing them and lowering the flow of blood to the retina. Patients must follow an exercise regimen and medication schedule to maintain healthy cholesterol levels because this impeded circulation might worsen the evolution of diabetic retinopathy (Kang and Yang 2020). In diabetic retinopathy, early management is essential since it can preserve eyesight and dramatically delay the disease's progression. Before symptoms appear, medical professionals can identify retinal abnormalities and treat the condition to stop additional harm. Effective control of blood pressure, cholesterol, and blood sugar can help lower the chance of problems and enhance general eye health. Since the incidence of diabetic retinopathy rises with age, especially in individuals who have suffered from type 2 diabetes for a long time, age is another important risk factor.

Additionally, poor regulation of blood sugar and the emergence of other problems like high blood pressure and cholesterol are closely associated with obesity and inactivity, all of which increase the risk for diabetic eye damage (Waly and Hamed 2018). Increased leakage and oedema may result from hypertension's further harm to the retina's already delicate blood vessels. This accelerates retinal tissue degradation and contributes to vision loss, worsening the course of diabetic retinopathy. To prevent diabetic retinopathy in people with diabetes, blood pressure must be effectively controlled with medication and lifestyle modifications. Another worrying element is the existence of kidney failure (diabetic nephropathy) since the primary causes of vascular damage are similar in both illnesses. In addition, smoking and ethnicity are known risk factors; smoking exacerbates vascular damage, and ethnic groups are more likely to experience difficulties. These risks are particularly prevalent in African American, Hispanic, and Native American populations (Nazar 2014). Diabetic nephropathy must be identified early to enable prompt treatments that can stop or slow the course of kidney damage.

Healthcare professionals can limit the progression of renal disease by detecting problems early on and putting blood pressure and blood sugar control measures into place. Additionally, early intervention can lower the likelihood of renal failure reaching its final stage, enhancing the individual's level of life, and lowering the need for kidney transplantation or dialysis. The risk of developing diabetic retinopathy may be influenced by genetic factors (Song et al. 2017). Even if blood sugar is properly managed, some people may be genetically predisposed to the illness. Ongoing eye tests are consequently essential for those with type II diabetes since diabetic retinopathy vision loss can be greatly avoided or postponed with early detection and prompt treatments. Diabetic retinopathy is influenced by genetic factors, including differences in genes that impact inflammatory responses and blood vessel integrity. Regardless of lifestyle choices or diabetes treatment, these genetic predispositions can raise a person's vulnerability to the illness (Moosavi-Movahedi et al. 2021). It is essential to comprehend these genetic impacts to create individualized treatment programs and to determine which patients would benefit from more regular monitoring. Genetic analysis can also be used to identify those who are in danger of diabetic retinopathy, permitting early therapies and closer monitoring.

Additionally, by identifying those who are susceptible to issues like vision loss associated with diabetic retinopathy, genetic analysis can assist in providing more individualized therapies and preventative measures (Patibandla et al. 2024).

6 Prevention and Treatment of Diabetic Retinopathy

In a determination to reduce the disease's prevalence and growth, new treatments for DR are now being investigated. Even individuals with advanced DR have found hope in the new salvation treatment options that have emerged; it is estimated that 95% of these people could maintain their eyesight if they receive treatment before the retina is seriously damaged. These methods of saving the eye include laser surgery, corticosteroid injections, vitrectomy, and anti-VEGF injections. Inhibiting the formation of aberrant blood circulation in the eye is how anti-VEGF medication reduces swelling and enhances vision. Numerous people have found that this medication is successful in maintaining their vision and reducing the course of diabetic retinopathy (Meharwade 2019). Corticosteroids are useful for the treatment of diabetic retinopathy because they assist in decreasing swelling and inflammation in the retina, which is associated with enhanced vision and reduced severity of the disease. Corticosteroids may additionally enhance the efficacy of additional therapies, such as anti-VEGF treatment, by offering a multifaceted approach for controlling DR. This makes anti-VEGF injections an intriguing possibility for maintaining vision for those with DR by targeting the fundamental cause of the disease. Treatment for DR is based on its type or stage (Simo and Hernandez 2015).

It is important to keep in mind that this condition can be categorized as either proliferative or non-proliferative macula oedema. A screening program is advised for mild moderate to severe non-proliferative DR to control blood sugar and maintain ongoing ophthalmological revision. The best way to manage diabetes and DR is to keep the HbA1c level between 6 and 7% and control the diabetes. Future developments in gene therapy could open up new avenues for the treatment of diabetic retinopathy by focusing on particular genetic variables that are implicated in the condition (Pei et al. 2024). Additionally, more reliable and durable therapeutic benefits may be possible with the development of novel drug delivery mechanisms, such as sustained-release implants. The possibility of regenerative medicine, such as stem cell therapy, to restore sight for individuals with advanced DR and regenerate damaged tissues in the retina is additionally being studied by researchers. Because it is unknown how long type 2 diabetes has existed, the initial retinal examination should be done as soon as the condition is diagnosed. When retinal alterations are not present, annual follow-up exams are advised; otherwise, shorter intervals are advised (Sontheimer 2004). Since one in two women with proliferative diabetic retinopathy may experience worsening retinal abnormalities, pan-retinal photocoagulation using a laser should be carried out if the condition occurs before or soon after conception. Retinal laser photocoagulation can achieve a 50% reduction in this risk. Stem cell therapy introduces healthy cells that can develop into different types of retinal cells, potentially regenerating damaged retinal cells. These embryonic stem cells can restore adequate retinal function and enhance vision by replacing damaged or absent cells.

Additionally, stem cell therapy might encourage the production of growth hormones that aid in the restoration and regeneration of underlying retinal tissue (Shadmani et al. 2021). The retina's stem cell differentiation mechanism guides stem cells to differentiate into particular retinal cell types, including light receptors, ganglion cells within the retina, and retinal pigment epithelium. Extrinsic inputs from the retinal environment and intrinsic genetic instructions combine to guide this differentiation. By this mechanism, stem cells progressively take on the traits and capabilities of mature cells in the retina, which eventually allows them to blend in with the remaining retinal tissue and aid in the restoration of vision. Undifferentiated cells called stem cells, which may grow and differentiate becoming any type of cell required, are available to aid in this healing process. Due to their ability to replace or repair damaged or diseased tissue, stem cells hold promise for application in regenerative medicine (Mahla et al. 2023).

7 Anti-angiogenic Therapy

7.1 Anti-VEGF Therapy

Superior outcomes from anti-VEGF medications that blocked all VEGF isoforms overshadowed their potential. The pathological situation, as opposed to the physiological, and the ocular neovascularization that results in PDR, is significantly influenced by VEGF, an angiogenic factor unique to endothelial cells. As a vasopermeable factor, VEGF also relaxes endothelial cell connections, increasing vascular permeability. It is important to block all VEGF isoforms since this completely blocks the mechanisms that promote aberrant blood vessel development and elevated vascular permeability. The abnormal angiogenesis and leakage that worsen disorders like PDR and DME are prevented by this method, which successfully slows their progression (Patel et al. 2023). Anti-VEGF treatments are therefore an effective means of protecting eyesight and halting more ocular damage. These effects are partially blocked in DR by VEGF inhibition, as evidenced by several new clinical studies and instances utilizing anti-VEGF agents. Despite the lack of a published level I uncontrolled clinical trial, Conferment may be an additional effective anti-VEGF drug (Liu et al. 2020).

The effectiveness of aflibercept, ranibizumab, and bevacizumab in eyes with centre-involved DME and VA was not different, according to a hypothetical, randomized, efficacy comparison evaluation. Which includes other anti-VEGF medications including bevacizumab, ranibizumab, and aflibercept, Conbercept seeks to prevent VEGF from causing aberrant proliferation and vascular permeability. Conbercept is unique in its structure of molecules, which may provide additional adsorption capabilities and perhaps increase its effectiveness, even though these well-established medicines have demonstrated comparable performance in treating disorders like DME (Peinado et al. 2023). However, it is still unclear whether Conferment is better than alternative anti-VEGF therapies in the absence of a level I randomized clinical trial. While all three drugs—bevacizumab, ranibizumab, and aflibercept—target and decrease VEGF activity, their precise molecular structures and processes vary. Bevacizumab is a completely monoclonal antibody, whereas ranibizumab is a shorter antibody fragment that allows for more effective tissue penetration.

Aflibercept functions as a VEGF retain by binding to VEGF using a fusion protein made up of parts of the VEGF receptors, potentially offering broader suppression of VEGF isoforms. If anti-VEGF therapy causes incomplete drying of the macula, focal laser added when compared to its utilization in a delayed way, the initial use does not enhance VA outcomes (Duker and Liang 2024). For diabetic retinopathy, intravitreal ranibizumab injections administered once a month for DME increase the percentage of eyes with a two- or three-gradual reduction in the intensity of the condition, decrease the percentage of eyes with a two- or three-step worsening of the condition, and decrease the percentage of eyes that progress to proliferation diabetic retinopathy. Additionally, linked to these injections is a lower chance of vision loss. Furthermore, these injections have little negative side effects and are well tolerated. Anti-VEGF treatment is currently recommended for DME that causes vision loss, while photocoagulation with lasers keeps proliferative DR eyes from experiencing significant vision loss (Pessoa 2022). For the treatment of DME, corticosteroids, which can be injected via implanted within the eye, can be used either alone or in conjunction with other medications or laser surgery.

Corticosteroids can aid in lowering vascular permeability and inflammation, two major factors that contribute to diabetic macular oedema (DME). By maintaining the blood-retinal barrier and preventing the production of inflammatory cytokines, they may lessen oedema and enhance visual acuity (Starace et al. 2021). Furthermore, individuals who are unresponsive to anti-VEGF medicines or who are contraindicated for them may benefit from corticosteroids. Diabetic macular oedema can be effectively treated with both corticosteroid and anti-VEGF treatments, although their modes of action differ. Anti-VEGF treatments reduce aberrant blood vessel development and leakage by focusing on the excessive production of a growth factor known as vascular endothelial growth contrarily, corticosteroids work to stabilize the blood-retinal barrier and reduce inflammation, which makes them a good substitute for patients who cannot or will not respond to anti-VEGF therapies (Cardones and Banez 2006).

7.2 Corticosteroids in Diabetic Retinopathy

Corticosteroids have demonstrated efficacy in the management of DME. The actual treatment for this diabetes consequence includes intravitreal injections of corticosteroids and implants, vascular endothelial growth factor (VEGF) inhibitors, and focused laser photocoagulation. Anti-VEGF antibodies have completely changed the way that DR is treated, but a sizable portion of patients do not improve with treatment, and there is growing evidence that chemokines and inflammatory cytokines in addition to VEGF might have a role in the disease process. The pathophysiology of DME is significantly influenced by inflammatory cytokines, which increase inflammation and vascular permeability in the retina (Mateos-Olivares et al. 2021). Fluid buildup and subsequent retinal oedema can result from these molecules' disruption of the blood-retinal barrier. Patients who don't react to conventional anti-VEGF medications may have other therapeutic alternatives if these inflammatory pathways are targeted. Several cytokines, including VEGF, are downregulated by steroid medications, dexamethasone, triamcinolone, along fluocinolone acetonide. Even though these medications raise intraocular pressure, dexamethasone implants have shown promise in treating this illness (Arrigo and Bandello 2021).

Intravitreal corticosteroids may be very helpful for pseudophakic patients, and intravitreal dexamethasone implants exhibited morphology and functional enhancement in eyes with DME as soon as one month of use versus a year after the treatment. For refractory DME, a sustained-release corticosteroid implant in conjunction with an anti-VEGF drug shows promise (Dugel et al. 2015). The blood-retinal barrier-stabilizing, anti-inflammatory, and anti-angiogenic properties of corticosteroids make them beneficial for treatment. Combining corticosteroids with anti-VEGF drugs provides a multifaceted treatment strategy for DME by tackling inflammation and angiogenesis. Improved visual results and more substantial decreases in retinal oedema may result from this dual process. Combination therapy may also lessen the need for frequent injections, improving patient adherence and treatment effectiveness generally (García-Pérez et al. 2013).

7.3 Integration of Laser Therapy and Pharmacotherapy

Whether combining medication with laser therapy-conventional, targeted, or subthreshold-can reduce the patient's treatment burden while still attaining the best possible therapeutic outcomes. Concerning visual progression and treatment load, several investigations have indicated that anti-VEGF and PRP combined therapy may be better than monotherapy in PDR. Combination therapy may improve treatment results by focusing on several pathways that contribute to the illness process. In patients with proliferative diabetic retinopathy (PDR), neovascularization (NV) may recede more quickly and significantly when laser therapy and medication are used together (Li Calzi et al. 2010). Furthermore, this strategy might reduce the number of treatments needed, enhancing patient adherence and general quality of life. Scatter (pre-retinal) photocoagulation may lower the risk of vision loss from PDR as compared to indefinite deferral, according to research from the Diabetic Retinopathy Study (DRS). Photocoagulation was found to drastically lessen severe sight loss after only 2 years of treatment. The precise process by which laser treatment causes neovascularization to regress and lower DME is still determined (Emerson and Lauer 2007).

Increased oxygenation and reduced retinal tissues are associated with decreased retinal blood flow, the direct repair of leaking microaneurysms, and stimulation of the pigment epithelium of the retina (RPE) are all considered to be implicated. Laser therapy continues to be a valuable adjuvant or rescue treatment, even if anti-VEGF therapy has progressively taken the lead in treating DR. Treatment options for proliferative diabetic retinopathy (PDR) include laser therapy and anti-VEGF therapy, both of which have advantages and disadvantages (Cook 2019). Peripheral vision loss may result after laser treatment including pre-retinal photocoagulation (PRP), which is very effective at lowering the possibility of severe visual loss by rupturing aberrant blood vessels. A non-invasive method with fewer side effects, anti-VEGF therapy inhibits neovascularization by targeting vascular endothelial growth factors; nonetheless, it may need more frequent treatments to remain effective. Combining anti-VEGF medication with laser therapy may improve treatment results by targeting several proliferative diabetic retinopathy (PDR) factors (Wu et al. 2019). Anti-VEGF therapy can supplement PRP by preventing neovascularization and minimizing fluid leakage, while PRP effectively breaks up aberrant blood vessels to avert severe vision loss. This dual strategy may reduce the number of treatments needed and the negative effects of each therapy, which will ultimately improve patient adherence and quality of existence (Maciel et al. 2016).

8 Diabetic Foot Ulcer

Diabetic foot ulcers are a serious health issue. In 2011, approximately 350 million individuals worldwide (6.6% of the population) had diabetes mellitus, and over 55 million in Europe. Estimates for 2025 indicate that there will be over 65 million patients with diabetic foot ulcers. Diabetic foot ulcers are mostly caused by neuropathy, poor circulation, and immunological deficiencies. Neuropathy, or damage to the nerves, can cause a loss of feeling in the feet, which makes it difficult for people to notice injuries, while inadequate circulation may decrease blood flow to the feet, impeding the healing process and raising the risk of infection. Diabetic foot ulcer complications are the main factor contributing to diabetic patients' hospitalization and amputation; in fact, diabetic foot complications account for 20–40% of healthcare resources used for diabetes (Wang et al. 2022). The population's ageing and growth, as well as lifestyle changes, are the primary elements contributing to the increase in the number of people with diabetes. Early identification of diabetic foot ulcers is critical in preventing serious complications and lowering the

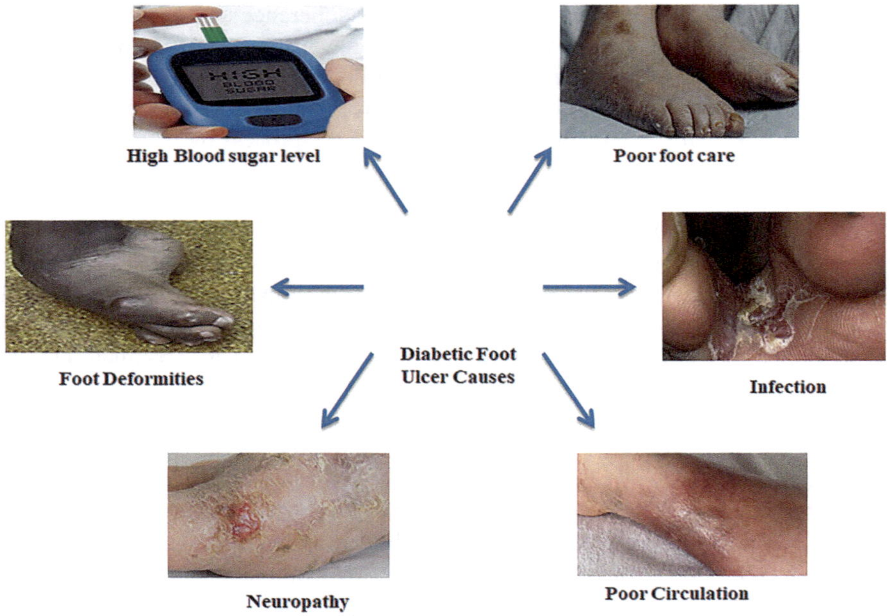

Fig. 3.2 Various complications of diabetes-associated foot ulcer complications

possibility of being hospitalized and amputations (Fig. 3.2). By detecting and treating ulcers early on, healthcare providers may implement more efficient measures to promote rehabilitation and prevent infections (Latlief et al. 2012). Self-care education and routine foot examinations are important methods in managing this component of diabetes care.

High serum glucose levels and abnormalities in the metabolism of carbohydrates, fats, and proteins are hallmarks of diabetes mellitus, a metabolic disease that can be caused via modifications to the actions, release, or both of insulin. Open sores or lesions that do not heal correctly on the feet are common symptoms of diabetic foot ulcers (DFUs). In addition to fluid or pus draining, symptoms may include warmth, redness, and swelling around the afflicted area (Ogoina 2011). Due to nerve injury, there may occasionally be a detectable odour and a decrease in foot sensation. These two elements may function alone, in concert, or conjunction with other illnesses including reduced joint mobility, biomechanical anomalies, microvascular disease, and heightened infection susceptibility.

According to certain research, the main causes of the challenges with DFU healing are either low levels of MMP inhibitors or the excessive and prolonged activity of metalloproteinases (MMPs). About 15% of people with diabetes generally develop these ongoing, non-healing neuropathic foot ulcers, which could lead to significant medical costs and a significant detrimental influence on their overall quality of life (Vileikyte 2001). Regular exfoliation for removing dead cells, the use of speciality dressings to keep the healing environment moist, and unloading techniques to relieve the pressure on the region that is damaged are some of the measures that may be used in conjunction to effectively manage DFUs. Furthermore, a thorough treatment strategy must include managing levels of blood glucose and treating any infections with the proper medications. In more difficult cases, advanced treatments such as growth factors, skin replacements, and hyperbaric oxygen treatment could also be taken into consideration to promote healing. Given that DFU accounts for over 85% of all patient foot amputations, there is a higher chance that a DFU may worsen and eventually result in amputation (Mohsin et al. 2024). To treat diabetic foot and preserve the leg, it is helpful to make clear how important it is to define key steps in diagnosing and therapeutic techniques. The number of foot amputations in the intended diabetic population can only be decreased by identifying and addressing the factors that can adversely affect prognosis, such as critical ischemia along with revascularization, osteomyelitis with its surgical treatments, compartmental syndrome, and emergency surgery.

Determining treatment approaches for the various forms of diabetic foot syndrome is, thus, the goal of this review. About 50% of diabetic foot ulcers develop infection, making infection and progressive gangrene the leading causes of amputation of the lower extremity (Vuorlaakso 2022). A lower extremity amputation occurs in 15 to 20% of hospitalized patients, and hospital admission is required for up to 20% of individuals with diabetic foot ulcers. The pathophysiology of foot ulcers is rather complicated and involves many elements. More research is necessary because there isn't yet an international consensus on the factors that affect DFU recurrence. This research is a case-control retrospective study. By comparing clinical information from two distinct categories of patients with DFU diagnoses, we were able to identify the pertinent characteristics that affect DFU recurrence and offer recommendations for preventing DFU recurrence in high-risk patients (Singh et al. 2013). Optimizing blood sugar levels to aid in healing, routine debridement to remove dead tissue, and early and vigorous infection control are all possible therapeutic approaches for diabetic foot syndrome. Using cutting-edge wound care methods like hyperbaric oxygen therapy and negative pressure therapy can also speed up healing. Long-term limb preservation and ulcer recurrence prevention also depend on patient education regarding appropriate foot care and routine professional monitoring. Bony prominence excision, reconstruction, tendon lengthening, and tenotomy are surgical options for healing ulcers and preventing recurrence. However, additional ulceration and other problems could arise from these operations (Kumar et al. 2024).

High rates of ulcer recurrence can be decreased by providing post-healing footwear, educating patients appropriately, and practicing good foot hygiene. Diabetic foot problems are most frequently caused by peripheral sensory and autonomic neuropathy, which raises the risk of ulcers by causing excessive foot pressure, foot abnormalities, and unstable gait. As a result, it's critical to educate patients on the need for good foot hygiene and to provide post-healing footwear to lessen ulcer pressure. In addition, regular foot care can help detect and treat any abnormalities or unstable gait that may be causing the ulcers, which lowers the chance of ulcer development. Foot ulcers are related to higher foot difficulties, as demonstrated by an abundance of recent studies (Alavi et al. 2014) (Table 3.1).

Table 3.1 Various bioactive substances and their function against diabetes by various experimentation

S. no	Bioactive substances		Models	Results	References
1	Growth factors	VEGF	db diabetic mice	Increased neovascularization and the release of cells produced from bone marrow that are damaged at the site hasten the healing process	Galiano et al. (2004)
		bFGF	Patients with DFUs	75% of the treated patients' wound area was reduced. Tissue granulation and epithelization were stimulated.	Galiano et al. (2004)
		SDF-1α	STZ diabetic mice	Enhanced homing, wound healing, and EPC mobilization	Wallner et al. (2015)
2	Gene therapy	Adenoviral PDGF-B and adenoviral-Hox B3	db diabetic mice	Enhanced neovascularization and healing of wounds significantly. Adenoviral PDGF-B also promoted the recruitment of EPCs to the injury site and sped up the healing of wounds in diabetic mice.	Wallner et al. (2015)
		Lentiviral-SDF-1a	STZ diabetic mice- mice obese NOD/Ltj mice	Enhanced granulation tissue and improved diabetes wound rehabilitation with full epithelialization	Pastar et al. (2014)
3	Stem cells	CD133+ cells	STZ diabetic mice	Stimulated endothelial cells to migrate and proliferate, which accelerated the healing process and encouraged angiogenesis	Zhang et al. (2023)
		Basics	STZ diabetic mice	Enhanced the ability to break wounds. Moreover, it raised collagen levels and the expression of TGF-b, KGF, EGF, and VEGF.	Mehrabani et al. (2023)

9 Neuropathy and Ulcers

The main contributory factor influencing the growth of ulcers in patients with diabetes is always thought to be peripheral nerve damage and sensory loss. Nevertheless, abnormal sensations beneath and around the feet and ankles, in addition to deformities at various levels within the foot due to arthritis and ageing, endanger the sensory disruptions even more. Slow healing, non-healing, and recurrent foot ulcers are all caused by the same pathophysiology. Diabetics' ability to sense pain, heat, or cold—all crucial for the early detection of ulcers—is compromised by nerve loss and aberrant pressure points in their feet (Ahmeti et al. 2011). The foot deformities also impair the healing process by compressing the peripheral nerves and blood vessels, which makes it harder for oxygen and vital nutrients to get to the wound. It is crucial to evaluate the condition of neuropathy, especially concerning the prognosis for healing and recurrence prevention.

Damage to nerve cells' structure and function is directly linked to the pathological alterations in the neurological system caused by diabetes's disrupted metabolic environment (Duarte 2015). Hyperglycaemia, elevated AGE, increased acylcarnitine, and oxidized LDLs, for instance, affect the form and function of nerve cells, hence affecting sensory, motor, and other functions. Due to the loss of blood flow and sensation in the feet, neuropathy can cause additional problems like persistent discomfort, diseases, and even gangrene if left untreated. If the ulcers worsen and don't heal properly, this may eventually lead to the need for amputations. Decreased coordination and balance might also raise the chance of accidents and injuries. Diabetes patients are at a much higher risk of experiencing trauma due to their inability to perceive pain signs (Steinberg et al. 2019). While AGEs alter or impair the functions of proteins, they also alter the expression of genes and regulate intracellular signal transduction by binding in AGE-specific receptors (RAGEs). Acetyl-CoA molecules are transformed into acylcarnitine due to an oversaturated delivery system and an excess of substrate. Sensory and motor capabilities may be compromised as a result of decreased nerve conduction velocity carried on by axonal degeneration. Patients with diabetes may have tingling, weakness, or numbness in the extremities, which can affect their everyday activities and overall quality of life (Zochodne and Toth 2014). Prolonged nerve fibre loss can worsen these symptoms over time, making it harder for patients to recognize wounds or infections and raising the possibility of serious consequences.

Lower extremity vasomotor dysfunction brought on by autonomic neuropathy frequently results in arteriovenous shifts of the epidermal vascular network (Smith and Singleton 2012). Autonomic neuropathy can also lead to the warming of the skin by increasing blood perfusion in the deep skin and causing malfunction in sweat gland production. The skin's protective function can be further compromised, and the danger of developing a foot ulcer is increased by excessive sweat evaporation caused by abnormalities in the sweat glands' output. Serious side effects, including delayed stomach evacuation and gastrointestinal conditions, including gastroparesis, can result from untreated autonomic neuropathy. Cardiovascular problems can also occur, such as orthostatic hypotension, which is distinguished by a significant decrease in blood pressure following standing, raising the risk of fainting and dizziness (Vinubharathi 2018). Diabetic patients may also experience bladder dysfunction, which can lead to incontinence or urine retention, making their lives even more difficult. The function of important nervous system plastic molecules, such as patterns of expression of neuromodulin, heat-shock protein, β-tubulin and poly-ADP-ribose polymerase in DRG, are altered by hyperglycaemia. Important linkages in diabetic neuropathy include aberrations in DRG function, such as altered metabolic function in the body, altered motor neuron protein expression, and upregulated GW-body (mRNA processing site).

Polyol and inositol metabolism abnormalities have been identified in certain investigations, including nonenzymatic glycoprotein of diabetic nerves' protein transporters and neurons, neurovascular defects, neuroprotective disorders, axonal transmission defects, and the breakdown of the Na/K-adenosine triphosphate (ATP) enzyme (Deng et al. 2023). Pathological alterations in these crucial processes can result in aberrant protein synthesis, oxidative stress, and neuronal mitochondrial dysfunction, which can impair peripheral nerve function. Because oxidative damage impairs nerve function and causes cellular stress, it plays a crucial role in the emergence of neuropathy. Excessive reactive oxygen species and free radicals can harm DNA, proteins, and lipids in cells, which can cause neuronal death. The deterioration of peripheral nerves and exacerbation of neuropathy symptoms is caused by an accumulation of oxidative damage, which intensifies the inflammatory response (Pop-Busui et al. 2006). Glutathione, the antioxidant vitamin C and vitamin E are examples of antioxidants that are crucial for shielding cells from the harmful effects of oxidative damage. Antioxidants can alleviate the symptoms of neuropathy by reducing inflammation and neutralizing free radicals. In diabetic neuropathy, prompt treatment is essential to avoiding irreparable nerve damage and enhancing outcomes for patients. Prompt care can preserve nerve function, improve quality of life, and decrease diseases. Patients can lessen consequences and the intensity of neuropathic symptoms by eliminating risk factors and changing their lifestyle (Mallet et al. 2020).

9.1 Pathogenesis of Diabetic Foot Ulcers

The foot is the intersection of numerous destructive processes in diabetes patients, involving nearly every lower extremity component, including blood vessels, nerves, muscles, bones, joints, skin, and subcutaneous tissue. Developing and implementing monitoring and preventative strategies requires an awareness of these processes. Diabetic foot problems are frequently accompanied by peripheral artery disease, which lowers the blood supply to the feet, and diabetic neuropathy, a condition which can cause a loss of feeling. These issues raise the possibility of foot ulcers, diseases, and, in extreme situations, gangrene and amputation. Preventing these consequences requires early detection and appropriate therapy (Zaib et al. 2021). The development of ulceration is a complex process. A significant proportion of diabetic foot ulcers are caused by the critical combination of neuropathy, moderate foot trauma, and foot deformity. Joint rigidities, unequal plantar pressures, and a compromised capacity for wound healing are additional risk factors.

Diabetes patients who sustain tissue damage first have a compromised immune response, with fewer chemotactic actions to attract inflammatory cells to the injured tissues. This slows the healing process and raises the probability of bacterial infection (Coffman and Eberhardt 2002). Diabetic patients with joint rigidities may experience aberrant pressure distribution when walking due to changes in the foot's typical biomechanics. Because of the increased pressure on specific foot parts, the skin may break down and develop calluses, increasing the risk of ulceration. The treatment of diabetic foot disorders can become even more challenging as a result of these pressure sites' potential to worsen tissue damage and slow healing. More particular mechanisms may be involved, according to current cellular research on chronic diabetic ulcers (Yan et al. 2024). After this initial phase, when the response of inflammatory cells is finally developed, the process shifts to an increase in inflammation along with proteolysis. Prolonged exposure to hyperglycaemia also causes glycation of proteins and disruptions of cell responses, further impeding the process of inflammation and tissue repair (Song et al. 2022).

9.2 Screening for Risk of Diabetic Foot Ulcers

A podiatrist or primary care physician will examine the feet once a year to determine the risk of ulcers in diabetics. The examination should involve checking for peripheral artery disease, skin breakdown, and neuropathy severe enough to impair protective feeling. Gangrene and infection are two serious consequences that can result from diabetic foot ulcers if untreated. Amputation may be required in severe situations, and hospitalization may be necessary as a result of these complications. Additionally, a patient's mobility and quality of life might be significantly affected by an untreated ulcer (Brem et al. 2004). Calluses, interdigital maceration, and thicker nails should all be checked during a physical examination since they could be signs of an infection from fungi and elevated pressure on the bed of the nail. In common ulceration sites, digital abnormalities such as hammer or claw toes manifest as greater prominence of the metatarsal head over the plantar region and the interphalangeal joints dorsally.

Ulceration also frequently occurs at the point of a toe that is exposed to elevated pressure when it comes into contact with a surface or shoe (Armstrong et al. 2023). Evaluation of the ankle's range of motion in both dorsiflexion and plantarflexion can detect equines deformity, which raises forefoot plantar pressure and is recognized as less than zero degrees of flexion at the ankle joint. Foot ulcer risk can be raised by arterial disease, which can be detected with the use of an ankle-brachial value or toe brachial index. Physicians can identify equinus deformity, which might end in elevated forefoot plantar pressure, by measuring the dorsiflexion and plantar flexion ankles' range of motion. The malformation of the equinus is important in ulcer formation because it restricts the ankle's dorsiflexion, which puts greater stress on the forefoot when walking (Kakwani 2017). This elevated pressure can cause skin deterioration and recurrent damage, increasing the forefoot's vulnerability to ulcers. Reducing this pressure and the chance of ulcer development can be achieved by treating equinus deformity with procedures like stretching, orthotics, or surgery. Orthotics are essential for preventing ulcers because they relieve high-pressure areas by transferring pressure across the foot. They lower the risk of skin deterioration and ulcer development by offering support and cushioning. In individuals who have peripheral artery disease or foot abnormalities, orthotics can greatly reduce the risk of pressure-induced injuries by providing additional stability and rectifying aberrant gait patterns (Arhos et al. 2022).

10 Therapy for Diabetic Foot Ulcers

Diabetes patients with limb-threatening blockage in the femoral or peroneal arteries who might have been forced to have their extremities amputation frequently have their limbs restored by improved peripheral blood flow following distal revascularization surgery. With long-term patency rates in the 80% range, popliteal through plantar-arch vein-graft bypasses justify vascular repair. By restoring blood flow without requiring open surgery, endovascular procedures like angioplasty and stenting are examples of alternative treatments (Jones and Wyers 2018). To enhance circulation and stop more arterial blockage, pharmaceutical strategies such as antiplatelet treatment and vasodilators may be implemented. To improve tissue salvage and wound healing, hyperbaric oxygen therapy may also be an option in certain circumstances. Gene therapy is one of the newest treatments for peripheral vascular disease. It works by supplying certain growth factors to increase angiogenesis and blood flow. Using progenitor cells to repair damaged blood arteries and improve circulation is another potential strategy: stem cell therapy (Müller et al. 2018). New drug-eluting technologies are also being developed to administer treatments directly to specific locations that are afflicted, perhaps lowering the danger of restenosis and increasing outcomes over time. In peripheral vascular disease, pentoxifylline—a hemorheologic drug that increases the deformation of stiff blood vessels and cells—has been tried to improve distal flow; however, no large-scale, randomized controlled prospective trials have demonstrated an impact in diabetic patients (Bowering 2001).

Pentoxifylline is effectively used as a last option, while anecdotal accounts indicate that it can occasionally be beneficial when taken in diabetic patients at greater dosages for longer periods. Their main responsibility is to preserve the current flow and lessen the chance of additional thrombotic blockage. This is because pentoxifylline's mechanism of action is unclear and it is not a very efficient vasodilator. It may also have dangerous adverse effects including anaemia and extreme hypotension (Banik and Berger 2014). A multifaceted approach that combines medication interventions with lifestyle changes like diet and exercise is frequently more beneficial for diabetic patients. Using bioengineered skin substitutes and other advanced wound care treatments can help manage diabetic foot ulcers. Furthermore, it is essential to regulate blood sugar levels through insulin therapy and continuous glucose monitoring to reduce the probability of complications of peripheral vascular disease (Jadav et al. 2024).

10.1 Topical Ulcer Therapy

The administration of suitable dressings and routine exfoliation to eliminate necrotic tissue from the ulceration site are fundamental components of optimal wound care. The most effective debridement is done using a scalpel and forceps. Whirlpool treatment and foot soaking are ineffective for debridement and may further worsen tissue damage due to high heat or maceration. Topical antiseptics, like hydrogen peroxide, should be avoided since they are harmful to the skin cells that are recovering. Acticoat, an antibacterial silver dressing, and cadexomer iodine (Iodosorb), two novel compounds, have demonstrated potential as therapeutic antibiotics which do not damage underlying tissue. Keeping the wound surroundings wet promotes healing for all types of wounds. The ideal topical treatment for ulcers should maintain a moist but not saturated ulcer bed to promote maceration (Martin et al. 2013).

10.2 Pressure Relief and Avoiding Further Trauma

Foot ulcers in individuals with peripheral nerve damage and loss of protective feeling are often triggered by unrecognized injuries from improperly fitted footwear. Many foot ulcers can be avoided by teaching patients how to buy shoes that are wide enough, deep enough, and supportive of the arch. Repeated damage from walking on the injured area keeps the ulcer from healing. When treating foot ulcers, early intervention is essential to avoiding complications and accelerating healing (Edmonds 2020). By treating the ulcer as soon as possible, medical professionals can lower the chance of infection and additional tissue damage. Additionally, those suffering from peripheral neuropathy can have much better results if offloading techniques and appropriate wound care are used. A complete contact cast is an option for a diabetic patient who has a foot ulcer. These casts need skilled technicians who have received specialized application training and are inappropriate for ischemia or infected ulcers (Vaidya 2020). With minimal cushioning outside of a few crucial locations, the casts encompass neuropathy feet and lower limbs in plaster. A patient who is not cooperative or an incompetent technician applying a cast to the wrong limb can have disastrous results (Foster 2006). Other alternate therapies for foot ulcers include, along with complete contact casting, the application of therapeutic footwear, such as orthotic inserts or custom-moulded shoes, which assist disperse pressure and lessen damage to the affected area. Vacuum-assisted closure therapy is an additional choice. It draws out extra fluid and encourages blood flow to the wound by using negative pressure. To stop infection and encourage the healing of ulcers in the feet, regular debridement—the removal of dead tissue—is also crucial (Dayya et al. 2022).

10.3 Diabetes-Related Foot Ulcers (DFUs) and Impaired Wound Repair

The intricate mechanism of wound healing includes the simultaneous activation of blood and parenchymal cells, the extracellular matrix (ECM), and soluble mediators. There are various stages to this process, including inflammation, granulation tissue development, homeostasis/coagulation, re-epithelialization, and remodelling.

These stages may overlap and are generally not linked to a strict and clearly defined time frame. Typically, the development and differentiation, along with the most important cells in the mechanism of wound healing-macrophages, mast cells, fibroblasts and keratinocytes, determines the transition between phases. Following tissue damage, a fibrin plug forms to restore homeostasis, and aggregating platelets release several growth factors and cytokines, including the monocyte chemoattractant protein 1 (MCP-1) and transforming growth factor beta (TGF-β), which draw monocytes and neutrophils to the wound site. The expression of platelet-derived growth factor (PDGF), colony-stimulating factor 1 (CSF-1), and tumour necrosis factor-α (TNF-α) is induced by these inflammatory cells and is crucial for the initial stage of the process of new tissue creation. In most cases, re-epithelialization starts a few hours after the injury (Farhangniya and Samadikuchaksaraei 2023). Activated fibroblasts and keratinocytes move from the outermost layers of the wound towards the wound site when exposed to such growth factors, where they multiply and create the extracellular matrix (ECM) that will promote wound closure.

Angiogenesis is induced by angiogenic factors like fibroblast growth factor (FGF), vascular endothelial growth factor (VEGF), and PDGF, which stimulate macrophages and endothelial cells to produce basic fibroblast and vascular endothelial growth factors. A common side effect of diabetes is chronic diabetic neuropathy, which is distinguished by an ongoing degradation of peripheral nerve impulses brought on by inadequate blood flow and elevated blood sugar levels (Honnegowda et al. 2015). It is defined as either long-lasting or temporary harm to nerve tissue. The degree of neuropathy is significantly influenced by the length of time and severity of the hyperglycaemia exposure. Depending on the body area impacted, diabetic neuropathy may be categorized as focal, proximal, autonomic, or peripheral. However, while some diabetes individuals may develop neuropathy at an early age, many may never get it. DFU is still difficult to treat medically. Enhancing our appreciation of the pathology and molecular mechanisms of diabetic wounds might contribute to enhanced and more efficient treatments. Their application method and/or efficacies are still insufficient to guarantee significant DFU recovery (Firmansyah et al. 2024).

10.4 Hydrocolloids and Hydrogel Treatments

Hydrocolloids are moist wound dressings that typically consist of hydrophilic/colloidal layer particles with a backing substance (e.g., foams, non-woven fibres of polyester, or partially permeable films). Biocompatible gels made of polysaccharides (such as cellulose and its metabolites) may be present in these particles or proteins (such as collagen and gelatin). These dressings will create a moist environment by absorbing wound fluids when they come into touch with wound exudate. Autolytic debridement is facilitated by this wet environment, which enables the body to eliminate dead tissue on its own. Because the hydrocolloid components cushion and protect nerve terminals, they also aid in pain reduction. By covering the wound with a barrier, they can also stop bacteria from invading (Baptista-Silva et al. 2022). Additionally, they can be semi-permeable to oxygen and water. The use of hydrophilic dressings in severe wounds with infections has been questioned; nevertheless, because of the potentially hypoxic and too-moist environment, which may intensify necrotic tissue autolysis and raise the risk of illness at the wound site. Diabetes foot wounds that are shallow show no symptoms of infection or have little to moderate wound exudates can be treated with hydrocolloid-type dressings. In wounds with a lot of exudate, hydrocolloid dressings shouldn't be used since too much moisture can cause the surrounding skin to deteriorate (Moura et al. 2013). Because they may make the infection worse, they ought to be avoided as well in wounds that exhibit clinical symptoms of infection, which include increasing irritation, warmth, bruising, or purulent discharge. Furthermore, these dressings should not be used for wounds that need to be checked often or for people who have known allergies to any of the substances found in the dressing (Massella et al. 2019).

The primary function of hydrogels is to preserve extremely moist wound settings. These systems are made up of single or combined hydrated polymers that provide at least 20% of their total weight in retained water, i.e., in the form of a gel. These materials are often classified as superabsorbent if their water content exceeds 95%. Hydrogels offer a barrier of defence that keeps infections at bay and encourages a moist environment that aids in healing. The natural healing process of wounds is aided by this wetness, which promotes tissue regeneration and cell migration. Additionally, by hydrating nerve endings and having a calming impact on the wound region, hydrogels can help lessen pain and discomfort. Hydrogels can enlarge (or contract) reversibly in water-based conditions with particular pH and ionic strength values. They can be cross-linked either covalently or non-covalently to control their expanding capacities and to retain their flexible structures (Lafifi et al. 2024).

10.5 Polymeric Wound Dressings for DFU Treatment

The effectiveness of wound healing is dependent on several variables, including the type and stage of the wound, the degree of the injury, the patient's health, the tissues involved, the dressing used, and the impact of therapeutic drugs and healing boosters. Hydroactive or passive dressings can be used to treat wounds. The former are typically used for sudden injuries because they can provide adequate protection and absorb a sufficient quantity of exudates, while the latter are typically used in chronic wounds because they can readily adapt to lesions and keep an environment moist, which helps promote healing. Growth regulators and antibiotics are examples of healing enhancers that can greatly speed up the healing process of wounds by encouraging cell proliferation and lowering the danger of infection (Negut et al. 2018). By promoting the production of collagen and tissue remodelling, these compounds can also raise the general calibre of the repaired tissue. They can also aid in lowering inflammation, which is essential for avoiding difficulties and guaranteeing a quicker recovery. Various constituent polymeric materials with unique chemical, physical, and biological characteristics can be used to create wound dressing systems with varying designs, sizes, and shapes. This allows for the creation of finished products with a variety of functional qualities (Powers et al. 2016). Considering the source of those polymeric materials—natural or synthetic—is one of the easiest methods to distinguish between copolymers and synthetic polymers. Copolymers combine the advantageous qualities of many monomers to produce materials with increased mechanical strength and flexibility, which improves the effectiveness of wound dressings. This enables the dressings to suit the patient more comfortably and adapt better to the wound site. Copolymers can also be designed to have particular permeability and biodegradability properties, which can enhance the release of medicinal substances and aid in the healing process (Saghazadeh et al. 2018).

10.6 The Notch Signalling Pathways Function in DFU Healing

Human tissue homeostasis is regulated by the Notch pathway, which is crucial for the differentiation of cells, proliferation, and angiogenesis. Proliferation issues, decreased angiogenesis, and suppression of cell differentiation are all consequences of Notch pathway malfunction in diabetic wounds that impede the healing process. Notch1 activation results in decreased keratinocyte proliferation; while a keratinocyte Notch1 signalling deficiency causes increased epidermal proliferation in mice, which may even lead to skin cancers. Notch1 signalling is essential for preserving the delicate equilibrium needed for successful wound healing. It controls the growth and development of keratinocytes, guaranteeing that these cells develop and migrate appropriately to seal wounds (Abouelezz et al. 2023). The intricate process of tissue regeneration and repair is facilitated by Notch1, which modifies cellular activity and the local environment. Hyperglycaemia triggers a positive feedback cycle involving the Notch1 receptor along with the Dll4 ligand, which inhibits diabetic wound healing and directly encourages the abnormal stimulation of Notch1 signalling in the skin of diabetics. Significant amounts of kallikrein binding protein, or DFUs, cause a marked inflammatory response, encourage macrophage differentiation to M1, activate the Notch signalling pathway, and enhance the total amount of macrophages in the wound, all of which delay down wound healing in diabetic patients (Wu et al. 2022a, b). The M1 phenotype of macrophage polarization creates an environment that is pro-inflammatory and may be harmful to wound healing. The inflammatory phase is prolonged by the cytokines along with additional inflammatory mediators produced by M1 macrophages, which also hinder the proliferative phase that is required for tissue healing. Diabetic patients may develop permanent wounds that do not heal as a result of the stopped wound-healing process (Attinger and Bulan 2001).

10.7 NF-κB Signalling Pathway's Function in DFU Healing

The NF-κB pathway regulates numerous genes that are implicated in several vital biological processes, such as inflammatory processes, proliferation, migration of cells, and death. Tumour necrosis factor (TNF)-α and IL-6 are among the proinflammatory enzymes and proteins that are stimulated whenever the activation NF-κB component penetrates the nucleus and binds to the DNA that contains the B-binding motif. Furthermore, NF-κB-mediated inflammatory problems can result in neurological illnesses and inflammation of wounds. NF-κB can be activated by high glucose, human endothelial cells can undergo apoptosis due to increased caspase-3 activity, and wound healing can be delayed in diabetics. One potential treatment approach to lessen the consequences of diabetes is to target the NF-κB pathway (Castano et al. 2005). NF-κB activation inhibitors may help diabetes people heal wounds more effectively and with less inflammation. New therapies that enhance endothelium cell proliferation and function might come from studies into natural substances and medications that can alter this pathway. By inhibiting the stimulation of NF-κB, which in turn prevents the decomposition of pro-inflammatory enzymes and cytokines, wound therapy can raise the rate of wound healing, shorten the healing period, and lower the possibility of amputation. By activating the NF-κB signal, hyperbaric oxygenation can promote wound healing and angiogenesis, regulate cell migration and proliferation, and activate fibroblast' stromal cells-derived factor-1 and VEGF expression (Capó et al. 2023).

Cell metabolism and ATP production can be enhanced by hyperbaric oxygen treatment, which increases the amount of oxygen available to tissues. This increased oxygen level promotes angiogenesis, the formation of fresh vascular tissue, which is necessary for effective wound healing, and reduces inflammation. This treatment can improve overall wound outcomes and encourage the repair of injured tissues in diabetes patients, hence counteracting the decreased healing process. Growth factors like platelet-derived growth factors (PDGF), which promote cell migration and proliferation and hasten wound healing, are another potential strategy (Shen et al. 2020). Through the introduction of cells that can develop into multiple cell types required for repair, stem cell treatment is also being studied for its efficacy in improving healing and rebuilding damaged tissues. Additionally, cutting-edge wound dressings which distribute antibacterial substances or keep an environment moist can be extremely important in preventing infections and promoting the healing process (Paladini and Pollini 2019).

11 Cardiovascular Disease in Diabetes

Diabetes mellitus significantly raises the risk of cardiovascular disease (CVD) in both before and after-menopausal women (Feng et al. 2008). Type II diabetes is more common in women than in men in the majority of ethnic groupings. Diabetes and CVD are associated with advancing age, boosted lipids, obesity, hereditary factors, and sedentary lifestyles. Diabetes-related cardiovascular consequences comprise atherosclerotic CAD, stroke, diabetic cardiomyopathy, and kidney failure. Several epidemiological studies have established a connection between diabetes and cardiovascular disease in the public density (Khawandanah 2019).

11.1 High-Density Lipoprotein (HDL) and Cardiovascular Disease (CVD) in Diabetes

The failure of therapeutic therapies and the lack of an apparent connection between HDL-regulating genes and CVD have prompted a reconsideration of the relationship between HDL and atherosclerosis. Consequently, it is unknown if HDL has a physiologic function in altering atherogenesis, even if HDL-C levels suggest CVD risk. Parameters that are connected to the risk of CVD include obesity, hypertriglyceridemia, and lower HDL-C levels. The composition and levels of HDL in diabetic individuals have been documented. A criterion for diagnosing metabolic syndrome, HDL-C levels are frequently lower in those with type 2 diabetes (Kaur 2014; Anupama 2015). Women with T1DM who had significantly higher HDL-C levels (>80 mg/dL) have a bigger risk of CVD compared to those with lower levels. High-density lipoprotein (HDL) facilitates to transfer the intracellular cholesterol from cultured macrophages, a process known as cholesterol efflux capacity (CEC), which serves as a surrogate marker for in vivo reverse cholesterol transport (Cochran et al. 2021). Elevated levels of altered lipoproteins may hasten the growth of cardiovascular problems in diabetes through proatherogenic, pro-inflammatory, and immunogenic mechanisms, even though large epidemiological studies have linked plasma HDL-C levels to a lower danger of atherosclerosis, and HDL-associated proteins have been shown to have cardioprotective qualities. HDL contains a large diversity of proteins (Marsche et al. 2020).

11.2 Type 1 Diabetes vs Type 2 Diabetes in CVD

CVD is also more likely to occur among people with type 2 diabetes (Shah et al. 2015). There were reports of comparable alterations in plasma lipids. However, metabolic syndrome is a significant confounding factor in individuals with type 2 diabetes. There is a greater prevalence of hypertension and a significant proportion of obese people. Most significantly, type 1 diabetes has a significantly longer exposure to elevated glucose levels than type 2 diabetes. Accordingly, both type 1 and type 2 diabetes are linked to an atherogenic lipoprotein discovery and an occurring risk for CVD; however, type 1 diabetes has a significantly longer exposure period, which causes the illness to manifest more rapidly (Dabelea et al. 2017).

11.3 *Risk Factors for Cardiovascular Complications in Diabetes Mellitus*

The population explosion and epidemiological investigation shifts are linked to several changes in the nations. First, the rising incidence of the native risk factors for coronary artery disease (CAD), such as smoking, a diet heavy in saturated fat, high blood pressure, obesity, diabetes, dyslipidaemia, and unproductive behaviour, as well as the frequently documented co-occurrence of several risk factors in a single person (Figs. 3.3 and 3.4). Diabetes patients are more likely to develop cardiovascular diseases (CVDs) and always have a worse prognosis following a heart attack than people without the disease. A group of lipid abnormalities, including decreased HDL cholesterol, smaller and denser LDLs, and raised plasma triglycerides, are linked to type 2 diabetes mellitus and have been linked to an increased risk of cardiovascular disease. Lipoprotein (a) is one of the new cardiovascular risk factors whose significance is slowly becoming apparent. However, there is still debate over its relevance in diabetes patients (Laakso 2001).

11.4 *Pathophysiology*

Diabetes contributes to CVD through various mechanisms including hyperglycaemia, dyslipidaemia, hypertension, and inflammation (Petrie et al. 2018).

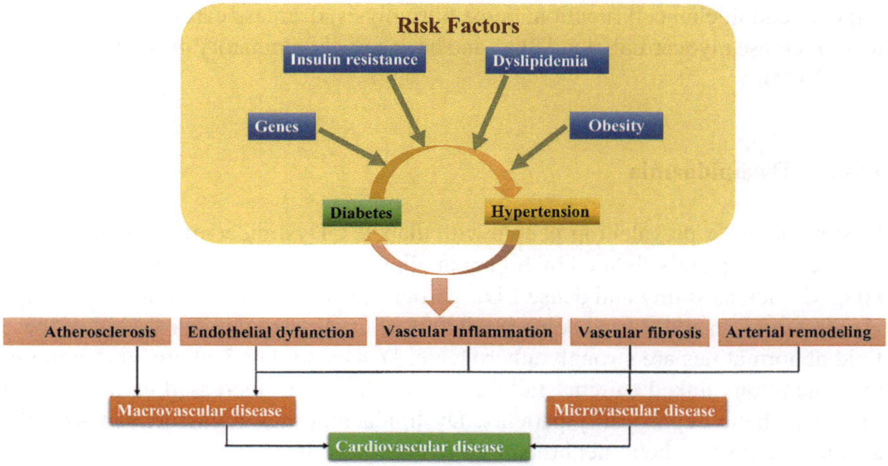

Fig. 3.3 Diabetes-associated risk factors towards CVD and hypertension

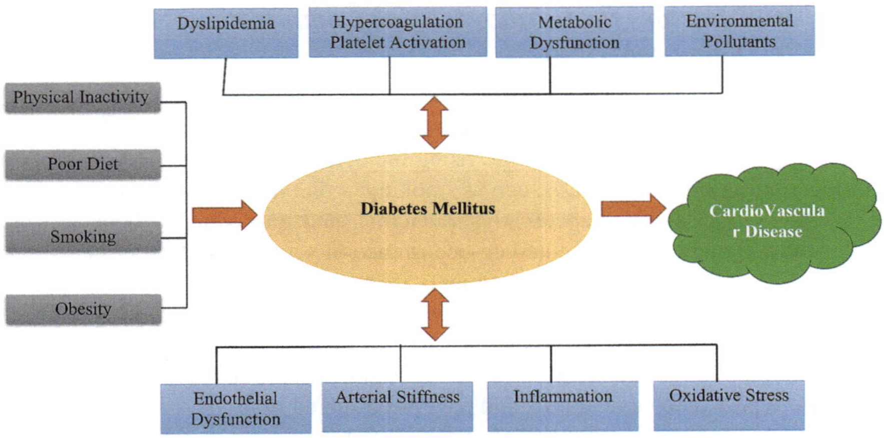

Fig. 3.4 Lifestyle-associated diabetes and its associated myocardial difficulties

11.4.1 Hyperglycaemia

Hyperglycaemia has been demonstrated to have a wide range of impacts on the immune system, inflammatory mediators, haemostasis, and endothelial cell function, all of which can injure numerous cells and organs. It has been postulated that pancreatic cell toxicity or desensitization caused by hyperglycaemia may result in relative hypoinsulinemia. These results in a persistent cycle of comparative hypoinsulinemia and hyperglycaemia spurred on by stress. Insulin has been exhibited to improve endothelial cell function, boost fibrinolysis, decrease cardiomyocyte apoptosis, increase myocardial blood flow, and have anti-inflammatory properties (Ghosh et al. 2023).

11.4.2 Dyslipidaemia

Dyslipidaemia is prevalent in people with diabetes. Dyslipidaemia associated with diabetes is frequently defined by higher apolipoprotein B (ApoB) levels, decreased HDL-C, increased tiny and dense LDL particles, fasting and postprandial hypertriglyceridemia. In addition to being predictors of cardiovascular disease, all of these lipid abnormalities are strongly atherogenic. Diabetes-related atherogenic lipoprotein alterations linked to renal failure may contribute to increased cardiovascular risk in diabetic nephropathy patients. Dyslipidaemia may be associated with the advancement of diabetic nephropathy in addition to its detrimental effects on the cardiovascular system (Kawanami et al. 2016) (Fig. 3.5).

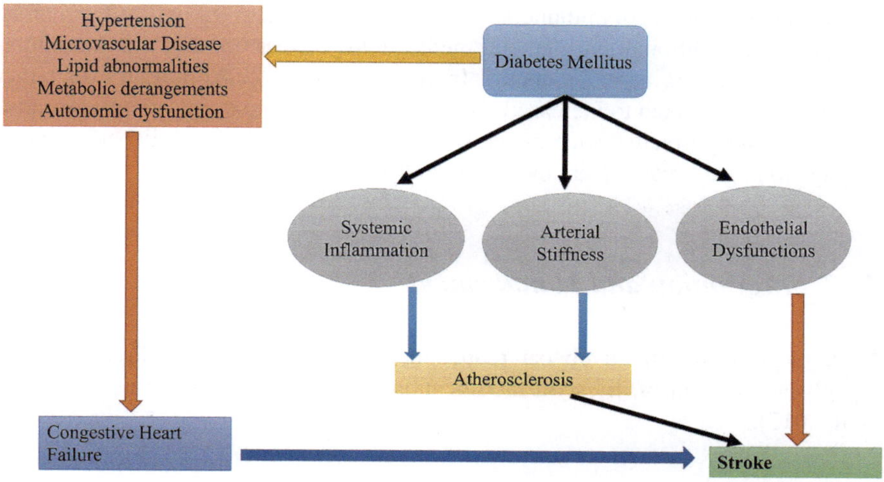

Fig. 3.5 Diabetes-associated nerve complications

11.5 Hypertension

The widespread presence of high blood pressure in diabetic patients has been substantially greater than that observed in non-diabetic individuals. These two diseases seem to usually coexist, even though there are many contributing factors to hypertension. Patients who have both diabetes and high blood pressure often develop renal insufficiency. People with diabetes had almost twice the prevalence of hypertension as age-matched adults without the illness, and those with hypertension are significantly likely to develop diabetes with normotension. In addition, hypertension is responsible for up to 75% of cardiovascular disease occurrences among diabetics. Hypertension is connected with a higher risk of microvascular complications (Strain and Paldánius 2018).

11.6 Inflammation

Experimental study shows inflammation and antibodies play a significant part in cardiovascular morbidity, hypertension, and metabolic dysfunction. Research has shown that T2D patients have a higher amount of total leukocytes, with neutrophils and lymphocytes especially higher than those without the disease. The inflammatory biomarkers in the setting of metabolic dysfunction can potentially provide a sound basis for developing tailored cardiovascular therapy in the setting of metabolic dysfunction, which is partially caused by inflammation of adipose tissue. The use of inflammatory biomarkers in developing customized cardiovascular therapy can potentially prove beneficial. The findings of genetic research and clinical trials supporting the benefits of immune-targeted therapy, as well as the anti-inflammatory properties of traditional anti-diabetic medications, further confirm that inflammation is connected to type 2 diabetes. Increased immune cell infiltration of target tissues has been linked to increased glucotoxicity and lipotoxicity, which can cause cardiovascular problems, such as metabolic cardiomyopathy, and diabetes-related damage to target organs (Daryabor et al. 2020).

11.7 Prevention and Management

Managing diabetes effectively can reduce the risk of cardiovascular disease including glycaemic control, lifestyle modifications, and regular monitoring (Newman et al. 2017).

11.7.1 Glycaemic Control

The variations in the outcomes can be partially explained by the clinical features of the populations under study concerning the existence of CVD, the length of diabetes and the kind of rigorous engagement used. This advantage persisted even after the groups' differences in glycaemic control disappeared over the additional five years of observational subsequent studies. The study's people had an elevated threat identify for heart disease when they started receiving strict glycaemic control. The most significant finding from the results presented is that, when started initially, people with instantaneous diabetes and minimal heart attack have positive cardiovascular effects. In elderly individuals with a greater cardiovascular risk profile who have been exposed to hyperglycaemia for years, the opposite cannot be accurate while searching for alleviated glycaemic goals (Davies et al. 2018).

11.7.2 Lifestyle Modifications

The foundation of clinical therapy for people with type 2 diabetes is lifestyle management, which includes food control and increased physical activity. Physical activity and diet are frequently assessed together as a component of an all-encompassing lifestyle intervention. More than 5145 T2D patients participated in the largest and most comprehensive study of the link between exercise and cardiovascular morbidity and death. Clinicians were paired accidentally with receiving usual care in diabetes education and support or an intensive lifestyle intervention that, by week 26, included $175 a moment each week of vigorous physical activity, meal replacement, calorie restriction, and a predetermined calorie intake of fats and proteins (Franz et al. 2002).

11.7.3 Exercise

A nutritional intervention, physical activity and exercise, and weight loss were among the lifestyle factors. Exercise-focused studies have also shown benefits in cardiovascular risk variables (BP, dyslipidaemia, and body structure) in individuals with type 2 diabetes. Nevertheless, no clinical study has shown that exercise lowers mortality or key CVD endpoints in T2D patients. Although all groups exhibited a propensity to have lower HbA1c, only the group that did both aerobic and strength training had a significant decline (Liu et al. 2019).

11.7.4 Weight Management

The main strategies for managing weight in people who have T2D include calorie-restrictive diets, increased energy expenditure via aerobic exercise and regular physical activity, and lifestyle modifications. For type 2 diabetes patients allocated to the rigorous lifestyle modification, the study failed to show a decrease in cardiovascular system events, even though the intensive treatment group maintained weight loss and improved CVD risk variables. Although they have not been demonstrated to enhance the outcomes of T2D, extensive lifestyle interventions, nutritional counselling, and calorie restriction have been shown in prior clinical trials to enhance controlling cardiovascular danger indicators in affected individuals (Joris et al. 2017).

11.7.5 Smoking Cessation

There is compelling evidence linking smoking cigarettes to several negative health outcomes, such as heart disease. To prevent cardiovascular disease, individuals with and without type 2 diabetes should have a frequent and complete examination of their tobacco use, as well as cessation counselling and medication when needed (Banks et al. 2019). While certain individuals could put on kg subsequently quitting cigarette usage, a study finds that this does not significantly affect the massive reduction in cardiovascular disease that smoking cessation delivers.

12 Obesity

Obesity and lack of physical activity are not the only elements that lead to the onset of diabetic complications. Numerous studies have connected a decreased incidence of diabetes to lifestyle changes, weight loss, and nonpharmacologic treatment. As a result, compared to a placebo, intervention caused a 58% drop in the incidence of diabetes, which continued for ten years of follow-up. Therefore, to combat the obesity epidemic, measures could be implemented to promote the use of nutritious lifestyles (Avenell et al. 2004).

13 Stroke in Diabetes

In the globe, stroke is the second most prevalent trigger of mortality and can cause serious, permanent impairment. In addition to a little over 80% of the people experiencing motor deficits necessitating rehabilitative therapy, over 75% of stroke survivors have a handicap because the situation is challenging to obtain work. Strokes can occur in patients who have had diabetes for 1.5-2 years, and the longer the diabetes has been present, the greater the risk and genetic inheritance connected (Mathisen 2018). Furthermore, these risks may differ depending on a person's sexuality, with women typically more vulnerable than men. Stroke is caused by many diabetes-related causes, such as cardiac embolism, cerebral small vessel disease, and major artery atherosclerosis (Donnelly and Horton 2008).

13.1 Why Diabetes Increases Stroke Risk

Diabetes, increasing a risk factor for stroke, includes atherosclerosis, high cholesterol, insulin resistance, and diabetic retinopathy (Alloubani et al. 2018).

13.1.1 Atherosclerosis

Major cerebral artery atherosclerosis is one of the primary causes of stroke worldwide, with changes ranging from modest wall thickening to hemodynamically significant luminal stenosis. Intracranial atherosclerosis disease (ICAD) can happen either by itself as well in conjunction with systemic atherosclerosis in different arterial beds, encompassing the extracranial, coronary, or peripheral arteries. Atherosclerosis is a major pathomorphological process that causes cerebrovascular illness by narrowing the artery walls inside the brain and throughout the body. Chronic inflammation and arterial wall destruction within the peripheral or CV system are thought to be the causes of atherosclerosis (Khanchi et al. 2020).

13.1.2 High Cholesterol

In the general population, an excellent rate of LDL insulin is thought to be a predictor of cardiovascular and cerebrovascular disorders (CVD). However, in recent years, the function of HDL cholesterol (high-density lipoprotein cholesterol) has been understood. There was conflicting evidence linking HDL cholesterol to the possibility of an ischemic stroke. Individuals suffering from type 2 diabetes have been reported greater danger of CVD while their HDL cholesterol levels are low, which is one of the main causes of lipoprotein abnormalities. Rarely, though, has research examined the potential link between low HDL cholesterol and a higher possibility of stroke in individuals with type 2 diabetes.

13.1.3 Increased Blood Pressure

The most prevalent danger component that can be changed is hypertension. Stroke affects more than 30% of people in developed nations. Hypertension is more common in older persons. It is designed to more precisely and rapidly assess stroke risk factors. Approximately 90% of women and men aged 55 to 65 were found to be non-hypertensive, with average life expectancy ranging from 80 to 85 years. In hypertensive people, including older patients with isolated systolic hypertension, antihypertensive medication has been shown in several intervention trials to successfully reduce the morbidity and mortality of stroke (Gadi and Samaha 2007). Stroke risk in treated individuals has not been adequately established, and the blood pressure level before the beginning of stroke is not as clearly defined. Many patients who have received treatment still have elevated blood pressure, and they frequently have other cardiovascular possibility factors and problems. The danger of stroke may be linked to inadequate management of hypertension. Conversely, a sharp drop in blood pressure might lead to more cardiovascular problems. Low diastolic blood pressure has been connected to an increased risk of stroke or myocardial infarction recurrence (Franklin et al. 2015).

13.1.4 Hyperglycaemia

A typical occurrence that manifests during the early acute stroke phase is hyperglycaemia. It could be connected to a condition of nonfasting and stress that results in a poor metabolism of glucose. Stress hormone levels that are elevated increase the rates of aerobic glycolysis, block insulin-mediated glycogenesis, and promote the release of glucose from gluconeogenesis and glycogenolysis. Increased platelet adhesion and aggregation to the endothelium are observed in people with acute ischemic stroke also have diabetes and hyperglycaemia. Since hyperglycaemia is linked to negative outcomes, controlling post-stroke hyperglycaemia effectively is essential to enhancing results. Prolonged or severe hypoglycaemia is particularly concerning while using insulin treatment since it might cause irreversible brain damage. Blood glucose levels below 60 mg/dL need to be immediately rectified. Numerous studies have examined the potential benefits of glycemia management interventions for these individuals. Serious hypoglycaemia (glucose <40 mg/dl) occurred in 2.6% of the intervention group but not in any of the patients to stable treatment (Cruz et al. 2017).

14 Types of Stroke in Diabetes

- Ischaemic stroke.
- Haemorrhagic stroke.

14.1 *Ischaemic Stroke*

Cerebrovascular atherosclerosis and thrombosis are the main causes of ischaemic stroke in patients with diabetes mellitus. The pathogenesis of atherothrombosis is believed to be primarily caused by platelet activation and aggregation, and antiplatelet medication is successful in avoiding ischaemic stroke in individuals with cardiovascular disorders. The diagnosis of acute ischemic stroke was made based on the patient's medical history, neurological examination, MRI, angiography, and laboratory results. Only stroke patients with big artery atherosclerosis were chosen for this investigation, and acute ischemic stroke was categorized by etiologic diagnosis. The World Health Organization's criteria for analysing diabetes mellitus were used. The research excluded those individuals having a history of hypertension, atrial fibrillation, myocardial infarction, malignancy, acute infectious illnesses, immune system disorders, blood diseases, renal or liver failure, haemorrhagic stroke, or recurrent stroke (Flora and Nayak 2019).

14.2 *Haemorrhagic Stroke*

Haemorrhagic stroke continues to account for 10–20% of stroke patients, despite a significant drop in the occurrence and death velocity of stroke with intracranial haemorrhage (ICH). Haemorrhagic stroke is becoming more and more common worldwide. Hypertension and warfarin usage were risk factors for ICH. Mortality, intracerebral haemorrhage recurrence, mental illnesses, and epilepsy are among the surviving outcomes following haemorrhagic stroke. This was the approximate yearly rate of stroke recurrence following ICH. Insufficient management of blood pressure raises the risk of ICH recurrence. Antiplatelet medication did not enhance the bleeding volume or the risk of ICH recurrence in ICH patients. Recurrences of stroke can affect everyday functioning and activities and raise mortality. The risk factors for ischemic and haemorrhagic stroke recurrence are yet unclear. Consequently, the purpose was to identify and examine the danger factors for ischemic or haemorrhagic stroke recurrence as well as the effect of antiplatelets on stroke repetition in Taiwanese people with ICH (Carpenter et al. 2016).

14.3 Symptoms of Stroke

The sudden onset of six-stroke symptoms are as: unilateral weakness, unilateral numbness, visual loss, half-vision loss, impaired interpersonal comprehension, and impaired verbal or written self-expression. When a neurologist diagnoses a stroke based on a positive response to some of the symptoms, we found that the sensitivity was strong (82%) and the specificity was fair (62%). Individuals with diabetes also had a greater frequency of stroke symptom clusters than those having normal blood sugar levels or prediabetes. Compared to those with normal glycemia, those with diabetes had a higher likelihood of experiencing two or many stroke symptoms after multivariable adjustment. Despite adjusting for factors like age, ethnicity, and sex, diabetes was similarly linked to clusters of stroke symptoms; however, these relationships were diminished and ceased to be statistically significant following further multivariable adjustment (Reich et al. 2006). Neither any of the clusters of stroke symptoms nor having two nor more symptoms were linked to prediabetes. These results have consequences for stroke prevention initiatives and raise the possibility that monitoring for stroke symptoms in diabetics may be necessary.

14.4 Prevention of Stroke

Stroke sufferers are at risk of another incident since the majority of strokes are mild to moderate. Patients with diabetes, regardless of age, are about twice as likely to suffer a stroke as those without the disease. This is the case for both ischemic stroke and haemorrhagic stroke (ICH & SAH), however ischemic stroke is more likely to occur than haemorrhagic stroke. People with diabetes are more likely to get lacunar stroke and exhibit signs of small vessel ischemia. Strokes are greater possibilities to happen in people with type 1 diabetes who also have diabetic nephropathy. In research on stroke prevention in patients with mild subcortical strokes, overt diabetes mellitus was present in 37% of the population, which is almost twice as high as in acute stroke trials (Kelly 2020). Randomized studies have not sufficiently examined other significant outcomes like dementia or moderate cognitive impairment, its precursor.

The direct correlation between intensive hyperglycaemia treatment and a lower risk of developing clinical dementia is just unknown. The fact that clinical dementia has a long-term consequence makes it difficult to research. It is immoral to refuse people appropriate medicines when dementia is preceded or arises more quickly than other significant microvascular and macrovascular consequences. Diabetes patients should prioritize preventing strokes by addressing both (1) risk factors and (2) risk states. Risk conditions include carotid artery stenosis and atrial fibrillation. Anticoagulation should be started and atrial fibrillation should be investigated. Carotid revascularization should be made available to patients who have symptomatic carotid artery stenosis. Stroke risk can be decreased by general stroke preventive lifestyle practices such as quitting smoking, keeping a healthy weight, leading an active lifestyle, abstaining from excessive alcohol use, and abstaining from all forms of sympathomimetic drug usage, including illegal and over-the-counter medications (Musialek et al. 2023).

15 Diabetic Nephropathy

The most prevalent diabetes condition associated with the consequences of cardiovascular disease is diabetic nephropathy (DN). It is a leading contributor to cardiovascular death (Samsu 2021). Patients with diabetes mellitus will experience complications with their kidneys. As chronic kidney disease (CKD) worsens and leads to end-stage renal disease (ESRD), more and more individuals throughout the world need renal replacement therapy. Over 20 million persons in the United States and over 13 million in Japan have chronic renal disease (Thurlow et al. 2021). To screen for diabetic nephropathy, albumin levels in urine are measured on-site. It should be confirmed in each of the three samples, which were obtained three to six months apart if it is abnormal. Because some individuals have a lower rate of glomerular filtration even if their albumin levels in the urine are within the typical range, it is therefore advised that the glomerular filtration rate be regularly evaluated for the proper diagnosis of nephropathy (Zsom et al. 2022).

15.1 Diabetic Nephropathy: A Public Health Problem

We know that type 2 diabetes is far more prevalent than type 1 diabetes , accounting for almost 90% of cases. Between 30 and 40% of individuals with type 1 diabetes have nephropathy, which is a highly expected prevalence. Race has a significant impact on the prevalence of acute renal issues in individuals with type 2 diabetes (Lawrence et al. 2021). It is very high among Pima Indians (60%) but relatively low among Caucasians (5%). One hundred thousand (0.7%) of the approximately 15 million diabetics in the United States are receiving dialysis. Similar issues exist in Europe, where 22.5 million people have diabetes, with 100,000–120,000 of them receiving dialysis (0.5%). Diabetes mellitus and its consequences were estimated to have cost the United States $100 billion in 1995. This exceeds a nation like Italy's whole health expenditure for 2001. It is astounding to learn that up to $2 billion is spent on dialysis in the United States (Caraballo et al. 2020). Prevention must be the first focus in a situation where health costs are concerning and the phenomena are still increasing. The primary goal should be to prevent diabetes mellitus by making healthy lifestyle changes, which should be the focus of comprehensive educational initiatives.

Diabetic nephropathy's natural progression has been thoroughly explained. The glomerular filtration rate (GFR) is high when the illness first appears. After 5 years, it begins to deteriorate. Although the deterioration is gradual, it inevitably results in ESRD throughout 20 to 25 years. Early renal impairment is indicated by the onset of microalbuminuria, which may arise after 10 years. The transition to frank proteinuria is an important step because it accelerates kidney destruction and results in end-stage renal disease, also known as ESRD, within a few years.

15.2 *Stages of Diabetic Nephropathy*

15.2.1 Stage 1: Early Hypertrophy-Hyperfunction

Several renal alterations are previously evident when diabetes mellitus first manifests clinically, according to systematic studies conducted in the past ten years. The early hypertrophy-hyperfunction stage and the formation of a new entity are justified by these alterations, which are dominating in terms of structure, biochemistry, and function. Early in the course of diabetes, both human and experimental animal diabetes have significantly larger kidneys, and both conditions are accompanied by correspondingly improved function Gupta, A., & Moazzam, 2022. According to a three-month study by Gupta, A., and Moazzam on 1719 participants, regular insulin treatment reduces both renal size and GFR, showing that the changes are caused by metabolic effects that can only be reversed gradually. It was discovered previously that the functional parameter is partially normalized. The development of expanded kidney size and related phenomena at the initial phases of diabetes mellitus is likely to be associated with possibly some characteristics of late diabetic glomeruli.

15.2.2 Stage 2: Glomerular Lesions Devoid of Symptoms

One of the defining characteristics of standard control for years following the initiation of insulin therapy is the normality of all standard clinical measures associated with renal function. The glomerular filtration rate has, however, been consistently elevated in many studies utilizing the traditional renal function test described (Homer Smith Schwaiger et al. 2021). While the increase in the untreated scenario is approximately 30–40%, this increase is approximately 20–30%. The poor metabolic control we get with our regular insulin regimens is probably the cause of the increased renal function (Rahman et al. 2021). When GFR exceeds extracellular volume in type 1 diabetes, it has been referred to as “real kidney hyperactivity.” Even while renal biopsy specimens may show noticeable structural changes, normal albumin excretion continues, demonstrating the kidneys’ well-maintained function commonly observed in moreover individuals, which is followed by changes in the glomerular mesangium’s volume.

15.2.3 Stage 3: Incipient Diabetic Nephropathy

This stage's main symptom is excessively high baseline excretion of urinary albumin as measured by radioimmunoassay. According to current research, diabetics with adequate metabolic control who are tested in a lab setting while sitting or lying down should employ a j.g./min range of 15 to 300. An additional characteristic of this crucial stage of renal impairment in diabetes is a gradually rising albumin excretion over several years (Adeva-Andany et al. 2022). Proteinuria is unquestionably a significant predictive factor during the progression of diabetes mellitus. While it isn't particularly dangerous on its own, it does indicate in blood pressure and a potential deterioration in kidney function. Numerous researches have validated this well-established truth (Kardalas et al. 2024).

15.2.4 Stage 4: Diabetes Nephropathy That Is Overt

Chronic proteinuria without any signs of kidney disease that is not diabetic is the main feature of this classic stage 9. If the concomitant high BP is not treated, kidney function deteriorates. A decrease in the rate of glomerular filtering is evident in all patients with severe and ongoing proteinuria. Our research has shown the significance between the diastolic Bp after the control period and the decline rate of GFR. The average drop was determined to be around 1 millilitre per minute each month. During pregnancy, patients with diabetic nephropathy also experienced a comparable reduction in renal function, as noted by other researchers (Zhao et al. 2020). A lengthy renal clearance collecting period and very high molecular weight dextran were used to investigate the process by which diabetic kidney disease progresses to proteinuria. Extremely large molecular weight absorption of dextran is abnormally enhanced in diabetic patients with severe proteinuria. Disorders could be detectable in diabetic renal disease that is less severe, and our study's longer clearance times and higher molecular weight of dextran seem to improve the test's sensitivity (Tillman et al. 2022).

15.2.5 End-Stage: Due to Diabetic Nephropathy, Kidney Damage

A significant medical issue, uraemia resulting from diabetes mellitus, each insulin and non-insulin-dependent, is present in up to 25% of the increasing currently enrolled in end-stage kidney damage programs in the United States. It is arguably the most important issue in clinical nephrology.11314 It may be helpful to group patients into these five groups for clinical and research purposes (Ramadan 2022).

15.3 *Risk Factors for Diabetic Nephropathy*

15.3.1 Increased Urinary Albumin Excretion

Both type 1 diabetes (T1D) and type 2 diabetes have risk factors for the development of diabetic nephropathy in most individuals, the first sign of diabetic kidney disease is moderately high renal albumin elimination or 40–400 mg per gram creatine in a spot sample of urine Renal function decrease is especially likely to occur in patients with severely elevated albuminuria, which is defined as >330 mg of albumin/g of urea in a urine sample taken on-site (Halminen et al. 2022). Clinical albumin levels or macroalbuminuria are other names for this illness. Therefore, the development of diabetic nephropathy does not necessarily require high urine albumin excretion (UAE). This discovery has implications for the disease's diagnosis, particularly that along with UAE, GFR should be assessed (Lalić et al. 2024).

15.3.2 Elevated Glucose Levels

A primary risk factor for the onset and advancement of insufficient control of glucose levels is a symptom of diabetic kidney disease. In both type 1 and type 2 diabetes patients, elevated blood glucose levels are associated with an increased risk of nephropathy Mayeda et al. 2020. Those with moderate albuminuria and lower HbA1c levels were less likely to develop severe albuminuria or end-stage renal disease. Similar results were found in T1D and T2D patients in randomized controlled trials. Intensive glycaemic management decreased the chance of moderate albuminuria developing into severe albuminuria or ESRD in DCCT (Chen et al. 2022).

15.3.3 Primary Prevention of Diabetic Nephropathy

Targets diabetics with normoalbuminuria and aims to prevent microscopic renal damage from developing. Hypertension and hyperglycemia are two risk factors that can be managed. Three randomized, placebo-controlled trials that address the primary prevention of incipient diabetic nephropathy, that is, studies conducted in individuals with normoalbuminuria and normotension, considering the value of blood pressure regulation (Alsalemi et al. 2022). Ravid et al.'s study involved 156 individuals with noninsulin-dependent type 2 diabetes, the most prevalent form of the disease. Albumin excretion in placebo-treated participants was 12 mg/24 hours at baseline and increased to 26 mg/24 hours after the follow-up. On the other hand, albumin excretion was fairly steady in the enalapril-treated group (Khan et al. 2022). Notably, only 6.5% of patients in the enalapril group experienced microalbuminuria, compared to 19.5% in the placebo group. This means that enalapril reduced the risk of microalbuminuria by 66%. Three more studies FACET, ABCD, and

UKPDS focused on primary prevention in normoalbuminuric diabetics with hypertension. According to these studies, the effects of ACE inhibitors were not better than those of other groups of antihypertensive medications, such as atenolol, nisoldipine, and amlodipine (Schrier et al. 2007).

16 Conclusion

According to current increases in mortality, diabetes is an inescapable disease for people of all ages, including youngsters. The key causes of rising morbidity and death include people's modern lifestyles, eating choices, and lack of information about diabetes care. Furthermore, inadequate diabetes treatment and its accompanying secondary complications are substantial, with severe consequences for numerous organs. Individuals can control their symptoms and enhance their quality of life by implementing early identification and a thorough treatment plan that includes medical, lifestyle, and alternative therapies. Ongoing research and clinical developments improve our knowledge and treatment of this difficult disorder.

Acknowledgement The authors are thankful to Thiruvalluvar University.

Conflict of Interest None to declare.

References

Abel, E. D., Gloyn, A. L., Evans-Molina, C., Joseph, J. J., Misra, S., Pajvani, U. B., & Drucker, D. J. (2024). Diabetes mellitus-Progress and opportunities in the evolving epidemic. *Cell, 187*(15), 3789–3820.

Abel, R. (2014). The Eye Care Revolution:: Prevent And Reverse Common Vision Problems, Revised And Updated. Kensington Books.

Abouelezz, H. M., El-Kashef, D. H., Abdelaziz, R. R., & Nader, M. A. (2023). Tenofovir alone or combined with doxorubicin abrogates DMBA-induced mammary cell carcinoma: An insight into its modulatory impact on oxidative/Notch/apoptotic signalling. *Life Sciences, 326*, 121798.

Adeva-Andany, M. M., Fernández-Fernández, C., Funcasta-Calderón, R., Ameneiros-Rodríguez, E., Adeva-Contreras, L., & Castro-Quintela, E. (2022). Insulin resistance is associated with clinical manifestations of diabetic kidney disease (glomerular hyperfiltration, albuminuria, and kidney function decline). *Current Diabetes Reviews, 18*(7), 64–85.

Agarwal, A. (2011). An Interventional Follow Up Study of Resolution of Steroid Responsive Macular Edema After Intravitreal Triamcinolone Acetonide (Master's thesis, Rajiv Gandhi University of Health Sciences (India)).

Ahmed, S., Azhar, S., & Grove, M. (2024). P11 A one-in-a-million case: familial cold autoinflammatory syndrome. Rheumatology Advances in Practice, 8 (Supplement_1), rkae117-042.

Ahmeti, I., Guceva, N. L., Jovanovska, B., Milenkovic, T., & Adamova, K. (2011). Diabetic foot with risk for ulceration associated with diabetic retinopathy in type 2 diabetes. In *Endocrine Abstracts*. European Society of Endocrinology.

Alam, M. M. (2021). Prevalence of Type 2 Diabetes Mellitus complications in human (Doctoral dissertation, Chattogram Veterinary & Animall Sciences University).

Alam, U., Asghar, O., Azmi, S., & Malik, R. A. (2014). General aspects of diabetes mellitus. *Handbook of clinical neurology*, *126*, 211–222.

Alavi, A., Sibbald, R. G., Mayer, D., Goodman, L., Botros, M., Armstrong, D. G., ... & Kirsner, R. S. (2014). Diabetic foot ulcers: Part I. Pathophysiology and prevention. *Journal of the American Academy of Dermatology*, *70*(1), 1–e1.

Alloubani, A., Saleh, A., & Abdelhafiz, I. (2018). Hypertension and diabetes mellitus as a predictive risk factors for stroke. *Diabetes & Metabolic Syndrome: Clinical Research & Reviews*, *12*(4), 577–584.

Alsalemi, N., Sadowski, C. A., Elftouh, N., Louis, M., Kilpatrick, K., Houle, S. K., & Lafrance, J. P. (2022). The effect of renin–angiotensin–aldosterone system inhibitors on continuous and binary kidney outcomes in subgroups of patients with diabetes: a meta-analysis of randomized clinical trials. *BMC nephrology*, *23*(1), 161.

Al Shehabi, N., Hallak, Y., Battistin, U., Faraji, H., Othman, M., Alkowatli, H., & Butt, M. (2024). Takotsubo Cardiomyopathy Complicating Diabetic Ketoacidosis, Hypothermia and Hypernatremia in a Comatose Patient. *Cureus*, *16*(7).

Antonetti, D. A., Silva, P. S., & Stitt, A. W. (2021). Current understanding of the molecular and cellular pathology of diabetic retinopathy. *Nature Reviews Endocrinology*, *17*(4), 195–206.

Anupama, B. (2015). Study of HsCRP Levels in Metabolic Syndrome with Special Reference to Cardiovascular Abnormalities (Doctoral dissertation, Rajiv Gandhi University of Health Sciences (India).

Arhos, E. K., Di Stasi, S., Hartigan, E. H., & Snyder-Mackler, L. (2022). Males and females have different muscle activity patterns during gait after ACL injury and reconstruction. *Journal of Electromyography and Kinesiology*, *66*, 102694.

Armstrong, D. G., Tan, T. W., Boulton, A. J., & Bus, S. A. (2023). Diabetic foot ulcers: a review. *Jama*, *330*(1), 62–75.

Arrigo, A., & Bandello, F. (2021). Molecular features of classic retinal drugs, retinal therapeutic targets and emerging treatments. *Pharmaceutics*, *13*(7), 1102.

Attinger, C. E., & Bulan, E. J. (2001). Debridement: The key initial first step in wound healing. *Foot and ankle clinics*, *6*(4), 627–660.

Avenell, A., Broom, J., Brown, T. J., Poobalan, A., Aucott, L., Stearns, S. C., & Grant, A. M. (2004). Systematic review of the long-term effects and economic consequences of treatments for obesity and implications for health improvement. *Health technology assessment*, *8*(21).

Babel, R. A., & Dandekar, M. P. (2021). A review on cellular and molecular mechanisms linked to the development of diabetes complications. *Current Diabetes Reviews*, *17*(4), 457–473.

Banik, R. K., & Berger, J. S. (2014). Peripheral vasodilators. In *Essentials of Pharmacology for Anesthesia, Pain Medicine, and Critical Care* (pp. 257–273). New York, NY: Springer New York.

Banks, E., Joshy, G., Korda, R. J., Stavreski, B., Soga, K., Egger, S., ... & Lopez, A. D. (2019). Tobacco smoking and risk of 36 cardiovascular disease subtypes: fatal and non-fatal outcomes in a large prospective Australian study. *BMC medicine*, *17*, 1–18.

Baptista-Silva, S., Alves, P., Guimarães, I., Borges, S., Tavaria, F., Granja, P., & Oliveira, A. L. (2022). A View on Polymer-Based Composite Materials for Smart Wound Dressings. *Polymeric and Natural Composites: Materials, Manufacturing and Biomedical Applications*, 425–456.

Bowering, C. K. (2001). Diabetic foot ulcers. Pathophysiology, assessment, and therapy. *Canadian Family Physician*, *47*(5), 1007–1016.

Brem, H., Sheehan, P., & Boulton, A. J. (2004). Protocol for treatment of diabetic foot ulcers. *The American journal of surgery*, *187*(5), S1–S10.

Browning, D. J., Stewart, M. W., & Lee, C. (2018). Diabetic macular edema: evidence-based management. *Indian journal of ophthalmology*, *66*(12), 1736–1750.

Busik, J. V., Mohr, S., & Grant, M. B. (2008). Hyperglycemia-induced reactive oxygen species toxicity to endothelial cells is dependent on paracrine mediators. *Diabetes*, *57*(7), 1952–1965.

Caldwell, R. B., Bartoli, M., Behzadian, M. A., El-Remessy, A. E., Al-Shabrawey, M., Platt, D. H., & Caldwell, R. W. (2005). Vascular endothelial growth factor and diabetic retinopathy: role of oxidative stress. *Current drug targets*, *6*(4), 511–524.

Caplan, L. R. (2013). *Navigating the Complexities of Stroke*. Oxford University Press.

Capó, X., Monserrat-Mesquida, M., Quetglas-Llabrés, M., Batle, J. M., Tur, J. A., Pons, A., ... & Tejada, S. (2023). Hyperbaric oxygen therapy reduces oxidative stress and inflammation, and increases growth factors favouring the healing process of diabetic wounds. *International journal of molecular sciences*, *24*(8), 7040.

Caraballo, C., Valero-Elizondo, J., Khera, R., Mahajan, S., Grandhi, G. R., Virani, S. S., ... & Nasir, K. (2020). Burden and consequences of financial hardship from medical bills among non-elderly adults with diabetes mellitus in the United States. *Circulation: Cardiovascular Quality and Outcomes*, *13*(2), e006139.

Cardones, A. R., & Banez, L. L. (2006). VEGF inhibitors in cancer therapy. *Current pharmaceutical design*, *12*(3), 387–394.

Carpenter, A. M., Singh, I. P., Gandhi, C. D., & Prestigiacomo, C. J. (2016). Genetic risk factors for spontaneous intracerebral haemorrhage. *Nature Reviews Neurology*, *12*(1), 40–49.

Castano, A. P., Demidova, T. N., & Hamblin, M. R. (2005). Mechanisms in photodynamic therapy: part two-cellular signaling, cell metabolism and modes of cell death. *Photodiagnosis and photodynamic therapy*, *2*(1), 1–23.

Chen, S. J., Chiang, H. Y., Chen, P. S., Chang, S. N., Chen, S. H., Wu, M. Y., & Kuo, C. C. (2022). Association of poorly controlled HbA1c with increased risk of progression to end-stage kidney disease and all-cause mortality in patients with diabetes and chronic kidney disease. *Plos one*, *17*(9), e0274605.

Chen, Y., Meng, Z., Li, Y., Liu, S., Hu, P., & Luo, E. (2024). Advanced glycation end products and reactive oxygen species: uncovering the potential role of ferroptosis in diabetic complications. *Molecular Medicine*, *30*(1), 141.

Chowdhury, T. (2014). *Diabetes management in clinical practice*. Springer.

Cochran, B. J., Ong, K. L., Manandhar, B., & Rye, K. A. (2021). High density lipoproteins and diabetes. *Cells*, *10*(4), 850.

Coffman, J. D., & Eberhardt, R. T. (Eds.). (2002). *Peripheral arterial disease: diagnosis and treatment*. Springer Science & Business Media.

Cook, C. A. (2019). *Phototherapeutic Devices for the Treatment of Diabetic Retinopathy*. California Institute of Technology.

Cruz, P., Blackburn, M. C., & Tobin, G. S. (2017). A systematic approach for the prevention and reduction of hypoglycemia in hospitalized patients. *Current diabetes reports*, *17*, 1–6.

Dabelea, D., Stafford, J. M., Mayer-Davis, E. J., D'Agostino, R., Dolan, L., Imperatore, G., & SEARCH for Diabetes in Youth Research Group. (2017). Association of type 1 diabetes vs type 2 diabetes diagnosed during childhood and adolescence with complications during teenage years and young adulthood. *Jama*, *317*(8), 825–835.

Daryabor, G., Atashzar, M. R., Kabelitz, D., Meri, S., & Kalantar, K. (2020). The effects of type 2 diabetes mellitus on organ metabolism and the immune system. *Frontiers in immunology*, *11*, 1582.

Davies, M. J., D'Alessio, D. A., Fradkin, J., Kernan, W. N., Mathieu, C., Mingrone, G., ... & Buse, J. B. (2018). Management of hyperglycemia in type 2 diabetes, 2018. A consensus report by the American Diabetes Association (ADA) and the European Association for the Study of Diabetes (EASD). *Diabetes care*, *41*(12), 2669.

Dayya, D., O'Neill, O. J., Huedo-Medina, T. B., Habib, N., Moore, J., & Iyer, K. (2022). Debridement of diabetic foot ulcers. *Advances in wound care*, *11*(12), 666–686.

De Deus, I. J., Martins-Silva, A. F., Fagundes, M. M. D. A., Paula-Gomes, S., Silva, F. G. D. E., Da Cruz, L. L., ... & De Queiroz, K. B. (2023). Role of NLRP3 inflammasome and oxidative stress in hepatic insulin resistance and the ameliorative effect of phytochemical intervention. *Frontiers in Pharmacology*, *14*, 1188829.

Deng, H., Li, B., Shen, Q., Zhang, C., Kuang, L., Chen, R., ... & Li, G. (2023). Mechanisms of diabetic foot ulceration: A review. *Journal of diabetes*, *15*(4), 299–312.

Dilmurodovna, T. D. (2024). DIABETES MELLITUS IN CENTRAL ASIA: PROBLEMS AND SOLUTIONS. *Лучшие интеллектуальные исследования*, *12*(4), 204-213.

Donnelly, R., & Horton, E. (Eds.). (2008). *Vascular complications of diabetes: current issues in pathogenesis and treatment*. John Wiley & Sons.

Dorrell, M., Uusitalo-Jarvinen, H., Aguilar, E., & Friedlander, M. (2007). Ocular neovascularization: basic mechanisms and therapeutic advances. *Survey of ophthalmology*, *52*(1), S3–S19.

Duarte, J. M. (2015). Metabolic alterations associated to brain dysfunction in diabetes. *Aging and disease*, *6*(5), 304.

Dugel, P. U., Bandello, F., & Loewenstein, A. (2015). Dexamethasone intravitreal implant in the treatment of diabetic macular edema. *Clinical Ophthalmology*, 1321–1335.

Duker, J., & Liang, M. (2024). *Anti-VEGF use in ophthalmology*. CRC Press.

Edmonds, M. (2020). Metabolic Disorders and Management of High-Risk Patients. *Neale's Disorders of the Foot and Ankle E-Book: Neale's Disorders of the Foot and Ankle E-Book*, 261.

Elgazzaz, M., Woodham, P. C., Maher, J., & Faulkner, J. L. (2024). Implications of pregnancy on cardiometabolic disease risk: preeclampsia and gestational diabetes. *American Journal of Physiology-Cell Physiology*, *327*(3), C646–C660.

Emerson, M. V., & Lauer, A. K. (2007). Emerging therapies for the treatment of neovascular age-related macular degeneration and diabetic macular edema. *BioDrugs*, *21*, 245–257.

Fakhruddin, S., Alanazi, W., & Jackson, K. E. (2017). Diabetes-induced reactive oxygen species: mechanism of their generation and role in renal injury. *Journal of diabetes research*, *2017*(1), 8379327.

Farhangniya, M., & Samadikuchaksaraei, A. (2023). A Review of Genes Involved in Wound Healing. *Medical Journal of the Islamic Republic of Iran*, *37*.

Feng, Y., Hong, X., Wilker, E., Li, Z., Zhang, W., Jin, D., ... & Xu, X. (2008). Effects of age at menarche, reproductive years, and menopause on metabolic risk factors for cardiovascular diseases. *Atherosclerosis*, *196*(2), 590–597.

Firmansyah, Y., Sidharta, V. M., Wijaya, L., & Tan, S. T. (2024). Unraveling the significance of growth factors (TGF-β, PDGF, KGF, FGF, Pro Collagen, VEGF) in the Dynamic of Wound Healing. *Asian Journal of Medicine and Health*, *22*(3), 49–61.

Flora, G. D., & Nayak, M. K. (2019). A brief review of cardiovascular diseases, associated risk factors and current treatment regimes. *Current pharmaceutical design*, *25*(38), 4063–4084.

Forrester, J. V., Kuffova, L., & Delibegovic, M. (2020). The role of inflammation in diabetic retinopathy. *Frontiers in immunology*, *11*, 583687.

Foster, A. V. (2006). *Podiatric assessment and management of the diabetic foot*. Elsevier Health Sciences.

Franklin, S. S., Gokhale, S. S., Chow, V. H., Larson, M. G., Levy, D., Vasan, R. S., ... & Wong, N. D. (2015). Does low diastolic blood pressure contribute to the risk of recurrent hypertensive cardiovascular disease events? The Framingham Heart Study. *Hypertension*, *65*(2), 299–305.

Franz, M. J., Bantle, J. P., Beebe, C. A., Brunzell, J. D., Chiasson, J. L., Garg, A., ... & Wheeler, M. (2002). Evidence-based nutrition principles and recommendations for the treatment and prevention of diabetes and related complications. *Diabetes care*, *25*(1), 148–198.

Gacche, R. N., & Meshram, R. J. (2014). Angiogenic factors as potential drug target: efficacy and limitations of anti-angiogenic therapy. *Biochimica et Biophysica Acta (BBA)-Reviews on Cancer*, *1846*(1), 161–179.

Gadi, R., & Samaha, F. F. (2007). Dyslipidemia in type 2 diabetes mellitus. *Current diabetes reports*, *7*(3), 228–234.

Galiano, R. D., Tepper, O. M., Pelo, C. R., Bhatt, K. A., Callaghan, M., Bastidas, N., ... & Gurtner, G. C. (2004). Topical vascular endothelial growth factor accelerates diabetic wound healing through increased angiogenesis and by mobilizing and recruiting bone marrow-derived cells. *The American journal of pathology*, *164*(6), 1935–1947.

Galvão, J. M. F. (2013). *Evaluation of the potential protective effects of adenosine A3 receptor in retinal ganglion cell degeneration* (Doctoral dissertation).

García-Pérez, L. E., Álvarez, M., Dilla, T., Gil-Guillén, V., & Orozco-Beltrán, D. (2013). Adherence to therapies in patients with type 2 diabetes. *Diabetes Therapy*, *4*, 175–194.

Ghosh, N., Chacko, L., Bhattacharya, H., Vallamkondu, J., Nag, S., Dey, A., ... & Dewanjee, S. (2023). Exploring the complex relationship between diabetes and cardiovascular complications: understanding diabetic cardiomyopathy and promising therapies. *Biomedicines*, *11*(4), 1126.

Gong, Q., Zhang, R., Wei, F., Fang, J., Zhang, J., Sun, J., ... & Wang, H. (2022). SGLT2 inhibitor-empagliflozin treatment ameliorates diabetic retinopathy manifestations and exerts protective effects associated with augmenting branched chain amino acids catabolism and transportation in db/db mice. *Biomedicine & Pharmacotherapy*, *152*, 113222.

Halminen, J., Sattar, N., Rawshani, A., Eliasson, B., Eeg-Olofsson, K., Bhatt, D. L., & Rawshani, A. (2022). Range of risk factor levels, risk control, and temporal trends for nephropathy and end-stage kidney disease in patients with type 1 and type 2 diabetes. *Diabetes Care*, *45*(10), 2326–2335.

Honnegowda, T. M., Kumar, P., Udupa, E. G. P., Kumar, S., Kumar, U., & Rao, P. (2015). Role of angiogenesis and angiogenic factors in acute and chronic wound healing. *Plastic and Aesthetic Research*, *2*, 243–249.

Jadav, R. K., Yee, K. C., Turner, M., & Mortazavi, R. (2024, August). Potential benefits of continuous glucose monitoring for predicting vascular outcomes in Type 2 diabetes: a rapid review of primary research. In *Healthcare* (Vol. 12, No. 15, p. 1542). MDPI.

Jenkins, A. J., Rowley, K. G., Lyons, T. J., Best, J. D., Hill, M. A., & Klein, R. L. (2004). Lipoproteins and diabetic microvascular complications. *Current pharmaceutical design*, *10*(27), 3395–3418.

Jones, D. W., & Wyers, M. C. (2018). Lower Extremity Arterial Reconstruction in Patients with Diabetes Mellitus: Principles of Treatment. *The Diabetic Foot: Medical and Surgical Management*, 327–343.

Joris, P. J., Plat, J., Kusters, Y. H., Houben, A. J., Stehouwer, C. D., Schalkwijk, C. G., & Mensink, R. P. (2017). Diet-induced weight loss improves not only cardiometabolic risk markers but also markers of vascular function: a randomized controlled trial in abdominally obese men. *The American journal of clinical nutrition*, *105*(1), 23–31.

Kakwani, R. (2017). Foot and ankle clinical cases. *Postgraduate Orthopaedics: The Candidate's Guide to the FRCS (Tr & Orth) Examination*, 174.

Kang, Q., & Yang, C. (2020). Oxidative stress and diabetic retinopathy: Molecular mechanisms, pathogenetic role and therapeutic implications. *Redox biology*, *37*, 101799.

Kannel, W. B. (1987). Status of risk factors and their consideration in antihypertensive therapy. *The American journal of cardiology*, *59*(2), A80–A90.

Kardalas, E., Paikopoulou, A., Vassiliadi, D. A., Kounatidis, D., Vallianou, N. G., Vourlakou, C., ... & Stratigou, T. (2024). The role of kidney biopsy in deciphering diabetic versus non-diabetic origin of kidney disease among patients with type 2 diabetes mellitus and nephrotic range proteinuria: A retrospective study. *Metabolism Open*, *23*, 100313.

Kaur, J. (2014). [Retracted] A Comprehensive Review on Metabolic Syndrome. *Cardiology research and practice*, *2014*(1), 943162.

Kawanami, D., Matoba, K., & Utsunomiya, K. (2016). Dyslipidemia in diabetic nephropathy. *Renal Replacement Therapy*, *2*, 1–9.

Kelly, D. (2020). *The impact of chronic kidney disease on stroke risk, mechanisms, and outcomes* (Doctoral dissertation, University of Oxford).

Khan, A. A., Rejnmark, L., Rubin, M., Schwarz, P., Vokes, T., Clarke, B., ... & Shu, A. D. (2022). PaTH forward: a randomized, double-blind, placebo-controlled phase 2 trial of TransCon PTH

in adult hypoparathyroidism. *The Journal of Clinical Endocrinology & Metabolism, 107*(1), e372–e385.
Khanchi, M., Zh, M. A., Demeuov, T. N., Baubekov, A. A., Zhakubayev, M. A., Tergeussizov, A. S., ... & Shamshiev, A. S. (2020). CLINICAL AND ACADEMIC ASPECTS OF PERIPHERAL ARTERY ANEURISM (LITERATURE REVIEW) PART ONE. *Вестник хирургии Казахстана*, (3 (64)), 36–47.
Khawandanah, J. (2019). Double or hybrid diabetes: a systematic review on disease prevalence, characteristics and risk factors. *Nutrition & diabetes, 9*(1), 33.
Khunti, K., & Millar-Jones, D. (2017). Clinical inertia to insulin initiation and intensification in the UK: a focused literature review. *Primary care diabetes, 11*(1), 3–12.
Khuu, L. A. (2015). *In-Vivo Biomarkers for Detection of Early Diabetic Retinopathy Using Aqueous Humor Cytokines and Functional Retinal Imaging Techniques*. University of Toronto (Canada).
Kowluru, R. A., & Chan, P. S. (2007). Oxidative stress and diabetic retinopathy. *Experimental diabetes research, 2007*.
Kumar, P., Hasan, F., Kumar, V., Chawla, R., & Goyal, S. K. (2024). Diabetic Wound Healing: Navigating Physiology, Advancements and Research Frontiers. *Journal of Diabetes Research Reviews & Reports. SRC/JDRR-202. DOI:* https://doi.org/10.47363/JDRR/2024 *(6), 181*, 2–11.
Laakso, M. (2001). Cardiovascular disease in type 2 diabetes: challenge for treatment and prevention. *Journal of internal medicine, 249*(3), 225–235.
Lafifi, Z., Bjij, I., & Soliman, M. E. (2024). Covalent Flexible Peptide Docking in Drug Discovery: Current Challenges and Potential Interventions. *Innovation Discovery, 1*.
Lalić, K., Popović, L., Lukač, S. S., Rasulić, I., Petakov, A., Krstić, M., ... & Lalić, N. M. (2024). Practicalities and importance of assessing urine albumin excretion in type 2 diabetes: A cutting-edge update. *Diabetes Research and Clinical Practice, 215*, 111819.
Latlief, G., Elnitsky, C., Hart-Hughes, S., Phillips, S. L., Adams-Koss, L., Kent, R., & Highsmith, M. J. (2012). Patient safety in the rehabilitation of the adult with an amputation. *Physical Medicine and Rehabilitation Clinics, 23*(2), 377–392.
Lawrence, J. M., Divers, J., Isom, S., Saydah, S., Imperatore, G., Pihoker, C., ... & SEARCH for Diabetes in Youth Study Group. (2021). Trends in prevalence of type 1 and type 2 diabetes in children and adolescents in the US, 2001-2017. *Jama, 326*(8), 717–727.
Lebovitz, H. E., & Banerji, M. A. (2004). Treatment of insulin resistance in diabetes mellitus. *European Journal of Pharmacology, 490*(1–3), 135–146.
Lenin, R., Thomas, S. M., & Gangaraju, R. (2018). Endothelial activation and oxidative stress in neurovascular defects of the retina. *Current pharmaceutical design, 24*(40), 4742–4754.
Li Calzi, S., Neu, M. B., Shaw, L. C., & Grant, M. B. (2010). Endothelial progenitor dysfunction in the pathogenesis of diabetic retinopathy: treatment concept to correct diabetes-associated deficits. *EPMA Journal, 1*, 88–100.
Liu, H., Ma, Y., Xu, H. C., Huang, L. Y., Zhai, L. Y., & Zhang, X. R. (2020). Updates on the management of ocular vasculopathies with VEGF inhibitor conbercept. *Current Eye Research, 45*(12), 1467–1476.
Liu, Y., Ye, W., Chen, Q., Zhang, Y., Kuo, C. H., & Korivi, M. (2019). Resistance exercise intensity is correlated with attenuation of HbA1c and insulin in patients with type 2 diabetes: a systematic review and meta-analysis. *International journal of environmental research and public health, 16*(1), 140.
Maciel, A. P. F., Pimenta, H. B., & Caldeira, A. P. (2016). Quality of life and medication adherence in hypertensive patients. *ACTA Paulista De Enfermagem, 29*, 542–548.
Mahla, R. S., Mukherjee, A. K., Amin, S., Jainarayanan, A., Mouroug-Anand, N., Nandakumar, A., & Prasad, A. D. (2023). Stem Cells Application in Eye Regeneration and Restoration of Vision. In *Handbook of Stem Cell Applications* (pp. 1–31). Singapore: Springer Nature Singapore.
Mallet, M. L., Hadjivassiliou, M., Sarrigiannis, P. G., & Zis, P. (2020). The role of oxidative stress in peripheral neuropathy. *Journal of molecular neuroscience, 70*, 1009–1017.

Mandal, S., Pal, S., Maiti, P., & Roy, S. (2022). Roles Of Reactive Oxygen Species In The Pathogenesis Of Type 2 Diabetes Mellitus. *Journal of Pharmaceutical Negative Results*, 5234–5245.

Marsche, G., Heine, G. H., Stadler, J. T., & Holzer, M. (2020). Current understanding of the relationship of HDL composition, structure and function to their cardioprotective properties in chronic kidney disease. *Biomolecules*, *10*(9), 1348.

Martin, C., Low, W. L., Amin, M. C. I. M., Radecka, I., Raj, P., & Kenward, K. (2013). Current trends in the development of wound dressings, biomaterials and devices. *Pharmaceutical patent analyst*, *2*(3), 341–359.

Massella, D., Giraud, S., Guan, J., Ferri, A., & Salaün, F. (2019). Manufacture techniques of chitosan-based microcapsules to enhance functional properties of textiles. *Sustainable Agriculture Reviews 35: Chitin and Chitosan: History, Fundamentals and Innovations*, 303–336.

Mateos-Olivares, M., García-Onrubia, L., Valentín-Bravo, F. J., González-Sarmiento, R., Lopez-Galvez, M., Pastor, J. C., ... & Pastor-Idoate, S. (2021). Rho-kinase inhibitors for the treatment of refractory diabetic macular oedema. *Cells*, *10*(7), 1683.

Mathisen, S. M. (2018). Risk factors influencing long-term survival and nursing home placement in stroke survivors and patients with carotid stenosis.

Mayeda, L., Katz, R., Ahmad, I., Bansal, N., Batacchi, Z., Hirsch, I. B., ... & De Boer, I. H. (2020). Glucose time in range and peripheral neuropathy in type 2 diabetes mellitus and chronic kidney disease. *BMJ Open Diabetes Research and Care*, *8*(1), e000991.

Meharwade, S. J. (2019). *Intravitreal Bevacizumab for Primary Treatment of Diabetic Macular Edema* (Master's thesis, Rajiv Gandhi University of Health Sciences (India)).

Mehrabani, M., Mohammadyar, S., Rajizadeh, M. A., Bejeshk, M. A., Ahmadi, B., Nematollahi, M. H., ... & Amirkhosravi, A. (2023). Boosting therapeutic efficacy of mesenchymal stem cells in pulmonary fibrosis: The role of genetic modification and preconditioning strategies. *Iranian Journal of Basic Medical Sciences*, *26*(9), 1001.

Mishra, K. C., Kumar, S., Jha, A. K., Banerjee, A., Banik, S., & Banerjee, P. (2024). Meta-analysis reveals an association of CAPN10 SNP 44 polymorphism with type 2 diabetes mellitus in Middle Eastern populations. *Human Gene*, 201305.

Mohsin, F., Javaid, S., Tariq, M., & Mustafa, M. (2024). Molecular immunological mechanisms of impaired wound healing in diabetic foot ulcers (DFU), current therapeutic strategies and future directions. *International Immunopharmacology*, *139*, 112713.

Moosavi-Movahedi, A. A., Taghavi, F., Rahban, M., & Moosavi-Movahedi, A. A. (2021). Lifestyle in the Regulation of Diabetic Disorders. *Rationality and Scientific Lifestyle for Health*, 129–153.

Morze, C. V., Allu, P. K., Chang, G. Y., Marco-Rius, I., Milshteyn, E., Wang, Z. J., ... & Pearce, D. (2018). Non-invasive detection of divergent metabolic signals in insulin deficiency vs. insulin resistance in vivo. *Scientific reports*, *8*(1), 2088.

Moura, L. I., Dias, A. M., Carvalho, E., & de Sousa, H. C. (2013). Recent advances on the development of wound dressings for diabetic foot ulcer treatment—A review. *Acta biomaterialia*, *9*(7), 7093–7114.

Müller, P., Lemcke, H., & David, R. (2018). Stem cell therapy in heart diseases–cell types, mechanisms and improvement strategies. *Cellular Physiology and Biochemistry*, *48*(6), 2607–2655.

Musialek, P., Bonati, L. H., Bulbulia, R., Halliday, A., Bock, B., Capoccia, L., ... & Lip, G. Y. (2023). Stroke risk management in carotid atherosclerotic disease: A Clinical Consensus Statement of the ESC Council on Stroke and the ESC Working Group on Aorta and Peripheral Vascular Diseases. *Cardiovascular Research*, cvad135.

Nair, P. V. (2018). *A Prospective Study of Exudative Diabetic Retinopathy and Its Association with Serum Lipid Profile and Haemoglobin* (Master's thesis, Rajiv Gandhi University of Health Sciences (India)).

Nazar, C. M. J. (2014). Diabetic nephropathy; principles of diagnosis and treatment of diabetic kidney disease. *Journal of nephropharmacology*, *3*(1), 15.

Neelam, K., Cheung, C. M. G., Ohno-Matsui, K., Lai, T. Y., & Wong, T. Y. (2012). Choroidal neovascularization in pathological myopia. *Progress in retinal and eye research, 31*(5), 495–525.
Negut, I., Grumezescu, V., & Grumezescu, A. M. (2018). Treatment strategies for infected wounds. *Molecules, 23*(9), 2392.
Newman, J. D., Schwartzbard, A. Z., Weintraub, H. S., Goldberg, I. J., & Berger, J. S. (2017). Primary prevention of cardiovascular disease in diabetes mellitus. *Journal of the American College of Cardiology, 70*(7), 883–893.
Ogoina, D. (2011). Fever, fever patterns and diseases called 'fever'–a review. *Journal of infection and public health, 4*(3), 108–124.
Osborne, N. N., Ugarte, M., Chao, M., Chidlow, G., Bae, J. H., Wood, J. P. M., & Nash, M. S. (1999). Neuroprotection in relation to retinal ischemia and relevance to glaucoma. *Survey of ophthalmology, 43*, S102–S128.
Paladini, F., & Pollini, M. (2019). Antimicrobial silver nanoparticles for wound healing application: progress and future trends. *Materials, 12*(16), 2540.
Pastar, I., Stojadinovic, O., Yin, N. C., Ramirez, H., Nusbaum, A. G., Sawaya, A., ... & Tomic-Canic, M. (2014). Epithelialization in wound healing: a comprehensive review. *Advances in wound care, 3*(7), 445–464.
Patel, S. A., Nilsson, M. B., Le, X., Cascone, T., Jain, R. K., & Heymach, J. V. (2023). Molecular mechanisms and future implications of VEGF/VEGFR in cancer therapy. *Clinical Cancer Research, 29*(1), 30–39.
Patibandla, R. L., Rao, B. T., & Murty, M. R. (2024). Revolutionizing Diabetic Retinopathy Diagnostics and Therapy through Artificial Intelligence: A Smart Vision Initiative. In *Transformative Approaches to Patient Literacy and Healthcare Innovation* (pp. 136–155). IGI Global.
Pedersen, Z. O., Ewers, B., Wimmelmann, C. L., Kofoed-Enevoldsen, A., Køster-Rasmussen, R., Couppé, C., ... & Dammeyer, J. (2024). A Cross-Sectional Study Investigating Associations between Personality Traits, Glycemic Control, and BMI in Persons with Diabetes: Lolland-Falster Health Study, Denmark. *International Journal of Environmental Research and Public Health, 21*(9), 1231.
Pei, X., Huang, D., & Li, Z. (2024). Genetic insights and emerging therapeutics in diabetic retinopathy: from molecular pathways to personalized medicine. *Frontiers in Genetics, 15*, 1416924.
Peinado, C., Liuzzi, D., Sluijter, S. N., Skorikova, G., Boon, J., Guffanti, S., ... & Rojas, S. (2023). Review and perspective: Next generation DME synthesis technologies for the energy transition. *Chemical Engineering Journal*, 147494.
Pessoa, B. B. D. S. T. (2022). *The Role of Vitreous and Vitreoretinal Interface in the Management of Diabetic Macular Edema* (Doctoral dissertation, Universidade do Porto (Portugal)).
Petrie, J. R., Guzik, T. J., & Touyz, R. M. (2018). Diabetes, hypertension, and cardiovascular disease: clinical insights and vascular mechanisms. *Canadian Journal of Cardiology, 34*(5), 575–584.
Pop-Busui, R., Sima, A., & Stevens, M. (2006). Diabetic neuropathy and oxidative stress. *Diabetes/metabolism research and reviews, 22*(4), 257–273.
Powers, J. G., Higham, C., Broussard, K., & Phillips, T. J. (2016). Wound healing and treating wounds: Chronic wound care and management. *Journal of the American Academy of Dermatology, 74*(4), 607–625.
Rahman, M. S., Hossain, K. S., Das, S., Kundu, S., Adegoke, E. O., Rahman, M. A., ... & Pang, M. G. (2021). Role of insulin in health and disease: an update. *International journal of molecular sciences, 22*(12), 6403.
Ramadan, A. B. (2022). Osteoinductive Factor and Diabetic Nephropathy in Type 2 Diabetes Mellitus Patients: A Literature Review. *Sohag Medical Journal, 26*(2), 25–50.
Ramasarma, T., & Rafi, M. M. (2016). A glucose-centric perspective of hyperglycemia.
Reich, L. M., Folsom, A. R., Key, N. S., Boland, L. L., Heckbert, S. R., Rosamond, W. D., & Cushman, M. (2006). Prospective study of subclinical atherosclerosis as a risk factor for venous thromboembolism. *Journal of Thrombosis and Haemostasis, 4*(9), 1909–1913.

Riddell, M. C., Gal, R. L., Bergford, S., Patton, S. R., Clements, M. A., Calhoun, P., ... & Sherr, J. L. (2024). The acute effects of real-world physical activity on glycemia in adolescents with type 1 diabetes: The type 1 diabetes exercise initiative pediatric (T1DEXIP) study. *Diabetes care*, *47*(1), 132–139.

Rudraraju, M., Narayanan, S. P., & Somanath, P. R. (2020). Regulation of blood-retinal barrier cell-junctions in diabetic retinopathy. *Pharmacological research*, *161*, 105115.

Saghazadeh, S., Rinoldi, C., Schot, M., Kashaf, S. S., Sharifi, F., Jalilian, E., ... & Khademhosseini, A. (2018). Drug delivery systems and materials for wound healing applications. *Advanced drug delivery reviews*, *127*, 138–166.

Sakini, A. S. A., Hamid, A. K., Alkhuzaie, Z. A., Al-Aish, S. T., Al-Zubaidi, S., Tayem, A. A. E., ... & Khunda, S. S. (2024). Diabetic macular edema (DME): dissecting pathogenesis, prognostication, diagnostic modalities along with current and futuristic therapeutic insights. *International Journal of Retina and Vitreous*, *10*(1), 83.

Samsu, N. (2021). Diabetic nephropathy: challenges in pathogenesis, diagnosis, and treatment. *BioMed research international*, *2021*(1), 1497449.

Savastano, M. C., Savastano, A., & Rizzo, S. (2018). Pharmacological adjuncts to vitrectomy surgery. *Current Pharmaceutical Design*, *24*(41), 4843–4852.

Schrier, R. W., Estacio, R. O., Mehler, P. S., & Hiatt, W. R. (2007). Appropriate blood pressure control in hypertensive and normotensive type 2 diabetes mellitus: a summary of the ABCD trial. *Nature Clinical Practice Nephrology*, *3*(8), 428–438.

Schwaiger, E., Krenn, S., Kurnikowski, A., Bergfeld, L., Pérez-Sáez, M. J., Frey, A., ... & Hecking, M. (2021). Early postoperative basal insulin therapy versus standard of care for the prevention of diabetes mellitus after kidney transplantation: a multicenter randomized trial. *Journal of the American Society of Nephrology*, *32*(8), 2083–2098.

Shadmani, A., Razmkhah, M., Jalalpoor, M. H., Lari, S. Y., & Eghtedari, M. (2021). Autologous activated omental versus allogeneic adipose tissue-derived mesenchymal stem cells in corneal alkaline injury: an experimental study. *Journal of Current Ophthalmology*, *33*(2), 136–142.

Shah, A. D., Langenberg, C., Rapsomaniki, E., Denaxas, S., Pujades-Rodriguez, M., Gale, C. P., ... & Hemingway, H. (2015). Type 2 diabetes and incidence of cardiovascular diseases: a cohort study in 1· 9 million people. *The lancet Diabetes & endocrinology*, *3*(2), 105–113.

Shen, S., Wang, F., Fernandez, A., & Hu, W. (2020). Role of platelet-derived growth factor in type II diabetes mellitus and its complications. *Diabetes and Vascular Disease Research*, *17*(4), 1479164120942119.

Shruti, B. (2019). *A Hospital Based Study to Evaluate the Role of Glycosylated Haemoglobin (HbA1c) on Sight Threatening Diabetic Retinopathy* (Master's thesis, Rajiv Gandhi University of Health Sciences (India)).

Simo, R., & Hernandez, C. (2015). Novel approaches for treating diabetic retinopathy based on recent pathogenic evidence. *Progress in Retinal and eye Research*, *48*, 160–180.

Singh, S., Pai, D. R., & Yuhhui, C. (2013). Diabetic foot ulcer–diagnosis and management. *Clin Res Foot Ankle*, *1*(3), 120.

Smith, A. G., & Singleton, J. R. (2012). Diabetic neuropathy. *Continuum: lifelong learning in neurology*, *18*(1), 60–84.

Smith, J. M., & Steel, D. H. (2015). Anti-vascular endothelial growth factor for prevention of postoperative vitreous cavity haemorrhage after vitrectomy for proliferative diabetic retinopathy. *Cochrane Database of Systematic Reviews*, (8).

Song, J., Hu, L., Liu, B., Jiang, N., Huang, H., Luo, J., ... & Chen, Q. (2022). The emerging role of immune cells and targeted therapeutic strategies in diabetic wounds healing. *Journal of Inflammation Research*, 4119–4138.

Song, M. K., Bischoff, D. S., Song, A. M., Uyemura, K., & Yamaguchi, D. T. (2017). Metabolic relationship between diabetes and Alzheimer's disease affected by Cyclo (His-Pro) plus zinc treatment. *BBA clinical*, *7*, 41–54.

Sontheimer, R. D. (2004). The management of dermatomyositis: current treatment options. *Expert opinion on pharmacotherapy*, *5*(5), 1083–1099.

Starace, V., Battista, M., Brambati, M., Cavalleri, M., Bertuzzi, F., Amato, A., ... & Cicinelli, M. V. (2021). The role of inflammation and neurodegeneration in diabetic macular edema. *Therapeutic Advances in Ophthalmology*, *13*, 25158414211055963.

Steinberg, N., Gottlieb, A., Siev-Ner, I., & Plotnik, M. (2019). Fall incidence and associated risk factors among people with a lower limb amputation during various stages of recovery–a systematic review. *Disability and rehabilitation*, *41*(15), 1778–1787.

Strain, W. D., & Paldánius, P. M. (2018). Diabetes, cardiovascular disease and the microcirculation. *Cardiovascular diabetology*, *17*, 1–10.

Tattersall, R. B., & Matthews, D. R. (2024). The history of diabetes mellitus. *Textbook of diabetes*, 1–21.

Thurlow, J. S., Joshi, M., Yan, G., Norris, K. C., Agodoa, L. Y., Yuan, C. M., & Nee, R. (2021). Global epidemiology of end-stage kidney disease and disparities in kidney replacement therapy. *American journal of nephrology*, *52*(2), 98–107.

Tillman, L., Tabish, T. A., Kamaly, N., Moss, P., El-Briri, A., Thiemermann, C., ... & Yaqoob, M. M. (2022). Advancements in nanomedicines for the detection and treatment of diabetic kidney disease. *Biomaterials and Biosystems*, *6*, 100047.

Tomic, D., Shaw, J. E., & Magliano, D. J. (2022). The burden and risks of emerging complications of diabetes mellitus. *Nature Reviews Endocrinology*, *18*(9), 525–539.

Vaidya, R. G. (2020). *Efficacy of low level laser therapy on wound healing in patients with diabetic foot ulcers* (Doctoral dissertation, BLDE (Deemed to be University)).

Vileikyte, L. (2001). Diabetic foot ulcers: a quality of life issue. *Diabetes/metabolism research and reviews*, *17*(4), 246–249.

Vinubharathi, B. (2018). *Preclinical and Comparative Clinical Study of Maruthampattai Kudineer and Yogam Therapy in the management of Madhumegam (Type II Diabetes)* (Doctoral dissertation, National Institute of Siddha, Chennai).

Vuorlaakso, M. (2022). Diabetic Foot Infection and Lower Extremity Amputations: Recent trends, outcomes and prognostic factors.

Wallner, C., Schira, J., Wagner, J. M., Schulte, M., Fischer, S., Hirsch, T., ... & Behr, B. (2015). Application of VEGFA and FGF-9 enhances angiogenesis, osteogenesis and bone remodeling in type 2 diabetic long bone regeneration. *PloS one*, *10*(3), e0118823.

Waly, E. H., & Hamed, M. S. (2018). Hypertension and dyslipidemia among type ii diabetic patients and related risk factors and complications. *Egypt J Commun Med*, *36*(1), 31–43.

Wang, X., Yuan, C. X., Xu, B., & Yu, Z. (2022). Diabetic foot ulcers: Classification, risk factors and management. *World journal of diabetes*, *13*(12), 1049.

Williams, G. (2009). *Increased hexosamine biosynthetic pathway flux impairs myocardial GLUT4 translocation* (Doctoral dissertation, Stellenbosch: University of Stellenbosch).

Wu, L., Acón, D., Wu, A., & Wu, M. (2019). Vascular endothelial growth factor inhibition and proliferative diabetic retinopathy, a changing treatment paradigm?. *Taiwan journal of ophthalmology*, *9*(4), 216–223.

Wu, X., He, W., Mu, X., Liu, Y., Deng, J., Liu, Y., & Nie, X. (2022a). Macrophage polarization in diabetic wound healing. *Burns & Trauma*, *10*, tkac051.

Wu, X., Zhou, X., Lai, S., Liu, J., & Qi, J. (2022b). Curcumin activates Nrf2/HO-1 signaling to relieve diabetic cardiomyopathy injury by reducing ROS in vitro and in vivo. *The FASEB Journal*, *36*(9), e22505.

Yan, L., Wang, Y., Feng, J., Ni, Y., Zhang, T., Cao, Y., ... & Zhao, C. (2024). Mechanism and application of fibrous proteins in diabetic wound healing: A literature review. *Frontiers in Endocrinology*, *15*, 1430543.

Yang, X., Yu, X. W., Zhang, D. D., & Fan, Z. G. (2020). Blood-retinal barrier as a converging pivot in understanding the initiation and development of retinal diseases. *Chinese Medical Journal*, *133*(21), 2586–2594.

Yusuf, S., Ibrahim, S. S., Musa, A., Salihu, Y. A., Nyako, F. T., & Wujika, N. D. (2024). Risk Factors for Lower Extremity Amputation among Diabetic Patients with Diabetic Foot Gangrene in ATBUTH, Bauchi. *Open Journal of Orthopedics*, *14*(9), 391–403.

Zaib, S., Talib, M. A., Ahmad, S., Arif, A., Mustafa, F., Junaid, M., & Khan, I. (2021). Diabetes and its Association with Peripheral Arterial Disease. *Biomedical Journal of Scientific & Technical Research, 37*(5), 29735–29746.

Zhang, C., & Luo, X. (2024). The increase in diabetes in children from underdeveloped countries. *Current Opinion in Pediatrics*, 10–1097.

Zhang, X., Saaddine, J. B., Chou, C. F., Cotch, M. F., Cheng, Y. J., Geiss, L. S., ... & Klein, R. (2010). Prevalence of diabetic retinopathy in the United States, 2005–2008. *Jama, 304*(6), 649–656.

Zhang, Y., Zhang, Y. Y., Pan, Z. W., Li, Q. Q., Sun, L. H., Li, X., ... & Yang, B. F. (2023). GDF11 promotes wound healing in diabetic mice via stimulating HIF-1α-VEGF/SDF-1α-mediated endothelial progenitor cell mobilization and neovascularization. *Acta Pharmacologica Sinica, 44*(5), 999–1013.

Zhao, Y., Cai, J., Zhu, X., van Donkelaar, A., Martin, R. V., Hua, J., & Kan, H. (2020). Fine particulate matter exposure and renal function: A population-based study among pregnant women in China. *Environment International, 141*, 105805.

Zhou, B., Rayner, A. W., Gregg, E. W., Sheffer, K. E., Carrillo-Larco, R. M., Bennett, J. E., ... & Can, G. (2024). Worldwide trends in diabetes prevalence and treatment from 1990 to 2022: a pooled analysis of 1108 population-representative studies with 141 million participants. *The Lancet*.

Zochodne, D. W., & Toth, C. (2014). Diabetes and the nervous system. In *Aminoff's Neurology and General Medicine* (pp. 351–368). Academic Press.

Zsom, L., Zsom, M., Salim, S. A., & Fülöp, T. (2022). Estimated glomerular filtration rate in chronic kidney disease: a critical review of estimate-based predictions of individual outcomes in kidney disease. *Toxins, 14*(2), 127.

Chapter 4
Antidiabetic Properties of Microalgae Compounds

Richard Luan Silva Machado, Darissa Alves Dutra, Adriane Terezinha Schneider, Rosangela Rodrigues Dias, Mariany Costa Deprá, Leila Queiroz Zepka, and Eduardo Jacob-Lopes

Abstract Due to rapid global urbanization and changes in population lifestyles, diabetes mellitus is currently the third leading cause of death on the planet. This metabolic disorder is characterized by hyperglycemia, resulting from insufficient insulin secretion or an inadequate response by the body to its action. In this sense, the growing demand for natural products has attracted the interest of the pharmaceutical industry in microalgae. These photosynthetic microorganisms stand out for the vast diversity of bioactive compounds capable of improving insulin sensitivity, attributed to their antioxidant and anti-inflammatory properties. However, further research on the action of these microalgal metabolites is needed for their full exploration and effective pharmaceutical and industrial applications. In this context, this chapter aims to provide a more comprehensive view of the diversity of microalgal biocompounds applicable in the prevention of this metabolic disorder, including (i) aspects related to the overall risk of the disease; (ii) microalgal biocompounds and mechanisms associated with antidiabetic effects; (iii) related scientific evidence, as well as (iv) future perspectives and existing challenges.

Keywords Antioxidant · Bioactive · Biocompounds · Biomass · Diabetes mellitus

1 Introduction

Diabetes mellitus (DM) is one of the most critical public health problems, expected to affect 700 million people by 2045, with approximately 90% of cases diagnosed as type 2 diabetes (Hossain et al. 2024). This disease is characterized by chronic hyperglycemia, which is associated with excessive oxidative stress, resulting in macromolecular damage to cells of great importance in the body, such as lipids and DNA. In the long term, these complications can lead to damage that includes

R. L. S. Machado · D. A. Dutra · A. T. Schneider · R. R. Dias · M. C. Deprá · L. Q. Zepka · E. Jacob-Lopes (✉)
Rural Sciences Center, Federal University of Santa Maria, Santa Maria, RS, Brazil

R. Raja et al. (eds.), *Algae in Diabetes Management*,
https://doi.org/10.1007/978-981-95-0571-5_4

neuropathy, nephropathy, retinopathy, cataracts, and cardiovascular diseases (Cardullo et al. 2020; de Alencar Alves et al. 2020; Yusuf et al. 2021). Considered a global public health risk, the International Diabetes Federation (IDF) reports that approximately 540 million adults, between 20 and 79 years old, are affected by the disease, representing approximately 10.5% of the adult population.

Furthermore, the World Health Organization (WHO) (2024) reported a 3% increase in the mortality rate from diabetes between 2000 and 2019, predicting that this disease will be the seventh leading cause of death in the world by 2030. In addition, the WHO highlighted that dietary patterns and lifestyle habits can contribute to the risk factors for the development of chronic diseases such as diabetes. In the state of the art, only a certain percentage of microalgae and derivatives are approved for inclusion in the human diet, even though they are highly recommended due to their high content of beneficial and bioactive compounds (Aboulthana et al. 2023). The importance of natural products derived from algae in diabetes research lies in their potential to target multiple aspects of the disease. Microalgae produce a wide range of bioactive compounds, such as polysaccharides, polyphenols, pigments, fatty acids, and sterols, which have demonstrated promising pharmacological properties in the control and treatment of diabetes. These compounds can modulate glucose metabolism, increase insulin sensitivity, improve lipid profiles, and exert antioxidant and anti-inflammatory effects, all of which are crucial factors in diabetes management (Bocanegra et al. 2021; Machado et al. 2024).

However, natural products derived from microalgae and the knowledge about their bioactive compounds remain largely underexplored. The identification of these compounds is a complex task that demands interdisciplinary approaches. The constant updating of analytical and molecular techniques is crucial in this process, being a prerequisite for the development of new products through high-throughput strategies. In recent years, the growing interest and public and private investments in marine biotechnology have significantly expanded the capacity to generate information and collect large volumes of data, promoting a deeper understanding of diverse cellular processes and biological phenomena. Furthermore, biotechnology uses advanced methodologies, such as heterologous expression and genetic engineering, to identify species with bioactive potential and increase the production of compounds of interest (Lauritano et al. 2019).

In this sense, the chapter will initially present a comprehensive overview of the antidiabetic effects of bioactive compounds present in microalgae biomass. Initially, the general panorama of diabetes mellitus will be discussed, including its global prevalence and its impact on public health. Next, the focus will be on the therapeutic potential of bioactive compounds derived from microalgae, such as pigments, proteins, and other metabolites of interest, with antidiabetic action. Subsequently, studies carried out in human models, in vitro and in vivo, which evaluate the efficacy of these compounds will be reviewed. Finally, future perspectives and promising directions for the development of this area of research will be discussed.

2 Diabetes Mellitus as a Global Public Health Risk

Diabetes mellitus is a chronic disease characterized by hyperglycemia and represents a serious global public health problem. This condition results from persistent metabolic dysfunction caused by deficiency in insulin production, resistance to its action, or both (Bodke et al. 2023). The two main types of diabetes are: type 1, an autoimmune condition that defeats the insulin-producing pancreatic cells, and type 2, associated with insulin resistance and insufficient production, strongly related to sedentary lifestyle and obesity (Kumar et al. 2020; Hossain et al. 2024). The prevalence of diabetes has grown alarmingly. In 2021, the International Diabetes Federation (IDF) estimated that 537 million adults were living with the disease, with a predicted number of 783 million by 2045.

Unfortunately, type 2 diabetes, which accounts for approximately 90% of cases, is particularly prevalent in low- and middle-income countries, where changes in dietary and lifestyle habits have increased risk factors (Hossain et al. 2024). Complications associated with diabetes are serious, including cardiovascular disease, kidney failure, blindness, and amputations (Al-Lawati 2017; Forray et al. 2023). Furthermore, diabetes is a leading cause of mortality worldwide (IDF 2021). The disease also generates high costs for health systems, with global expenditures estimated at over US$850 billion in 2021 (Quarti Machado Rosa et al. 2018; IDF 2021). The main approaches to controlling diabetes mellitus involve the use of oral medications and insulin therapy, although their side effects, such as hypoglycemia and gastrointestinal problems, are well documented (Blahova et al. 2021). However, in recent years, there has been a growing interest in new sources of bioactive substances, such as alternatives with fewer adverse effects (Barrea et al. 2023; Cotas et al. 2024).

Regarding the treatment of patients with diabetes, glycemic control in patients with type 1 diabetes is commonly achieved with the administration of intravenous insulin, while in type 2 diabetes, control can be managed with diet, exercise, and oral medications (Sugandh et al. 2023). Among the therapeutic approaches are the inhibition of the enzymes α-amylase and α-glucosidase, which reduce the release of glucose into the bloodstream after meals (Munawaroh et al. 2020). Another strategy includes inhibition of the enzyme dipeptidyl peptidase IV (DPP-IV), which prolongs the action of incretin hormones, promoting insulin secretion. However, drugs such as acarbose, biguanide, and DPP-IV inhibitors, although effective, can cause side effects and are expensive (Jia et al. 2021; Odenthal et al. 2024).

Microalgae contain bioactive compounds, such as pigments, proteins, fatty acids, and fibers, which can neutralize oxidative stress and inflammation, thus improving insulin sensitivity and reducing the risk of diabetes-related complications. In addition, these bioactive compounds have demonstrated therapeutic properties, including antioxidant, anti-inflammatory, and immunomodulatory actions (Melloni et al. 2023; Tamel Selvan et al. 2023). A schematic representation of the action of biocompounds of microalgae on α-glucosidase is shown in Fig. 4.1.

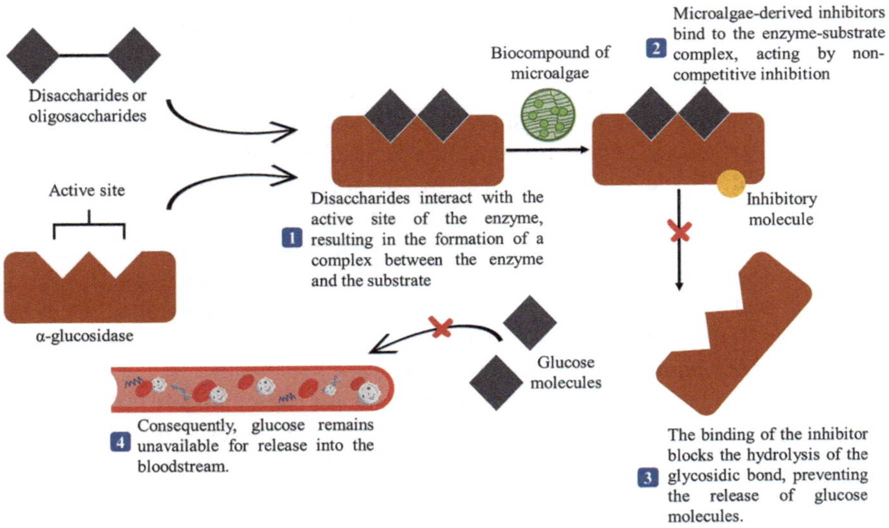

Fig. 4.1 Inhibition of α-glycosidase enzyme activity by microalgae extracts in the control of glucose absorption

Some microalgae such as *Arthrospira platensis*, *Nannochloropsis oculata*, *Porphyridium purpureum,* and *Chlorella vulgaris* have antidiabetic potential by inhibiting enzymes related to diabetes and alleviating oxidative stress (Vieira et al. 2021). Since the 1970s, Eastern Europe has led the production of microalgae-based medicines, consolidating these sources as a sustainable alternative for obtaining drugs (Zhuang et al. 2022). Recent studies, such as that by Kang et al. (2024), have demonstrated the promising use of the microalgae *Haematococcus* sp. in the treatment of chronic diabetic wounds, through the development of hydrogels that promote healing by improving cell proliferation and blood vessel formation.

Due to these advances, it is suggested that microalgae can reduce the dependence on conventional drugs produced from synthetic chemicals, offering more natural alternatives and playing a fundamental role in medical and pharmaceutical biotechnology (Mehariya et al. 2021). These unicellular organisms with a high capacity for biomass production represent a new frontier for the chemical industry, offering a sustainable alternative to traditional processes (Mehariya et al. 2021). The discovery of microalgae-based drugs opens new perspectives for health. Compounds extracted from microalgae have demonstrated significant therapeutic activity in scientific studies, such as using soft gel capsules, which can help reduce brain damage by showing immunostimulatory activity and improving the immune defense system (Sathasivam et al. 2019).

In addition, microalgae have also been used in the production of nutraceuticals, showing promise in the control of diabetes mellitus through the production of supplements and functional foods. Microalgae-based supplements and functional foods have shown promise in the management of diabetes mellitus. These value-added products can be integrated into diets to modulate carbohydrate digestion, glucose absorption, and lipid metabolism, thereby helping to control blood sugar levels (Bocanegra et al. 2021; Tamel Selvan et al. 2023). In silico studies have identified specific microalgae metabolites, such as fucoxanthin, zeaxanthin, and lutein that interact with key proteins in glucose metabolism and insulin signaling. This suggests potential molecular targets for the treatment of diabetes.

In this sense, microalgae extract, as a natural bioproduct, has vast therapeutic potential and can replace or complement the use of conventional chemicals, minimizing side effects and promoting a more sustainable approach to human health (Silva et al. 2024). The ability of these biocompounds to regulate metabolic pathways involved in insulin resistance and oxidative stress highlights their relevant role in the prevention and treatment of diabetes (Pereira and Valado 2023). Therefore, microalgae biocompounds and their antidiabetic mechanisms emerge as areas of great interest for the advancement of biotechnology and medicine, opening new perspectives for more natural and effective therapies.

3 Microalgae Biocompounds and Their Relationship with Antidiabetic Effects

Microalgae biomass is a powerful source of natural pigments, such as carotenoids, chlorophylls, and phycobiliproteins (Dutra et al. 2024). Among them, microalgal carotenoids, including β-carotene, α-carotene, zeaxanthin, lutein, violaxanthin, neoxanthin, myxoxanthophyll, echinenone, and canthaxanthin, stand out for their biological activities, intrinsically linked to their chemical structure. This structure is characterized by long hydrocarbon chains with multiple conjugated double bonds, which give them antioxidant, anti-inflammatory, and other biological benefits (Saide et al. 2021). These compounds neutralize free radicals and reduce oxidative stress, which is essential for cellular protection and overall health (Jacob-Lopes et al. 2019).

These biochemical alterations and oxidative stress are observable clinical and pathological factors frequently associated with diabetes mellitus. Among the parameters of biochemical alterations are the enzymes ALT (alanine aminotransferase) and AST (aspartate aminotransferase), involved in the central metabolism of all organisms. AST is found mainly in the heart, liver, skeletal muscle, and kidney, while ALT is found mainly in the liver and kidney (Martins et al., 2020). ALT and AST are characterized as biomarkers of liver function, since their increased levels in the blood indicate damage to liver cells, and are often used to diagnose diseases such as hepatic steatosis, in addition to being associated with metabolic conditions such as diabetes mellitus. Research suggests that bioactive compounds present in some species of microalgae, especially antioxidants such as carotenoids, chlorophylls, phycobiliproteins, and omega-3 fatty acids, can reduce serum levels of these enzymes, signaling a possible hepatoprotective action (Altinok-Yipel et al. 2020; Hassaan et al. 2021; Eldessouki et al. 2024). These compounds help mitigate oxidative stress, one of the main factors responsible for cellular damage that results in increased ALT and AST.

Table 4.1 Antidiabetic properties of peptides derived from various algae species

Species	Enzyme	Peptide sequence	Bioactivities (IC_{50})	References
Arthrospira platensis	Alcalase		5.257 mg/mL	Liu et al. (2022)
	Bromelain		4.195 mg/mL	
	Papain		3.819 mg/mL	
	Pepsin		4.059 mg/mL	
	Trypsin		5.603 mg/mL	
Chlorela pyrenoidosa	Pepsin	FLKPLGSGK	63.7% inhibition at 5 mg/mL	Li et al. (2021)
Palmaria palmata	Alcalase + Flavourzyme		0.70 mg/mL	McLaughlin et al. (2021)
Porphyra dioica	Alcalase + Flavourzyme	YLVA	439 μM	Cermeño et al. (2019)
Ulva sp.	Alkaline protease + Flavourzyme	SLAVSVH	1.21 mg/mL	Cian et al. (2022)

Protein content not only helps diagnose and treat diseases involving the kidneys, liver, bone marrow, and other nutritional or metabolic problems. The development of products containing protein extracts from microalgae biomass may be able to inhibit enzymes such as dipeptidyl peptidase IV (DPP-IV), α-amylase, and α-glucosidase (Table 4.1). DPP-IV, a serine peptidase enzyme involved in metabolism, is an important therapeutic target in the treatment of type 2 diabetes (T2D) because it inactivates incretin hormones such as glucagon-like peptide 1 (GLP-1) and glucose-dependent insulinotropic polypeptide (GIP). These hormones play essential roles in the release of insulin after meals and in the control of blood glucose levels. In addition, an inhibitor of carbohydrate-digesting enzymes such as α-amylase and α-glucosidase can reduce glucose absorption in the intestine. Bioactive peptides obtained from several algae species have demonstrated antidiabetic properties, with *Arthorpira platensis*, *Chlorella* sp., and *Palmaria palmat*a being the most studied. Other algae investigated include *Porphyradioca* (Cermeño et al. 2019) and *Ulva* sp. (Cian et al. 2022).

Dipeptidyl peptidase IV (DPP-IV) can be inhibited by compounds with different chemical structures, largely due to the easy accessibility to its active site, located in a long cavity (Cian et al. 2022). Peptides that act as DPP-IV inhibitors are typically composed of short sequences of 2 to 5 amino acid residues. The position of the amino acids within the sequence, especially at the N-terminus (the initial end of the amino acid chain), directly influences the efficacy of DPP-IV inhibition. Peptides that have the proline (Pro) residue in specific positions of the N-terminus, such as in the first, second, third, or fourth position (with emphasis on the penultimate one), generally demonstrate a strong ability to inhibit DPP-IV. Furthermore, the presence of residues such as glycine (Gly), alanine (Ala), phenylalanine (Phe), valine (Val), and leucine (Leu) in the penultimate and last positions of the C-terminus (the end of the amino acid chain) is also associated with excellent inhibitory activity (Li et al. 2021). However, the presence of isoleucine (Ile) in the penultimate position of the

N-terminus can decrease the efficacy of DPP-IV inhibition (Harnedy et al. 2015). Thus, this demonstrates that the position and composition of amino acids within the peptide sequence are determining factors in its ability to inhibit DPP-IV.

Another class of compounds present in microalgae biomass is lipids, which include phospholipids, glycolipids, and apolarglycerolipids, as well as betaine and some lipids specific to certain genera or species. They are generally present in approximately 20 to 50% of microalgae biomass, although this proportion varies according to the species and cultivation conditions (Barkia et al. 2019). Among these lipids, polyunsaturated fatty acids (PUFAs) and pigments (carotenoids) are the most notable functional components for health (Conde et al. 2021). In addition, microalgae contain a wide variety of fatty acids and sterols, in addition to other unsaponifiable fractions, such as terpenoids, carotenoids, and tocopherols. Regarding fatty acids, omega-3 PUFAs are the most abundant, reaching between 10% and 50% of the total fatty acid content of microalgae. Incorporated into lipid fractions, carotenoids have significant nutritional value. Some studies have highlighted the importance of microalgal fatty acids—including omega-3 PUFAs, phytosterols (β-sitosterol), antioxidants (α-tocopherol)—and polyphenols in reducing inflammation and oxidation, in addition to contributing to improvements in micro- and macrovascular endothelial function. These effects play a preventive role in diabetes mellitus (Murray et al. 2018; Bocanegra et al. 2021; Valado et al. 2023).

It is noteworthy that lipid quality, rather than quantity, may influence insulin sensitivity and diabetes risk (Telle-Hansen et al. 2019). Prospective studies associate high intake of saturated fatty acids with increased diabetes risk, while polyunsaturated fatty acids (PUFAs) reduce this risk. Forouhi et al. (2016) found that high blood levels of α-linolenic acid and linoleic acid were associated with a lower long-term risk of diabetes, while other omega-6 fatty acids increased the risk. Marine omega-3 fatty acids, such as DHA and EPA, present in fish oil and microalgae, have antioxidant and anti-inflammatory properties, essential in reducing inflammation (Wang et al. 2019; Bocanegra et al. 2021). Compounds present in microalgae biomass, such as omega-3 fatty acids, may demonstrate the ability to improve insulin sensitivity, modulating its signaling and facilitating glycemic control in patients with diabetes (Abo-Shady et al. 2023). In addition, these compounds are based on chronic inflammation, related to insulin resistance and the development of diabetes (Sinha et al. 2023). Microalgae lipids, such as EPA and DHA, have anti-inflammatory properties that help reduce cytokines, improving disease control (Conde et al. 2023).

Dietary fibers present in microalgae biomass, composed of polysaccharides such as cellulose, hemicellulose, and pectin, can be classified as soluble and insoluble (Deprá et al. 2023). These can contribute to the prevention of insulin resistance, balancing the intestinal microbiota (prebiotic effect), and acting directly on the epithelial and immune cells that regulate the intestinal barrier and immune function. In addition, they help reduce cholesterol and control blood glucose levels in diabetics (Cai et al. 2020). From a centesimal point of view, microalgae can contain dietary fiber ranging from 36 to 60% of dry weight, making them an excellent source of health-beneficial fiber (Peñalver et al. 2020). The antioxidant effect of polysaccharides from natural products derived from microalgae in controlling diabetes is also noteworthy. Hyperglycemia associated with diabetes promotes the production of reactive oxygen species (ROS), which contributes to oxidative stress and inflammation. Natural compounds present in microalgae have important biological activities in neutralizing ROS, preventing cellular damage, and acting on processes involved in diabetes (Pereira and Valado 2023). An example of this is the study by Guo et al. (2021) who found that high molecular weight polysaccharides obtained from *Chlorella pyrenoidosa* (CPS) and *Spirulina platensis* (SPS) reduced obesity in high-fat-induced obese mice (C57BL/6). These polysaccharides influence hepatic lipolysis or lipogenesis, protecting against energy imbalance, systemic inflammation, liver fat accumulation, and dyslipidemia.

4 Scientific Evidences

Without adequate scientific evidence, any and all studies can be discredited. Thus, providing solid foundations for claims may be the way to bridge the gap between basic research and successful development. Therefore, research aimed primarily at the food and pharmaceutical industries must apply in vivo, in vitro, and human studies (Shegokar 2020; Ramalho and Petrica 2023). In this sense, diabetes mellitus, which is, in fact, a public health problem, highlights the need for including emerging therapies. Thus, approaches to research based on the use of microalgae for the control of diabetes mellitus include in vitro, in vivo, and human trials. Table 4.2 presents some of these studies that are available in databases.

Several species of microalgae have been used in in vitro studies. Of course, the microalgae *Arthrospira platensis* stands out for being used in in vivo and human studies. This microalgae is already classified as "generally recognized as" (GRAS) by the FDA (US Food and Drug Administration) (Schneider et al. 2023). In this way, its use is already more established and reaches more sectors. Research related to the in vitro application of microalgae for diabetes mellitus has achieved positive results, focusing on the first category of antidiabetic agents. Achieving inhibition of α- and β-glucosidase and α-amylase is essential, as it can be an effective strategy for controlling blood glucose, especially in periods following food intake (Kaushik et al. 2023).

Regarding in vivo studies, Qiu and collaborators (2022) highlight in their research that when supplementing mice with *C. pyrenoidosa* polysaccharides, they observed a decrease in insulin. Through this result, they concluded that, just like the application of metformin, *C. pyrenoidosa* polysaccharides could also be applied to optimize glucose metabolism while maintaining insulin production. In a study by Sadek and collaborators (2017), diabetes was induced in rats using streptozotocin (STZ). Afterward, when supplementing with *A. platensis*, a significant reduction was observed in serum glucose and glycated hemoglobin (HbA1c), and a relevant increase in serum insulin and the activity of antioxidant enzymes. These studies portray a little of the efficient application of microalgae, as well as the great expansion of the possibility of using different strains.

Table 4.2 In vitro, in vivo, and human studies using microalgae to control diabetes mellitus

Experiments	Microalgae	Effects on diabetes mellitus	References
In vitro studies			
α-amylase inhibition	*Arthrospira platensis, Nannochloropsis oculata, Porphyridium purpureum, Chlorella vulgaris*	Obtained inhibition of α-amylase	Vieira et al. (2021)
α- and β-glucosidase inhibition	*Synechococcus* sp.	High α- and β-glucosidase inhibition	Gradíssimo et al. (2021)
In vivo studies			
Male Wistar rats	*Nannochloropsis oculata (NOM) powder (0, 10, 20, mg/kg BW/day) for 21 days*	↑BW ↓Serum glucose level ↑ Serum insulin level	Fereidouni et al. (2022)
Male Kunming mice (6 weeks old)	*Chlorella pyrenoidosa* Polysaccharide powder (150 and 300 mg/kg BW/day) for 4 weeks	↑BW ↑Insulin level ↑ Glucose uptake	Qiu et al. (2022)
Male Wistar rats (2 months old)	10% *Nannochloropsis gaditana* powder/day for 2 months, administered orally	↑BW ↓Serum glucose, HbA1c, TG, cholesterol	Nacer et al. (2020)
Male Wistar rats	*Nannochloropsis oculata (10, 20 mg/kg BW/day) for 3 weeks*	↑BW ↓Serum glucose, TG, cholesterol, and LDL-C ↑Serum insulin and HDL-C	Nasirian et al. (2019)
Male Wistar rats (2.5 months old)	*Arthrospira plantensis* extract (20, 30 mg/kg BW/day) for 3 weeks	↓Plasma glucose, TG, cholesterol, LDL-C	Nasirian et al. (2019)
Male Albino rats	*Arthrospira platensis* powder (500 mg/kg BW) twice weekly for 2 months	↓Serum glucose ↓HbA1C ↑ Serum insulin	Sadek et al. (2017)
Male Wistar rats or Swiss mice	*Arthrospira platensis* (Spi, 25, 50, and 100 mg/kg BW/day) for 5 and 10 days	↓Serum TG and cholesterol ↓Serum glucose	Joventino et al. (2012)
Male Sprague–Dawley rats (12 ± 2 weeks old)	*Dunaliella salina* (150 mg/kg BW/day) at 72, 64, 48, 40, and 24 h before sacrifice for 3 days	↑BW ↓TG	Ruperez et al. (2009)
Human studies			
160 male patients with T2DM (45–60 years old)	*Arthrospira platensis* capsule (1 g/day) for 12 weeks	↓FBG ↓HbA1C ↓TG, TC, LDL-C ↑ HDL-C	Anitha and Chandralekha (2010)
40 males and females with T2DM (30–60 years old)	*Arthrospira platensis* powder (14 g/day) for 45 days	↓FBG ↓Postprandial blood glucose level	Alam et al. (2016)

(continued)

Table 4.2 (continued)

Experiments	Microalgae	Effects on diabetes mellitus	References
50 male and female patients with T2DM (30–60 years old)	*Arthrospira platensis* powder (8 g/day) for 3 months	↓FBG	Beihaghi and Taherzadeh (2017)
60 male and female patients with T2DM (25–50 years old)	*Arthrospira platensis* powder (2 g/day) for 3 months	↓FBG ↓HbA1C ↓TG, TC, LDL-C ↑ HDL-C	Karizi et al. (2023)

BW body weight, *FBG* fasting blood glucose, *HbA1C* hemoglobin A1c, *HDL-C* high-density lipoprotein cholesterol, *LDL-C* low-density lipoprotein cholesterol, *TC* total cholesterol, *TG* triglyceride, *T2DM* type 2 diabetes mellitus

While both in vitro and in vivo research have their value, human studies are an important part of the overall process. A study with 60 patients was carried out regarding supplementation with *A. platensis* for 3 months, as a complementary form of treatment already carried out with metformin. During this period, patients who were assigned to supplement treatment with microalgae achieved a significant reduction in hemoglobin A1C levels, fasting blood sugar, triglycerides, total cholesterol, LDL-C, and an increase in HDL-C levels. In conclusion of this study, the combination of *A. platensis* with metformin treatment was favorable for long-term glucose control (Karizi et al. 2023). In comparison, one study evaluated treatments for diabetes with metformin and *A. platensis* administered individually. The authors of this study, Alam et al. (2016), observed that both had a similar effect in terms of controlling the reduction of fasting blood sugar and postprandial glycemia.

Finally, the scientific literature presents several possible paths to include the use of microalgae for the treatment of diabetes mellitus. Furthermore, it is still clear that toxicological studies need to cover more microalgal species and compounds. Ensuring the safety of consuming microalgae-based products would open up an even wider range of possibilities. However, there is still a clear need for more robust research to confirm its real effectiveness and achieve a more comprehensive understanding of its effects in different clinical population.

5 Future Prospects and Limitations

Microalgae biomass presents a wide spectrum of applications, and a growing number of researchers are dedicated to fully exploiting its potential. Emerging technologies are propelling algae research into a new era, enabling innovative approaches to the discovery and exploitation of bioactive compounds. In this sense, genetic engineering of microalgae offers promising opportunities to enhance their commercial applications, particularly in the production of pharmaceutical products (Grama et al. 2022). Recent advances in omics technologies, genome editing tools, and high-throughput screening methodologies have accelerated progress in improving microalgae strains (Kumar et al. 2020). Furthermore, the use of microalgae in the formulation of supplements and functional foods is an area of great interest, allowing a preventive approach to diabetes, especially in the control of postprandial glycemia (Golovinskaia and Wang 2023).

Although the outlook is promising, there are significant limitations that need to be overcome. Research on sales availability indicates that the actual proportion of microalgae in the market is still at a low level (Zhuang et al. 2022). Another challenge that microalgae present is their variability in biochemical composition, which, depending on cultivation conditions, can affect the consistency of therapeutic effects (Barkia et al. 2019). One of the limitations highlighted in the pharmaceutical sector is the scarcity of clinical trials in humans that validate the antidiabetic effects observed in preclinical studies. Furthermore, the cost of producing microalgae on a commercial scale is also a limiting factor, especially when compared to more affordable synthetic therapies.

In any case, both the advances and the difficulties in the application of microalgae biomass in the treatment of diabetes mellitus are recognized, but the need to deepen scientific understanding is reinforced. It is essential to continue investing in research and in the improvement of these microorganisms, which have the potential to replace chemical compounds that, although effective, have side effects on human health. Looking to the future, microalgae have great potential for the development of new drugs and functional foods, contributing to improving quality of life and promoting human health (Ibrahim et al. 2023).

6 Conclusion

There is a long way to go, but it does not seem to be that long. As a result, significant antidiabetic properties of microalgal biomass extracts may mark an intriguing advance in the search for innovative functional uses on an industrial scale, including functional foods and pharmaceuticals. The wide range of compounds present, such as natural pigments (carotenoids, chlorophylls, and phycobiliproteins), proteins, lipids, and dietary fibers, raises expectations for useful and commercially viable products. Although they are marketed as a functional food and drug, their market performance may still fall short of expectations. However, consumer interest in using natural bioactive compounds as medicines has recently increased. Furthermore, the numerous biological processes of microalgae have the potential to increase their value as a health-benefiting ingredient in the pharmaceutical and food industries. In conclusion, the use of microalgae biomass as biofactories for biocompounds and application in bioproducts with antidiabetic action presents a promising avenue for further research and development. Although the existing literature highlights the significant potential of microalgae biomass in providing inhibitors for enzymes associated with diabetes, additional efforts are needed to optimize the extraction and isolation conditions.

References

Abo-Shady, A. M., Gheda, S. F., Ismail, G. A., Cotas, J., Pereira, L., & Abdel-Karim, O. H. (2023). Antioxidant and antidiabetic activity of algae. Life, 13(2), 460.

Aboulthana, W. M., Bekheit, R. A. S., & Serag, W. M. (2023). Bio-and phyto-chemical study on *Nannochloropsis oculata* algal extract incorporated with gold nanoparticles, *in vitro* study. Frontiers in Scientific Research and Technology, 5(1).

Alam, A., Siddiqui, M. A., Quamri, A., Fatima, S., Roqaiya, M., & Ahmad, Z. (2016). Effica-cy of *Spirulina* (Tahlab) in patients of type 2 diabetes mellitus (ZiabetusShakri): A ran-domized controlled trial. Journal of Diabetes & Metabolism, 7(10), 1–5.

Al-Lawati, J. A. (2017). Diabetes mellitus: a local and global public health emergency. Oman medical journal, 32(3), 177.

Altinok-Yipel, F., Tekeli, I. O., Ozsoy, S. Y., Guvenc, M., Sayin, S., &Yipel, M. (2020). Investigation of hepatoprotective effect of some algae species on carbon tetrachloride-induced liver injury in rats. Archives of Physiology and Biochemistry, 126(5), 463–467.

Anitha, L., & Chandralekha, K. (2010). Effect of supplementation of spirulina on blood glu-cose, glycosylated hemoglobin and lipid profile of male non-insulin dependent diabetics.

Barkia, I., Saari, N., & Manning, S. R. (2019). Microalgae for high-value products towards human health and nutrition. Marine drugs, 17(5), 304.

Barrea, L., Vetrani, C., Verde, L., Frias-Toral, E., Ceriani, F., Cernea, S., &Muscogiuri, G. (2023). Comprehensive approach to medical nutrition therapy in patients with type 2 diabetes mellitus: from diet to bioactive compounds. Antioxidants, 12(4), 904.

Beihaghi, M., &Taherzadeh, Z. (2017). The effects of oral administration of spirulina platen-sis (cultured iranian) on blood glucose and glycosylated hemoglobin blood in type 2 diabetes mellitus patients. Iranian Journal of Diabetes and Metabolism, 16(3), 183–190.

Blahova, J., Martiniakova, M., Babikova, M., Kovacova, V., Mondockova, V., & Omelka, R. (2021). Pharmaceutical drugs and natural therapeutic products for the treatment of type 2 diabetes mellitus. Pharmaceuticals, 14(8), 806.

Bocanegra, A., Macho-González, A., Garcimartín, A., Benedí, J., & Sánchez-Muniz, F. J. (2021). Whole alga, algal extracts, and compounds as ingredients of functional foods: composition and action mechanism relationships in the prevention and treatment of type-2 diabetes mellitus. International Journal of Molecular Sciences, 22(8), 3816.

Bodke, H., Wagh, V., & Kakar, G. (2023). Diabetes mellitus and prevalence of other comorbid conditions: a systematic review. Cureus, 15(11).

Cai, Y., Folkerts, J., Folkerts, G., Maurer, M., & Braber, S. (2020). Microbiota-dependent and-independent effects of dietary fibre on human health. British journal of pharmacology, 177(6), 1363–1381.

Cardullo, N., Muccilli, V., Pulvirenti, L., Cornu, A., Pouységu, L., Deffieux, D., & Tringali, C. (2020). C-glucosidic ellagitannins and galloylated glucoses as potential functional food ingredients with anti-diabetic properties: A study of α-glucosidase and α-amylase inhibition. Food chemistry, 313, 126099.

Cermeño, M., Stack, J., Tobin, P. R., O'Keeffe, M. B., Harnedy, P. A., Stengel, D. B., & FitzGerald, R. J. (2019). Peptide identification from a *Porphyra dioica* protein hydrolysate with antioxidant, angiotensin converting enzyme and dipeptidyl peptidase IV inhibitory activities. Food & function, 10(6), 3421–3429.

Cian, R. E., Nardo, A. E., Garzón, A. G., Añon, M. C., & Drago, S. R. (2022). Identification and *in silico* study of a novel dipeptidyl peptidase IV inhibitory peptide derived from green seaweed Ulva spp. hydrolysates. Lwt, 154, 112738.

Conde, T. A., Zabetakis, I., Tsoupras, A., Medina, I., Costa, M., Silva, J., & Domingues, M. R. (2021). Microalgal lipid extracts have potential to modulate the inflammatory response: A critical review. International Journal of Molecular Sciences, 22(18), 9825.

Conde, T., Neves, B., Couto, D., Melo, T., Lopes, D., Pais, R., & Domingues, M. R. (2023). Polar Lipids of Marine Microalgae *Nannochloropsis oceanica* and *Chlorococcumamblystomatis*

Mitigate the LPS-Induced Pro-Inflammatory Response in Macrophages. Marine Drugs, 21(12), 629.

Cotas, J., Lomartire, S., Pereira, L., Valado, A., Marques, J. C., & Gonçalves, A. M. (2024). Seaweeds as Nutraceutical Elements and Drugs for Diabetes Mellitus: Future Perspectives. Marine Drugs, 22(4), 168.

de Alencar Alves, M. F., de Almeida Barreto, F. K., de Vasconcelos, M. A., do Nascimento Neto, L. G., Carneiro, R. F., da Silva, L. T., & Teixeira, E. H. (2020). Antihyperglycemic and antioxidant activities of a lectin from the marine red algae, *Bryothamnionseaforthii*, in rats with streptozotocin-induced diabetes. International journal of biological macromolecules, 158, 773–780.

Deprá, M. C., Dias, R. R., Maroneze, M. M., Nascimento, T. C., Severo, I. A., Zepka, L. Q., & Jacob-Lopes, E. (2023). Bioactive polysaccharides from microalgae: A close look at the biomedical applications. Recent Patents on Biotechnology, 17(4), 296–311.

Dutra, D. A., Deprá, M. C., Dias, R. R., Zepka, L. Q., & Jacob-Lopes, E. (2024). Ensuring nutrition and food safety within planetary boundaries: The role of microalgae-based ingredients in sustainable food chain. Journal of Applied Phycology, 1–20.

Eldessouki, E. A., Elshopakey, G. E., Elbahnaswy, S., Shakweer, M. S., Abdelwarith, A. A., Younis, E. M., & Eissa, E. S. H. (2024). Influence of astaxanthin-enriched *Haematococcus pluvialis* microalgae on the growth efficacy, immune response, antioxidant capacity, proinflammatory cytokines, and tissue histomorphology of hybrid red tilapia. Aquaculture International, 1–22.

Fereidouni, A., Khaleghian, A., Mousavi-Niri, N., &Moradikor, N. (2022). The effects of supplementation of *Nannochloropsis oculata* microalgae on biochemical, inflammatory and antioxidant responses in diabetic rats. Biomolecular Concepts, 13(1), 314–321.

Forouhi, N. G., Imamura, F., Sharp, S. J., Koulman, A., Schulze, M. B., Zheng, J., & Wareham, N. J. (2016). Association of plasma phospholipid n-3 and n-6 polyunsaturated fatty acids with type 2 diabetes: the EPIC-InterAct case-cohort study. PLoS medicine, 13(7), e1002094.

Forray, A. I., Coman, M. A., Simonescu-Colan, R., Mazga, A. I., Chereches, R. M., & Borzan, C. M. (2023). The Global Burden of Type 2 Diabetes Attributable to Dietary Risks: Insights from the Global Burden of Disease Study 2019. Nutrients, 15(21), 4613.

Golovinskaia, O., & Wang, C. K. (2023). The hypoglycemic potential of phenolics from functional foods and their mechanisms. Food Science andHumanWellness, 12(4), 986–1007.

Gradíssimo, D. G., Oliveira da Silva, V. C., Xavier, L. P., do Nascimento, S. V., Valadares, R. B. D. S., Faustino, S. M. M., & Santos, A. V. (2021). Glucosidase inhibitors screening in microalgae and cyanobacteria isolated from the amazon and proteomic analysis of inhibitor producing Synechococcus sp. GFB01. Microorganisms, 9(8), 1593.

Grama, S. B., Liu, Z., & Li, J. (2022). Emerging trends in genetic engineering of microalgae for commercial applications. Marine Drugs, 20(5), 285.

Guo, W., Zhu, S., Li, S., Feng, Y., Wu, H., & Zeng, M. (2021). Microalgae polysaccharides ameliorates obesity in association with modulation of lipid metabolism and gut microbiota in high-fat-diet fed C57BL/6 mice. International journal of biological macromolecules, 182, 1371–1383.

Harnedy, P. A., O'Keeffe, M. B., & FitzGerald, R. J. (2015). Purification and identification of dipeptidyl peptidase (DPP) IV inhibitory peptides from the macroalga *Palmaria palmata*. Food Chemistry, 172, 400–406.

Hassaan, M. S., Mohammady, E. Y., Soaudy, M. R., Sabae, S. A., Mahmoud, A. M., & El-Haroun, E. R. (2021). Comparative study on the effect of dietary β-carotene and phycocyanin extracted from Spirulina platensis on immune-oxidative stress biomarkers, genes expression and intestinal enzymes, serum biochemical in Nile tilapia, *Oreochromis niloticus*. Fish & Shellfish Immunology, 108, 63–72.

Hossain, M. J., Al-Mamun, M., & Islam, M. R. (2024). Diabetes mellitus, the fastest growing global public health concern: Early detection should be focused. Health Science Reports, 7(3), e2004.

Ibrahim, T. N. B. T., Feisal, N. A. S., Kamaludin, N. H., Cheah, W. Y., How, V., Bhatnagar, A., & Show, P. L. (2023). Biological active metabolites from microalgae for healthcare and pharmaceutical industries: a comprehensive review. Bioresource Technology, 372, 128661.

IDF – International Diabetes Federation. (2021). Diabetes facts & figures. Available online: https://idf.org/about-diabetes/diabetes-facts-figures/. (accessed on 2 October 2024).

Jacob-Lopes, E., Maroneze, M. M., Deprá, M. C., Sartori, R. B., Dias, R. R., &Zepka, L. Q. (2019). Bioactive food compounds from microalgae: An innovative framework on industrial biorefineries. Current Opinion in Food Science, 25, 1–7.

Jia, Y., Xue, Z., Wang, Y., Lu, Y., Li, R., Li, N., ... & Chen, H. (2021). Chemical structure and inhibition on α-glucosidase of polysaccharides from corn silk by fractional precipitation. Carbohydrate polymers, 252, 117185.

Joventino, I. P., Alves, H. G., Neves, L. C., Pinheiro-Joventino, F., Leal, L. K. A., Neves, S. A., & Viana, G. B. (2012). The microalga Spirulina platensis presents anti-inflammatory action as well as hypoglycemic and hypolipidemic properties in diabetic rats. Journal of complementary and integrative medicine, 9(1).

Kang, Y., Xu, L., Dong, J., Yuan, X., Ye, J., Fan, Y., & Ji, X. (2024). Programmed microalgae-gel promotes chronic wound healing in diabetes. Nature communications, 15(1), 1042.

Karizi, S. R., Armanmehr, F., Azadi, H. G., Zahroodi, H. S., Ghalibaf, A. M., Bazzaz, B. S. F., &Taherzadeh, Z. (2023). A randomized, double-blind placebo-controlled add-on trial to assess the efficacy, safety, and anti-atherogenic effect of *Spirulina platensis* in patients with inadequately controlled type 2 diabetes mellitus. Phytotherapy Research, 37(4), 1435–1448.

Kaushik, A., Sangtani, R., Parmar, H. S., & Bala, K. (2023). Algal metabolites: Paving the way towards new generation antidiabetic therapeutics. Algal Research, 69, 102904.

Kumar, G., Shekh, A., Jakhu, S., Sharma, Y., Kapoor, R., & Sharma, T. R. (2020). Bioengineering of microalgae: recent advances, perspectives, and regulatory challenges for industrial application. Frontiers in Bioengineering and Biotechnology, 8, 914.

Lauritano, C., Ferrante, M. I., & Rogato, A. (2019). Marine natural products from microalgae: An-omics overview. Marine Drugs, 17(5), 269.

Li, Y., Aiello, G., Fassi, E. M. A., Boschin, G., Bartolomei, M., Bollati, C., & Lammi, C. (2021). Investigation of *Chlorella pyrenoidosa* protein as a source of novel angiotensin I-converting enzyme (ACE) and dipeptidyl peptidase-IV (DPP-IV) inhibitory peptides. Nutrients, 13(5), 1624.

Liu, J., Bai, X., & Fu, P. (2022). In silico and in vitro assessment of bioactive peptides from *Arthrospira platensis* phycobiliproteins for DPP-IV inhibitory activity, ACE inhibitory activity, and antioxidant activity. Journal of Applied Phycology, 34(3), 1497–1511.

Machado, R. L. S., Dutra, D. A., Schneider, A. T., Dias, R. R., Zepka, L. Q., & Jacob-Lopes, E. (2024). Recent Advances in Algal Nexus for Circular Economy. Algae as a Natural Solution for Challenges in Water-Food-Energy Nexus: Toward Carbon Neutrality, 61–77.

McLaughlin, C. M., Harnedy-Rothwell, P. A., Lafferty, R. A., Sharkey, S., Parthsarathy, V., Allsopp, P. J., &O'Harte, F. P. (2021). Macroalgal protein hydrolysates from Palmariapalmata influence the 'incretin effect *in vitro via* DPP-4 inhibition and upregulation of insulin, GLP-1 and GIP secretion. European Journal of Nutrition, 60(8), 4439–4452.

Mehariya, S., Goswami, R. K., Karthikeysan, O. P., & Verma, P. (2021). Microalgae for high-value products: A way towards green nutraceutical and pharmaceutical compounds. Chemosphere, 280, 130553.

Melloni, M., Sergi, D., Simioni, C., Passaro, A., & Neri, L. M. (2023). Microalgae as a Nutraceutical Tool to Antagonize the Impairment of Redox Status Induced by SNPs: Implications on Insulin Resistance. Biology, 12(3), 449.

Munawaroh, H. S. H., Gumilar, G. G., Nurjanah, F., Yuliani, G., Aisyah, S., Kurnia, D., & Show, P. L. (2020). In-vitro molecular docking analysis of microalgae extracted phycocyanin as an anti-diabetic candidate. Biochemical Engineering Journal, 161, 107666.

Murray, M., Dordevic, A. L., Ryan, L., & Bonham, M. P. (2018). An emerging trend in functional foods for the prevention of cardiovascular disease and diabetes: Marine algal polyphenols. Critical reviews in food science and nutrition, 58(8), 1342–1358.

Nacer, W., Baba Ahmed, F. Z., Merzouk, H., Benyagoub, O., & Bouanane, S. (2020). Evaluation of the anti-inflammatory and antioxidant effects of the microalgae *Nannochloropsis gaditana* in streptozotocin-induced diabetic rats. Journal of Diabetes & Metabolic Dis-orders, 19, 1483–1490.

Nasirian, F., Sarir, H., &Moradi-Kor, N. (2019). Antihyperglycemic and antihyperlipidemic activities of *Nannochloropsis oculata* microalgae in streptozotocin-induced diabetic rats. Biomolecular concepts, 10(1), 37–43.

Odenthal, K., Nunes, E., Nunes, N., Fernandes, T., Fernandes, I. A., & de Carvalho, M. A. P. (2024). Microalgae and cyanobacteria as natural sources of antioxidant enzymes and enzyme inhibitors for Alzheimer's and diabetes. AlgalResearch, 82, 103610.

Peñalver, R., Lorenzo, J. M., Ros, G., Amarowicz, R., Pateiro, M., &Nieto, G. (2020). Seaweeds as a functional ingredient for a healthy diet. Marine Drugs, 18(6), 301.

Pereira, L., & Valado, A. (2023). Algae-Derived Natural Products in Diabetes and Its Complications-Current Advances and Future Prospects. Life, 13(9), 1831.

Qiu, Y., Gao, X., Chen, R., Lu, S., Wan, X., Farag, M. A., & Zhao, C. (2022). Metabolomics and biochemical insights on the regulation of aging-related diabetes by a low-molecular-weight polysaccharide from green microalga *Chlorella pyrenoidosa*. Food Chemistry: X, 14, 100316.

Quarti Machado Rosa, M, dos Santos Rosa, R., Correia, M. G., Araujo, D. V., Bahia, L. R., & Toscano, C. M. (2018). Disease and economic burden of hospitalizations attributable to diabetes mellitus and its complications: a nationwide study in Brazil. International journal of environmental research and public health, 15(2), 294.

Ramalho, A., & Petrica, J. (2023). Knowledge in motion: a comprehensive review of evi-dence-based human kinetics. International Journal of Environmental Research and Public Health, 20(11), 6020. DOI: https://doi.org/10.3390/ijerph20116020.

Ruperez, F. J., Garcia-Martinez, D., Baena, B., Maeso, N., Vallejo, M., Angulo, S., & Bar-bas, C. (2009). *Dunaliella salina* extract effect on diabetic rats: metabolic fingerprinting and target metabolite analysis. Journal of pharmaceutical and biomedical analysis, 49(3), 786–792.

Sadek, K. M., Lebda, M. A., Nasr, S. M., & Shoukry, M. (2017). *Spirulina platensis* prevents hyperglycemia in rats by modulating gluconeogenesis and apoptosis *via* modification of oxidative stress and MAPK-pathways. Biomedicine&Pharmacotherapy, 92, 1085–1094.

Saide, A., Martínez, K. A., Ianora, A., &Lauritano, C. (2021). Unlocking the health potential of microalgae as sustainable sources of bioactive compounds. International Journal of Molecular Sciences, 22(9), 4383.

Sathasivam, R., Radhakrishnan, R., Hashem, A., & Abd-Allah, E. F. (2019). Microalgae metabolites: A rich source for food and medicine. Saudi journal of biological sciences, 26(4), 709–722.

Schneider, A. T., Deprá, M. C., Dias, R. R., Zepka, L. Q., & Jacob-Lopes, E. (2023). Microalgae as superfood. In Algae Materials (pp. 93–102). Academic Press.

Shegokar, R. (2020). Preclinical testing-Understanding the basics first. In Drug Delivery Aspects (pp. 19–32). Elsevier.

Silva, M., Avni, D., Varela, J., & Barreira, L. (2024). The Ocean's pharmacy: Health discoveries in marine algae. Molecules, 29(8), 1900.

Sinha, S., Haque, M., Lugova, H., & Kumar, S. (2023). The effect of omega-3 fatty acids on insulin resistance. Life, 13(6), 1322.

Sugandh, F. N. U., Chandio, M., Raveena, F. N. U., Kumar, L., Karishma, F. N. U., Khuwaja, S., & Kumar, S. (2023). Advances in the management of diabetes mellitus: a focus on personalized medicine. Cureus, 15(8).

Tamel Selvan, K., Goon, J. A., Makpol, S., & Tan, J. K. (2023). Therapeutic potentials of microalgae and their bioactive compounds on diabetes mellitus. Marine Drugs, 21(9), 462.

Telle-Hansen, V. H., Gaundal, L., & Myhrstad, M. C. (2019). Polyunsaturated fatty acids and glycemic control in type 2 diabetes. Nutrients, 11(5), 1067.

Valado, A., Ponedilok, P., Caseiro, A., & Pereira, L. (2023). Macroalgae as a Source of Functional Foods in the Prevention of Cardiovascular Diseases. Recent Progress in Nutrition, 3(4), 1–27.

Vieira, M. V., Turkiewicz, I. P., Tkacz, K., Fuentes-Grünewald, C., Pastrana, L. M., Fuciños, P., & Nowicka, P. (2021). Microalgae as a potential functional ingredient: evaluation of the phytochemical profile, antioxidant activity and *in-vitro* enzymatic inhibitory effect of different species. Molecules, 26(24), 7593.

Wang, M., Ma, L. J., Yang, Y., Xiao, Z., & Wan, J. B. (2019). n-3 Polyunsaturated fatty acids for the management of alcoholic liver disease: A critical review. Critical reviews in food science and nutrition, 59(sup1), S116–S129.

World Health Organization. (2024). Diabetes. Available online: https://www.who.int/news-room/fact-sheets/detail/diabetes. (accessed on 20 September 2024).

Yusuf, B. O., Yakubu, M. T., & Akanji, M. A. (2021). Chromatographic fractions from *Chrysophyllum albidum* stem bark boost antioxidant enzyme activity and ameliorate some markers of diabetes complications. Journal of Traditional and Complementary Medicine, 11(4), 336–342.

Zhuang, D., He, N., Khoo, K. S., Ng, E. P., Chew, K. W., & Ling, T. C. (2022). Application progress of bioactive compounds in microalgae on pharmaceutical and cosmetics. Chemosphere, 291, 132932.

Chapter 5
Macroalgae with Anti-diabetic Properties

S. Bharathi, Yolin Angel P. A. Soundara Rajan, S. Prakash, Grasian Immanuel, and R. Ramasubburayan

Abstract Macroalgae, also known as seaweeds, are vital to aquatic environments. Seaweeds exhibit numerous bio-cellular activities, including anti-diabetic properties, due to their complex chemical components, together with primary and secondary metabolites. The physiological active compounds present in seaweeds could manage digestive effects and maintain intestinal permeability. Also, they regulate glucose metabolism and insulin sensitivity, control postprandial blood glucose levels, and oxidative stress, thus demonstrating anti-diabetic properties. Bioactive compounds from seaweed have been characterised, and their actions have been partially elucidated. The anti-diabetic advantages of seaweeds have been proved by various in vitro, in vivo, and human studies. While oral seaweed intake remains common in many countries, additional research on the bioactivities and possible adverse effects of functional chemicals in edible seaweeds is essential for developing seaweed-based anti-diabetic medications. The main objective of this chapter is to give readers a basic understanding of the nutritional and anti-diabetic properties of macroalgae. The forthcoming chapters of this book include more comprehensive details about the mechanism of action and research studies in this area.

S. Bharathi
Centre for Modern Organic Agriculture Research (CMOAR), Sathyabama Institute of Science and Technology, Chennai, Tamil Nadu, India

Yolin Angel P. A. Soundara Rajan
Centre for Nanoscience and Nanotechnology, Sathyabama Institute of Science and Technology, Chennai, Tamil Nadu, India

S. Prakash
Marine Biotechnology Research Laboratory, Department of Basic Sciences, Institute of Fisheries Post Graduate Studies (IFPGS), Tamil Nadu Dr. J. Jayalalithaa Fisheries University, Vaniyanchavadi, Chennai, Tamil Nadu, India

G. Immanuel
Centre for Marine Science & Technology, Manonmaniam Sundaranar University, Rajakkamangalam, Kanyakumari District, Tamil Nadu, India

R. Ramasubburayan (✉)
Centre for Marine and Aquatic Research, Department of Prosthodontics, Saveetha Dental College and Hospitals, SIMATS, Saveetha University, Chennai, Tamil Nadu, India

R. Raja et al. (eds.), *Algae in Diabetes Management*,
https://doi.org/10.1007/978-981-95-0571-5_5

Keywords Anti-diabetic · Seaweeds · Glucose metabolism · Insulin sensitivity · Digestive enzymes

1 Introduction

Macroalgae are recognised as essential components of our planet, serving as primary producers in aquatic systems and acting as bioindicators of water quality for the bioremediation process (Lourenço-Lopes et al. 2020). Macroalgae provide oxygen, nutrition, and shelter to a wide range of living beings in the ocean, and their role in mitigating ocean acidification has highlighted them as a potential solution for global warming. These multi-cellular macroalgae are commonly referred to as seaweeds. In East Asian nations such as China, Japan, and Malaysia, seaweeds have been cultivated and incorporated into traditional meals for generations. Besides culinary applications, the demand for seaweed has been rising for non-food purposes, including supplements in nutraceutical preparations, cosmetics, cattle feed, fertilizers (especially in horticulture), bio-packaging, and biofuels (Shannon and Abu-Ghannam 2019). Macroalgae have received increased interest in recent years because they offer numerous health-promoting qualities, which may minimise the risk of several chronic diseases and extend the lifespan (Ganesan et al. 2019). Due to its potential anti-coagulation, anti-oxidation, anti-tumour, anti-viral, and anti-bacterial properties, as well as non-toxic, edible, cost-effective, and easily cultivable qualities, seaweeds can be exploited in the development of novel pharmaceuticals and functional foods, exhibiting minimal or no negative consequences (Lomartire and Gonçalves 2022).

Seaweeds provide USD 6 billion to the world economy through diverse industry applications (Horn 2009). Records show that over 10,000 unique species of macroalgae inhabit various saltwater environments globally (Barbier et al. 2019; FAO 2021), although approximately 220 species are being utilised for commercial purposes (Cai et al. 2021). About 10 species are intensively farmed to meet demand; among these, the brown seaweed *Saccharina japonica* makes up 33%, followed by *Kappaphycus alvarezii* at 17%. The remaining seaweed species, such as *Undaria pinnatifida*, *Porphyra* sp., *Sargassum fusiforme*, *Eucheuma*, *Gracilaria* sp., *Enteromorpha clathrata*, *Monsostroma nitidium*, and *Caulerpa* sp. are commonly accounted as cultivable species (Choudhary et al. 2021).

2 Types of Seaweeds and Their Global Distribution

These multi-cellular macroalgae exhibit colour variations in their thallus, categorised depending on morphology: Rhodophyta (red algae) (e.g. *Porphyra, Gelidium amansii*), Chlorophyta (green algae) (*Enteromorpha compressa, Codium fragile*), and Phaeophyta (brown algae) (e.g. *Saccharina japonica, U. pinnatifida,* and

Sargassum fusiform) (Hamid et al. 2019). The broad group of brown algae exhibit pigments varied from yellow to dark brown. The pigments found in red algae include chlorophylls a and d, carotenoids, phycocyanin, phycoerythrin, and allophycocyanin (Denis et al. 2009). Green algae predominantly comprise chlorophyll, essential for photosynthesis and other biological processes (Aryee et al. 2018). The presence of biochemical compounds in seaweeds mainly relies on species type, harvest season, and geographical location (Abomohra and Almutairi 2020). De Melo et al. (2021) observed that the carbohydrate content of *Sargassum vulgare* was lower during rainfall compared to the dry season. Therefore, a thorough examination of seaweeds is crucial for effectively utilising value-added compounds, as it may elucidate their potential in manufacturing such compounds. Proximate and ultimate studies, coupled with FTIR and TGA/DTG analysis, are conducted to characterise the seaweeds based on the bioavailability of carbohydrates, proteins, and lipids contained within them (Yang et al. 2021). Due to their physiologically active and regenerative properties, algae represent a viable and cost-effective biomass source of significant chemicals with potential applications in the food, chemical, pharmaceutical, and cosmetic industries (Biris-Dorhoi et al. 2020).

3 Nutritional and Metabolite Composition of Seaweeds

The nutritional composition of seaweeds exceeds that of terrestrial plants due to the absence of energy expenditure on roots, stems, leaves, or circulatory systems, allowing for energy conservation (Choudhary et al. 2023). Seaweeds are low-calorie food sources, abundant in nutrition, such as vitamins, minerals, essential trace elements, polyunsaturated fatty acids, bioactive metabolites, proteins, polysaccharides, and dietary fibres (Cotas et al. 2024). The substantial quantity of many physiologically active compounds, including complex organic molecules and primary and secondary metabolites, renders them important (Biris-Dorhoi et al. 2020). The bioactive compounds, including nutritive metabolites, phenolics, pigments, etc., are given in Table 5.1. The health advantages of seaweed supplements, combined with a balanced diet, have been shown beyond standard consumption (Cotas et al. 2024).

Seaweeds contain more amino acids and unsaturated fatty acids than vegetables (Jimenez-Escrig and Cambrodon 1999). Seaweeds are a considerable source of macro minerals, including potassium, magnesium, calcium, and phosphorus, as well as microminerals such as iron, manganese, molybdenum, nickel, selenium, and zinc (Baghel et al. 2023). Secondary metabolites are primarily generated in seaweeds in response to stress conditions induced by temperature, salinity, ultraviolet radiation, or pollutants. Seaweeds produce a range of bioactive components, including sterols, terpenes, polyphenols, peptides, and halogenated compounds. Therefore, secondary metabolite production in seaweeds is essential for protection against structural and photodynamic damage incurred in extreme growth conditions (Tanna and Mishra 2019).

Table 5.1 Nutrients and bioactive secondary metabolites reported in selected seaweeds

Nutrients/ bioactive secondary metabolites	Quantity of contents in dry weight	Subtypes	Seaweeds
Proteins and amino acids	3–50%	Glutamic acid, aspartic acid, alanine, glycine, lectins, alkaline phosphatase and oxidases, fibrinolytic enzymes, rubisco	Species of the genera *Enteromorpha, Palmaria, Vertebrata, Sargassum, Codium, Solieria, Halimeda, Caulerpa, Kappaphycus*
Lipids and fatty acids	< 5%	Eicosapentaenoic, arachidonic, hexadecatetraenoic, oleic, palmitic, linoleic, linolenic, and docosahexaenoic, stearidonic acids, glycolipids, phospholipids, omega-3 and omega-6	Species of the genera *Ulva, Fucus, Sargassum, Porphyra, Neopyropia, Cystoseira, Gracilariopsis, Cladophora*
Carbohydrate	32–60%	Porphyran Sulphated Fucose-Fucoidans Sulphated Galactan-Carrageenans Sulphated Heteroglycans-Ulvans	Species of the genera *Laminaria, Undaria, Saccharina, Gracilaria, Ulva, Kappaphycus, Eucheuma, Pyropia, Sargassum, Cladosiphon, Fucus, Hypnea, Tichocarpus, Eucheuma, A scophyllum, Porphyra, Palmaria, Gigartina, Chondrus*
Fibre	36–60%	Agars, alginates, fucoidan, carrageenan, laminaran, porphyrin, cellulose, hemicelluloses, mannans, xylan	Species of the genera *Chondrus, Porphyra, Fucus, Caulerpa, Enteromorpha Ulva, Garateloupia, Durvillaea, Himanthalia, Saccharina, Sargassum, Undaria*
Vitamins	Green and brown algae: 500–3000 mg/kg; Red algae: 100–800 mg/kg	Vitamins A, E, B1, B2, B12, C Thiamine, riboflavin, β-carotene, tocopherols	Species of the genera Udotea, *Turbinaria, Polycladia, Colpomenia, Ascophyllum, Fucus, Ulva, Gracilaria, Palmaria, Porphyra*

(continued)

Table 5.1 (continued)

Nutrients/ bioactive secondary metabolites	Quantity of contents in dry weight	Subtypes	Seaweeds
Mineral	40%	Sodium, calcium, magnesium, potassium, chloride, phosphorus, sulphur, chromium, copper, fluoride, iodine, iron, manganese, molybdenum, selenium, and zinc	Species of the genera *Fucus, Laminaria, Porphyra, Gracilaria, Centroceras, Caulerpa, Monostroma, Hypnea, Ulva, Kappaphycus, Acanthophora, Macrocystis, Nereocystis, Eisenia, Codium, Sargassum, Chnoospora, Mastocarpus, Ecklonia, Durvillaea, Ascophylum, Palmaria, Gelidium, Laurencia, Chondrus, Enteromorpha*
Pigments	–	Carotenoids, chlorophyll, phycobiliproteins, Beta-carotene, astaxanthin, phycocyanin, fucoxanthin, phycobilins, violaxanthin, lutein, neoxanthin, antheraxanthin, zeaxanthin	Species of the genera *Caulerpa, Gracilaria, Eucheuma, Saccharina, Undaria, Himanthalia, Myagropsis, Dictyota, Sargassum, Padina, Turbinaria*
Phenolic compounds	–	Bromophenols, morin, monobromophenols, 2,4-bromophenol, 2,6-dibromophenol, 2,4,6-tribromophenol, myricetin, phlorotannins, quercetin, phloroglucinol, hesperidin, catechins, benzoic, salicylic, vanillic, gallic, p-hydroxybenzoic, gentisic, protocatechuic, syringic, caffeic, chlorogenic, ferulic, and cinnamic acids, phlorotannins, flavonoids, hydroxy benzoic and hydroxy cinnamic acids	Species of the genera *Ecklonia, Gracilaria, Eisenia, Caulerpa, Cystoseira, Acanthophora. Himanthalia*
Terpenes	–	Halogenated monoterpenes, sesquiterpenoid caulerpenyne, sesquiterpene	Species of the genera *Locamium, Portieria, Ochtodes, Laurencia, Caulerpa*

Baghel et al. (2023), Biris-Dorhoi et al. (2020), Salido et al. (2023), Aknaf et al. (2024), and Lu and Chen (2022)

4 Bioactive Compounds in Seaweeds with Anti-diabetic Properties

Polysaccharides from seaweed enhance insulin sensitivity, improve glucose absorption, and reduce blood glucose levels. *S. horneri*, a seaweed, contains higher concentrations of advantageous polysaccharides and fucoxanthin than commonly consumed brown seaweeds such as Wakame (*U. pinnatifida*) and Japanese tangle (*S. japonica*) (Kim et al. 2021). Dietary fibres, such as alginate, carrageenan, fucoidan, and phenolic compounds, facilitate the sensation of fullness and satiety, thus assisting in weight management. Besides carbohydrates and dietary fibres, seaweeds encompass various vitamins and minerals, such as iodine, calcium, magnesium, iron, and vitamin C (Lomartire and Gonçalves 2022).

Seaweeds possess a variety of non-nutritive bioactive substances, such as fucoxanthin, phlorotannins, and sulphated polysaccharides, recognised for their anti-diabetic benefits without inducing adverse effects (Kim et al. 2021). Seaweed peptides exhibit potential anti-diabetic properties, and some peptides demonstrate insulin-mimetic action (Gunathilaka et al. 2020). Mannitol, a sugar alcohol common in seaweeds, is not metabolised by the body and is expelled intact in the urine. Mannitol is occasionally utilised as a sweetener in low-calorie dishes due to its mild laxative properties (Grembecka 2015). Iodine contributes to the enhancement of reducing fasting blood glucose levels, while antioxidants play a crucial role in neutralising reactive oxygen species. Both of these elements found in seaweeds are beneficial for managing type 2 diabetes mellitus (T2DM) (Lee and Han 2018).

Dieckol, a compound in the brown algae *Ecklonia cava*, can improve glycaemia and insulin resistance; it also exhibits antioxidant properties that alleviate oxidative stress damage associated with hyperglycaemia. Likewise, phytosterol in seaweed lowers blood glucose levels and exhibits antioxidant, anti-inflammatory, anti-obesity, and anti-diabetic characteristics (Sohn et al. 2021). The fatty acid profiles of seaweeds vary among brown, red, and green varieties, with green seaweeds exhibiting higher levels of omega-3 fatty acids, especially alpha-linolenic acid (ALA). *Ulva* sp. is considered the most promising candidate for the formulation of low-fat meals enhanced with PUFA-rich nutraceuticals (Al-Adilah et al. 2021). Indian Phaeophytes and Rhodophytes are rich in omega-3 polyunsaturated fatty acids (PUFAs) and omega-6 PUFAs, respectively, while Chlorophytes are abundant in monounsaturated fatty acids (MUFAs) (Kumar et al. 2011).

Marine algae supply plentiful sources of non-animal polysaccharides. Porphyran, a sulphated carbohydrate derived from the genus *Porphyra,* has exhibited anti-inflammatory, antioxidant, and antihyperlipidemic activities. Ulvan derived from green algae exhibits immunostimulatory properties (Lomartire and Gonçalves 2022). Carrageenan is predominantly derived from red algae. Fucoidan, alginate, and laminarin are polysaccharides found solely in brown seaweeds. Seaweeds contain polyphenols that have been shown to possess antioxidant, anti-diabetic, anti-inflammatory properties, and inhibitory properties on α-glucosidase, the enzyme responsible for reducing postprandial blood glucose levels (Valko et al. 2007). Among the different types of algal polyphenols, phlorotannins from brown algae are the most prominent due to their established pharmacological potential (Lopes et al. 2012). The health benefits of phytosterols derived from marine algae, particularly fucosterol, include anti-diabetic properties.

The intake of bioactive substances from seaweed enhances the makeup of gut microbiota (Agarwal et al. 2022). The presence of fibre components, including fucoidan, laminarin, algin, ulvan, porphyran, polyphenols, and peptides, demonstrates beneficial effects on gut microbiota (Ojo et al. 2021). The gut microbiota plays a crucial role in regulating immune markers, interacting with food constituents, and effectively managing glucose and lipid metabolism, insulin sensitivity, energy balance, and intestinal permeability. These functions are consistently associated with a reduced incidence of T2DM (Gurung et al. 2020).

5 Mechanisms of Anti-diabetic Action of Bioactive Compounds in Seaweeds

Numerous epidemiological, clinical, and meta-analytical studies relate a decreased incidence of various chronic diseases, including diabetes mellitus (DM), with a dietary profile of seaweed consumption (Deleris et al. 2016). It has been proposed that the biotechnological significance of seaweeds arises from their ability to adapt to extreme conditions. The compounds associated with the aforementioned unique property may enhance their defence mechanisms, thereby combating diseases (Dixit et al. 2018). However, the mechanism underlying the anti-diabetic efficacy and regulating blood glucose levels remains questionable. The symptoms of diabetes, characterised by elevated blood glucose levels resulting from either inadequate insulin production or inadequate insulin utilisation, are managed through several natural plant-based substances, yet the search for a definitive cure persists (Rachdaoui 2020). The glycaemic index can be reduced by altering nutrient composition to decrease energy derived from available carbohydrates, inhibiting digestive enzyme activity to slow digestion, prolonging gastric emptying and glucose absorption by increasing digest viscosity, and improving insulin sensitivity by anti-inflammatory effects (Fig. 5.1). Seaweeds, which are abundant in high-value nutrients, can diminish glycaemic potency by interfering with essential processes. This section analyses the current uses of seaweeds and the macro and micronutrients they possess that may exhibit glycaemic-lowering effects, thereby demonstrating anti-diabetic properties.

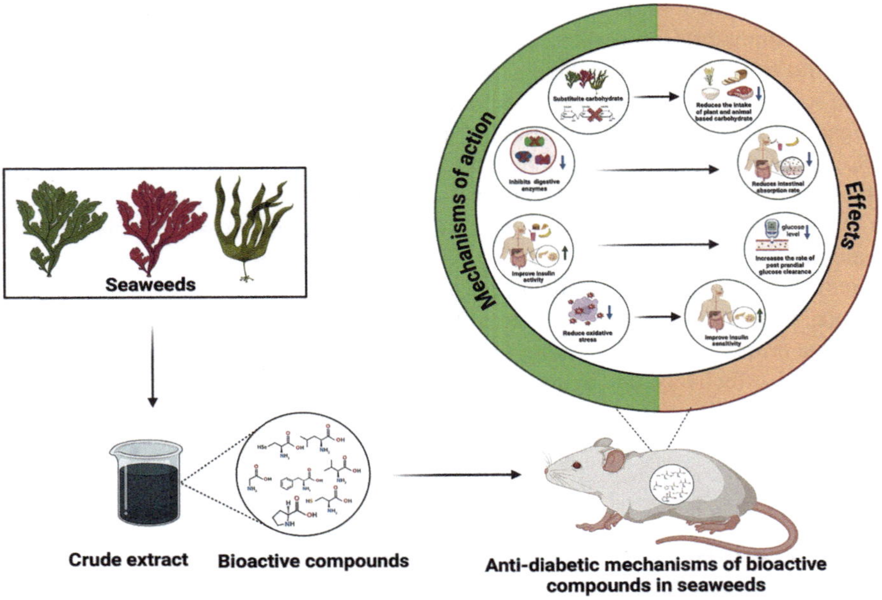

Fig. 5.1 In vivo model showing the mechanism of action of bioactive compounds from seaweeds in controlling diabetes

5.1 Seaweeds: Alternate Carbohydrate Source

Reducing the available carbohydrate content can be accomplished by substituting these carbohydrates with alternative macronutrients sourced from seaweeds. Seaweeds contain nutrients that enhance carbohydrate quality and reduce the glycaemic potential of foods. Seaweeds provide significant amounts of dietary fibre, omega-3 fatty acids, polyphenols, and phytochemicals, including carotenoids (Mendis and Kim 2011). The health benefits and potential for developing functional ingredients of these nutrients have been acknowledged and confirmed (Cai et al. 2021).

The red seaweed species, such as Fucus, Ascophyllum, Porphyra, and *Palmaria* have the highest concentrations of accessible carbohydrates, such as glucose, mannose, and galactose, which exist in minimal amounts. Dietary fibres in seaweeds are crucial in regulating glycaemic levels, as they are neither utilised nor absorbed by the human body (Holdt and Kraan 2011). Seaweeds comprise two primary categories of dietary fibres: water-soluble fibres, accounting for 25% (Lu and Chen 2022), and water-insoluble fibres, constituting 21.8% (Penalver et al. 2020). Soluble fibres, including agars, alginates, fucoidan, carrageenan, laminaran, and porphyran, can enhance satiety in the human gastrointestinal tract, improve intestinal microbiota, and slowdown the digestion and absorption of other nutrients (Holdt and Kraan 2011). Insoluble fibres, such as cellulose, hemicelluloses, mannans, and xylan, are not fermentable in the large intestine, yet they maintain water content in the structural matrix, aiding in large intestine transit and increasing faecal mass (Holdt and Kraan 2011).

The amino acid profile of seaweed proteins can serve as an effective substitute for flours in large quantities to decrease the available carbohydrates. The seaweed proteins incorporated into the cereal products can restrict starch availability and digestibility by forming a barrier around starch granules (Yang et al. 2019). Fucoidan is a complex sulphated polysaccharide predominantly found in the cell walls of brown seaweeds (Skriptsova et al. 2010). Over the past decade, the anti-diabetic potential of fucoidan has generated interest due to its therapeutic effects on several stages of glucose metabolism. The image (Fig. 5.2) illustrates the diverse anti-diabetic characteristics of fucoidan. Mannitol, a sugar alcohol derived from the brown seaweeds *Laminaria* and *Ecklonia* sp., serves as a substitute for sucrose in sugar-free compounds (Holdt and Kraan 2011).

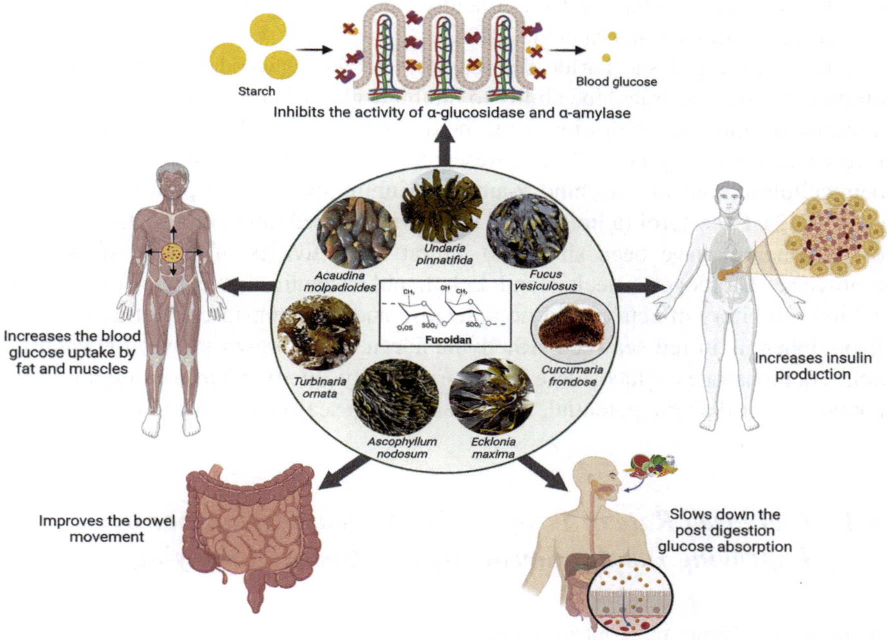

Fig. 5.2 Anti-diabetic properties of fucoidan reported from various seaweeds

5.2 Inhibition of Digestive Enzymes and Reduction in the Intestinal Absorption Rate

Following food consumption, the breakdown of carbohydrates is mostly facilitated by two principal enzymes: α-amylase and α-glucosidase. Alpha-amylase facilitates the digestion of long-chain carbohydrates, whereas alpha-glucosidase catalyses the breakdown of starch or disaccharides into monosaccharide glucose units. Inhibitors of these enzymes may hinder glucose absorption, reducing postprandial blood glucose levels and presenting a viable strategy for diabetes management (Lu et al. 2022).

Soluble fibres, including alginate and fucoidan, significantly reduce the rate of glucose release during digestion (Dhital et al. 2015). Alginate, a significant dietary fibre in brown algae belonging to the genera *Laminaria, Undaria*, and *Hizikia* (Holdt and Kraan 2011), demonstrates a pronounced effect in suppressing postprandial blood glucose levels. Huang et al. (2019) reported that calcium alginate possesses the ability to block the brush border enzyme α-glucosidase. Agar and carrageenans derived from red seaweeds function as thickening agents, capable of forming a continuous hydrated network that envelops the surface of starch granules, thereby establishing a barrier. Consequently, the barriers impede the amylose leaching during gelatinisation (Ai et al. 2013).

Laminarin, a polysaccharide derived from brown seaweed, such as Laminaria digitata, has demonstrated the ability to inhibit α-glucosidase activity, consequently diminishing glucose absorption from the digestive tract and lowering blood sugar levels (Karuppusamy et al. 2022). Research indicates that seaweed cellulose and hemicellulose content may bind α-amylase, inhibiting starch depolarisation (Jung et al. 2012). Fucosterol (a lipid found in brown seaweed) and phlorotannins (a type of polyphenol) have been shown to inhibit the activities of α-glucosidase and α-amylase. Moreover, Dieckol and Eckol, phlorotannins have been reported to exhibit inhibitory effects on aldose reductase and angiotensin-converting enzyme. Bromophenols in red seaweed (Rhodymeniaceae) and brown seaweed (*Leathesia nana*) demonstrate α-glucosidase inhibitory activity. Fucoxanthin, a carotenoid with abundant anti-diabetic potential, is found in brown seaweeds (Lu and Chen 2022).

5.3 Increased Rate of Postprandial Glucose Clearance by Improving Insulin Sensitivity and Gastric Emptying

Inadequate glycaemic control could result from elevated postprandial glucose (PPG) levels. The PPG profile is influenced by carbohydrate absorption, insulin and glucagon release, and their synergistic effects on glucose metabolism in the liver and peripheral organs. PPG plays a distinctive role in the aetiology of diabetes complications and should be a therapeutic target. Marine macroalgae contain bioactive compounds with the potential to inhibit dipeptidyl peptidase IV (DPP-IV), also known as adenosine deaminase complexing protein 2, an enzyme found in the human body (Yang et al. 2019), encoded by the DPPIV gene. DPPH-IV enzyme degrades incretin, which may play a role in promoting a reduction in blood glucose levels (Murakami et al. 2021). Incretins are a metabolic hormone class that stimulates insulin secretion from the Langerhans islet cells. This hormone is responsible for producing insulin only when necessary. Hence, the glucose production by the liver can be controlled (Jia et al. 2020). The bioactive compounds in seaweeds reduce the action of DPPH-IV, thereby enhancing incretin activity and decreasing glucose production. Fucoidan extracted from the brown algae *Cucumaria frondose* and *Acaudina molpadioides* may improve blood glucose clearance through two mechanisms: boosting insulin secretion via the activation of the PI3K/PKB pathway and inhibiting DPP IV, as well as promoting GLUT4 translocation.

Consuming proteins with carbohydrates during a meal could boost insulin secretion (Desterke et al. 2020). Peptides obtained from the red seaweed *Palmaria palmate* demonstrate a notable inhibitory effect on DPP-IV, thereby raising insulin production (Xia et al. 2017). The peptide SLAVSVH, extracted from *Ulva* sp., exhibited inhibitory action against the DDP-IV enzyme (Cian et al. 2022). Protein hydrolysates obtained from the macroalgae *P. palmata* exhibit inhibitory action against DPP-IV (Mc Laughlin et al., 2021). Polyphenols, such as phlorotannins and bromophenols, enhance insulin-mediated glucose clearance in skeletal muscle by activating the PI3K/PKB pathway (Roy et al. 2011; Lee and Jeon 2013). Fucoxanthin has shown the capacity to modulate GLUT4 mRNA in skeletal muscle tissues and improve blood glucose clearance. Fucoxanthin has been found to reduce insulin resistance and significantly lower blood glucose and plasma insulin levels in diabetic rats via downregulating TNFα mRNA (Park et al. 2011).

The rate of post-digestion glucose absorption can be attenuated by delaying the gastric emptying time. The fluidity of food increases due to dietary fibres, which slows the gastric emptying rate and subsequently reduces the rate of glucose absorption and release into the bloodstream (Lau et al. 2016). The addition of exogenous lipids and proteins has been proposed to delay gastric emptying time (Henry et al. 2008). Hence, the co-ingestion of these compounds and diet can significantly reduce the glycaemic impact. The negative charge of fucoidan may play a role in regulating and modulating SGLT1 (odium Na-glucose transporter 1), thereby slowing the post-digestion absorption of glucose in the lumen (Kim et al. 2015). Post-ingestion, alginate can create solid intragastric gels within digesta, leading to stomach distension and delayed gastric emptying and glucose absorption (Clegg et al. 2012). In addition to these, numerous other dietary fibres are directly and indirectly associated with the regulation of postprandial blood glucose levels. These include laminarin, agar, ulvan, carrageenan, and mannitol (Lu and Chen 2022).

5.4 Improving Insulin Sensitivity by Reducing Oxidative Stress and Inflammation

In addition to the potential benefits of polysaccharides in regulating blood sugar levels, proteins, lipids, phenolic compounds, and pigments, all of which also exhibit anti-inflammatory properties (Kim et al. 2021). Cellular and endothelial dysfunctions caused by the proinflammatory cytokines in the diabetic-associated obese state further impair insulin sensitivity and glucose homeostasis (Ahmed et al. 2021). The antioxidant properties of seaweed polysaccharides are more effective in alleviating oxidative stress, which contributes to the onset of diabetic problems (Loayza-Gutiérrez et al. 2022). Numerous polysaccharides derived from red seaweeds, including agar, carrageenan, porphyran, agar-oligosaccharide, and carrageenan oligosaccharide, have been documented to possess antioxidant properties. Brown seaweeds contain alginate, fucoidan, and laminarin oligosaccharides, all of which have significant antioxidant capacity (Baghel et al. 2023). Seaweeds inhabit intertidal to subtidal zones, exposing themselves to various stressors, which leads to increased buildup of reactive oxygen species. Seaweeds produce antioxidant enzymes, including ascorbate peroxidase, catalase, and other substances to manage stress (VijaySankar et al. 2023).

6 Anti-Diabetic Potential of Seaweeds—An Overview of Study Reports

At present, a variety of alternative medicines are sourced from the bioactive compounds present in a range of biological entities, such as microbes, fungi, seaweeds, plants, and animals. Alternative treatments are not provided to patients solely based on conventional knowledge or theoretical frameworks. Numerous in vitro and in vivo research utilising microorganisms and animal models are performed to investigate the potential anti-diabetic capabilities and to identify the bioactive compounds found within them (Table 5.2). As a primary screening method, in vitro research is crucial for clarifying the roles of bioactive materials, thereby reducing the detrimental effects of any unidentified substances present. Varying techniques for preparing seaweed extract, including the solvent type and factors employed, may differ based on the types of seaweed and bioactive material that needs to be collected.

The methanol extract of *Chaetomorpha aerea* and *Chlorodesmis* sp. collected from Indian coastal regions exhibited α-amylase inhibition in in vitro investigations. GC-MS analysis was used to identify the bioactive chemicals responsible for suppressing α-amylase activity (Unnikrishnan et al. 2015). In vitro study conducted by

Table 5.2 Study reports on the anti-diabetic properties exhibited by the different seaweeds

Seaweed	Bioactive compounds	Anti-diabetic properties	Type of study	References
Ecklonia cava	Dieckol	Reduces the levels of blood glucose, serum insulin, and body weight Improves the activities of antioxidant enzymes and AMPK and Akt phosphorylation in the muscle tissues	Diabetic rats (IP injected—14 days of intervention)	Kang et al. (2013)
E. cava	–	Decreases the postprandial glucose levels	Double-blind, placebo-controlled trial (12 weeks of supplementation)	Lee and Jeon (2015)
Sargassum coreanum	–	Reduces blood glucose, glycosylated haemoglobin levels, and the insulin resistance and increases the hepatic glucose metabolism	Diabetic rats (6 weeks of supplementation)	Park et al. (2015)
Kappaphycus alvarezii Doty ex P. Silva	–	Reduces the blood glucose level Restores the glycosylated haemoglobin and plasma insulin	Diabetic rats (28 days of supplementation)	Cyriac and Eswaran (2016)
Gracilaria opuntia	Sulphated galactopyran motif	Inhibits the activities of digestive enzymes and dipeptidyl peptidase-4 (DPP4)	In vitro study	Makkar and Chakraborty (2017)
Sargassum oligocystum	–	Reduces the levels of fasting blood glucose, triglyceride and HOMA-IR Reconstitute damaged pancreatic β-cells	Diabetic rats (30 days of intervention)	Akbarzadeh et al. (2018)

(continued)

Table 5.2 (continued)

Seaweed	Bioactive compounds	Anti-diabetic properties	Type of study	References
Dendrobium loddigesii	–	Increase the insulin level and decrease the blood glucose, body weight, low density lipoprotein cholesterol levels Possesses anti-oxidative, anti-inflammatory effects Maintains the gut microbiome	In vivo (diabetic mice)	Li et al. (2018)
Saccharina latissima	–	Lowers the bodyweight and improves the HbA1c and insulin levels	Human trial (10 weeks of intervention)	Sorensen et al. (2019)
Turbinaria decurrens	Cyclotrisiloxane Hexamethyl	Possesses antioxidant activity Inhibits α-amylase and α-glucosidase Non-toxic against fibroblast normal cell lines	In vitro study	Ismail et al. (2019)
Sargassum confusum	Oligosaccharides	Regulates the insulin receptor substrate 1 and reduces the levels of fasting blood glucose Improves the levels of good bacteria count in intestine	In vitro study	Yang et al. (2019)
Sargassum fusiforme Setch, Macrocystis pyrifera	Polysaccharides	Hypoglycaemic and hypolipidemic activities	In vivo (diabetic mice)	Jia et al. (2020)
Kappaphycus sp.	–	Ameliorates insulin resistance and influences on glucose homeostasis	In vitro study	Chin et al. (2020)
Cladophora dalmatica, Ectocarpus siliculosus, Padina boergesenii, Padina gymnospora	–	Inhibit α-amylase and α-glucosidase activities	In vitro study	Landa-Cansigno et al. (2020)

(continued)

Table 5.2 (continued)

Seaweed	Bioactive compounds	Anti-diabetic properties	Type of study	References
Ascophyllum nodosum, Fucus vesiculosus, Undaria pinnatifida	Carbohydrate and polyphenolic compounds	Inhibit maltase, sucrase, and α-glucosidase enzymes	In vitro study	Attjioui et al. (2021)
Palmaria palmata	Protein hydrolysate	Inhibits the DPP4 activity and directly influences the incretin effect, thus improves the insulin secretion	In vivo (diabetic mice)	McLaughlin et al. (2021)
Cystoseira compressa	Phlorotannins	Improves the levels of total antioxidant capacity, serum insulin, hepatic glutathione, and AMPKα2 expression in skeletal muscle Reduces the damage of β cells of pancreas Reduces the level of serum glucose, α-amylase, and glucosidase	In vivo (diabetic mice—4 weeks of treatment)	Gheda et al. (2021)
Sargassum kjellmanianum	Alginate	Attenuate the oxidative stress-induced cell damage, thereby accelerate diabetic wound healing	In vivo (diabetic mice)	Lu et al. (2022)
Padina tetrastromatica	Sulphated galactofucan	Inhibits the activity of carbolytic enzymes	In vitro study	Antony et al. (2022)
Caulerpa verticillate	Flavonoid	Inhibits α-amylase and α-glucosidase Possesses antioxidant activity	In vitro study	Maheswari and Salamun (2022)
Ascophyllum nodosum	Phlorotannins	Inhibits the activity of α-amylase and α-glucosidase Possesses antioxidant activity	In vitro study	Gisbert et al. (2023)
Sargassum wightii	–	Inhibits the activity of diabetic enzymes Possesses the anti-oxidant activity	In vitro study	Deepak et al. (2024)

Deepak et al. (2024) revealed that 9-hydroxyhexadecanoic acid, found in the seaweed *S. wightii*, exhibits inhibitory effects on α-amylase and α-glucosidase as well as it possesses antioxidant activity. Recent research has compared the inhibitory efficacy of seaweed extracts or compounds with existing pharmaceutical agents utilised for diabetes management through dietary enzyme inhibition. Extracts from the Mexican seaweeds *C. dalmatica, E. siliculosus, P. boergesenii*, and *P. gymnospora* show significant inhibition of α-glucosidase in comparison to acarbose, a synthetic α-glucosidase and α-amylase inhibitor (>1000 μg/mL) (Landa-Cansigno et al. 2020).

Peptides (sulphated galactofucan polysaccharide) isolated from brown seaweed, *P. tetrastromatica*, enhance the intake of glucose and insulin sensitivity (Nyakundi and Yang 2023). The antihyperglycemic activity of the purified polysaccharide was evaluated for its effectiveness in inhibiting the proteolytic enzyme DPP-4 and the carbohydrate-digesting enzymes α-amylase and α-glucosidase. The inhibitory value was recorded as IC_{50} less than 1 mg/mL, indicating that *P. tetrastromatica* may be a novel, naturally active agent in mitigating hyperglycaemia-related conditions (Antony et al. 2022). Crude fucoidan extracts from *U. pinnatifida*, administered at a dosage of 45 mg/kg body weight, significantly inhibited hyperglycaemia in genetically diabetic (db/db) mice. The preventive effect was attained by enhancing glucose utilisation and reducing fasting blood glucose and serum insulin levels (Kim et al. 2012). Sporophylls, in addition to seaweed, have also been studied for their anti-diabetic activity. The consumption of Mekabu, the sporophylls of *U. pinnatifida*, with a breakfast of white rice, reduced postprandial glucose concentration. It has been proposed that this effect is caused by the chemical fucoxanthin, which is present in Mekabu (Tanemura et al. 2014).

Raw seaweed biomass is rich in nutrients and bioactive substances, yet its inconsistent inherent composition may affect its biological efficacy in the human body. Laboratory-based models are effective for mimicking scenarios of the human digestive system. In vivo investigations precisely elucidate the digestion and fermentation processes of seaweed-derived proteins, polysaccharides, and lipids by gut bacteria (Monteiro et al. 2021). Various animal studies have demonstrated that the oral administration of seaweed peptides lowers the blood glucose levels in diabetic rats (Siddiqui et al. 2022). Administration of dieckol from seaweeds to diabetic mice led to a notable reduction in blood glucose and insulin levels relative to the control group (Kang et al. 2013). Alginate significantly decreased the fasting blood glucose levels in male ICR mice. Polyphenols extracted from the green seaweed *E. prolifera* enhanced the proliferation of beneficial Alistipes intestinal bacteria in the gastrointestinal tract of diabetic mice following 28 days of administration. Meanwhile, fasting blood glucose levels were reduced, and glucose tolerance was enhanced relative to the model group. Moreover, polyphenol exhibited an impact on the proteins involved in glucose metabolism, which include the enhancement of mRNA expression of PI3K and downregulation of JNK1 expression has been observed (Lin et al. 2019).

It was demonstrated that the extracts from *F. vesiculosus* and *A. nodosum* could completely block the enzymes α-amylase and α-glucosidase at the concentrations of 30 μg/mL and 2 μg/mL, respectively (Gabbia et al. 2017). Extracts of *E. prolifera* reduced fasting blood glucose levels, enhanced oral glucose tolerance, and provided protection against liver and kidney injury while decreasing inflammation in diabetic mice. The hypoglycaemic effect of *E. prolifera* activates the IRS1/PI3K/AKT pathway and inhibits the JNK1/2 insulin pathway in the liver. The treatment significantly increased the abundance of *Lachnospiraceae* and *Alisties*, which exhibited a positive correlation with metabolic phenotypes (Yan et al. 2019). Treatment with the seaweed *Dictyopteris divaricata* improved oral glucose tolerance, insulin secretion, β-cell mass rejuvenation, and pro-inflammatory cytokine levels in Type 1 diabetic-induced mice. The population of beneficial bacteria, including *Firmicutes*, *Bacteroidetes*, and *Lactobacillus* sp., has increased in the gut of mice. Immunohistological examination confirmed the reduction in inflammation and restored the structural morphology of the colon while increasing levels of insulin receptor substrate-1, Mucin-2, and tight-junction proteins, thereby maintaining gut structures and barrier permeability. *D. divaricata* promises a safe treatment and management for T1DM disorders, suggesting its potential as a dietary supplement alternative to conventional diabetes medications (Siddiqui et al. 2022). Fucoxanthin derived from *S. angustifolium* significantly reduced the weight gain rate and fasting blood glucose levels. Elevated plasma insulin concentrations were reported, comparable to the effects of metformin. Moreover, total cholesterol, triglycerides, and low-density lipoprotein levels were reduced in the treated groups. The histopathological assessment of pancreatic tissue in diabetic mice demonstrated that oral administration of fucoxanthin induced the regeneration of pancreatic beta cells (Oliyaei et al. 2021). Likewise, the fucoidan compound derived from *S. japonica* stimulated the release of insulin from the pancreas and lowered the plasma glucose concentrations in diabetic rats (Wang et al. 2013). Edible traditional Japanese seaweed, *Gloiopeltis furcate* (Rhodophyta), diet has shown a positive effect on obesity-related metabolic diseases, such as insulin resistance, hyperglycaemia, hepatic steatosis, and dyslipidaemia in obese mice (Murakami et al. 2023).

In a meta-analysis of randomised controlled trials, seaweed consumption decreased the risk of post-meal hyperglycaemia and enhanced blood sugar regulation and metabolic health. Haemoglobin glycation results in the formation of HbA1c, which reflects average blood glucose levels, thus serving as a standard biomarker for diabetes. HOMA-IR identifies early-stage diabetes, underscoring its significance for primary prevention. Seaweed and its extracts have been shown to lower postprandial blood glucose levels, HbA1c, and HOMA-IR, thereby serving as a potential therapy for diabetes (Sherwani et al. 2016). A crossover meal trial study demonstrated that extracts of brown algae (*L. digitata* and *U. pinnatifida*) potentially affect blood glucose levels in participants weighing less than 63 kg (Zaharudin et al. 2021). The Korea National Health and Nutrition Examination Survey 2005 revealed that the prevalence of T2DM was significantly lower in men with the highest seaweed intake (Lee et al. 2010).

7 Safety and Toxicology of Seaweed Consumption

Despite the nutritional advantages of seaweed, limited research indicates the detrimental effects of trace elements present in seaweeds. Seaweeds possess high concentrations of minerals, exceeding the quantities seen in meat and spinach for iron and copper (Barandiaran et al. 2024). Certain brown seaweeds, particularly from the *Laminaria* and *Saccharina* genera (Kombu), exhibit high iodine concentrations, exceeding the adult upper intake limit of 600 μgas established by the European Scientific Committee (EFSA 2006). Roleda et al. (2019) identified the presence of heavy metals in *Saccharina latissima* and *Alaria esculenta* (brown algae) as well as *P. palmata* (red alga). Studies indicate that brown seaweeds (e.g. *S. fusiforme*) possess higher quantities of total arsenic than red and green seaweeds (Huang et al. 2022). Substantially elevated iodine levels have been observed following Kombu eating (Barandiaran et al. 2024), whereas the highest total arsenic levels have been recorded after Wakame and Nori consumption. The quantities of trace elements, including iodine and arsenic, can significantly fluctuate depending on the processing or cooking method applied to the seaweed (Laparra et al. 2004).

Naidu et al. (1993) observed that hemagglutination induced by lectin compounds found in *U. fasciata*, *C. taxifolia*, *E. linza*, and *S. johnstonii* resulted in acute toxicity in Wistar-CFT rats. This may cause growth retardation in animals. The majority of mortality and morbidity cases are documented for the species *G. verrucosa* and *G. edulis* (Cheney 2016). Prostaglandin, particularly prostaglandin E2, found in *G. vermiculophylla*, *G. lichenoides*, and *G. verrucose*, is associated with diarrhoea, nausea, headache, vomiting, bleeding, and hypertension, all of which elevate cardiovascular risk. Polycarvenoside A and polycarvenoside B are the two most lethal toxins identified in *G. edulis*, resulting in symptoms such as vomiting, diarrhoea, and muscle spasms (Kumar and Sharma 2021). However, regulations governing seaweed intake limits and quality are essential, including detailed legislation that indicates the maximum permissible concentration of heavy metals in seaweeds to guarantee safe consumption and safeguard public health.

8 Conclusion

Seaweeds are increasingly recognised globally for their exceptional nutritional value and medicinal properties. Despite the predominance of aquatic areas on our planet, there exists considerable opportunity for seaweed cultivation and commercialisation. Furthermore, there are no concerns regarding cultivation challenges comparable to those seen in agricultural crop production. As a result, extensive research on the composition and bioactivities of seaweeds globally has resulted in a significant increase in seaweed production over the past two decades. Further, thorough research is essential to understand the effects and fundamental bioactivities associated with seaweeds. This would enable the complete use of physiologically active substances derived from seaweeds to develop diabetic therapies and generate functional food products that can sustain cellular functions in the human body, hence preventing the onset of diabetes. This research should prioritise precise documentation of the dangerous and toxic qualities of seaweed uses across many sectors, which is crucial. Extensive preclinical and clinical research is necessary to establish a correlation between seaweed consumption and potential benefits for diabetes management.

References

Abomohra AE and Almutairi AW 2020. A close-loop integrated approach for microalgae cultivation and efficient utilization of agar-free seaweed residues for enhanced biofuel recovery. *Bioresource Technology*, 317, 1–11.

Agarwal S, Singh V and Chauhan 2022. Antidiabetic Potential of Seaweed and Their Bioactive Compounds: A Review of Developments in Last Decade.*Critical Reviews in Food Science and Nutrition,* 63, 5739–5770.

Ahmed B, Sultana R and Greene MW 2021. Adipose tissue and insulin resistance in obese. *Biomedicine & Pharmacotherapy,* 137, 1–13.

Ai Y, Hasjim J and Jane J 2013. Effects of Lipids on Enzymatic Hydrolysis and Physical Properties of Starch.*Carbohydrate Polymers,* 92, 120–127.

Akbarzadeh S, Gholampour H, Farzadinia P, Daneshi A, Ramavandi B, Moazzeni A, Keshavarz M and Bargahi A 2018. Anti-diabetic effects of *Sargassum oligocystum* on Streptozotocin-induced diabetic rat. *Iranian Journal of Basic Medical Sciences*, 3, 342–346.

Aknaf A, Ahsan U, Baghour M, Rahhou A, Imperl J, Petrova P and Kolar M 2024. Minerals and trace element compositions of some seaweeds from the Marchica lagoon, North-East Mediterranean coast of Morocco. *Marine Pollution Bulletin*, 206, 1–8.

Al-Adilah H, Al-Sharrah TK, Al-Bader D, Ebel R, Küpper FC and Kumari P 2021. Assessment of Arabian Gulf seaweeds from Kuwait as sources of nutritionally important polyunsaturated fatty acids (Pufas). *Foods*, 10, 1–17

Antony T, Chakraborty K and Dhara S 2022. SulfatedGalactofucan from Seaweed *Padina tetrastromatica*Attenuates Proteolytic Enzyme Dipeptidyl-Peptidase-4: A Potential Anti-Hyperglycemic Lead. *Natural Product Research,* 36, 6240–6250.

Aryee AN, Agyei D, and Akanbi TO 2018. Recovery and utilization of seaweed pigments in food processing. *Current Opinion in Food Science*, 19:113–119.

Attjioui M, Ryan S, Ristic AK, Higgins T, Goñi O, Gibney ER, Tierney J and O'Connell S 2021. Comparison of edible brown algae extracts for the inhibition of intestinal carbohydrate digestive enzymes involved in glucose release from the diet. *Journal ofNutritional Science*, 10, 1–10.

Baghel RS, Choudhary B, Pandey S, Pathak PK, Patel MK and Mishra A 2023. Rehashing our insight of seaweeds as a potential source of foods, nutraceuticals, and pharmaceuticals. *Foods*, 19, 1–39.

Barandiaran LN, Taylor VF and Karagas MR 2024. Exposure to iodine, essential and non-essential trace element through seaweed consumption in humans. *Scientific Reports*, 14, 1–8.

Biris-Dorhoi ES, Michiu D, Pop CR, Rotar AM, Tofana M, Pop OL, Socaci SA and Farcas AC 2020. Macroalgae—A sustainable source of chemical compounds with biological activities. *Nutrients*, 12, 1–25.

Cai J, Lovatelli A, Aguilar-Manjarrez J, Cornish L, Dabbadie L, Desrochers A, Diffey S, Garrido Gamarro E, Geehan J, Hurtado A and Lucente D 2021. Seaweeds and microalgae: an overview for unlocking their potential in global aquaculture development. *FAO Fisheries and Aquaculture Circular*, 1229, 1–48.

Cheney, D., 2016. Toxic and harmful seaweeds. In. J. Fleurence, & I. Levione (Eds.). *Seaweed in health and disease prevention*, pp. 407–421. Amsterdam: Elsevier.

Chin YX, Chen X, Cao WX, Sharifuddin Y, Green BD, Lim PE, Xue CH and Tang QJ 2020. Characterization of seaweed hypoglycemic property with integration of virtual screening for identification of bioactive compounds. *Journal of functional foods*, 64, 1–11.

Choudhary B, Khandwal D, Gupta NK, Patel J and Mishra A 2023. Nutrient composition, physic-biochemical analyses, oxidative stability and antinutritional assessment of abundant tropical seaweeds from the Arabian Sea. *Plants*, 12, 1–27.

Choudhary P, Khade M, Savant S, Musale A, Chelliah MS and Dasgupta S 2021. Empowering blue economy: From underrated ecosystem to sustainable industry. *Journal of environmental management*, 291, 1–16.

Cian RE, Nardo AE, Garzón AG, Anon MC and Drago SR 2022. Identification and In Silico Study of a Novel Dipeptidyl Peptidase IV Inhibitory Peptide Derived from Green Seaweed Ulva Spp. Hydrolysates. *LWT-Food Science and Technology*, 154, 1–8.

Clegg ME, Pratt M, Markey O, Shafat A and Henry CJK 2012. Addition of Different Fats to a Carbohydrate Food: Impact on Gastric Emptying, Glycaemic and Satiety Responses and Comparison with in Vitro Digestion. *Food Research International*, 48, 91–97.

Cotas J, Lomartire S, Pereira L, Valado A, Marques JC and Gonçalves AM 2024. Seaweeds as Nutraceutical Elements and Drugs for Diabetes Mellitus: Future Perspectives. *Marine Drugs*, 168, 1–22.

Cyriac B and Eswaran K 2016. Anti-hyperglycemic effect of aqueous extract of *Kappaphycusalvarezii* (Doty) Doty ex. P. Silva in alloxan-induced diabetic rats. *Journal of Applied Phycology*, 28, 2507–2513.

de Melo NS, Cardoso LG, de Castro Nunes JM, Brito GB, Caires TA, de Souza CO, Portz L and Druzian JI 2021. Effects of dry and rainy seasons on the chemical composition of *Ulva fasciata, Crassiphycuscorneus,* and *Sargassum vulgare* seaweeds in tropical environment.*Brazilian Journal of Botany*, 44, 331–344.

Deepak P, Perumal P, Balakrishnan R, Balasubramanian B and Velmurugan P 2024. Assessment of antioxidant and digestive enzyme inhibition by phyco-molecules isolated from marine brown alga *Sargassum wightii*. *Food and Humanity*, 100226, 1–11.

Deleris P, Nazih H and Bard JM 2016. Seaweeds in Human Health. In. J. Fleurence, & I. Levione (Eds.). *Seaweed in health and disease prevention*, pp. 319–367. Amsterdam: Elsevier.

Denis C, Ledorze C, Jaouen P and Fleurence J 2009. Comparison of different procedures for the extraction and partial purification of R-phycoerythrin from the red macroalga *Grateloupiaturuturu.Botanica Marina*, 52, 278–281.

Desterke C, Turhan AG, Bennaceur-Griscelli A and Griscelli F 2020. PPAR Cistrome Repression during Activation of Lung Monocyte-Macrophages in Severe COVID-19. *Science* 23 (10), 101611.

Dhital S, Gidley MJ and Warren FJ 2015. Inhibition of Amylase Activity by Cellulose: Kinetic Analysis and Nutritional Implications. *Carbohydrate Polymers*, 123, 305–312.

Dixit DC, Reddy CRK, Balar N, Suthar P, Gajaria T and Gadhavi, DK 2018. Assessment of the Nutritive, Biochemical, Antioxidant and Antibacterial Potential of Eight Tropical Macro Algae Along Kachchh Coast, India as Human Food Supplements.*Journal of aquatic food product technology*, 27, 61–79.

European Food Safety Authority (EFSA) 2006. Statement on a request from the Commission related to iodine in seaweed. *EFSA Journal,*4(10), 1046–1047.

FAO. 2021. *FAO Yearbook. Fishery and Aquaculture Statistics 2019/FAO annuaire. Statistiques des pêches et de l'aquaculture 2019/FAO anuario. Estadísticas de pesca y acuicultura 2019.* pp.110, Rome.

Gabbia D, Dall'Acqua S, DiGangi IM, Bogialli S, Caputi V, Albertoni L, Marsilio I, Paccagnella N, Carrara M, Giron MC and De Martin S 2017. The Phytocomplex from *Fucus vesiculosus* and *Ascophyllum nodosum* Controls Postprandial Plasma Glucose Levels: An *in vitro* and *in vivo*study in a Mouse Model of NASH. *Marine Drugs*, 15, 41–53.

Ganesan AR, Tiwari U and Rajauria G 2019. Seaweed nutraceuticals and their therapeutic role in disease prevention. *Food Science and Human Wellness*, 8, 252–63.

Gheda S, Naby MA, Mohamed T, Pereira L and Khamis A 2021. Antidiabetic and antioxidant activity of phlorotannins extracted from the brown seaweed *Cystoseiracompressa* in streptozotocin-induced diabetic rats.*Environmental Science and Pollution Research*, 28, 22886–22901.

Gisbert M, Franco D, Sineiro J and Moreira R 2023. Antioxidant and antidiabetic properties of phlorotannins from *Ascophyllum nodosum* seaweed extracts. *Molecules*, 28, 4937–4967.

Grembecka M, 2015. Sugar alcohols—their role in the modern world of sweeteners: a review. *European Food Research and Technology*, 241, 1–14.

Gunathilaka TL, Samarakoon K, Ranasinghe P and Peiris LDC 2020. Antidiabetic potential of marine brown algae—a mini review. *Journal of Diabetes Research*, 1, 1230218.

Gurung M, Li Z, You H, Rodrigues R, Jump DB, Morgun A and Shulzhenko N 2020. Role of gut microbiota in type 2 diabetes pathophysiology. *eBioMedicine*, 51, 102590.

Hamid SS, Wakayama M, Ichihara K,Sakurai K, Ashino Y, Kadowaki R, Soga T and Tomita M 2019. Metabolome profiling of various seaweed species discriminates between brown, red, and green algae. *Planta*, 249, 1921–1947.

Henry CJ, Lightowler HJ, Newens KJ and Pata N 2008. The influence of adding fats of varying saturation on the glycaemic response of white bread.*International journal of food sciences and nutrition*, 59, 61–69.

Holdt SL and Kraan S 2011. Bioactive Compounds in Seaweed: Functional Food Applications and Legislation.*Journal of applied phycology*, 23, 543–597.

Horn SJ 2009. Seaweed biofuels: production of biogas and bioethanol from brown macroalgae. VDM, Verlag Dr. Müller. Saarbrucken, Germany.

Huang Z, Bi R, Musil S, Pétursdóttir ÁH, Luo B, Zhao P, Tan X and Jia Y 2022.Arsenic species and their health risks in edible seaweeds collected along the Chinese coastline.*Science of the Total Environment*,847, 157429.

Huang Z, Wang Y, Shafer R, Winn NC, Kanaley JA and Vardhanabhuti B 2019. Glycemic Effects Following the Consumption of Mixed Soy Protein Isolate and Alginate Beverages in Healthy Adults.*Food & function*, 10, 1718–1725.

Ismail GA, Gheda SF, Abo-Shady AM and Abdel-Karim OH 2019. *In vitro* potential activity of some seaweeds as antioxidants and inhibitors of diabetic enzymes. *Food Science and Technology*, 40, 681–91.

Jia RB, Wu J, Li ZR, Ou ZR, Lin L, Sun B and Zhao M 2020. Structural characterization of polysaccharides from three seaweed species and their hypoglycemic and hypolipidemic activities in type 2 diabetic rats. *International journal of biological macromolecules*, 155, 1040–1049.

Jimenez-Escrig AN and Cambrodon G 1999. Nutritional evaluation and physiological effects of edible seaweeds. *Archivoslatinoamericanos de nutricion*, 49, 114–120.

Jung HA, Islam MN, Lee CM, Jeong HO, Chung HY, Woo HC and Choi JS 2012. Promising antidiabetic potential of fucoxanthin isolated from the edible brown algae *Eiseniabicyclis* and *Undariapinnatifida*.*Fisheries Science*. 78, 1321–1329.

Kang MC, Wijesinghe WA, Lee SH, Kang SM, Ko SC, Yang X, Kang N, Jeon BT, Kim J, Lee DH and Jeon YJ 2013. Dieckol isolated from brown seaweed *Eclonia cava* attenuates type II diabetes in db/db mouse model.*Food and chemical toxicology*, 53, 294–298.

Karuppusamy S, Rajauria G, Fitzpatrick S, Lyons H, McMahon H, Curtin J, Tiwari BK and O'Donnell C 2022. Biological Properties and Health-Promoting Functions of Laminarin: A Comprehensive Review of Preclinical and Clinical Studies. *Marine Drugs*, 20, 772–798.

Kim E, Cui J, Kang I, Zhang G and Lee Y 2021. Potential Antidiabetic Effects of Seaweed Extracts by Upregulating Glucose Utilization and Alleviating Inflammation in C2C12 Myotubes. *International Journal of Environmental Research and Public Health*, 18, 1367–1370.

Kim KJ, Yoon KY, and Lee BY 2012. Fucoidan Regulate Blood Glucose Homeostasis in C57BL/KSJ M+/+db and C57BL/KSJ Db/Db Mice. *Fitoterapia*, 83, 1105–1109.

Kim KT, Rioux LE and Turgeon S 2015. Molecular Weight and Sulfate Content Modulate the Inhibition of α-Amylase by Fucoidan Relevant for Type 2 Diabetes Management. *Pharma Nutrition*, 3,108–114.

Kumar M, Kumari P, Trivedi N, Shukla MK, Gupta V, Reddy CR and Jha B 2011. Minerals, PUFAs and antioxidant properties of some tropical seaweeds from Saurashtra coast of India.*Journal of Applied Phycology*, 23, 797–810.

Kumar MS and Sharma SA 2021. Toxicological effects of marine seaweeds: a cautious insight for human consumption. *Critical Reviews in Food Science and Nutrition*. 61, 500–521.

Landa-Cansigno C, Hernández-Domínguez EE, Monribot-Villanueva JL, Licea-Navarro AF, Mateo-Cid LE, Segura-Cabrera A and Guerrero-Analco JA 2020. Screening of Mexican tropical seaweeds as sources of α-amylase and α-glucosidase inhibitors. *Algal Research*, 49, 1–11.

Laparra JM, Velez D, Montoro R, Barbera R and Farre R 2004. Bioaccessibility of inorganic arsenic species in raw and cooked *Hizikiafusiforme* seaweed. *Applied Organometallic Chemistry*,18, 662–669.

Lourenço-Lopes C, Fraga-Corral M, Jimenez-Lopez C, Pereira AG, Garcia-Oliveira P, Carpena M, Prieto MA and Simal-Gandara J 2020. Metabolites from macroalgae and its applications in the cosmetic industry: A circular economy approach. *Resources*, 9, 101–131.

Lau E, Zhou W and Henry CJ 2016. Effect of Fat Type in Baked Bread on Amylose-Lipid Complex Formation and Glycaemic Response.*British Journal of Nutrition*, 115, 2122–2129.

Lee GY and Han SN 2018. The role of vitamin E in immunity. *Nutrients* 10, 1614–1627.

Lee HJ, Kim HC, Vitek L and Nam CM 2010. Algae consumption and risk of type 2 diabetes: Korean national health and nutrition examination survey in 2005. *Journal of nutritional science and vitaminology*, 56, 13–18.

Lee SH and Jeon YJ 2015. Efficacy and safety of a dieckol-rich extract (AG-dieckol) of brown algae, *Eclonia cava*, in pre-diabetic individuals: A double-blind, randomized, placebo-controlled clinical trial. *Food & function*, 6, 853–858.

Lee SH and Jeon YJ 2013. Anti-Diabetic Effects of Brown Algae Derived Phlorotannins, Marine Polyphenols through Diverse Mechanisms. *Fitoterapia*, 86, 129–136.

Li XW, Chen HP, He YY, Chen WL, Chen JW, Gao L, Hu HY and Wang J 2018. Effects of rich-polyphenols extract of *D. loddigesii* on anti-diabetic, anti-inflammatory, anti-oxidant, and gut microbiota modulation in db/db mice. *Molecules*, 23, 3245–3565.

Lin G, Liu X, Yan X, Liu D, Yang C, Liu B, Huang Y and Zhao C 2019. *Role of green macroalgae Enteromorpha prolifera polyphenols in the modulation of gene expression and intestinal microflora profiles in type 2 diabetic mice.*International Journal of Molecular Sciences, *20, 25–36.*

Loayza-Gutiérrez LT, Apumayta-Suárez EV, Abdala R, Aguilar-Mendoza LÁ, Chávez-Pérez JA and Decara J 2022. Anti-Hyperglycemic and Antioxidant Effect of Fucoidan Extract from *Lessonia trabeculata* in Alloxan-Induced Diabetes Rats.*Journal of Applied Phycology*, 34, 3247–3261.

Lomartire S and Gonçalves AM 2022. An overview of potential seaweed-derived bioactive compounds for pharmaceutical applications. *Marine Drugs*, 20 141–173.

Lopes G, Sousa C, Silva LR, Pinto E, Andrade PB, Bernardo J, Mouga T and Valentão P 2012. Can phlorotannins purified extracts constitute a novel pharmacological alternative for microbial infections with associated inflammatory conditions? *PLoS ONE*, 7, e31145.

Lu LW and Chen JH 2022. Seaweeds as ingredients to lower glycemic potency of cereal foods synergistically—A perspective. *Foods*, 11, 714–738.

Lu X, Qin L, Guo M, Geng J, Dong S, Wang K, Xu H, Qu C, Miao J and Liu M 2022. A novel alginate from Sargassum seaweed promotes diabetic wound healing by regulating oxidative stress and angiogenesis.*Carbohydrate Polymers*, 289, 119437–11941.

Maheswari A and Salamun DE 2022. *In vitro* correlation studies of antidiabetic, antioxidant activity and HPLC-ESI-MS/MS analysis of marine seaweeds from Gulf of Mannar. *Regional Studies in Marine Science*, 56, 102682–102695.

Makkar F and Chakraborty K 2017. Antidiabetic and anti-inflammatory potential of sulphated polygalactans from red seaweeds *Kappaphycusalvarezii* and *Gracilaria opuntia*. *International Journal of Food Properties*, 20,1326–1337.

McLaughlin CM, Harnedy-Rothwell PA, Lafferty RA, Sharkey S, Parthsarathy V, Allsopp PJ, McSorley EM, Fitzgerald RJ and O'Harte FP 2021. Macroalgal Protein Hydrolysates from *Palmariapalmata*influence the 'Incretin Effect' *in vitro* via DPP-4 Inhibition and Upregulation of Insulin, GLP-1 and GIP Secretion.*European Journal of Nutrition*, 60, 4439–4452

Mendis E and Kim SK 2011. Present and future prospects of seaweeds in developing functional foods. *Advances in food and nutrition research*, 64, 1–15.

Monteiro P, Lomartire S, Cotas J, Pacheco D, Marques JC, Pereira L and Gonçalves AM 2021. Seaweeds as a Fermentation Substrate: A Challenge for the Food Processing Industry. *Processes*, 9, 1953–1968.

Murakami S, Hirazawa C, Ohya T, Yoshikawa R, Mizutani T, Ma N, Moriyama M, Ito T, and Matsuzaki C 2021. The edible brown seaweed *Sargassum horneri*(Turner) C. Agardh ameliorates high-fat diet-induced obesity, diabetes, and hepatic steatosis in mice. *Nutrients* 3, 551–565.

Murakami S, Hirazawa C, Mizutani T, Yoshikawa R, Ohya T, Ma N, Owaki Y, Owaki T, Ito T and Matsuzaki C 2023. The anti-obesity and anti-diabetic effects of the edible seaweed *Gloiopeltis furcata* (Postels et Ruprecht) J. Agardh in mice fed a high-fat diet. *Food Science & Nutrition*,11, 599–610.

Naidu KA, Tewari A, Joshi HV, Viswanath S, Ramesh HP and Rao SV, 1993. Evaluation of nutritional quality and food safety of seaweeds of India. *Journal of food safety*, 13, 77–90.

Nyakundi BB and Yang J 2023. Uses of Papaya Leaf and Seaweed Supplementations for Controlling Glucose Homeostasis in Diabetes. *International Journal of Molecular Sciences*, 24, 6846–6861.

Ojo O, Ojo OO, Zand N and Wang X 2021. The effect of dietary fibre on gut microbiota, lipid profile, and inflammatory markers in patients with type 2 diabetes: a systematic review and meta-analysis of randomised controlled trials. *Nutrients,* 13, 1805–1823.

Oliyaei N, Moosavi-Nasab M, Tamaddon AM and Tanideh N 2021. Antidiabetic effect of fucoxanthin extracted from *Sargassum angustifolium* on streptozotocin-nicotinamide-induced type 2 diabetic mice.*Food Science & Nutrition*, 9, 3521–3529

Park HJ, Lee MK, Park YB, Shin YC and Choi MS 2011. Beneficial effects of *Undaria pinnatifida* ethanol extract on diet-induced-insulin resistance in C57BL/6J mice. *Food and Chemical Toxicology*, 49, 727–733.

Park MH, Nam YH, Han JS 2015. *Sargassumcoreanum* extract alleviates hyperglycemia and improves insulin resistance in db/db diabetic mice. *Nutrition Research and Practice*, 9, 472–479.

Penalver R, Lorenzo JM, Ros G, Amarowicz R, Pateiro M and Nieto G 2020. Seaweeds as a functional ingredient for a healthy diet. *Marine Drugs*, 18, 301–328.

Rachdaoui, N 2020. Insulin: The Friend and the Foe in the Development of Type 2 Diabetes Mellitus.*International journal of molecular sciences*, 21, 1770–1791.

Roleda MY, Marfaing H, Desnica N, Jónsdóttir R, Skjermo J, Rebours C and Nitschke U 2019. Variations in polyphenol and heavy metal contents of wild-harvested and cultivated seaweed bulk biomass: Health risk assessment and implication for food applications. *Food Control* 95, 121–134

Roy MC, Anguenot R, Fillion C, Beaulieu M, Bérubé J and Richard D 2011. Effect of a commercially-available algal phlorotannins extract on digestive enzymes and carbohydrate absorption *in vivo*. *Food research international*, 44, 3026–3029.

Salido M, Soto M and Seoane S 2023. Seaweed: Nutritional and gastronomic perspective. A review. *Algal Research*, 77, 103357–103368.

Shannon E and Abu-Ghannam N, 2019. Seaweeds as nutraceuticals for health and nutrition. *Phycologia*, 58, 563–577.

Sherwani SI, Khan HA, Ekhzaimy, A, Masood, A, Sakharkar, MK 2016. Significance of HbA1c Test in Diagnosis and Prognosis of Diabetic Patients. *Biomark Insights*, 11, 95–104.

Siddiqui NZ, Rehman AU, Yousuf W, Khan AI, Farooqui NA, Zang S, Xin Y and Wang L 2022. Effect of crude polysaccharide from seaweed, *Dictyopterisdivaricata* (CDDP) on gut microbiota restoration and anti-diabetic activity in streptozotocin (STZ)-induced T1DM mice. *Gut Pathogens*, 14, 39–65.

Skriptsova AV, Shevchenko NM, Zvyagintseva TN and Imbs TI 2010. Monthly Changes in the Content and Monosaccharide Composition of Fucoidan from *Undaria pinnatifida* (Laminariales, Phaeophyta).*Journal of Applied Phycology*, 22, 79–86.

Sohn SI, Rathinapriya P, Balaji S, Jaya Balan D, Swetha TK, Durgadevi R, Alagulakshmi S, Singaraj P and Pandian S 2021. Phytosterols in seaweeds: an overview on biosynthesis to biomedical applications. *International Journal of Molecular Sciences*, 22, 12691–12718.

Sorensen LE, Jeppesen PB, Christiansen CB, Hermansen K and Gregersen S 2019. Nordic seaweed and diabetes prevention: Exploratory studies in KK-Ay mice. *Nutrients*, 11, 1435–1445.

Tanemura Y, Yamanaka-Okumura H, Sakuma M, Nii Y, Taketani Y and Takeda E 2014. Effects of the intake of *Undaria pinnatifida* (Wakame) and its Sporophylls (Mekabu) on Postprandial Glucose and Insulin Metabolism.*The Journal of Medical Investigation*, 61, 291–297.

Tanna B and Mishra A 2019. Nutraceutical potential of seaweed polysaccharides: Structure, bioactivity, safety, and toxicity. *Comprehensive Reviews in Food Science and Food Safety*, 18, 817–831.

Unnikrishnan PS, Suthindhiran K and Jayasri MA 2015. Alpha-amylase inhibition and antioxidant activity of marine green algae and its possible role in diabetes management. *Pharmacognosy Magazine*, 11, 511–515.

Valko M, Leibfritz D, Moncol J, Cronin MT, Mazur M and Telser J 2007. Free radicals and antioxidants in normal physiological functions and human disease. *The international journal of biochemistry & cell biology*, 39, 44–84.

VijaySankar NP, Jagtap AS, Baghel RS, Imchen T and Manohar CS 2023. Elucidation of the antioxidant potential of marine macroalgal biomolecules for healthcare applications: Current status and future prospects. Se-Kwon Kim, S.-K., Shin, KH, Venkatesan, J.(Eds.), Marine Antioxidants: Preparations, Syntheses, and Applications. Elsevier, Amsterdam, Netherlands, 2023.

Wang J, Jin W, Zhang W, Hou Y, Zhang H and Zhang Q 2013. Hypoglycemic Property of Acidic Polysaccharide Extracted from *Saccharina japonica*and its Potential Mechanism.*Carbohydrate polymers*, 95, 143–147.

Xia EQ, Zhu SS, He MJ, Luo F, Fu CZ and Zou TB 2017. Marine Peptides as Potential Agents for the Management of Type 2 Diabetes Mellitus—A Prospect.*Marine drugs*, 15, 88–105.

Yan X, Yang C, Lin G, Chen Y, Miao S, Liu B and Zhao C 2019. Antidiabetic potential of green seaweed *Enteromorpha prolifera* flavonoids regulating insulin signaling pathway and gut microbiota in type 2 diabetic mice. *Journal of food science*, 84, 165–73.

Yang CF, Lai SS, Chen YH, Liu D, Liu B, Ai C, Wan XZ, Gao LY, Chen XH and Zhao C 2019. Anti-diabetic effect of oligosaccharides from seaweed *Sargassum confusum* via JNK-IRS1/PI3K signalling pathways and regulation of gut microbiota. *Food and Chemical Toxicology*, 131, 110562.

Yang Y, Zhang M, Alalawy AI, Almutairi FM, Al-Duais MA, Wang J and Salama ES 2021. Identification and characterization of marine seaweeds for biocompounds production. *Environmental Technology & Innovation*, 24, 101848–101860.

Zaharudin N, Tullin M, Pekmez CT, Sloth JJ, Rasmussen RR and Dragsted LO 2021. Effects of brown seaweeds on postprandial glucose, insulin and appetite in humans–A randomized, 3-way, blinded, cross-over meal study. *Clinical Nutrition*, 40, 830–838

Chapter 6
Antidiabetic Secondary Metabolites and Active Molecules from Algae

P. Madhan Kumar, R. Parameswari, R. Babujanarthanam, and J. Lesitha Jeeva Kumari

Abstract Diabetes mellitus is an inescapable long-term global health burden that necessitates treatment methods based on new creative agents that are simple, straightforward, economical, and successful in managing and lowering various problems through their pharmacological effects. In recent years, algae have attracted interest for their possible bioactive chemicals and antidiabetic action. As a result, the current chapter concentrates on a complete assessment of natural compounds produced from algae with antidiabetic properties and their treatment. Notable algae derivatives from polysaccharides, polyphenols, proteins and peptides, and vitamins, as well as their contributions, addressed specific antidiabetic features. However, cohorts with problems using microalgae from natural sources should be managed, with a focus on pharmacokinetic factors in innovative antidiabetic therapies. Furthermore, continued research into tailoring bioactive algal molecules from therapies aimed at reducing the global diabetes burden, which leads to effective individual recovery.

Keywords Natural algae · Diabetes mellitus · Polyphenols · Insulin · Global burden

1 Introduction—Diabetes

Diabetes mellitus is a rapidly developing disease that affects the health and wealth of individuals of all ages and genders. Cancer and other metabolic problems follow diabetes mellitus. Diabetes had the highest recorded death and morbidity worldwide, and this trend continued year after year. By 2050, around 50% of the

---

P. M. Kumar · R. Parameswari · R. Babujanarthanam (✉)
Department of Biotechnology, Thiruvalluvar University, Vellore, Tamil Nadu, India

J. L. J. Kumari
Department of Biotechnology, Marudhar Kesari Jain College for Women, Thirupathur, Tamil Nadu, India

R. Raja et al. (eds.), *Algae in Diabetes Management*,
https://doi.org/10.1007/978-981-95-0571-5_6

population will have type II diabetes mellitus. Furthermore, it is overruling and affecting millions of people, causing a crisis in both the economic and healthcare sectors. Type 1 diabetes is caused by lifestyle and sedentary activities. In general, diabetes and its consequences are linked and create substantial health difficulties, such as cardiovascular abnormalities and renal failure, which lead to major organ degradation. Diabetes is the inability of insulin secretion by the Langerhans of beta cells in the pancreas to produce insulin (Kokil et al. 2015). Hyperglycemia, commonly known as type II diabetes mellitus, is caused by inadequate insulin production and lack of glucose oxidation in blood circulation. Middle-aged adults with no gender differences were the ones affected. Even though type II diabetes is common, it accounts for 90% of all diabetic cases. Obesity, smoking, drinking, junk food, a sedentary lifestyle, systemic infections, viral infections, and a poor diet associated with BMI are all key risk factors for diabetes.

Type I diabetes mellitus/neonatal diabetes mellitus/autoimmune is characterized by the absence of the insulin-secreting gland throughout infancy owing to inheritance (Sen et al. 2016). In the early stages of this kind of diabetes, insulin injections were most commonly recommended. Furthermore, gestational diabetes is a form of diabetes that develops during pregnancy in women who are on an unregulated diet and recover after birth, according to studies from specialists. Various factors, including multiple organ damage or failure, cause most difficulties. The most prominent cause is hyperglycemia, which causes macro- and microvascular damage. Microvascular damage begins with the eyelids and small blood vessels, known as retinopathy, nephropathy, and neuropathy and is caused by diabetes. Macrovascular injury affects important organs by linking big blood arteries and contributing significantly to organ loss, such as CAD, stroke, and peripheral artery damage (Souza et al. 2012).

2 Complications of Diabetes

Diabetes, a long-term metabolic disease marked by decreased insulin synthesis or usage, has become an epidemic on a global scale. Millions of people suffer from the illness, which places a heavy strain on economies and healthcare systems. Diabetes is linked to several complications that might significantly lower life satisfaction and increase the disease's premature death and morbidity rates in individuals impacted by illness. Diabetes mellitus type 2 (T2DM) is a metabolic disease that is triggered by hyperglycemia and has grown to be a major issue globally (Trikkalinou et al. 2017). T2DM is becoming more and more common worldwide due to several causes, including obesity, ageing populations, urbanization, and a decline in physical activity. A major contributor to postprandial hyperglycemia and a major factor in the morbidity and prognosis of T2DM patients is a dysfunctional rise in the digestion of glucose, an identifiable characteristic of the disease. Numerous organ systems are impacted by the wide range of problems associated with diabetes. Microvascular and macrovascular abnormalities arise due to diabetes's chronic hyperglycemia (Sen et al. 2016). Cardiovascular disorders such as peripheral arterial disease, coronary artery disease, and stroke are made more likely by macrovascular problems, which affect the body's major blood vessels. These issues can lead to serious health outcomes, such as cardiovascular events, lower limb amputations, end-stage renal disease, and visual impairment. The overall burden of diabetes is further increased by its association with an increased risk of other illnesses like dyslipidemia and hypertension, along with non-alcoholic fatty liver disease.

Natural substances and related compounds have drawn much attention recently as potential treatments for hyperglycemia and its consequences. These substances, which come from a wide range of natural sources like fungi, microbes, plants, and animals, provide a wide range of bioactive components with complex pharmacological actions (Souza et al. 2012). By utilizing natural products, diabetes research can benefit from a wealth of new chemicals and conventional medicine successfully passed along through the years. The wide range of chemical composition of natural products is one of their main benefits. For example, plants—including algae—produce a remarkable variety of secondary metabolites, such as phenolic chemicals, alkaloid compounds, flavonoids, and terpenoids, which have a broad spectrum of biological actions. These varied chemical structures provide a useful tool for creating targeted treatments, enabling the modification of several molecular targets related to insulin production, glucose homeostasis, and insulin sensitivity (Kokil et al. 2015). Microvascular and macrovascular disputes might arise due to chronic hyperglycemia's production of oxidative stress and activation of inflammatory pathways. Diabetes treatment may benefit greatly from using natural ingredients and their derivatives. Their pleiotropic effects, variety of chemical ingredients, and potential for reducing problems make them appealing candidates for the creation of innovative therapeutic strategies. We can increase our therapeutic arsenal and enhance the condition of those with diabetes by utilizing the potential of natural products (Ríos et al. 2015). The present chapter focuses on the promising results of algal as bioactive compounds from various derivatives and acts as antidiabetic agents in diabetes management.

3 Algal Bioactive Compounds with Potentially Antidiabetic Applications

As photosynthetic organisms inhabiting various watery environments, algae are remarkably adept at creating a wide range of distinct and physiologically active substances. These natural compounds made from algae present fresh opportunities to study the biology of diabetes and create cutting-edge treatments for its control. Natural compounds made from algae are important for diabetes research since they can target several components of the condition. Many bioactive substances, including polysaccharides, polyphenols, dyes and pigments, fatty acids, and sterols, are produced by algae and have shown a variety of pharmacological characteristics that are pertinent to diabetes (Bocanegra et al. 2021). These substances can improve lipid profiles, increase insulin sensitivity, modify glucose metabolism, and exhibit anti-inflammatory and antioxidant properties—all critical for managing diabetes. Natural compounds made from algae show potential in treating the particular issues correlated with diabetes. Inflammation and oxidative stress in the retina are indicators of retinopathy caused by diabetes, a chronic microvascular condition. The anti-inflammatory and antioxidant qualities of substances produced via algae have demonstrated promise in safeguarding retinal cells and slowing the advancement of retinopathy (Popović-Djordjević et al. 2021). By lowering albuminuria, eliminating renal fibrosis, and modifying the inflammatory pathways, chemicals produced from algae have also shown reno-protective properties, offering a possible treatment option for diabetic nephropathy. The ability to bridge therapeutic gaps while enhancing current diabetic treatments. These agents have the potential to address therapeutic gaps and enhancecurrent diabetes treatments. When administered as adjuvants, they may increase the efficacy of standardantidiabetic medications or produce synergistic effects. They can improve traditional antidiabetic medications' effectiveness or synergistic effects when used as adjuvants.

Additionally, natural products made from algae may provide an alternate therapeutic treatment for people who are resistant to or have an intolerance for existing drugs. Natural compounds made from algae thus constitute an important and unexplored resource for the study and treatment of diabetes (Leena et al. 2020). Algae produce a wide range of bioactive chemicals that provide a wealth of possible therapeutic agents that can target various components of the ailment and its implications. Genetic variety and adaptability to various ecological niches are the main causes of algae's significant chemical diversity. Algae belonging to different taxonomic groups have unique metabolic processes and biosynthesis capacities, which enable the production of chemicals unique to that species. Red algae, or Rhodophyta, are widely recognized for producing sulphated polysaccharides, such as agar and carrageenan, which have been shown to have anti-inflammatory, anti-viral, and anticoagulant properties (Popper et al. 2011).

3.1 Polysaccharides

Many research investigations have demonstrated the anticancer, antibacterial, antidiabetic, and anti-obesity activities of polysaccharides extracted from mushrooms. As a result, they have enormous promise for both functional foods and medicine. Polysaccharides derived from fungi, primarily β-glucans, are an intriguing class of chemicals because of their therapeutic qualities—Fungi's immunostimulatory and antitumor characteristics. In mice with Sarcoma 180 tumors, he demonstrated that extracts from boletus mushrooms (Boletus edulis) improved their survival chances (James 2023). Lucas also extracted the compound calvacin, which increases the efficacy of sarcoma and leukemia treatments, through the berries of the enormous puffball Calvatia gigantea. Their molecular weight, water solubility, branch count, and tertiary structure affect enzyme activity. The human body does not have any negative impacts from these polysaccharides.

However, not all glucans have anticancer qualities. Their degree of branching, size and molecular weight, solubility in water, and mode of occurrence all affect this characteristic. β-glucans with high molecular weights primarily composed of β (1 → 3) linkages have the strongest anticancer effects (Amin et al. 2022). The antitumor properties of polysaccharides are extensive and multivalent. They restrict damage to cell DNA, lower the levels of carcinogens in the body, prevent them from activating, and stop cancer cells from growing. Their capacity to stimulate the immune system allows them to aid in the treatment of infectious disorders as well as cancer. One significant benefit of naturally occurring polysaccharides is that they do not harm the human body and can mitigate the negative effects of chemotherapy. Polysaccharides, an anticancer medication generated from fungi, have effectively treated lung, breast, cervical, and gastrointestinal tract malignancies (Samanta 2022). The bioactive components have anticancer qualities and may also help lessen the signs and consequences of Parkinson's and Alzheimer's diseases.

Additionally, they aid in the therapeutic management of some types of cancer, stimulate the body, and enhance digestion. The mushroom known as lion's mane generates a variety of physiologically active compounds, such as extracellular polysaccharides (EPS) and intracellular polysaccharides (IPS), which have anticancer and immunostimulatory effects. Breast, prostate gland, lung, and cervical cancer cells may all be successfully inhibited by barbituric acid-containing Usnea longissima extract. Human melanoma cells UACC-62 were susceptible to the cytotoxic effects of norstictic acid, which was isolated from Ramalina sp. Numerous research has documented lichens' anticancer properties (Priori 2021). Utilizing a variety of mechanisms, including cytotoxic action, regulating the Lichen's metabolites exhibits anticancer properties, including cell cycle, anti-proliferation, anti-migration and angiogenesis, telomere inhibitory, and inhibition of endothelium tube formation. Lichen thalli secondary metabolites affect various cancer cell types, includingthose from skin, leukemia, and brain tumors such as astrocytoma. They also influence cells from the prostate, lung, gastrointestinal tract, breast, liver, cervix, rectum, pancreas, ovary, and lymphatic system (Ślusarczyk et al. 2021).

3.2 Vitamins and Sterols

Among the few foods containing cholesterol, ergosterol is an intermediate to vitamin D2 found in mushrooms. Ergocalciferol (D2) and cholecalciferol (D3) are the primary physiological metabolites of active vitamin D. Under ultraviolet light, ergosterol can be transformed into vitamin D2. UV light with a wavelength between 280 and 320 nm photolyzes ergosterol to produce a variety of compounds. Provitamin D2 and its isomers, tachysterol and lumisterol, are the primary ones. Heat causes provitamin D2 to reorganize into vitamin D2 spontaneously. Vitamin D2 is metabolized after consumption, passing via an intermediate derivative

25-hydroxyvitamin D2 to reach the physiologically functioning form 1α, 25-dihydroxyvitamin D2 (Wei et al. 2024). In addition to the north and the south, mushrooms are the primary source of vitamin D. Children, the elderly, and post-menopausal women can all benefit from vitamin D's ability to improve calcium digestion. Osteoporosis in both adults and children can result from a lack of vitamin D. Prostate, colorectal, and breast cancers, as well as autoimmune illnesses, particularly type I diabetes and Crohn's disease, along with cardiovascular conditions, have all been found to be significantly decreased by vitamin D. Chanterelle, or Cantharellus cibarius, is one of the most highly important and often harvested types of mushrooms. Along with its eye-pleasing orange color, fragrant aroma, and high nutritional content, it is also resistant to infestations of insects (Raman et al. 2011).

3.3 Carotenoids and Flavonoids

Carotenoids are substances produced by bacteria, fungi, algae, and plants. Carotenoids that have thirteen double bonds and four keto groups with a monocyclic structure can be produced by fungi. These conjugated bonds give these compounds their distinctive yellow, orange, or purple appearance. Carotenoids are essential for the human body because they defend cells against the potentially lethal combination of damaging light, ultraviolet damage, and oxidizing-antioxidation by neutralizing individual oxygen molecules, which might give off additional electrical energy through the process of oxidation process (Alcaíno et al. 2016). These substances function as photoreceptors and stabilize cell membranes. In addition to preventing sun-induced irritation, they slow the skin's ageing process. Atherosclerotic plaques are prevented from forming due to the carotenoid complex's ability to prevent oxidizing of its components of LDL lipoprotein. Compared to a diet deficient in these substances, increasing your intake of carotenoids reduces your chance of developing cataracts. Tricholoma equestre, Agaricus bisporus, Pleurotus ostreatus, Boletus edulis, Suillus bovinus, and Cantharellus cibarius are among the varieties of mushrooms which are edible and contain β-carotene. Fruiting bodies of A. bisporus, B. edulis, P. ostreatus, S. bovinus, T. equestre, and C. cibarius (Lin and Xu 2020) have been found to contain lycopene. C. cibarius fruiting bodies contain lutein, α-carotene, and xanthotoxin, whereas B. edulis fruiting bodies contain γ-carotene, auroxanthin, and neurosporin. The reduction of peroxides along with hydroxides, rapid transfer of hydrogen fuel into its carboxyl compound, and inhibiting oxidizing enzymes like lipoxygenase are the mechanisms by which flavonoids limit the generation of reactive oxygen species and free oxygen radicals in cells. The chelation of ions made up of transition metals Cu and Fe, which stops harmful hydroxyl radicals from forming in cells, is the basis of the secondary mechanism of action. Additionally, during lipid peroxidation, flavonoids can interrupt the chain of free radical reactions. These substances protect cell membranes, enzymes that break down, and genetic material (Goyal et al. 2022; Alcaíno et al. 2016).

3.4 Phenolic Compounds

In recent decades, the discovery of several bioactive compounds in fungi, particularly macrofungal ones, along with observing their health-promoting qualities, has led to an increase in interest in fungi, notably lichen-forming fungi. Researchers are especially interested in antioxidants because they can eliminate free radicals produced within the human body. This enables them to combat a wide range of prevalent ailments faced by society, such as cancer, diabetes, atherosclerosis, cardiovascular disease, and neurological diseases. To incorporate antioxidants into a daily diet, researchers are now looking for new sources of antioxidants (Al-Tamimi et al. 2024). Numerous studies' findings show that both spontaneous and cultivated mushrooms' fruiting bodies are abundant in antioxidants. The majority of the phenolic chemicals present in mushrooms are phenolic acids. They exhibit a variety of biological activities because of their potent antioxidant impact and capacity to shield important structures: nucleic acids, protection of cell membranes, protein scaffolding, enzymes, and lipids against oxidative damage. The chemicals that are derivatives of cinnamon, vanillic acid, and caffeic acid exhibit the highest antioxidant properties, including protecting cells against hydrogen peroxide (Holger et al. 2014). Protocatechuic, gallic, and p-hydroxybenzoic acids—found in fungi—have been shown to have anti-inflammatory, antibacterial, antiviral, and antioxidant properties both in vitro and in vivo. They also stimulate the production of gastric secretions. The effects of protocatechuic acid include immunomodulation, spasmolysis, cardioprotection, anticoagulation, and chemoprevention. The existence of protocatechuic acid, gallic acid, homogentisic acid, and chlorogenic acid in P. ostreatus (Ślusarczyk et al. 2021). Differing genetic characteristics or the different areas that originated mushroom fruiting bodies for variations in the qualitative and contractions of phenolic compounds reported by various investigators. The extent of damage to the fruiting individuals or shifting environmental factors, such as air pollution, may also cause tumors.

A great source of important antioxidants is edible mushrooms. Lichenized fungi (lichens) are another material that can be used to produce therapeutic medicines because of their secondary metabolites' antioxidant, antibacterial, and anticancer capabilities (Montano et al. 2022). For instance, Lichen extracts from the Usnea genus showed strong antioxidant capacity and may preserve human lymphocytes by controlling enzymatic activity. Acid dibenzofuran, separated from Usnea species and other lichens, has shown various biological activity, including anticancer, antibacterial, and cytoprotective properties. Additionally, because of their capacity to scavenge free radicals, several lichen species like B. fuscescens, P. tilapia, and U. decussate were considered possible sources of natural antioxidants. Lichen secondary metabolites and their full biological capabilities were examined in conventional medicine. Phenolic compounds can also be produced in high concentrations by lichens, including species of the genera Peltigera, Solorina, Nephroma, Cetraria, Flavocetraria, and Alectoria (Rahman et al. 2021; Ślusarczyk et al. 2021).

4 Algae and Its Bioactive Compounds' Mechanisms of Action in Enhancing Medicine

The ability to influence the key pathways linked to insulin sensitivity, glucose homeostasis, and the avoidance or reduction of consequences from diabetes is illustrated by these bioactive components. The full therapeutic potential of organic substances obtained from algae depends on a thorough understanding of these mechanisms. Algae extracts show different bioactive component profiles, and each one may be able to address different elements of diabetes. While certain extracts may be better at improving insulin sensitivity or reducing problems, others may have better effects on glucose regulation. The most promising algal extracts require careful research, including in vitro and in vivo trials and comparisons of their therapeutic effects. High concentrations of particular bioactive substances, like peptides, polyphenols, or polysaccharides, found in algae extracts may be the secret to improved antidiabetic effectiveness (Menaa et al. 2021; Tiwari et al. 2021) (Table 6.1).

5 Controlling Glucose Metabolism via Increasing Hormone Insulin

Natural products made from algae have been shown to alter the activity of enzymes like α-glucosidase and α-amylase that are involved in the cellular breakdown of glucose. Compounds derived from algae can lower postprandial glucose levels by slowing down the decomposition and absorption of carbohydrates by blocking these enzymes. These substances usually attach to the enzymes' active sites or interact with necessary amino acid residues to stop the enzymes' catalyzing the

Table 6.1 Algae and its bioactive compounds

S. no	Compounds	Microalgae/Macroalgae	References
1	Lipids	Trachydiscus minutus, Phaeodactylum tricornutum/fatty acids (Phaeophyta)	Alsenani et al. (2020)
2	Polysaccharides	Ulvales, Chlorophyta/(Laminaria sp.), Phaeophyta, Gelidiales, Gigartinales, Gracilariales	Máximo et al. (2018) and Shahidi and Rahman (2018)
3	Bioflavonoids	Quercetin, kaempherol	Wells et al. (2017)
4	Polyphenols	Bacillariophyceae, Eustigmatophyceae, Haematococcus pluvialis/catechins, tannins, flavonoids, mycosporine, lignans	Chojnacka et al. (2012)
5	Sterols	Monoraphidium, Scenedesmus/Rhodophyta, Phaeophyta	Shalaby (2011)
6	Pigments	Chlorophyll b (green algae)/dinoflagellate, phycoerythrin (Rhodophyta)	Thomas and Kim (2013)

decomposition of carbohydrates. This competitive or non-competitive inhibition effectively slows the rate at which carbohydrates are converted to glucose (Wong et al. 1995). Consequently, the total amount of glucose released into the blood is reduced, aiding blood sugar regulation. Natural sources, including specific fruits, vegetables, and herbs, have shown similar enzyme-inhibiting qualities. Compounds in fenugreek, green tea, and cinnamon, for example, have been shown to block α-glucosidase and α-amylase, which helps to control blood sugar levels. Combining these natural substitutes with items made from algae offers greater possibilities for controlling glucose metabolism.

A distinctive characteristic of type 2 diabetes is insulin resistance, typified by a decreased cell absorption of glucose and a decreased response to insulin (Pereira and Valado 2023). Natural compounds made from algae have the potential to increase insulin sensitivity. They can promote glucose absorption and cell utilization by activating signaling pathways, including the PI3K/Akt pathway, implicated in insulin action. By promoting insulin sensitivity, these substances help improve metabolic regulation and control glycemic levels. The PI3K/Akt pathway plays a critical role in glucose metabolism by facilitating glucose absorption into cells. This pathway increases the efficiency of insulin's entry into cells by promoting the translocation of insulin transporter proteins towards the cell membrane (Świderska et al. 2018). Ensuring that cells have enough energy for their metabolic processes and preserving adequate blood sugar levels depend on this process. Products made from algae have many advantages over controlling blood sugar. Omega-3 fatty acids, antioxidants, and minerals are essential for general health and well-being. These products also have antioxidant qualities that can shield cells from oxidative stress, lowering the risk of long-term illnesses and increasing longevity (Zhao et al. 2019).

6 Abiotic and Biotic Elements in the Synthesis of Algal Biomolecules

Enhancing the yield of important bio compounds from algal cultures and optimizing their synthesis. Many medicinal substances, including ash (43.6%), high fiber (52.3%), lacking in protein (37.8%), and trace levels of FAs (5.2%), can be synthesized by the majority of algae. The accessibility of essential nutrients (N, P, K, etc.), humidity, temperature, oxygen consumption, the intensity of light, CO2, seawater pH, and sampling variability are some of the environmental factors that affect algal evolution, the total biomass production yield, in addition to the formation of therapeutic compounds including lipids, amino acids, polymers, vitamin and mineral deficiencies, pigments (Geada et al. 2021). Pharmaceuticals are one of the industries where these compounds may be used in developing medications and dietary supplements. They can be used as natural colorants or nutritional additions in food processing. They also show potential in the cosmetics sector for developing anti-ageing and antioxidant products. Determining appropriate levels of mineral substances, polysaccharides, and tiny substances present in algae thus remains extremely difficult. The growth of marine algae may be adversely affected by abiotic stressors. Different algae species have varying pH needs for growth and food absorption, and pH variations negatively impact algal cells (Gomez-Zavaglia et al. 2019). Reduced algal productivity can result from abiotic stressors such as temperature, pH, and salinity variations because they interfere with cellular functions and stunt growth. These pressures can impact the production of beneficial compounds, which can also result in oxidative damage and nutritional imbalances. Thus, it is crucial to comprehend and control these stressors to increase the synthesis of compounds and optimize algal cultivation. Chlorella vulgaris, for example, is a green microalga that can thrive in a wide range of pH values, although its pH 8–10 had the fastest growth rate and maximum biomass efficiency. Algal development efficiency is impacted by light, which negatively affects cell reproduction (Abideen et al. 2022). Abiotic stress in algae production can be reduced by controlling salinity, preserving stable pH levels, and optimizing nutrient concentrations. Algal stress can be lessened using bioreactors with controlled environmental parameters, such as temperature and light intensity adjustments.

Furthermore, selective breeding and genetic engineering of algae variants may improve their resistance to harsh environments, guaranteeing long-term growth and the production of compounds. Higher light levels will accelerate the pace of photosynthetic activity; additionally, photorespiration and photoinhibition balance the rate of increase until it reaches a threshold, after which the cell is disrupted. While the material of nearly all FAs associated with the brown seaweed Sargassum piluliferum dropped with increasing light, there was an increase in the green seaweed Ulva pertusa, indicating that the type of algae determines the number of FAs produced by light (Basin et al. 1975). Algae have vital photoprotective systems that preserve their photosynthetic machinery from harm brought on by prolonged exposure to light. These processes include the production of pigments such as carotenoids, which can release surplus energy as heat and stop the production of dangerous reactive oxygen species (Zia-Ul-Haq 2021).

Furthermore, algae can modify the structure and composition of their light-harvesting complexes to maximize light absorption and reduce photodamage under different light circumstances. Because they absorb and dissipate additional light energy as heat, carotenoids are essential for photoprotection because they stop the production of dangerous reactive oxygen species. By acting as antioxidants, they guard against oxidative damage to the photosynthetic mechanism. Additionally, carotenoids help light-harvesting complexes maintain their structural stability, which minimizes photodamage and ensures effective energy transfer (Sohrabi et al. 2024).

7 Algae Extracts' Impact on Insulin Sensitivity and Secretion

The potential of algal extract for glycemic control and diabetes treatment has been shown by considerable research on its effects on insulin secretion and sensitivity. Various bioactive substances in algae-derived natural products have shown promise in regulating insulin production and improving insulin sensitivity. The insulin release by pancreatic β-cells has been reported to be influenced by algae extracts. According to studies, several chemicals derived from algae can increase insulin release in reaction to stimulation with glucose. Among the many ways these substances work is by blocking ATP-sensitive potassium channels, which allows calcium ions to enter and cause the release of insulin vesicles (Oluwamodupe and Babalola 2024). Algal extracts can regulate blood glucose and maintain glucose homeostasis by encouraging insulin production. Animal studies and many cell types have shown that algae extracts can improve insulin sensitivity. These extracts activate intracellular signaling pathways essential for insulin-mediated glucose absorption, including the PI3K/Akt pathway. Adipose tissue has a significant role in controlling glucose metabolism and insulin sensitivity. Adipocyte activity has been demonstrated to be modulated by algae extracts, which affect adipocyte differentiation and adipokine release (Guru et al. 2021). Adiponectin and other adipokines are essential for the metabolism of fats and carbohydrates. Algae extracts can raise adiponectin secretion, improving insulin sensitivity and cell uptake of glucose. Extracts from algae have also been shown to stimulate the discoloration of white adipose tissue and prevent adipocyte hypertrophy, further enhancing metabolism.

Algae extracts have been shown to enhance insulin sensitivity and cellular energy metabolism by improving mitochondrial activity. These extracts can improve ATP synthesis, raise mitochondrial respiration, and encourage mitochondrial biogenesis (Bocanegra et al. 2021). Algal extracts enhance mitochondrial activity, which maximizes cellular energy use and facilitates glucose uptake and metabolism. Insulin resistance and poor glucose metabolism are caused by oxidative damage and chronic inflammation. Because algae extracts are antioxidants and anti-inflammatory, they can lessen oxidative stress and inflammation in tissues that respond to insulin. Algal extracts enhance insulin sensitivity and encourage improved glycemic management by addressing these underlying causes. The effect of algae compounds on insulin sensitivity and secretion offers important new information on their possible medical uses in diabetes treatment and glycemic control (Rahman et al. 2022). Numerous bioactive chemicals that target the major pathways involved in glucose metabolism and insulin control are present in these extracts. Additional research is necessary to identify the specific pathways that regulate the activity of algal extracts and pinpoint the active ingredients that benefit them. The creation of innovative diabetes treatment approaches could greatly benefit from utilizing the potential of natural compounds obtained from algae to enhance insulin secretion and sensitivity (Wang et al. 2024).

8 Inhibition of Carbohydrate-Digesting Enzymes via Algae

Natural products made from algae have demonstrated encouraging inhibitory effects on the enzymes that break down carbohydrates, which makes them viable options for managing diabetes and glycemic control. Enzymes that break down carbohydrates, such as α-glucosidase and α-amylase, are essential for converting complex carbs into glucose. By inhibiting these enzymes, blood glucose levels after meals can be controlled, and glucose spikes are preventable. These enzymes' active sites can be bound by bioactive chemicals found in algae-based products, preventing them from catalyzing the breakdown of polysaccharides. Because of this competitive inhibition, complex carbs take longer to convert to glucose, which causes blood sugar levels to rise more gradually (Crapo et al. 1976). Algal-based solutions provide a natural way to maintain glycemic control by adjusting the activity of these enzymes. The impact of natural compounds derived from algae on enzymes that break down carbohydrates is examined here, along with the consequences. Natural compounds derived from algae have been shown to block α-glucosidase, an enzyme that is in charge of the last stage of breaking down complex carbs into glucose. Algal-based substances slow down the breakdown and absorption of carbs by blocking α-glucosidase, which causes glucose to enter the bloodstream more slowly (Wylie-Rosett et al. 2004). The enzyme α-amylase, which starts the breakdown of starch via smaller carbohydrate units, has also been shown to be inhibited by natural compounds derived from algae. Algal-based chemicals slow down starch digestion by inhibiting α-amylase, lowering the proportion of glucose delivered into the bloodstream.

Algal-based substances can help control postprandial blood sugar levels and avoid hyperglycemia by blocking α-amylase. Certain algae species, such as Chlorella vulgaris and Spirulina platensis, are known to inhibit α-amylase. Bioactive substances found in these species efficiently reduce enzyme activity, providing a natural method of blood sugar regulation (Sharma et al. 2024). Scientists are still investigating how this and other algae could be used to create nutraceuticals and functional foods. Compared to conventional crops, algae can thrive in various environments and require less water and land, making their cultivation for these goods extremely beneficial to the environment. Because algae absorb CO2 during photosynthesis, growing algae also lowers carbon dioxide levels. Integrating algae farming with water treatment facilities can improve ecological sustainability by cleaning water and generating useful bioactive substances (Murray et al. 2013).

9 AMPK Enzyme Phosphorylation Enhancement

The enzyme AMP-activated protein kinase (AMPK) controls how lipids are metabolized and carbohydrates and is a critical player in metabolic pathways. AMPK is an essential component of the cell's energy production system. This enzyme produces adenosine triphosphate (ATP) in the cell. Cellular activity depends on ATP, which is a typical fuel for cells. Its essential energy is released when ATP is changed into ADP or AMP. AMP phosphorylates and resynthesizes ATP by stimulating the AMPK enzyme. One significant target in treating metabolic diseases is AMPKs and insulin resistance because of these functions and the control of insulin signaling

(Bullon et al. 2016). Hepatic fatty acid oxidation is stimulated in tandem with AMPK activation. The role that AMPKs play in treating diabetes-related conditions is demonstrated by their inhibition of the metabolism of cholesterol, lipogenesis, and triglycerides; their upregulation of insulin production on pancreatic β-cells; and their stimulation in the metabolic process of muscle fat as well and absorption of glucose by tissues. By stimulating AMPK in muscle cells, researchers expect to increase diabetes type 2 treatment and sensitivity to insulin. It is hypothesized that diabetes can be cured if we can figure out how to increase AMPK activity (Jeon 2016).

9.1 Reduction in Oxidative Stress

Antioxidant compounds prevent the cell from being destroyed by neutralizing free radicals and boosting immune system performance—deterioration of antioxidant activity results in oxidative stress. The definition of oxidative stress is, in other words, an imbalance between antioxidant defense and free radical production. While performing the glycation process, free radicals such as superoxide and hydroxyl radicals change the structure and function of enzymes, leading to diabetes (Jakus 2000). Conversely, the prevalence of oxidants leads to insulin resistance factors, particularly in the skeletal muscle. Angiotensin receptors contribute to the process by boosting the activity level of cellular NADPH oxidase, which produces hydrogen peroxide and superoxide ions in conditions of high energy. The auto-oxidation of glucose is one consequence of hyperglycemia, which also typically raises the production of reactive oxygen species (ROS) and alters the oxidative DNA, proteins and lipids in different tissues.

Furthermore, by altering the quantity of arginine residues in proteins, glycogen auto-oxidation might impair functionality (Hatziagapiou et al. 2019). Hyperglycemia boosts nitric oxide synthesis, which causes high amounts of superoxide generation in the mitochondria. Inflammatory responses of ROS and other oxidizing agents are produced when superoxide levels are high. This results in reduced insulin production in muscular and skeletal tissue, insulin resistance, and malfunction of pancreatic beta cells. These elements imply that one of the most crucial objectives in the fight against diabetes may be to investigate and clarify the elements that lessen oxidative stress (Mauricio and Alonso 2024).

10 Therapeutic Potential

Algal-based natural products, such as Ascophyllum nodosum, Fucus vesiculosus, and Undaria pinnatifida in the Phaeophyceae, inhibit the enzymes that break down carbohydrates. This is an appealing therapeutic strategy for managing diabetes and glycemic control. These natural compounds provide a natural and safer substitute for synthetic enzyme inhibitors by slowing down the digestion of carbohydrates and decreasing glucose absorption. Utilizing organic compounds made from algae as enzyme inhibitors also fits nicely with the increased interest in plant-based treatments and creating food or vitamin supplements to treat diabetes (Khursheed et al. 2019). Rate-digesting enzymes demonstrate their promise for managing diabetes and glycemic control. These natural compounds have inhibitory effects on the role of α-glucosidase and α-amylase, which regulate postprandial levels of glucose and prevent glucose production (Jugran et al. 2021).

11 Substances Derived from Marine Sources That Are Beneficial for Controlling the Effects of Diabetes

11.1 Macroalgae

Marine resources include seaweed, macroalgae, fish, invertebrates, and microbes that produce microalgae. Marine resources contain a range of bioactive substances that may be helpful in the management or treatment of diabetes. Unsaturated fatty acids, polysaccharides, polyphenols, vitamins, minerals, and dietary fiber are all found in seaweed. These substances exhibit numerous biological actions, including antidiabetic, anti-inflammatory, antioxidant, antibacterial, and antiviral properties. Seaweeds contain bioactive substances that, by controlling oxidative stress and inhibiting the enzymes that break down starch, can be utilized to treat long-term conditions like diabetes (Biris-Dorhoi et al. 2020). Polysaccharides found in seaweed are also crucial for neutralizing free radicals and shielding organisms via oxidative stress-related harm. Seaweed contains polyunsaturated fatty acids, namely omega-3 fatty acids and omega-6 fatty acids, essential for a balanced diet. Omega-3 fatty acids have a hypotensive effect, inflammatory indicators, and triglyceride levels in people with type 2 diabetes. Plant-based omega-6 fatty acids increase insulin sensitivity (Das et al. 2024).

According to a 2012 study by Pereira and colleagues, for example, which looked at 17 different kinds of red, brown seaweed, and green seaweed found that if the amounts of omega-6 and omega-3 in the phytoplankton are high enough, they can be considered an advantageous source of nutritional supplements polyunsaturated fatty acid (PUFAs). Consuming dietary fiber results in decreased feed intake and greater wear and tear since it takes longer to digest. According to several studies, eating too much fiber at meals is linked to a lower risk of developing diabetes (Pereira et al. 2012). Ascophyllum nodosum, Palmaria palmata, and Laminaria digitalata are examples of elongated Himanthalia. Glycemia is decreased when the fiber is used in algae properly.

Additionally, the biochemical indicators of insulin resistance typically decrease when dietary fiber intake rises. The polyphenolic chemicals found in red algae are among the other significant substances. By influencing the digestive enzymes involved in the breakdown of starch, these substances serve as an efficient component in treating diabetes. Laminarin, fucoidan, and alginate are among the polysaccharide molecules found in brown algae that have antidiabetic effects (Costas et al., 2024).

12 Effects of Chemicals and Extracts Derived from Algae on Diabetes

Ascophyllum nodosum brown algae extracts, both aqueous and ethanolic, have antidiabetic qualities and can inhibit the α-glucosidase enzyme. The phenolic chemicals in this algae extract have antioxidant properties and inhibit α-glucosidase and α-amylase. Phenolic extracts from brown algae belonging to the Ascophyllum and Alaria species have been shown to inhibit α-amylase activity. In addition, extracts from Ascophyllum also inhibit α-glucosidase. Research has indicated that dietary supplements made from marine algae such as Saccharina japonica and Undaria pinnatifida may lower blood lipids, regulate blood glucose, and enhance the activity of antioxidant enzymes like CAT and GSH-px in individuals with diabetes (Abo-Shady et al. 2023). A red algae known as Rhodomela confervoides was used to create a novel and promising combination for treating diabetes. The enzyme PTP1B, known as HPN, was inhibited by this mixture. Analysis has shown that HPN's antihyperglycemic effects are similar to those of rosiglitazone, an antidiabetic drug. The potential for HPN as a novel medication to treat type 2 diabetes is thus significant. The possible existence of α-glucosidase enzyme inhibiting effects shows that two bromophenols obtained from Grateloupia elliptica, a redalgae, had a stronger inhibitory impact than the other species. The Caulerpa racemosa extract of microalgae ethyl acetate was found to have a substantial inhibitory impact on α-amylase by Senthil et al. (2013). The MUFA extracts and derivatives separated from Ulva lactuca's algal extract activated several antioxidants ARE genes by acting as Nrf2-ARE pathway activators (Senthil et al. 2013) (Table 6.2).

13 Antioxidant Properties and Antihyperglycemic Characteristics of Different Algae-Derived Extracts

Numerous studies have connected the medicinal potential of microalgae to substances such as alkaloid compounds, carotenoids and terpenoids, as well as steroids, a substance called phenolic elements, halogen-containing ketones, alkaline compounds, cyclic polysulfides, and various other biological compounds. These components are also prevalent in vitamins, amino acids, pigments, fatty acids, and other vital minerals. Butanol was derived via the cyanobacterium Arthrospira platensis.

Table 6.2 Extracts derived from algae on diabetes metabolism

S. no	Species	Compounds	Mechanism of implementation	References
1	Pelvetia siliquosa	Fucosterol	Inhibiting the breakdown of glycogen and levels of blood sugar concentrations	Lee et al. (2004)
2	Ishige foliacea	Octaphlorethol A	GLUT4's increased absorption of glucose via AMPK signaling pathways Increasing the antioxidant enzymes' activity	Lee et al. (2012)
3	Undaria pinnatifida	Supplementary fibers	Enhanced antioxidant enzyme activity Lower lipids and blood sugar levels	Kim et al. (2008)
4	Ecklonia cava	Dieckol	Decreasing hepatic cholesterol levels and enhancing glucose tolerance Mitigation of oxidative stress-induced glucose-induced beta cell damage in the INS	Lee et al. (2010)
5	Ecklonia stolonifera	Dioxinodehydroeckol 7-Phloroeckol Dieckol	Blocking aldose reductase	Moon et al. (2011)
6	Gracilaria lemaneiformis	polysaccharide	Restoring the pancreatic tissues after they have deteriorated	Liao et al. (2015)
7	Symphyocladia latiuscula	Bromophenols: 1,2-Bis(2,3,6-tribromo-4,5-dihydroxyphenyl)-ethane	Restriction of the PTP1B enzyme	Harnedy and FitzGerald (2013)
8	Gelidium amansii	Ethanolic extract	Notable drop plasma glucose	Choi et al. (2015)
9	Rhodomela confervoides	Bromophenol derivatives: Bis(2,3-dibromo-4,5-dihydroxybenzyl) ether	Restriction of the PTP1B enzyme	Shi et al. (2008)
10	Kappaphycus striatus	Ethanolic extract	Lipase and α-amylase enzyme activity reduction	Balasubramaniam et al. (2013)
11	Caulerpa racemosa	Ethyl acetate extract	Restriction of α-amylase enzyme	Senthil et al. (2013)
12	Ulva fasciata	Sulfated polysaccharides	Inhibit the metabolism of carbohydrates and enhance the metabolism of glucose	Abirami and Kowsalya (2013)

(continued)

Table 6.2 (continued)

S. no	Species	Compounds	Mechanism of implementation	References
13	Ulva Lactuca	Seaweed extracts, unsaturated fatty acid constituents	Enhance antioxidant activity and activate a large number of ARE antioxidant genes	Tian et al. (2015)
14	Cladophora rupestris	Raw extracts	Suppression of the enzymes α-amylase & α-glucosidase	Suthindhiran Krish and Abhinab Das (2014)
15	Derbesia marina	Methanolic extracts carotenoids: Neoxanthin	Boost the action of antioxidants suppression of the PTP1B enzyme	Wang et al. (2018))

The highest antioxidant activity was measured using nitric oxide scavenging activity, hydroxyl radical, decreasing power, and DPPH assays. The methanolic extract of the microalgae Nannochloropsis sp. showed potential antioxidant activity, and increasing levels of vital fatty acids indicate the substance may be utilized as an inherent antioxidant for reducing oxidative stress (Abo-Shady et al. 2023). Because of their capacity to reduce the detrimental effects of ROS-induced oxidative damage, seaweeds are therefore acknowledged as renewable, high-quality antioxidant resources. Phenolic chemicals can act as antioxidants by strengthening the body's natural antioxidant system and chelating metal ions like Cu and Fe that catalyze reactions that produce free radicals. Protein ultrafiltrate gathered through Fucus spiralis-Ochrophyta showed a strong FRAP and a much higher scavenging activity of DPPH radical in ferrous ion-chelating assay, according to the antioxidant analysis. Enzymatic hydrolysis generates bioactive peptides and polyphenols responsible for this action (Abdullah et al. 2017). Seaweed is also considered one of the best natural antioxidant sources due to its high phenolic molecule concentration. Seaweed extracts were suggested as potential antidiabetic medication options by numerous investigations because of their capacity to lessen the damaging effects of excessive ROS-induced oxidative damage. U. lactuca's ethyl acetate extract has antidiabetic effects by blocking the enzymes that hydrolyze carbohydrates because of the activity of its bioactive ingredients (Ojo et al. 2023). Additionally, it has been demonstrated that the acetone-based extract of S. weight and an ethyl extract of S. polycystin can inhibit α-glucosidase (IC50 290.7 μg/mL) and α-amylase (IC50 440.5 μg/mL), hence preventing postprandial hyperglycemia. Additionally, significant α-glucosidase inhibitory effects were demonstrated by the flavonoids and tannins present in the methanol extract of Spatoglossum asperum (Martins-Gomes et al. 2024).

14 Antihyperglycemic Effects of Wide-Ranging Algal Extracts In Vivo

Cyanobacteria treatment restores liver damage by restoring normal serum hepatic marker enzymes and causes streptozotocin (STZ)-induced diabetic mice to regain body weight. Cyanobacteria exhibit anti-hypoglycemic properties by enhancing pancreatic insulin production from the islets' undamaged β-cells. According to a recent study, feeding powdered A. platensis cyanobacterium to diabetic rabbits reduced the animals' blood glucose levels, demonstrating antihyperglycemic action (Abo-Shady et al. 2023). Diabetes's negative effects on blood parameters and the inflammatory phase can be mitigated or prevented with the help of the antioxidant components found in A. platensis extract. The liver, kidneys, total bilirubin, lipid profile, and high glucose levels were significantly reduced by the methanol extract of A. platensis.

Additionally, the extract alleviated the histological cellular damage caused by diabetes development in the treated rats' pancreatic and liver tissues without adverse effects. Being non-toxic to the vital tissues, the extract had the potential to reverse the difficulties associated with diabetes. Aqueous extracts from U. reticulata show promise in preventing diabetes problems brought on by oxidative stress mediated by STZ (Gheda et al. 2021). Comparing this extract to a control group with diabetes showed a decrease in glutathione (GSH) level and increased plasma insulin production, vitamin E, and vitamin C activity. Lowering blood glucose and other physiological alterations exacerbated by diabetes demonstrated the antihyperlipidemic and antidiabetic properties of Turbinaria ornates sodium alginate. Recent investigations have indicated that, in addition to its hepatoprotective potential, the water-soluble extract of Padina boergesenii (Ochrophyta) has a critical influence on lowering increased blood glucose levels, renal biomarker levels, along with liver function lipid profile (Pereira and Cotas 2023).

15 Impact of Various Algal Extracts on the Therapy of Animal Hyperglycemia

A. platensis powder was given to diabetic rabbits in a recent study, and by reducing their blood glucose levels, the rabbits demonstrated antihyperglycemic action. A. platensis extract (400 mg/kg) may also have negative effects in hyperglycemic rats with diabetes induced by alloxan. Arthrospira, often known as spirulina, may have antihyperglycemic effects by boosting the ability of the β-cell islet to secrete insulin or by facilitating the movement of blood glucose to peripheral tissues (Gheda et al. 2021). The higher insulin levels in those with diabetes treated with Arthrospira, as shown in the results of the previous study, strongly suggested this. Elevated glucose levels were significantly reduced when S. wightii and U. lactuca ethyl acetate extracts were administered orally to diabetic rats. Accordingly, one possible way these extracts prevent hyperglycemia in diabetic rats is by improving glucose transport throughout cell membranes and increasing glycogen production, encouraging the breakdown of the glycolysis pathway via releasing the hormone insulin through the degranulation of pancreatic β-cells. Blood glucose levels impact hemoglobin levels (Ahmed 2006). Reducing blood glucose levels, which inevitably results in lower Hb values, may cause improved Hb levels in animals given varying dosages of Arthrospira. Another possibility is that arthrosis, a reliable iron supply, may help increase Hb levels. A platensis study used a recent powder to raise hemoglobin levels in diabetic rabbits. U. reticulata, the color green seaweed, has a therapeutic effect on restoring iron deficit and raising hemoglobin levels.

Treatment with Turbinaria conoides ethanolic extract and S. wightii & T. conoides methanolic extracts increased Hb levels, which were linked to decreased blood glucose levels. Additionally, diabetic rats treated with P. boergesenii aquatic extract had a considerable increase in Hb levels (Abdel-Karim et al. 2022). The level of bilirubin may indicate the condition and damage to the liver. Reducing blood levels of free fatty acids and related peroxides oxidation and hepatic inflammation improved the hepatic process. Insulin-like protein and ethanolic extract from Aphanizomenon flos-aqua (Cyanobacteria) were administered to diabetic rats, demonstrating the potential for liver-excretory function restoration. Increased gluconeogenesis and ketogenesis in diabetic diseases were formerly attributed to increased aminotransferase enzyme activity under insulin deprivation (Abo-Shady et al., 2021). Serum enzyme levels varied in diabetic rats, and these changes were strongly associated with changes in the metabolic processes of both the AST and ALT enzymes. The mechanism underlying the raised plasma concentrations of rats with diabetes showed that both aminotransferases might be related to liver dysfunction and increased excretion of these enzymatic processes in organs, especially the liver, due to oxidative damage and the production of final products that are increasingly glycosylated. Comparing the dietary supplementation of A. platensis powder to the induced diabetic rats, the former showed a noteworthy positive impact in lowering serum hepatic AST and ALT levels (Tiwari et al. 2013). A. platensis powder's strong ability to lower diabetic rabbits' elevated hepatic enzyme levels. A reduction in blood ALT and AST levels following therapy with A. platensis extract may lead to a subsequent improvement in liver disease. The treatment of 15 mg/kg b.w of A. platensis material significantly reduced hepatic transaminase expression compared to alloxan-diabetic rats as controls (Aissaoui et al. 2017).

16 Algal Compounds and Associated Nano-theranostics for Diabetes

Diabetes mellitus (DM) is a non-transmissible disease that is rapidly spreading around the world. According to recent statistics from the International Diabetes Federation (IDF) and the World Health Organization (WHO), there were over 382 million diabetic patients worldwide in 2013. By 2035, that figure might rise to almost 600 million cases. Diabetes mellitus is now acknowledged as a significant worldwide health issue that impacts both adults and children, leading to elevated rates of morbidity and mortality. DM is described by the ADA and WHO as an endocrine metabolic disease that is typified by persistent and sustained hyperglycemia brought on by either insufficient insulin action or production or a combination of both factors (Sen et al. 2016). An autoimmune condition known as diabetes mellitus type 1 is brought via an interaction of inherited and environmental variables. According to some research, the disease is brought on by a series of autoimmune reactions that target pancreatic β-cells. Pancreatic β-cells undergo systemic lymphocytic infiltration due to this interaction, which activates the adaptive immune system through CD4+ and CD8+ T-lymphocyte types. The invasion of pancreatic β-cells results in insulitis, which raises the expression of important inflammatory cytokines like IL-1β, TNF-α, along with INF-γ. These cytokines cause apoptosis, the destruction of pancreatic β-cells, and diabetes (Szablewski 2014). A multifactorial heterogenic disease, type 2 diabetes mellitus (T2DM), is defined by peripheral insulin resistance, inadequate insulin secretion by pancreatic β-cells, or both, which raises the synthesis of glucose in the liver.

Reduced sensitivity and signaling of peripheral insulin receptors cause insulin resistance. As a result, insulin concentrations become abnormally elevated due to impaired cellular signaling, not increased insulin production (Astorri 2011). Insulin signaling and pancreatic β-cell activity are genetically defective in type 2 diabetes, a complex genetic illness. According to genome-wide association (GWA) research, over 40 loci have been connected to the pathophysiology of type 2 diabetes. These include PPARG, IRS1, IGF1, IGF2, and KLF14, which are associated with insulin resistance, and KCNJ11, TCF7L2, and KCNQ1, which are linked to pancreatic β-cell activity. Insufficient insulin production and hyperglycemia are the results of a genetic mutation or deficiency in a single gene on separate chromosomes that occurs autosomally dominantly. This condition is known as diabetes of maturity of the young (MODY), which is a hereditary abnormality in pancreatic β-cell functioning (Azarova et al. 2023). Brown algae produce the most significant secondary bioactive metabolites, including alginic acids, phycoxanthin, fucosterols, fucoidan, and chlorotanins. They also exhibit potent biological actions, such as cytotoxic, antioxidant, anti-inflammatory, and antidiabetic effects. Most studies with chemicals derived from algae demonstrated that blood glucose levels could be controlled by blocking the enzymes that hydrolyze carbohydrates and protein tyrosine phosphatase 1B and by reducing insulin sensitivity, glucose absorption, and other protective effects against diabetes.

The marine particles found in the algae can be used to directly extract these chemicals (Menaa et al. 2021). The bioactive substances known as phlorotannins, which are abundant in brown algae, have been found to have potential as a treatment for several human illnesses, including diabetes. Phlorotannin subgroups exist in many brown algae species and have important antidiabetic properties through various mechanisms. Fucosterol, extracted from the brown algae Pteris siliquosa, decreased blood glucose levels and prevented the breakdown of glycogen in rats with diabetes induced by streptozotocin (STZ) (Menaa et al. 2021). Using an extract derived from the brown algae Pelvetia babingtonii, postprandial hyperglycemia was effectively suppressed and showed strong α-glucosidase inhibitory activity. New generations of medications based on selective nanoparticles are being developed and customized for particular metabolic situations, such as the development of age-related diseases, which is an important stage (Holte 2024). Because of the use of subpar animal models, the effects of age-related variables, such as frailty in very young kids and metabolic and physiologic modifications during old age, are still not well understood in nanomedicine or pharmacology in general. Additionally, the failure to translate promising animal ensues into human therapy may be due to factors like altered intestinal permeability, as defined in type 1 diabetes, or other metabolic process peculiarities, which include insulin resistance in type 2 diabetes. These factors may also impact the production of unconventional nanoparticulated drug preparations (Mancabelli et al. 2024).

17 Conclusion

Diabetes is already recognized as a chronic metabolic condition with the greatest death and morbidity rates in the globe. Furthermore, due to inadequate diabetes care, Asian countries are the most developed compared to European and Western countries. Clinical trials show that diabetes can be managed but not cured. Various drugs are recommended to control diabetes, with both beneficial and bad consequences. However, it is cost-dependent; bioavailability is important in determining drug management. Furthermore, medicine made from natural/green-synthesized materials properly treats glycemic and glycated hemoglobin conditions without problems. As a result, algae obtained from natural sources is a better alternative to treating diabetes and its consequences. Various tests have shown that natural algae, such as polysaccharides, polyphenols, pigments, vitamins, and peptides, have superior antidiabetic characteristics. In addition, blocking carbohydrate-digesting enzymes has a clear correlation with blood glucose homeostasis related to hormonal and molecular enzymatic sensitivity. Furthermore, encouraging findings in microvascular consequences of retinopathy and nephropathy at advanced stages were decreased by about 70% in severe diabetic patients. The fundamental mechanism and innovative algal biomedical isolation with maximum bioavailability will be investigated further to improve diabetes control.

Acknowledgement The authors are thankful to the Department of Biotechnology, Thiruvalluvar University, Tamil Nadu, India.

Conflict of Interest None to declare.

References

Abdel-Karim, O. H., Abo-Shady, A. M., Ismail, G. A., & Gheda, S. F. (2022). Potential effect of Turbinaria decurrens acetone extract on the biochemical and histological parameters of alloxan-induced diabetic rats. *International Journal of Environmental Health Research, 32*(7), 1447–1468.

Abdullah, M. A., Shah, S. M. U., Shanab, S. M. M., & Ali, H. E. A. (2017). Integrated algal bioprocess engineering for enhanced productivity of lipid, carbohydrate and high-value bioactive compounds. *Res Rev J Microbiol Biotechnol, 6*, 61–92.

Abideen, Z., Waqif, H., Munir, N., El-Keblawy, A., Hasnain, M., Radicetti, E., ... & Haider, G. (2022). Algal-mediated nanoparticles, phycochar, and biofertilizers for mitigating abiotic stresses in plants: A review. *Agronomy, 12*(8), 1788.

Abirami, R. G., & Kowsalya, S. (2013). Antidiabetic activity of Ulva fasciata and its impact on carbohydrate metabol-ism enzymes in alloxan induced diabetic rats. *International Journal of Research in Phytochemistry and Pharmacology, 3*(3), 136–141.

Abo-Shady, A. M., Gheda, S. F., Ismail, G. A., Cotas, J., Pereira, L., & Abdel-Karim, O. H. (2023). Antioxidant and antidiabetic activity of algae. *Life, 13*(2), 460.

Ahmed, S. M. (2006). *Study of the effect of Nicorandil on Insulin production and its influence on hypoglycemic action of glibenclamide on Alloxan induced diabetic rats* (Doctoral dissertation, Rajiv Gandhi University of Health Sciences (India)).

Aissaoui, O., Amiali, M., Bouzid, N., Belkacemi, K., & Bitam, A. (2017). Effect of Spirulina platensis ingestion on the abnormal biochemical and oxidative stress parameters in the pancreas and liver of alloxan-induced diabetic rats. *Pharmaceutical biology, 55*(1), 1304–1312.

Alcaíno, J., Baeza, M., & Cifuentes, V. (2016). Carotenoid distribution in nature. *Carotenoids in nature: biosynthesis, regulation and function*, 3–33.

Alsenani, F., Tupally, K. R., Chua, E. T., Eltanahy, E., Alsufyani, H., Parekh, H. S., & Schenk, P. M. (2020). Evaluation of microalgae and cyanobacteria as potential sources of antimicrobial compounds. *Saudi Pharmaceutical Journal, 28*(12), 1834–1841.

Al-Tamimi, H. M. A., Al-Sahlanee, B. J. A., Dawood, A. H., & Rahi, F. A. (2024). BIOLOGICALLY ACTIVE SUBSTANCES OF FUNGI AND ALGAE AS POTENTIAL PHARMACOLOGICAL AGENTS. *ББК 52.8 А 43*, 299.

Amin, Z. S., Abbas, S., Munir, A., Qadri, M. M. A., Raza, M., Danish, M., ... & Kashif, H. (2022). Role of Medicinal Mushroom Lentinula edodes in Nutrition, Nutraceutics and Ethnopharmacology. *Journal of Pharmaceutical Research International, 34*(30A), 18–35.

Astorri, E. (2011). Interplay between inflammation, autoimmunity and regeneration in the NOD mouse model of type 1 diabetes and Sjogren's Syndrome.

Azarova, I., Polonikov, A., & Klyosova, E. (2023). Molecular genetics of abnormal redox homeostasis in type 2 diabetes mellitus. *International Journal of Molecular Sciences, 24*(5), 4738.

Balasubramaniam, V., Mustar, S., Mustafa Khalid, N., Abd Rashed, A., Mohd Noh, M. F., Wilcox, M. D., ... & Pearson, J. P. (2013). Inhibitory activities of three Malaysian edible seaweeds on lipase and α-amylase. *Journal of applied phycology, 25*, 1405–1412.

Basin, M. M., Anderson DLT and Gill AE, 1975. Spin-up of a stratified ocean, with applications to upwelling. Deep-Sea Res., 22 (9): 583· 596.

Biris-Dorhoi, E. S., Michiu, D., Pop, C. R., Rotar, A. M., Tofana, M., Pop, O. L., ... & Farcas, A. C. (2020). Macroalgae—A sustainable source of chemical compounds with biological activities. *Nutrients, 12*(10), 3085.

Bocanegra, A., Macho-González, A., Garcimartín, A., Benedí, J., & Sánchez-Muniz, F. J. (2021). Whole alga, algal extracts, and compounds as ingredients of functional foods: composition and action mechanism relationships in the prevention and treatment of type-2 diabetes mellitus. *International Journal of Molecular Sciences, 22*(8), 3816.

Bullon, P., Marin-Aguilar, F., & Roman-Malo, L. (2016). AMPK/mitochondria in metabolic diseases. *AMP-activated Protein Kinase*, 129–152.

Choi, S., Oh, H., Jung, J., Park, S., Park, Y. I., Bak, S., & Lee, M. (2015). Effect of Agar-free Gelidium Amansii on Obesity in DIO C57BL/6J Mice Model. *The FASEB Journal, 29*, 750–2.

Chojnacka, K., Saeid, A., Witkowska, Z., & Tuhy, L. (2012, August). Biologically active compounds in seaweed extracts-the prospects for the application. In *The open conference proceedings journal* (Vol. 3, No. 1).

Crapo, P. A., Reaven, G., & Olefsky, J. (1976). Plasma glucose and insulin responses to orally administered simple and complex carbohydrates. *Diabetes, 25*(9), 741–747.

Das, G., Kameswaran, S., Ramesh, B., Bangeppagari, M., Nath, R., Talukdar, A. D., ... & Patra, J. K. (2024). Anti-Aging Effect of Traditional Plant-Based Food: An Overview. *Foods, 13*(23), 3785.

Geada, P., Moreira, C., Silva, M., Nunes, R., Madureira, L., Rocha, C. M., ... & Teixeira, J. A. (2021). Algal proteins: Production strategies and nutritional and functional properties. *Bioresource Technology, 332*, 125125.

Gheda, S. F., Abo-Shady, A. M., Abdel-Karim, O. H., & Ismail, G. A. (2021). Antioxidant and antihyperglycemic activity of arthrospira platensis (spirulina platensis) methanolic extract: in vitro and in vivo study. *Egyptian Journal of Botany, 61*(1), 71–93.

Gomez-Zavaglia, A., Prieto Lage, M. A., Jimenez-Lopez, C., Mejuto, J. C., & Simal-Gandara, J. (2019). The potential of seaweeds as a source of functional ingredients of prebiotic and antioxidant value. *Antioxidants, 8*(9), 406.

Goyal, R., Singh, O., Agrawal, A., Samanta, C., & Sarkar, B. (2022). Advantages and limitations of catalytic oxidation with hydrogen peroxide: from bulk chemicals to lab scale process. *Catalysis Reviews, 64*(2), 229–285.

Guru, A., Issac, P. K., Velayutham, M., Saraswathi, N. T., Arshad, A., & Arockiaraj, J. (2021). Molecular mechanism of down-regulating adipogenic transcription factors in 3T3-L1 adipocyte cells by bioactive anti-adipogenic compounds. *Molecular Biology Reports, 48*(1), 743–761.

Harnedy, P. A., & FitzGerald, R. J. (2013). In vitro assessment of the cardioprotective, anti-diabetic and antioxidant potential of Palmaria palmata protein hydrolysates. *Journal of applied phycology, 25*, 1793–1803.

Hatziagapiou, K., Kakouri, E., Lambrou, G. I., Bethanis, K., & Tarantilis, P. A. (2019). Antioxidant properties of Crocus sativus L. and its constituents and relevance to neurodegenerative diseases; focus on Alzheimer's and Parkinson's disease. *Current Neuropharmacology, 17*(4), 377–402.

Holger, T., Marie-Laurence, A., David, D., & Joël, B. (2014). Oxidative stress regulation in lichens and its relevance for survival in coastal habitats. In *Advances in Botanical Research* (Vol. 71, pp. 467–503). Academic Press.

Holte, C. F. (2024). Novel insights into the fenestrated scavenger endothelium of the liver sinusoid.

Jakus, V. (2000). The role of free radicals, oxidative stress and antioxidant systems in diabetic vascular disease. *Bratislavske lekarske listy, 101*(10), 541–551.

James, J. (2023). *Medicinal Mushrooms: The Extraction and Analysis of Bioactive Compounds from Hericium erinaceus and Ganoderma lucidum Mushrooms* (Master's thesis, University of Windsor (Canada)).

Jeon, S. M. (2016). Regulation and function of AMPK in physiology and diseases. *Experimental & molecular medicine, 48*(7), e245-e245.

Jugran, A. K., Rawat, S., Devkota, H. P., Bhatt, I. D., & Rawal, R. S. (2021). Diabetes and plant-derived natural products: From ethnopharmacological approaches to their potential for modern drug discovery and development. *Phytotherapy Research, 35*(1), 223–245.

Khursheed, R., Singh, S. K., Wadhwa, S., Kapoor, B., Gulati, M., Kumar, R., ... & Dua, K. (2019). Treatment strategies against diabetes: Success so far and challenges ahead. *European journal of pharmacology, 862*, 172625.

Kim, M. S., Kim, J. Y., Choi, W. H., & Lee, S. S. (2008). Effects of seaweed supplementation on blood glucose concentration, lipid profile, and antioxidant enzyme activities in patients with type 2 diabetes mellitus. *Nutrition research and practice, 2*(2), 62–67.

Kokil, G. R., Veedu, R. N., Ramm, G. A., Prins, J. B., & Parekh, H. S. (2015). Type 2 diabetes mellitus: limitations of conventional therapies and intervention with nucleic acid-based therapeutics. *Chemical Reviews*, *115*(11), 4719–4743.

Lee, S. H., Han, J. S., Heo, S. J., Hwang, J. Y., & Jeon, Y. J. (2010). Protective effects of dieckol isolated from Ecklonia cava against high glucose-induced oxidative stress in human umbilical vein endothelial cells. *Toxicology in vitro*, *24*(2), 375–381.

Lee, S. H., Kang, S. M., Ko, S. C., Lee, D. H., & Jeon, Y. J. (2012). Octaphlorethol A, a novel phenolic compound isolated from a brown alga, Ishige foliacea, increases glucose transporter 4-mediated glucose uptake in skeletal muscle cells. *Biochemical and biophysical research communications*, *420*(3), 576–581.

Lee, Y. S., Shin, K. H., Kim, B. K., & Lee, S. (2004). Anti-diabetic activities of fucosterol from Pelvetia siliquosa. *Archives of pharmacal research*, *27*, 1120–1122.

Leena, M. M., Silvia, M. G., Vinitha, K., Moses, J. A., & Anandharamakrishnan, C. (2020). Synergistic potential of nutraceuticals: Mechanisms and prospects for futuristic medicine. Food & function, 11(11), 9317–9337.

Liao, X., Yang, L., Chen, M., Yu, J., Zhang, S., & Ju, Y. (2015). The hypoglycemic effect of a polysaccharide (GLP) from Gracilaria lemaneiformis and its degradation products in diabetic mice. *Food & function*, *6*(8), 2542–2549.

Lin, L., & Xu, J. (2020). Fungal pigments and their roles associated with human health. *Journal of Fungi*, *6*(4), 280.

Mancabelli, L., Milani, C., De Biase, R., Bocchio, F., Fontana, F., Lugli, G. A., ... & Ventura, M. (2024). Taxonomic and metabolic development of the human gut microbiome across life stages: a worldwide metagenomic investigation. *Msystems*, *9*(4), e01294–23.

Martins-Gomes, C., Nunes, F. M., & Silva, A. M. (2024). Natural Products as Dietary Agents for the Prevention and Mitigation of Oxidative Damage and Inflammation in the Intestinal Barrier. *Antioxidants*, *13*(1), 65.

Mauricio, D., & Alonso, N. (Eds.). (2024). *Chronic complications of diabetes mellitus: current outlook and novel pathophysiological insights*. Elsevier.

Máximo, P., Ferreira, L. M., Branco, P., Lima, P., & Lourenço, A. (2018). Secondary metabolites and biological activity of invasive macroalgae of southern Europe. *Marine drugs*, *16*(8), 265.

Menaa, F., Wijesinghe, U., Thiripuranathar, G., Althobaiti, N. A., Albalawi, A. E., Khan, B. A., & Menaa, B. (2021). Marine algae-derived bioactive compounds: a new wave of nanodrugs?. *Marine drugs*, *19*(9), 484.

Montano, L., Maugeri, A., Volpe, M. G., Micali, S., Mirone, V., Mantovani, A., ... & Piscopo, M. (2022). Mediterranean diet as a shield against male infertility and cancer risk induced by environmental pollutants: a focus on flavonoids. *International Journal of Molecular Sciences*, *23*(3), 1568.

Moon, H. E., Islam, M. N., Ahn, B. R., Chowdhury, S. S., Sohn, H. S., Jung, H. A., & Choi, J. S. (2011). Protein tyrosine phosphatase 1B and α-glucosidase inhibitory phlorotannins from edible brown algae, Ecklonia stolonifera and Eisenia bicyclis. *Bioscience, biotechnology, and biochemistry*, *75*(8), 1472–1480.

Murray, P. M., Moane, S., Collins, C., Beletskaya, T., Thomas, O. P., Duarte, A. W., ... & Walsh, D. J. (2013). Sustainable production of biologically active molecules of marine based origin. *New Biotechnology*, *30*(6), 839–850.

Ojo, O. A., Ogunlakin, A. D., Gyebi, G. A., Ayokunle, D. I., Odugbemi, A. I., Babatunde, D. E., ... & Ojo, A. B. (2023). Profiling the antidiabetic potential of GC–MS compounds identified from the methanolic extract of Spilanthes filicaulis: experimental and computational insight. *Journal of Biomolecular Structure and Dynamics*, 1–22.

Oluwamodupe, C., & Babalola, O. O. (2024). Ion Channels and Regulation of Insulin Secretion in Beta-Cells. *Current Molecular Biology Reports*, *10*(4), 65–77.

Pereira, L. M., Obara, K., Dias, J. M., Menacho, M. O., Guariglia, D. A., Schiavoni, D., ... & Cardoso, J. R. (2012). Comparing the Pilates method with no exercise or lumbar stabilization for pain and functionality in patients with chronic low back pain: systematic review and meta-analysis. *Clinical rehabilitation*, *26*(1), 10–20.

Pereira, L., & Cotas, J. (2023). Therapeutic potential of polyphenols and other micronutrients of marine origin. *Marine drugs*, *21*(6), 323.

Pereira, L., & Valado, A. (2023). Algae-Derived Natural Products in Diabetes and Its Complications—Current Advances and Future Prospects. *Life*, *13*(9), 1831.

Popović-Djordjević, J. B., Katanić Stanković, J. S., Mihailović, V., Pereira, A. G., Garcia-Oliveira, P., Prieto, M. A., & Simal-Gandara, J. (2021). Algae as a source of bioactive compounds to prevent the development of type 2 diabetes mellitus. *Current Medicinal Chemistry*, *28*(23), 4592–4615.

Popper, Z. A., Michel, G., Hervé, C., Domozych, D. S., Willats, W. G., Tuohy, M. G., ... & Stengel, D. B. (2011). Evolution and diversity of plant cell walls: from algae to flowering plants. *Annual review of plant biology*, *62*(1), 567–590.

Priori, E. C. (2021). Oxidative stress and neuroinflammation effects in hippocampal development, aging, and disease.

Rahman, M. M., Islam, M. R., Rabbi, F., Islam, M. T., Sultana, S., Ahmed, M., ... & Behl, T. (2022). Bioactive compounds and diabetes mellitus: prospects and future challenges. *Current Pharmaceutical Design*, *28*(16), 1304–1320.

Rahman, M. M., Rahaman, M. S., Islam, M. R., Rahman, F., Mithi, F. M., Alqahtani, T., ... & Uddin, M. S. (2021). Role of phenolic compounds in human disease: current knowledge and future prospects. *Molecules*, *27*(1), 233.

Raman, M., Milestone, A. N., Walters, J. R., Hart, A. L., & Ghosh, S. (2011). Vitamin D and gastrointestinal diseases: inflammatory bowel disease and colorectal cancer. *Therapeutic Advances in Gastroenterology*, *4*(1), 49–62.

Ríos, J. L., Francini, F., & Schinella, G. R. (2015). Natural products for the treatment of type 2 diabetes mellitus. *Planta medica*, *81*(12/13), 975–994.

Samanta, S. (2022). Potential impacts of prebiotics and probiotics on cancer prevention. *Anti-Cancer Agents in Medicinal Chemistry (Formerly Current Medicinal Chemistry-Anti-Cancer Agents)*, *22*(4), 605–628.

Sen, S., Chakraborty, R., & De, B. (2016). *Diabetes mellitus in 21st century* (pp. 55–67). Singapore:: Springer.

Senthil, S. L., Kumar, T. V., Geetharamani, D., & Maruthupandi, T. (2013). Screening of seaweeds collected from southeast coastal area of India for α-amylase inhibitory activity, antioxidant activity and biocompatibility. *Int. J. Pharm. Pharm. Sci*, *5*, 240–244.

Shahidi, F., & Rahman, M. J. (2018). Bioactives in seaweeds, algae, and fungi and their role in health promotion. *Journal of Food Bioactives*, *2*, 58–81.

Shalaby, E. (2011). Algae as promising organisms for environment and health. *Plant signaling & behavior*, *6*(9), 1338–1350.

Sharma, S., Sahu, S., & Arya, S. K. (2024). Pretreatment and Fractionation of Algae Biomass for Value-Added Extraction. In *Value Added Products From Bioalgae Based Biorefineries: Opportunities and Challenges* (pp. 95–115). Singapore: Springer Nature Singapore.

Shi, D., Xu, F., He, J., Li, J., Fan, X., & Han, L. (2008). Inhibition of bromophenols against PTP1B and anti-hyperglycemic effect of Rhodomela confervoides extract in diabetic rats. *Chinese Science Bulletin*, *53*, 2476–2479.

Ślusarczyk, J., Adamska, E., & Czerwik-Marcinkowska, J. (2021). Fungi and algae as sources of medicinal and other biologically active compounds: A review. *Nutrients*, *13*(9), 3178.

Sohrabi, F., Saed-Moucheshi, A., Saed-Moucheshi, Z., & Pessarakli, M. (2024) Effects of Abiotic Stresses on the 26 Photosynthesis Apparatus in Plants. *energy*, *680*(680), 700.

Souza, M. R. D. A., Diniz, M. D. F. F. D. M., Medeiros-Filho, J. E. M. D., & Araújo, M. S. T. D. (2012). Metabolic syndrome and risk factors for non-alcoholic fatty liver disease. *Arquivos de gastroenterologia*, *49*, 89–96.

Suthindhiran Krish, S. K., & Abhinab Das, A. D. (2014). In-vitro bioactivity of marine seaweed, Cladophora rupestris.

Świderska, E., Strycharz, J., Wróblewski, A., Szemraj, J., Drzewoski, J., & Śliwińska, A. (2018). Role of PI3K/AKT pathway in insulin-mediated glucose uptake. *Blood glucose levels, 1*, 1–18.
Szablewski, L. (2014). Role of immune system in type 1 diabetes mellitus pathogenesis. *International immunopharmacology, 22*(1), 182–191.
Thomas, N. V., & Kim, S. K. (2013). Beneficial effects of marine algal compounds in cosmeceuticals. *Marine drugs, 11*(1), 146–164.
Tian, H., Yin, X., Zeng, Q., Zhu, L., & Chen, J. (2015). Isolation, structure, and surfactant properties of polysaccharides from Ulva lactuca L. from South China Sea. *International Journal of Biological Macromolecules, 79*, 577–582.
Tiwari, A., Melchor-Martínez, E. M., Saxena, A., Kapoor, N., Singh, K. J., Saldarriaga-Hernández, S., ... & Iqbal, H. M. (2021). Therapeutic attributes and applied aspects of biological macromolecules (polypeptides, fucoxanthin, sterols, fatty acids, polysaccharides, and polyphenols) from diatoms—A review. *International Journal of Biological Macromolecules, 171*, 398–413.
Tiwari, B. K., Pandey, K. B., Abidi, A. B., & Rizvi, S. I. (2013). Markers of oxidative stress during diabetes mellitus. *Journal of biomarkers, 2013*(1), 378790.
Trikkalinou, A., Papazafiropoulou, A. K., & Melidonis, A. (2017). Type 2 diabetes and quality of life. *World journal of diabetes, 8*(4), 120.
Wang, N., Manabe, Y., Sugawara, T., Paul, N. A., & Zhao, J. (2018). Identification and biological activities of carotenoids from the freshwater alga Oedogonium intermedium. *Food Chemistry, 242*, 247–255.
Wang, Y., Liu, X., & Li, Q. (2024). Plant Extracts for Type 2 Diabetes: Mechanisms, Clinical Implications and Future Directions—A Systematic Review. *Journal of Biobased Materials and Bioenergy, 18*(5), 771–794.
Wei, X., Pandohee, J., & Xu, B. (2024). Recent developments and emerging trends in dietary vitamin D sources and biological conversion. *Critical Reviews in Food Science and Nutrition, 64*(28), 10121–10137.
Wells, M. L., Potin, P., Craigie, J. S., Raven, J. A., Merchant, S. S., Helliwell, K. E., ... & Brawley, S. H. (2017). Algae as nutritional and functional food sources: revisiting our understanding. *Journal of applied phycology, 29*, 949–982.
Wong, C. H., Halcomb, R. L., Ichikawa, Y., & Kajimoto, T. (1995). Enzymes in organic synthesis: application to the problems of carbohydrate recognition (part 1). *Angewandte Chemie International Edition in English, 34*(4), 412–432.
Wylie-Rosett, J., Segal-Isaacson, C. J., & Segal-Isaacson, A. (2004). Carbohydrates and increases in obesity: does the type of carbohydrate make a difference?. *Obesity research, 12*(S11), 124S-129S.
Zhao, C., Yang, C., Wai, S. T. C., Zhang, Y., P. Portillo, M., Paoli, P., ... & Cao, H. (2019). Regulation of glucose metabolism by bioactive phytochemicals for the management of type 2 diabetes mellitus. *Critical Reviews in Food Science and Nutrition, 59*(6), 830–847.
Zia-Ul-Haq, M. (2021). Historical and introductory aspects of carotenoids. *Carotenoids: structure and function in the human body*, 1–42.

Chapter 7
Algae to the Rescue: Unveiling the Potential Role of Neuroprotection in Diabetic Neuropathy

Manikandan Kathirvel

Abstract Diabetes mellitus is a worldwide metabolic disorder that poses significant healthcare and financial issues due to its severe consequences and rising occurrence. Of them, diabetic neuropathy (DN) is a significant microvascular consequence characterized by damage to the nerves, resulting in incapacitating symptoms as tingling, numbness, and persistent pain. Morbidity rises and quality of life is severely reduced by DN. Recent data indicates that gut microbiota dysbiosis plays a crucial part in the development of DN and points to intricate relationships between inflammation, immunological response, and host metabolism. Pharmaceuticals and nutraceuticals are part of the current treatment approaches for diabetes and DN, although they are frequently expensive and ineffective. As a result, natural products especially marine algae have drawn more attention as possible sources of bioactive substances with neuroprotective and antidiabetic effects. Carotenoids, phlorotannins, and astaxanthin are examples of algae metabolites that have anti-inflammatory, antioxidant, and glucose-regulating properties. They may also alter metabolic pathways linked to the pathophysiology of DN. Notably, Symphyocladia latiuscula and other red algae have demonstrated promise in neuroprotection. This chapter explores the pathophysiology of DN, risk factors, conventional management approaches, and the emerging role of marine algae in combating diabetes-induced neuropathy, while addressing current challenges and limitations in translating these findings into clinical applications.

Keywords Diabetic neuropathy · Marine algae · Neuroprotection · Algal bioactive compounds · Antidiabetic agents

---

M. Kathirvel (✉)
Department of Life Sciences, School of Biological and Forensic Sciences, Kristu Jayanti (Deemed-to-be University), Bengaluru, Karnataka, India
e-mail: manikandan@kristujayanti.com

R. Raja et al. (eds.), *Algae in Diabetes Management*,
https://doi.org/10.1007/978-981-95-0571-5_7

1 Introduction

Globally, diabetes, a chronic metabolic disease marked by decreased insulin synthesis or usage, has become an epidemic. Millions of people are impacted by the illness, which places a heavy strain on economies and healthcare systems. Numerous problems linked to diabetes have a substantial negative influence on quality of life and raise rates of morbidity and death for those who have the disease (Pereira and Valado 2023; Wu et al. 2014; Tomic et al. 2022). Nearly half a billion people globally suffer from diabetes, making it a pandemic-level condition. By 2045, diabetes healthcare costs are predicted to reach $845 billion. The World Health Organization has divided it into two main categories: non-insulin-dependent diabetes mellitus (NIDDM)/type 2 diabetes mellitus (T2DM), which results from non-insulin-dependent elevated blood sugar levels, and insulin-dependent diabetes mellitus (IDDM)/type 1 diabetes mellitus (T1DM), which is caused by an insulin shortage. T2DM is the most common of them, accounting for around 90% of all cases of diabetes (Galicia-Garcia et al. 2020; Kaushik et al. 2023). Insulin resistance and insufficient insulin secretion work together to cause it. Obesity, poor eating habits, and sedentary activity are lifestyle variables that are frequently linked to type 2 diabetes (T2DM) (Kyrou et al. 2020). On the other hand, type 1 diabetes (T1DM) is an autoimmune disease that causes the pancreatic beta cells that produce insulin to be destroyed, leading to complete insulin insufficiency (Popoviciu et al. 2023).

Diabetes has a wide range of consequences that impact different organ systems. Diabetes-related chronic hyperglycemia causes both macrovascular and microvascular problems. Small blood arteries, such as those in the kidneys (diabetic nephropathy), nerves (diabetic neuropathy), and eyes (diabetic retinopathy), are the main site of microvascular problems. Large blood vessels in the body are affected by macrovascular problems, which raise the risk of cardiovascular conditions like peripheral arterial disease, diabetic neuropathy, retinopathy, nephropathy, coronary artery disease, and stroke (Cade 2008; Chawla et al. 2016). Diabetes complications that impact the neurological system include diabetic neuropathy (DN) (Thaifa et al. 2017). Sensory impairment brought on by DN includes burning, tingling, and numbness. DN alters the blood flow in the legs, raising the possibility of diabetic foot issues. With symptoms like tingling in the leg, numbness, muscle weakness, digestive problems (such as diarrhea or constipation), and urinary tract infections (UTI), diabetic nephropathy (DN) is a significant danger of hyperglycemia that harms nerves and results in severe disease, reducing patients' quality of life (Alkholifi et al. 2023). The intricate metabolic process that results in DN is brought on by persistently elevated blood glucose levels.

A new aspect of the gut microbiota in the pathophysiology of DN has been revealed by recent research. Neuropathy is closely linked to changes in the gut microbiota, which have been demonstrated to play a major influence in host metabolism, immune system function, and systemic inflammation. Dysbiosis is seen in people with diabetes or neuropathic symptoms, suggesting a link between these conditions and alterations found in gut microbial populations. Intestinal microbiota diversity and pathogens are increased by DN (Alkholifi et al. 2023). Although traditional diabetes treatment has advanced, there is increasing interest in investigating alternative therapy approaches that provide better efficacy, less adverse effects, and the ability to transform the illness (Williams et al. 2022; Davies et al. 2022). Natural products and their derivatives have drawn a lot of interest as possible diabetes control tools. These substances have a variety of chemical ingredients and a broad spectrum of pharmacological effects. They come from a variety of sources, such as plants, fungi, marine organisms, and algae (Abo-Shady et al. 2023; Quitério et al. 2021). Using algae products to their full potential in diabetes research could lead to the creation of innovative treatments and therapies that can successfully regulate blood sugar levels and lessen the consequences of the condition (Salehi et al. 2019; Liu et al. 2022). Because of their proven pharmacological qualities that are pertinent to the management of diabetes and their capacity to address many aspects of the disease, bioactive chemicals derived from algae, including polysaccharides, polyphenols, pigments, fatty acids, and sterols, are important in the study of diabetes. These substances have the ability to alter lipid profiles, increase insulin sensitivity, modify glucose metabolism, and have anti-inflammatory and antioxidant properties—all of which are critical components of diabetic care (Bocanegra et al. 2021; Pradhan et al. 2020).

Finding pharmacological or nutraceutical candidates that can be utilized to treat or prevent neurodegenerative disorders is a challenge. Although there are many targeted therapeutic modalities for diabetes treatment, including biguanides, α-glucosidase inhibitors, thiazolidinediones, non-sulphonylureas, secretagogues, incretin-based therapies, sulphonylureas, sulphonylureas, and direct insulin therapy, as well as natural remedies from nutraceuticals like curcumin, capsaicin, β-caryophyllene, chlorogenic acid (CGA), crocetin and crocin, Diosmin, Berberine (BR), and tocopherol (TOC), there is currently no cure for diabetes. Additionally, because diabetes is a chronic condition, treatments have been linked to significant financial costs. Algal metabolites are thus opening the door for the development of new antidiabetic treatments with improved clinical outcomes, which is urgently needed (Kaushik et al. 2023; Alkholifi et al. 2023; Mishra et al. 2022).

2 Diabetic Neuropathy

About 30 million people worldwide suffer from neuropathy, the most prevalent microvascular consequence and a major cause of both death and morbidity. One of the main causes of neuropathy is hyperglycemia (Bhadada et al. 2001). The absorption of glucose in neurons is largely dependent on its extracellular concentration, and their demand for it is consistently high. Their inability to go through anaerobic and glycolytic processes results in a fourfold increase in neuronal glucose. When such occurrences occur frequently or persistently, the intracellular metabolism of glucose may cause neuronal damage; this phenomenon is known as glucose neurotoxicity (Tomlinson and Gardiner 2008). All peripheral nerves, including sensory and motor neurons, are affected by diabetic peripheral neuropathy, a condition linked to chronic hyperglycemia. It is characterized by pain, paresthesia, hyperesthesia, dysesthesia, proprioceptive defect, loss of sensation, muscle weakness, and atrophy (Brown and Asbury 1984).

3 Classification of Diabetic Neuropathy

About 30% of diabetic individuals have neuropathy, and up to 50% of diabetics will eventually acquire neuropathy as their condition worsens. Diabetes destroys the peripheral nervous system's nerves; these conditions are categorized as localized and diffuse neuropathy. Diabetic autonomic neuropathy (DAN) and diabetic distal symmetrical polyneuropathy (DPN) are examples of diffuse neuropathies. According to the study, DPN is more common in people with T2DM than in T1DM, and its incidence rises with age. Small-fiber predominant neuropathy, large-fiber neuropathy, and mixed type of neuropathy are among the subclasses of DPN (Edwards et al. 2008; Callaghan et al. 2012). DPN patients frequently experience tingling, discomfort, weakness, numbness, and balance issues. They are frequently susceptible to ulcer development (Gordois et al. 2003; Van Acker et al. 2009). Stocking and glove distribution is the term used to describe how DPN pain spreads from the foot in a length-dependent manner. Small-fiber predominant neuropathy is the initial sign of peripheral nerve damage that gradually develops into DPN. It causes axonal loss in the tiny sensory nerves.

In individuals with diabetes, diabetic autonomic neuropathy (DAN) is a subtype of diffuse diabetic neuropathy and small-fiber predominant neuropathy. Any sympathetic or parasympathetic autonomic function of the autonomic nerve system (ANS) may be compromised by DAN. DAN can impact every bodily system because the ANS provides input to every organ. Heart arrhythmias, dry skin, heat intolerance, anhidrosis, erectile dysfunction, female sexual dysfunction, constipation, urine retention, and gastroparesis are among the prevalent symptoms. According to reports, DAN is very common, and autonomic neuropathy sufferers have a high chance of dying (Freeman 2005; Soedamah-Muthu et al. 2008). Diabetic amyotrophy (radiculoplexopathy), mononeuropathy, mononeuritis multiplex, and treatment-induced neuropathy are among the subtypes of focal neuropathy (Edwards et al. 2008; Callaghan et al. 2012). Either the cervical plexus or the lumbosacral area of the spine is impacted by diabetic radiculoplexopathy. Therapy-induced neuropathy is the term for neuropathy that can result from diabetes therapy, even if poor glucose control is linked to an increased risk of neuropathy.

4 Risk Factors for the Development of Neuropathy

Patients with diabetes who have metabolic syndrome, which includes hyperglycemia, hypertriglyceridemia, dyslipidemia, obesity, and hypertension are more likely to develop neuropathy.

4.1 Hyperglycemia

It is probable that biochemical pathways play a significant role in the emergence of problems from diabetes. It is obvious that hyperglycemia is a major factor in the onset and advancement of diabetic neuropathy as well as the other microvascular problems associated with diabetes. Increased flux via numerous glucose metabolism pathways, including glycolysis, polyol, and hexosamine, processes excess intracellular glucose.

4.1.1 Glycolysis Pathway

According to Vincent et al. (2004), excessive glycolysis produces too much NADH, which can overwhelm the mitochondrial electron transport chain. Additionally, the production of reactive oxygen species (ROS) damages mitochondria and activates the poly (ADP-ribose) polymerase pathway (PARP). ROS-induced PARP activation causes inflammation and neuronal impairment via interacting with the hexosamine and PKC pathways (Vincent et al. 2011).

4.1.2 Polyol Pathway

One alternative pathway for the metabolism of glucose that has been found to play a significant role in the development of neuropathy is the polyol pathway (Gabbay 1973). Oxidative stress, decreased NADPH levels, and increased cellular osmolarity can result from the increased flow through the polyol pathway (Obrosova 2005). Diabetic neuropathy has been linked to increased non-enzymatic glycation (Brownlee et al. 1988; Yagihashi 1995; Yagihashi et al. 2007) and polyol pathway overactivity (Greene et al. 1993; Cameron et al. 1997). Sorbitol dehydrogenase converts glucose to fructose via the polyol route, which is catalyzed by the enzyme aldose reductase. Since sorbitol is impermeable to cell membranes and cannot diffuse out, it accumulates in AR-containing tissues and peripheral nerves due to hyperglycemia, which has been linked to aldose reductase overactivity. This leads to hyperosmotic stress on the cell, which ultimately results in neuropathic pain, a feature of diabetic neuropathy (Gabbay et al. 1966; Kinoshita and Nishimura 1988). Aldose reductase inhibitors have been proven in animal models to avoid a number of problems, such as cataract, neuropathy, and nephropathy (Oates and Mylari 1999).

Other organic electrolytes that control cellular osmolality, such as taurine and myo-inositol, are depleted when intracellular sorbitol and fructose build up (Stevens et al. 1993). Peripheral nerve myo-inositol depletion disrupts phosphoinositide synthesis, resulting in insufficient diacylglycerol to maintain protein kinase C (PKC) levels, which are necessary for $Na^{+}K^{+}$-ATPase activation (Greene et al. 1987; Zhu and Eichberg 1990). Peripheral polyneuropathy is the result of these anomalies, which cause long-term, progressive damage and loss to both myelinated and unmyelinated peripheral nerve fibers (Low et al. 1997; Nagamatsu et al. 1995).

Long-term hyperglycemia causes advanced glycation end products (AGEs) to build up, which tends to hinder proteins' biological function and impact cellular function. The extracellular and intracellular AGE biomolecules trigger the production of NADPH oxidases, and influence the long-term inflammatory signaling cascades and oxidative stress. When extracellular AGEs, like plasma and matrix proteins, attach to the cell surface receptor for AGE (RAGE), the transcription factor like nuclear factor kappa B (NFκB) gets activated by AGE–RAGE interaction (Vincent et al. 2007; Sugimoto et al. 2008; Bierhaus and Nawroth 2009).

4.2 Modulation of PKC Family of Protein Kinases

A condition of complication in diabetic peripheral neuropathy is hyperglycemia, which is a crucial signaling event in the activation of the protein kinase C (PKC) family of protein kinases (Sheetz and King 2002; Ramana et al. 2004). Researchers have looked into how the polyol pathway contributes to PKC activation brought on by high glucose. Through neurovascular processes including blood flow and conduction velocity, PKC contributes to diabetic neuropathy. It was discovered that PKC overexpression lowers Na^+K^+-ATPase activity, which in turn reduces neuronal transmission and regeneration. After receiving a PKC inhibitor, the symptoms returned to normal (Lehning et al. 1994; Hermenegildo et al. 1992). PKC isoforms including α, −β1, −β2, −γ, −δ, and -ε have been shown to undergo membrane translocation in response to high glucose, which is followed by an increase in their phosphorylation (Borghini et al. 1994; Roberts and McLean 1997). High glucose-induced membrane translocation and PKC isoform phosphorylation were both inhibited by AR inhibitor treatment. Aldose reductase inhibition plays a crucial role in the study's findings, which show a significant decrease in the expression of PKC and AR proteins in the sciatic nerves of diabetic animals treated with red algae compounds compared to the diabetic animals (Samaddar and Koneri 2019). Thus accordingly, the AGE, polyol, hexosamine, protein kinase C (PKC), and PARP pathways all work together to cause neuronal damage, which includes impaired motor and sensory nerve transmission, reduced blood flow in nerves, resulting in death of the individual.

4.3 Dyslipidemia

A metabolic condition of aberrant blood lipids, dyslipidemia arises when the blood's lipid or lipoprotein levels are either high or low. Patients with type 2 diabetes are more likely to have dyslipidemia, which is linked to diabetic neuropathy. In vitro, it has been demonstrated that free fatty acids directly harm Schwann cells, which encourages adipocytes and macrophages to release inflammatory cytokines. Bad cholesterol called low-density lipoproteins (LDLs) undergo modification and attach to extracellular receptors like the LDL receptor LOX1, the toll-like receptor, and RAGE. This sets off signaling cascades that activate NADPH oxidase, which results in oxidative stress and neuronal apoptosis (Callaghan et al. 2012).

4.4 Impact of Cytokines (IL-6, IL-1β, and TNF-α) on Diabetic Neuropathy

Pro-inflammatory cytokines (IL-6, IL-1β, and TNF-α) are overexpressed in diabetic neuropathy, which directly raises nerve excitability and damages myelin, resulting in edema and further immune cell infiltration (Alexandraki et al. 2008). It has been discovered that the production of IL-1β in damaged nerves directly sensitizes primary afferent neurons' nociceptors (Binshtok et al. 2008). Since STAT3 is the primary mediator of signal transduction of pain in glial cells following peripheral injury, IL-6 can also exacerbate neuropathic pain in the dorsal horn by activating the STAT3 signaling pathway. Moreover, IL-6 directly causes pain by increasing the sensitivity of nerve endings (Xu et al. 1997). There is evidence that neuropathic pain and neuroinflammation can be effectively managed by inhibiting the production of cytokines (Ayepola et al. 2013). According to research by Samaddar and Koneri (2019), neuropathic animals treated with red algae Symphyocladia latiuscula polyphenols (SLPP) showed a dose-dependent decrease in the production of all three pro-inflammatory cytokines (IL-6, IL-1β, and TNF-α) in the sciatic nerve. To evaluate the treatment efficacy in human subjects, however, clinical trials must be conducted.

5 Management of Diabetes-Induced Neuropathy

A comprehensive strategy that includes lifestyle changes, glucose-lowering drugs, and focused therapies to address certain issues is needed to manage diabetes and associated complications. In the treatment of diabetic peripheral neuropathy, nutraceuticals derived from a variety of sources are essential. Nutraceutical treatment has proven to be an effective approach for stopping the progression. Numerous nutraceuticals have been shown to have strong antioxidant and anti-inflammatory qualities, which may prevent the early diabetes-driven molecular pathways that cause DPN, according to both in vitro and in vivo research. Nutraceuticals that come from a range of sources, such as plants, animals, and marine life, have been utilized effectively to cure DPN and safeguard health. The phrase "neuro-nutraceuticals" describes a wide variety of naturally occurring compounds found in food, herbal products, and dietary supplements that have medicinal properties. These substances are intended to improve nerve function, treat diabetic neuropathy, and help manage and prevent diseases.

6 Marine Algae as Potential Source of Bioactive Compounds

The oceans contain 90% of the world's living biomass, and marine species account for almost half of all biodiversity worldwide. Over the past few decades, people have become interested in marine algae as potential renewable resources. Globally, marine algae have been found to belong to about 8000 different classes (Naik et al. 2024). Microalgae, which include blue-green algae, dinoflagellates, and Bacillariophyta (diatoms), and macroalgae, which include seaweed, are divided into three classes based on their pigmentation: brown, red, and green algae. These groups are called Phaeophyceae, Rhodophyceae, and Chlorophyceae, respectively (Guardado Yordi et al. 2024). The majority of the estimated 1–10 million algal species on earth are microalgae. In addition to producing beneficial compounds, they can develop in a variety of environments, including rivers, lakes, wastes, and seas, and because they can withstand extreme environmental circumstances, microalgae have amassed a great deal of diversity (Kaushik et al. 2023). Algae offer a wide range of potential uses in medicine, including reducing inflammation, treating diabetes, fighting cancer, preventing obesity, and acting as antioxidants (Saddiqa et al. 2024). It is well known that marine algae are abundant sources of physiologically active, structurally varied chemicals with significant potential for use in medicine and medicine.

Scientists have found that marine algal-derived compounds are abundant in minerals (sodium, potassium, calcium, magnesium), vitamins, proteins, amino acids, dietary fiber, essential fatty acids (docosahexaenoic acid (DHA), eicosapentaenoic acid (EPA)), polysaccharides (fucoidan, laminarin, carrageenan), pigments, and a variety of phytochemicals (phytosterols, polyphenols, terpenoids, carotenoids, and tocopherols) that have antibacterial, anticoagulant, antiviral, antioxidant, anti-allergic, anti-cancer, anti-inflammatory, anti-obesity, anthelminthic, antifungal, neuroprotective, antifungal, and cytostatic properties (Pangestuti and Kim 2011; Martins et al. 2022).

7 Neuroprotective Effects of Marine Algae

According to estimates, by the 2040s, neurodegenerative illnesses would overtake cancer as the second leading cause of death for older adults (Pangestuti and Kim 2011; Bjarkam et al. 2001; Ansari et al. 2010). Nervous system failure and progressive death of neuronal cells are hallmarks of a diverse set of illnesses known as human neurodegenerative diseases. In order to prevent and treat neurodegenerative disorders, neuroprotective substances are crucial. Neuroprotective medications include antioxidants, protein aggregation inhibitors, anti-inflammatory, anti-excitotoxicity, matrix metalloproteinase (MMP) inhibitors, and enzyme inhibitors (Brown et al. 2005; Jomova et al. 2010; Lai et al. 2014; Baune 2015).

Through a variety of mechanisms, including cholinesterase inhibitors, antioxidant and anti-neuroinflammatory actions, and the prevention of neuron death, marine algae are currently playing a significant role in the field of neuroprotection. Antioxidants obtained from marine algae may be employed as preventative neuroprotective agents, according to studies. Because of their structural uniqueness and functional diversity, algal metabolites such as phenolics, alkaloids, terpenoids, carotenoids, phytosterols, and polysaccharides have garnered a lot of interest in medicinal chemistry. In preclinical models of neurodegenerative disorders, diabetes, and other conditions, these biofunctional molecules have demonstrated neuroprotection. Research points to algal metabolites—specifically, fucoxanthin, fucosterol, and fucoidan—as possible therapeutic targets for central nervous system disorders (Hannan et al. 2020).

The ability of marine red, brown, and green algae to effectively suppress cholinesterase (ChE) activity makes them potentially useful as functional neuroprotective agents. They are credited with biological functions such as preventing T2DM. While species like *Ecklonia stolonifera* and *Ishigeokamurae* have IC_{50} ranging from 17.11 to 389.1 μM (Stirk et al. 2007; Yoon et al. 2008; Yoon et al. 2009; Suganthy et al. 2010), notable marine algae species like *Caulerpa racemosa, Codium capitatum, Ulva fasciata, Ulva reticulate, Halimeda cuneata, Amphioroa ephedraea, Amphiora bowerbankii, Dictyota humifusa, Hypnea valentiae, Padina gymnospora,* and *Gracilaria edulis* have the potential to exhibit acetylcholinesterase inhibitory activities with IC_{50} ranging from 2.6 to 10 mg/mL.

7.1 Algal Molecules as Antidiabetic Agents

7.1.1 Carotenoids

Recently, carotenoids have been suggested as possible neuroprotective substances. Numerous carotenoids, including lutein, astaxanthin, fucoxanthin, and β-carotene, have been associated with antioxidant qualities. It has been observed that extracts from the microalgae D. salina that are rich in β-carotene and carotenoids may have the ability to inhibit AChE. Similarly, extracts from various carotenoids, including *Nannochloropsis oceanica, Tisochrysis lutea, Porphyridium cruentum,* and *Haematococcus pluvialis* were assessed as possible inhibitors of acetylcholinesterase (AChE) and butyrylcholinesterase (BChE) (Bueno et al. 2020). With an IC_{50} value of 47.17 μg/mL, *T. lutea* extract showed the strongest anti-AChE activity when compared to the other microalgae extracts under study. N. oceanic extract came in second with an IC_{50} of 66.29 μg/mL (Gallego et al. 2022).

7.1.2 Phlorotannins

Marine species, particularly those belonging to the brown algae groups *Fucaceae, Sargassaceae*, and *Araliaceae,* are rich in phenoltannins, a class of polyphenolic metabolites (Imbs and Zvyagintseva 2018). They can be found alone or in combination with other algal substances, such as alginic acid. In essence, phlorotannins are phloroglucinol polymers with varying degrees of polymerization. Because of their potential applications in medicine and health benefits, phlorotannins have been highly sought after in studies. These have been demonstrated to possess a broad range of biological activities, such as inhibitory effects on metalloproteinase and hyaluronidase enzymes, anticancer, antidiabetic, antioxidant, antiallergic, antihypertensive, anti-HIV, and radioprotective properties (Thomas and Kim 2011; Wijesekara et al. 2010). The study might pave the way for further research into phlorotannins' potential to reduce neuropathic pain and diabetes.

7.1.3 Astaxanthin (AX)

According to a study done on mice, giving the powerful antioxidant astaxanthin (AX) intraperitoneally restored blood glucose, oxidative stress, and mitochondrial dysfunction. It also made the behavioral abnormalities in animals with diabetic neuropathy (DN) worse. Additionally, after four weeks of administration, astaxanthin improved antioxidant defenses such as superoxide dismutase (SOD), catalase (CAT), glutathione reductase (GSH), glutathione peroxidase (GPx), thioredoxin reductase (TRR), and acetylcholinesterase (AChE) and significantly decreased blood glucose, reactive oxygen species (ROS), malondialdehyde (MDA), and nitric oxide (NO) levels (Gaur et al. 2022).

7.2 *Neuroprotective Role of the Red Alga* Symphyocladia latiuscula

One of the most abundant sources of structurally varied natural products is marine algae. Novel compounds have been identified from marine algae in greater numbers in recent years, and several of these compounds have been shown to have various biological functions (El Gamal 2010; Güven et al. 2010; Liu et al. 2011a, b; Wijesekara et al. 2011). It has been discovered that polyphenols that were separated from different marine macroalgae exhibit anti-diabetic properties (Bocanegra et al. 2009). High levels of bromophenols have been found in the red alga *Symphyocladia latiuscula*, which has been shown to have antibacterial (Kurata and Amiya 1980), antiviral (Park et al. 2005), anticancer (Lee et al. 2007), α-glucosidase inhibition (Kurihara et al. 1999), and free-radical scavenging properties (Duan et al. 2007; Xu et al. 2013). Activity of α-glucosidase is inhibited by the bromophenols. The interaction's primary driving forces are hydrogen and hydrophobic bonds, which cause the enzyme to undergo slight conformational changes (Liu et al. 2011a, 2011b). It has been observed that bromophenols have the ability to inhibit α-glucosidase and aldose reductase, which can help regulate hyperglycemia and alleviate neuropathic pain in patients having diabetic peripheral neuropathy (Kurihara et al. 1999; Kwon et al. 2008; Lee et al. 2009).

The polyphenolic components of *Symphyocladia latiuscula* (Choi et al. 2000; Jin et al. 2008) were demonstrated to have neuroprotective action in diabetic peripheral neuropathy in the study (Samaddar and Koneri 2019). *Symphyocladia latiuscula* polyphenols (SLPP) were studied for their protective effects on aldose reductase activity, compound muscle action potential (CMAP), nerve conduction velocity (NCV), and intraneural sorbitol buildup in peripheral nerves (sciatic nerve). NCV has been shown to drop dramatically in diabetic neuropathy (Sima and Sugimoto 1999; Coste et al. 1999). The neuromuscular electrophysiology (NCV of the sciatic nerve and CMAP of the gastrocnemius muscle) of diabetic mice is improved in experimental animals treated with *Symphyocladia latiuscula* polyphenols (SLPP), suggesting that it has a mitigating effect. In neuron physiology, a decrease in Na^+K^+-ATPase activity has major ramifications. Reduced NCV has been seen as a result of altered normal membrane axon repolarization following action potential-induced repolarization caused by decreased sciatic nerve Na^+K^+-ATPase activity (Greene et al. 1988; Kowluru et al. 1989). According to the study, sciatic nerve Na^+K^+-ATPase activity was greatly restored after SLPP treatment. Similarly, when pre-germinated brown rice was given to STZ-induced diabetic rats, decreased Na^+K^+-ATPase activity returned to normal (Usuki et al. 2007). Animals with diabetes have been shown to develop reduced conduction velocity in their motor and sensory nerves (Eliasson 1964). Aldose reductase inhibitor therapy has prevented or reversed the conduction impairment (Tomlinson et al. 1982).

Reduced Na^+K^+-ATPase activity and motor nerve conduction velocity (MNCV) of the caudal nerves in STZ-diabetic rats were also observed to be associated with sorbitol buildup, a result of Aldose reductase overactivity, under hyperglycemia conditions. An AR inhibitor administered strategically increased Na^+K^+-ATPase activity and restored the decreased MNCV (Hirata and Okada 1990). After administering SLPP to neuropathic rats, researchers observed a notable decrease in the production of Aldose reductase and a decrease in sciatic nerve AR activity (Samaddar and Koneri 2019). Therefore, the fact that SLPP, which is produced by the red algae *Symphyocladia latiuscula*, attenuates the two targets of sorbitol buildup and aldose reductase overactivity in the polyol pathway implies that SLPP may be utilized as a possible AR inhibitor in diabetic neuropathy treatment.

8 Challenges and Limitations

Given the antioxidant, anti-inflammatory, and neuroprotective qualities of marine algae, the utilization of algal bioactive compounds as possible neuroprotective agents in diabetic-induced neuropathy (DPN) offers a viable treatment option.

However, the effective development and use of these chemicals in the treatment of DPN is hampered by a number of obstacles and restrictions. The absence of strong clinical data is one of the main obstacles. There aren't many human therapeutic trials, despite the fact that a large number of in vitro and animal research indicate that algae bioactive chemicals can reduce oxidative stress, inflammation, and neuronal damage, all of which are important elements in the pathophysiology of DPN. More extensive, randomized clinical trials are required to determine the efficacy and safety because physiological variations may cause the results seen in animal models to not fully transfer to human patients.

The bioavailability of these bioactive substances is still another important obstacle. Numerous substances included in marine algae, including peptides, polyphenols, and polysaccharides, have low bioavailability, which means the body finds it difficult to absorb or use them. Many of these substances significantly degrade in the gastrointestinal tract after oral administration, which restricts their potential for therapeutic use. Their efficacy in treating neurodegenerative diseases like DPN is further diminished by the fact that, even if absorbed, they frequently have little capacity to pass through the blood-brain barrier (BBB), a crucial barrier that protects the central nervous system. Furthermore, because these chemicals can break down when exposed to light, heat, or oxygen, their stability during storage and during the body's metabolism is also an issue. The clinical use of algae bioactive chemicals is also significantly hampered by safety and toxicity issues. Although these substances might have short-term neuroprotective effects, little is known about their long-term safety. Pre-existing diseases like kidney or liver dysfunction are common in diabetic individuals and may impact how algal-derived chemicals are metabolized, potentially resulting in toxicity or negative reactions. Additional safety concerns are raised by the possibility of drug interactions with other prescriptions that are frequently given to diabetic patients.

In summary, whereas marine algae present a promising source of bioactive chemicals with neuroprotective qualities, there are still several obstacles in the way of their advancement as effective therapies for diabetic-induced neuropathy. To fully realize the therapeutic potential of algal bioactive chemicals in DPN, challenges pertaining to clinical validation, bioavailability, standardization, safety, and regulation must be resolved.

References

Abo-Shady, A. M., Gheda, S. F., Ismail, G. A., Cotas, J., Pereira, L., & Abdel-Karim, O. H. (2023). Antioxidant and antidiabetic activity of algae. Life, 13(2), 460.

Alexandraki, K. I., Piperi, C., Ziakas, P. D., Apostolopoulos, N. V., Makrilakis, K., Syriou, V. & Kalofoutis, A. (2008). Cytokine secretion in long-standing diabetes mellitus type 1 and 2: associations with low-grade systemic inflammation. *Journal of clinical immunology*, *28*, 314–321.

Alkholifi, F. K., Aodah, A. H., Foudah, A. I., & Alam, A. (2023). Exploring the Therapeutic Potential of Berberine and Tocopherol in Managing Diabetic Neuropathy: A Comprehensive Approach towards Alleviating Chronic Neuropathic Pain. *Biomedicines*, *11*(6), 1726.

Ansari, J. A., Siraj, A., & Inamdar, N. N. (2010). Pharmacotherapeutic approaches of Parkinson's disease. *International Journal of Pharmacology*, *6*(5), 584–590.

Ayepola, O. R., Chegou, N. N., Brooks, N. L., & Oguntibeju, O. O. (2013). Kolaviron, a Garcinia biflavonoid complex ameliorates hyperglycemia-mediated hepatic injury in rats via suppression of inflammatory responses. *BMC Complementary and Alternative Medicine*, *13*, 1–9.

Bhadada, S. K., Sahay, R. K., Jyotsna, V. P., & Agrawal, J. K. (2001). Diabetic neuropathy: current concepts. *J Indian Acad Clin Med*, *2*(4), 305–18.

Baune, B. T. (2015). Inflammation and neurodegenerative disorders: is there still hope for therapeutic intervention?. *Current opinion in psychiatry*, *28*(2), 148–154.

Bierhaus, A., & Nawroth, P. P. (2009). Multiple levels of regulation determine the role of the receptor for AGE (RAGE) as common soil in inflammation, immune responses and diabetes mellitus and its complications. *Diabetologia*, *52*, 2251–2263.

Binshtok, A. M., Wang, H., Zimmermann, K., Amaya, F., Vardeh, D., Shi, L & Samad, T. A. (2008). Nociceptors are interleukin-1β sensors. *Journal of Neuroscience*, *28*(52), 14062–14073.

Bjarkam, C. R., Sørensen, J. C., Sunde, N. Å., Geneser, F. A., & Østergaard, K. (2001). New strategies for the treatment of Parkinson's disease hold considerable promise for the future management of neurodegenerative disorders. *Biogerontology*, *2*, 193–207.

Bocanegra, A., Macho-González, A., Garcimartín, A., Benedí, J., & Sánchez-Muniz, F. J. (2021). Whole alga, algal extracts, and compounds as ingredients of functional foods: composition and action mechanism relationships in the prevention and treatment of type-2 diabetes mellitus. *International Journal of Molecular Sciences*, *22*(8), 3816.

Bocanegra, A., Bastida, S., Benedi, J., Rodenas, S., & Sanchez-Muniz, F. J. (2009). Characteristics and nutritional and cardiovascular-health properties of seaweeds. *Journal of medicinal food*, *12*(2), 236–258.

Borghini, I., Ania-Lahuerta, A., Regazzi, R., Ferrari, G., Gjinovci, A., Wollheim, C. B., & Pralong, W. F. (1994). α, βI, βII, δ, and ε protein kinase C isoforms and compound activity in the sciatic nerve of normal and diabetic rats. *Journal of neurochemistry*, *62*(2), 686–696.

Brown, M. J., & Asbury, A. K. (1984). Diabetic neuropathy. *Annals of Neurology: Official Journal of the American Neurological Association and the Child Neurology Society*, *15*(1), 2–12.

Brownlee, M., Cerami, A., & Vlassara, H. (1988). Advanced glycosylation end products in tissue and the biochemical basis of diabetic complications. *New England Journal of Medicine*, *318*(20), 1315–1321.

Brown, R. C., Lockwood, A. H., & Sonawane, B. R. (2005). Neurodegenerative diseases: an overview of environmental risk factors. *Environmental health perspectives*, *113*(9), 1250–1256.

Bueno, M., Vitali, C., Sánchez-Martínez, J. D., Mendiola, J. A., Cifuentes, A., Ibáñez, E., & Herrero, M. (2020). Compressed CO_2 technologies for the recovery of carotenoid-enriched extracts from *Dunaliella salina* with potential neuroprotective activity. *ACS Sustainable Chemistry & Engineering*, *8*(30), 11413–11423.

Cade, W. T. (2008). Diabetes-related microvascular and macrovascular diseases in the physical therapy setting. *Physical therapy*, *88*(11), 1322–1335.

Callaghan, B. C., Cheng, H. T., Stables, C. L., Smith, A. L., & Feldman, E. L. (2012). Diabetic neuropathy: clinical manifestations and current treatments. *The Lancet Neurology*, *11*(6), 521–534.

Cameron, N. E., Cotter, M. A., Basso, M., & Hohman, T. C. (1997). Comparison of the effects of inhibitors of aldose reductase and sorbitol dehydrogenase on neurovascular function, nerve conduction and tissue polyol pathway metabolites in streptozotocin-diabetic rats. *Diabetologia*, *40*, 271–281.

Chawla, A., Chawla, R., & Jaggi, S. (2016). Microvasular and macrovascular complications in diabetes mellitus: distinct or continuum?. *Indian Journal of Endocrinology and Metabolism*, *20*(4), 546–551.

Choi, J. S., Park, H. J., Jung, H. A., Chung, H. Y., Jung, J. H., & Choi, W. C. (2000). A cyclohexanonyl bromophenol from the red alga *Symphyocladia latiuscula*. *Journal of Natural Products*, *63*(12), 1705–1706.

Coste, T., Pierlovisi, M., Leonardi, J., Dufayet, D., Gerbi, A., Lafont, H. & Raccah, D. (1999). Beneficial effects of gamma linolenic acid supplementation on nerve conduction velocity, Na+, K+ ATPase activity, and membrane fatty acid composition in sciatic nerve of diabetic rats. *The Journal of Nutritional Biochemistry*, *10*(7), 411–420.

Davies, M. J., Aroda, V. R., Collins, B. S., Gabbay, R. A., Green, J., Maruthur, N. M. & Buse, J. B. (2022). Management of hyperglycemia in type 2 diabetes, 2022. A consensus report by the American Diabetes Association (ADA) and the European Association for the Study of Diabetes (EASD). *Diabetes Care*, *45*(11), 2753–2786.

Duan, X. J., Li, X. M., & Wang, B. G. (2007). Highly brominated mono-and bis-phenols from the marine red alga *Symphyocladia latiuscula* with radical-scavenging activity. *Journal of Natural Products*, *70*(7), 1210–1213.

Edwards, J. L., Vincent, A. M., Cheng, H. T., & Feldman, E. L. (2008). Diabetic neuropathy: mechanisms to management. *Pharmacology & Therapeutics*, *120*(1), 1–34.

Eliasson, S. G. (1964). Nerve conduction changes in experimental diabetes. *The Journal of Clinical Investigation*, *43*(12), 2353–2358.

El Gamal, A. A. (2010). Biological importance of marine algae. *Saudi Pharmaceutical Journal*, *18*(1), 1–25.

Freeman, R. (2005). Autonomic peripheral neuropathy. *The Lancet*, *365*(9466), 1259–1270.

Gabbay, K. H. (1973). The sorbitol pathway and the complications of diabetes. *New England Journal of Medicine*, *288*(16), 831–836.

Gabbay, K. H., Merola, L. O., & Field, R. A. (1966). Sorbitol pathway: presence in nerve and cord with substrate accumulation in diabetes. *Science*, *151*(3707), 209–210.

Galicia-Garcia, U., Benito-Vicente, A., Jebari, S., Larrea-Sebal, A., Siddiqi, H., Uribe, K. B. & Martín, C. (2020). Pathophysiology of type 2 diabetes mellitus. *International Journal of Molecular Sciences*, *21*(17), 6275.

Gallego, R., Valdés, A., Suárez-Montenegro, Z. J., Sánchez-Martínez, J. D., Cifuentes, A., Ibáñez, E., & Herrero, M. (2022). Anti-inflammatory and neuroprotective evaluation of diverse microalgae extracts enriched in carotenoids. *Algal Research*, *67*, 102830.

Gaur, S., Gaur, S., Mishra, R., Singh, R. K., & Bajpai, S. (2022). Astaxanthin reduces oxidative stress and alleviates diabetic neuropathy in STZ-induced diabetic mice. *International Journal of Diabetes in Developing Countries*, 1–11.

Gordois, A., Scuffham, P., Shearer, A., Oglesby, A., & Tobian, J. A. (2003). The health care costs of diabetic peripheral neuropathy in the US. *Diabetes care*, *26*(6), 1790–1795.

Greene, D. A., Lattimer, S. A., & Sima, A. A. (1987). Sorbitol, phosphoinositides, and sodium-potassium-ATPase in the pathogenesis of diabetic complications. *New England Journal of Medicine*, *316*(10), 599–606.

Greene, D.A., Lattimer, S.A., Sima, A.A.(1988) Are disturbances of sorbitol, phosphoinositide, and Na+-K+-ATPase regulation involved in pathogenesis of diabetic neuropathy, *Diabetes* 37, 688–693.

Greene, D. A., Sima, A. A. F., Stevens, M. J., Feldman, E. L., Killen, P. D., Henry, D. N. & Lattimer, S. A. (1993). Aldose reductase inhibitors: an approach to the treatment of diabetic nerve damage. *Diabetes/ Metabolism Reviews*, *9*(3), 189–217.

Guardado Yordi, E., Pérez Martínez, A., Radice, M., Scalvenzi, L., Abreu-Naranjo, R., Uriarte, E & Matos, M. J. (2024). Seaweeds as Source of Bioactive Pigments with Neuroprotective and/or Anti-Neurodegenerative Activities: Astaxanthin and Fucoxanthin. *Marine Drugs*, *22*(7), 327.

Güven, K. C., Percot, A., & Sezik, E. (2010). Alkaloids in marine algae. *Marine Drugs*, *8*(2), 269–284.

Hannan, M. A., Dash, R., Haque, M. N., Mohibbullah, M., Sohag, A. A. M., Rahman, M. A & Moon, I. S. (2020). Neuroprotective potentials of marine algae and their bioactive metabolites: Pharmacological insights and therapeutic advances. *Marine Drugs*, *18*(7), 347.

Hermenegildo, C., Felipo, V., Miñana, M. D., & Grisolía, S. (1992). Inhibition of protein kinase C restores Na+, K+-ATPase activity in sciatic nerve of diabetic mice. *Journal of Neurochemistry*, *58*(4), 1246–1249.

Hirata, Y., & Okada, K. (1990). Relation of Na+, K+-ATPase to delayed motor nerve conduction velocity: Effect of aldose reductase inhibitor, ADN-138, on Na^+,K^+-ATPase activity. *Metabolism, 39*(6), 563–567.

Imbs, T. I., & Zvyagintseva, T. N. (2018). Phlorotannins are polyphenolic metabolites of brown algae. *Russian Journal of Marine Biology, 44*, 263–273.

Jin, H. J., Oh, M. Y., Jin, D. H., & Hong, Y. K. (2008). Identification of a Taq DNA polymerase inhibitor from the red seaweed Symphyocladia latiuscula. *J. Environ. Biol, 29*, 475–478.

Jomova, K., Vondrakova, D., Lawson, M., & Valko, M. (2010). Metals, oxidative stress and neurodegenerative disorders. *Molecular and Cellular Biochemistry, 345*, 91–104.

Kaushik, A., Sangtani, R., Parmar, H. S., & Bala, K. (2023). Algal metabolites: Paving the way towards new generation antidiabetic therapeutics. *Algal Research, 69*, 102904.

Kinoshita, J. H., & Nishimura, C. (1988). The involvement of aldose reductase in diabetic complications. *Diabetes/metabolism Reviews, 4*(4), 323–337.

Kowluru, R., Bitensky, M. W., Kowluru, A., Dembo, M., Keaton, P. A., & Buican, T. (1989). Reversible sodium pump defect and swelling in the diabetic rat erythrocyte: effects on filterability and implications for microangiopathy. *Proceedings of the National Academy of Sciences, 86*(9), 3327–3331.

Kurata, K., & Amiya, T. (1980). Disodium 2, 3, 6-Tribromo-5-Hydroxybenzyl 1′, 4-Distjlfate, A New Bromophenol from the Red Alga, *Symphyocladia latiuscula*. *Chemistry Letters, 9*(3), 279–280.

Kurihara, H., Mitani, T., Kawabata, J., & Takahashi, K. (1999). Two new bromophenols from the red alga *Odonthalia corymbifera*. *Journal of Natural Products, 62*(6), 882–884.

Kwon, Y. I., Apostolidis, E., & Shetty, K. (2008). Inhibitory potential of wine and tea against α-amylase and α-glucosidase for management of hyperglycemia linked to type 2 diabetes. *Journal of Food Biochemistry, 32*(1), 15–31.

Kyrou, I., Tsigos, C., Mavrogianni, C., Cardon, G., Van Stappen, V., Latomme, J., & Manios, Y. (2020). Sociodemographic and lifestyle-related risk factors for identifying vulnerable groups for type 2 diabetes: a narrative review with emphasis on data from Europe. *BMC Endocrine Disorders, 20*, 1–13.

Lai, T. W., Zhang, S., & Wang, Y. T. (2014). Excitotoxicity and stroke: identifying novel targets for neuroprotection. *Progress in Neurobiology, 115*, 157–188.

Lee, J. H., Park, S. E., Hossain, M. A., Kim, M. Y., Kim, M. N., Chung, H. Y. & Kim, N. D. (2007). 2, 3, 6-Tribromo-4, 5-dihydroxybenzyl methyl ether induces growth inhibition and apoptosis in MCF-7 human breast cancer cells. *Archives of Pharmacal Research, 30*, 1132–1137.

Lee, S. H., Karadeniz, F., Kim, M. M., & Kim, S. K. (2009). α-Glucosidase and α-amylase inhibitory activities of phloroglucinal derivatives from edible marine brown alga, *Ecklonia cava*. *Journal of the Science of Food and Agriculture, 89*(9), 1552–1558.

Lehning, E. J., LoPachin, R. M., Mathew, J., & Eichberg, J. (1994). Changes in Na^+,K^+-ATPase and protein kinase C activities in peripheral nerve of acrylamide-treated rats. *Journal of Toxicology and Environmental Health, Part A Current Issues, 42*(3), 331–342.

Liu, Y., Zeng, S., Ji, W., Yao, H., Lin, L., Cui, H & Pan, G. (2022). Emerging theranostic nanomaterials in diabetes and its complications. *Advanced Science, 9*(3), 2102466.

Liu, M., Hansen, P. E., & Lin, X. (2011a). Bromophenols in marine algae and their bioactivities. *Marine Drugs, 9*(7), 1273–1292.

Liu, M., Zhang, W., Wei, J., & Lin, X. (2011b). Synthesis and α-glucosidase inhibitory mechanisms of bis (2, 3-dibromo-4, 5-dihydroxybenzyl) ether, a potential marine bromophenol α-glucosidase inhibitor. *Marine Drugs, 9*(9), 1554–1565.

Low, P. A., Yao, J. K., Kishi, Y., Tritschler, H. J., Schmelzer, J. D., Zollman, P. J., & Nickander, K. K. (1997). Peripheral nerve energy metabolism in experimental diabetic neuropathy. *Neuroscience Research Communications, 21*(1), 49–56.

Martins, B., Vieira, M., Delerue-Matos, C., Grosso, C., & Soares, C. (2022). Biological potential, gastrointestinal digestion, absorption, and bioavailability of algae-derived compounds with neuroprotective activity: A comprehensive review. *Marine Drugs, 20*(6), 362.

Mishra, G., Singh, P., Molla, M., Yimer, Y. S., Ewunetie, A., Tadesse, T. Y & Kefale, B. (2022). Nutraceuticals: A source of benefaction for neuropathic pain and fibromyalgia. *Journal of Functional Foods*, *97*, 105260.

Nagamatsu, M., Nickander, K. K., Schmelzer, J. D., Raya, A., Wittrock, D. A., Tritschler, H., & Low, P. A. (1995). Lipoic acid improves nerve blood flow, reduces oxidative stress, and improves distal nerve conduction in experimental diabetic neuropathy. *Diabetes Care*, *18*(8), 1160–1167.

Naik, B., Richa, S., Bharadwaj, S., Mishra, S., Kumar, V., Kumar, V.& Preet, M. S. (2024). Utilizing marine algal metabolites to fight neurodegenerative diseases. Frontiers in Marine Science, 11, 1370839.

Oates, P. J., & Mylari, B. L. (1999). Aldose reductase inhibitors: therapeutic implications for diabetic complications. *Expert Opinion on Investigational Drugs*, *8*(12), 2095–2119.

Obrosova, I. G. (2005). Increased sorbitol pathway activity generates oxidative stress in tissue sites for diabetic complications. *Antioxidants & Redox Signaling*, *7*(11–12), 1543–1552.

Pangestuti, R., & Kim, S. K. (2011). Neuroprotective effects of marine algae. *Marine Drugs*, *9*(5), 803–818.

Park, H. J., Kurokawa, M., Shiraki, K., Nakamura, N., Choi, J. S., & Hattori, M. (2005). Antiviral activity of the marine alga *Symphyocladia latiuscula* against herpes simplex virus (HSV-1) *in vitro* and its therapeutic efficacy against HSV-1 infection in mice. *Biological and Pharmaceutical Bulletin*, *28*(12), 2258–2262.

Pereira, L., & Valado, A. (2023). Algae-Derived Natural Products in Diabetes and Its Complications-Current Advances and Future Prospects. *Life*, *13*(9), 1831.

Popoviciu, M. S., Kaka, N., Sethi, Y., Patel, N., Chopra, H., & Cavalu, S. (2023). Type 1 diabetes mellitus and autoimmune diseases: a critical review of the association and the application of personalized medicine. *Journal of Personalized Medicine*, *13*(3), 422.

Pradhan, B., Nayak, R., Patra, S., Jit, B. P., Ragusa, A., & Jena, M. (2020). Bioactive metabolites from marine algae as potent pharmacophores against oxidative stress-associated human diseases: A comprehensive review. *Molecules*, *26*(1), 37.

Quitério, E., Soares, C., Ferraz, R., Delerue-Matos, C., & Grosso, C. (2021). Marine health-promoting compounds: Recent trends for their characterization and human applications. *Foods*, *10*(12), 3100.

Ramana, K. V., Friedrich, B., Srivastava, S., Bhatnagar, A., & Srivastava, S. K. (2004). Activation of nulcear factor-κB by hyperglycemia in vascular smooth muscle cells is regulated by aldose reductase. *Diabetes*, *53*(11), 2910–2920.

Roberts, R. E., & McLean, W. G. (1997). Protein kinase C isozyme expression in sciatic nerves and spinal cords of experimentally diabetic rats. *Brain Research*, *754*(1–2), 147–156.

Saddiqa, A., Faisal, Z., Akram, N., Afzaal, M., Saeed, F., Ahmed, A. & Hailu, G. G. (2024). Algal pigments: Therapeutic potential and food applications. *Food Science & Nutrition.*

Salehi, B., Ata, A., V. Anil Kumar, N., Sharopov, F., Ramírez-Alarcón, K., Ruiz-Ortega, A. & Sharifi-Rad, J. (2019). Antidiabetic potential of medicinal plants and their active components. *Biomolecules*, *9*(10), 551.

Samaddar, S., & Koneri, R. (2019). Polyphenols of marine red macroalga *Symphyocladia latiuscula* ameliorate diabetic peripheral neuropathy in experimental animals. *Heliyon*, *5*(5).

Sheetz, M. J., & King, G. L. (2002). Molecular understanding of hyperglycemia's adverse effects for diabetic complications. *Jama*, *288*(20), 2579–2588.

Sima, A. A. F., & Sugimoto, K. (1999). Experimental diabetic neuropathy: an update. *Diabetologia*, *42*, 773–788.

Soedamah-Muthu, S. S., Chaturvedi, N., Witte, D. R., Stevens, L. K., Porta, M., Fuller, J. H., & Eurodiab Prospective Complications Study Group. (2008). Relationship between risk factors and mortality in type 1 diabetic patients in Europe: the Eurodiab Prospective Complications Study (PCS). *Diabetes Care*, *31*(7), 1360–1366.

Stevens, M. J., Lattimer, S. A., Kamijo, M., Van Huysen, C., Sima, A. A. F., & Greene, D. A. (1993). Osmotically-induced nerve taurine depletion and the compatible osmolyte hypothesis in experimental diabetic neuropathy in the rat. *Diabetologia*, *36*, 608–614.

Stirk, W. A., Reinecke, D. L., & van Staden, J. (2007). Seasonal variation in antifungal, antibacterial and acetylcholinesterase activity in seven South African seaweeds. *Journal of Applied Phycology*, *19*, 271–276.

Suganthy, N., Pandian, S. K., & Devi, K. P. (2010). Neuroprotective effect of seaweeds inhabiting South Indian coastal area (Hare Island, Gulf of Mannar Marine Biosphere Reserve): Cholinesterase inhibitory effect of *Hypnea valentiae* and *Ulva reticulata*. *Neuroscience Letters*, *468*(3), 216–219.

Sugimoto, K., Yasujima, M., & Yagihashi, S. (2008). Role of advanced glycation end products in diabetic neuropathy. *Current Pharmaceutical Design*, *14*(10), 953–961.

Thaifa, M. S., Roshna, S., Arya, U. S., & Babu, A. G. (2017). A review on diabetes mellitus and diabetic neuropathy: A plant based approach. *Journal of Pharmacognosy and Phytochemistry*, *6*(3), 506–510.

Thomas, N. V., & Kim, S. K. (2011). Potential pharmacological applications of polyphenolic derivatives from marine brown algae. Environmental Toxicology and Pharmacology, 32(3), 325–335.

Tomlinson, D. R., & Gardiner, N. J. (2008). Glucose neurotoxicity. *Nature Reviews Neuroscience*, *9*(1), 36–45.

Tomlinson, D. R., Holmes, P. R., & Mayer, J. H. (1982). Reversal, by treatment with an aldose reductase inhibitor, of impaired axonal transport and motor nerve conduction velocity in experimental diabetes mellitus. *Neuroscience Letters*, *31*(2), 189–193.

Tomic, D., Shaw, J. E., & Magliano, D. J. (2022). The burden and risks of emerging complications of diabetes mellitus. *Nature Reviews Endocrinology*, *18*(9), 525–539.

Usuki, S., Ito, Y., Morikawa, K., Kise, M., Ariga, T., Rivner, M., & Yu, R. K. (2007). Effect of pre-germinated brown rice intake on diabetic neuropathy in streptozotocin-induced diabetic rats. *Nutrition & Metabolism*, *4*, 1–11.

Van Acker, K., Bouhassira, D., De Bacquer, D., Weiss, S., Matthys, K., Raemen, H & Colin, I. M. (2009). Prevalence and impact on quality of life of peripheral neuropathy with or without neuropathic pain in type 1 and type 2 diabetic patients attending hospital outpatients clinics. *Diabetes & Metabolism*, *35*(3), 206–213.

Vincent, A. M., Callaghan, B. C., Smith, A. L., & Feldman, E. L. (2011). Diabetic neuropathy: cellular mechanisms as therapeutic targets. *Nature Reviews Neurology*, *7*(10), 573–583.

Vincent, A. M., Perrone, L., Sullivan, K. A., Backus, C., Sastry, A. M., Lastoskie, C., & Feldman, E. L. (2007). Receptor for advanced glycation end products activation injures primary sensory neurons *via* oxidative stress. *Endocrinology*, *148*(2), 548–558.

Vincent, A. M., Russell, J. W., Low, P., & Feldman, E. L. (2004). Oxidative stress in the pathogenesis of diabetic neuropathy. *Endocrine Reviews*, *25*(4), 612–628.

Wijesekara, I., Pangestuti, R., & Kim, S. K. (2011). Biological activities and potential health benefits of sulfated polysaccharides derived from marine algae. *Carbohydrate Polymers*, *84*(1), 14–21.

Wijesekara, I., Yoon, N. Y., & Kim, S. K. (2010). Phlorotannins from Ecklonia cava (Phaeophyceae): Biological activities and potential health benefits. BioFactors, 36(6), 408–414.

Williams, D. M., Jones, H., & Stephens, J. W. (2022). Personalized type 2 diabetes management: an update on recent advances and recommendations. *Diabetes, Metabolic Syndrome and Obesity: Targets and Therapy*, 281–295.

Wu, Y., Ding, Y., Tanaka, Y., & Zhang, W. (2014). Risk factors contributing to type 2 diabetes and recent advances in the treatment and prevention. *International Journal of Medical Sciences*, *11*(11), 1185.

Xu, X. J., Hao, J. X., Andell-Jonsson, S., Poli, V., Bartfai, T., & Wiesenfeld-Hallin, Z. (1997). Nociceptive responses in interleukin-6-deficient mice to peripheral inflammation and peripheral nerve section. *Cytokine*, *9*(12), 1028–1033.

Xu, X., Yin, L., Gao, L., Gao, J., Chen, J., Li, J., & Song, F. (2013). Two new bromophenols with radical scavenging activity from marine red alga *Symphyocladia latiuscula*. *Marine Drugs*, *11*(3), 842–847.

Yagihashi, S. (1995). Pathology and pathogenetic mechanisms of diabetic neuropathy. *Diabetes/metabolism Reviews*, *11*(3), 193–225.
Yagihashi, S., Yamagishi, S. I., & Wada, R. (2007). Pathology and pathogenetic mechanisms of diabetic neuropathy: correlation with clinical signs and symptoms. *Diabetes Research and Clinical Practice*, *77*(3), S184-S189.
Yoon, N. Y., Chung, H. Y., Kim, H. R., & Choi, J. E. (2008). Acetyl-and butyrylcholinesterase inhibitory activities of sterols and phlorotannins from *Ecklonia stolonifera*. *Fisheries Science*, *74*, 200–207.
Yoon, N. Y., Lee, S. H., & Kim, S. K. (2009). Phlorotannins from Ishige okamurae and their acetyl- and butyrylcholinesterase inhibitory effects. *Journal of Functional Foods*, *1*(4), 331–335.
Zhu, X., & Eichberg, J. (1990). 1, 2-Diacylglycerol content and its arachidonyl-containing molecular species are reduced in sciatic nerve from streptozotocin-induced diabetic rats. *Journal of Neurochemistry*, *55*(3), 1087–1090.

Chapter 8
Diabetic-Induced Nephropathy and Its Mitigation by Algae

Priya Josson Akkara, Yajushi Yashitha Vinnakota, R. Esther Shoba, and M. Sonia Angeline

Abstract Diabetes is a major worldwide health issue that calls for original treatment ideas. Diabetic nephropathy (DN) or diabetic kidney disease is a condition marked by pathological amounts of urine albumin excretion, diabetic glomerular lesions, and loss of glomerular filtration rate (GFR) in diabetics. Managing diabetes and its consequences calls for a thorough approach that includes lifestyle changes, glucose-lowering drugs, and focused treatments for particular complications. The varied compositions and pharmacological properties of natural products and their derivatives have become interesting contenders for diabetes control. Algae have attracted interest as a possible source of bioactive chemicals with anti-diabetic qualities. Numerous bioactive substances, including polysaccharides, polyphenols, pigments, fatty acids, and sterols, are produced by algae and have shown a variety of pharmacological characteristics such as ability to improve lipid profiles, increase insulin sensitivity, modify glucose metabolism, have anti-inflammatory and antioxidant properties, all of which are critical for managing diabetes. Studies suggest that these algae-derived bioactive compounds will help in preventing complications from the disease. Hence proving that there is great potential for developing new antidiabetic treatments by utilizing the diverse pharmacological properties of algae.

Keywords Diabetes · Natural products · Algae · Antidiabetic · Diabetic nephropathy · Therapeutic potential

P. J. Akkara · Y. Y. Vinnakota · R. E. Shoba · M. S. Angeline (✉)
Department of Life Sciences, School of Biological and Forensic Sciences, Kristu Jayanti (Deemed-to-be University), Bangalore, India
e-mail: sonia.m@kristujayanti.com

R. Raja et al. (eds.), *Algae in Diabetes Management*,
https://doi.org/10.1007/978-981-95-0571-5_8

1 Introduction

Diabetic nephropathy is one of the most severe complications of diabetes, which poses a significant global health challenge. Extended periods of high blood sugar can lead to kidney filtration failure, ultimately resulting in progressive renal failure. If not addressed, this condition can progress to end-stage renal disease, characterized by increased albumin levels and a decreased glomerular filtration rate. Ongoing high blood sugar levels worsen oxidative stress, inflammation, and fibrosis, which are the primary mechanisms behind diabetic nephropathy (DN). Algae have been identified as potential agents to help mitigate the effects of DN. They are abundant in bioactive compounds, including antioxidants, polyunsaturated fatty acids, and essential vitamins, which have been shown to enhance insulin sensitivity, reduce inflammation, and stabilize blood glucose levels—all of which play significant roles in the progression of diabetic nephropathy. Additionally, compounds derived from algae have demonstrated protective effects on the kidneys by decreasing renal fibrosis, lowering albumin levels, and altering inflammatory pathways. Such compounds could serve as natural medicinal options to prevent or reverse kidney damage in individuals with diabetes. As research into algae continues to grow, it shows great promise as a complementary therapy for diabetes and diabetic nephropathy (DN), offering a natural alternative to traditional treatments that may come with fewer side effects. Studies indicate that algae can mitigate the impacts of DN. Packed with bioactive compounds like antioxidants, polyunsaturated fatty acids, and essential vitamins, algae can improve insulin sensitivity, reduce inflammation, and help regulate blood sugar levels—all crucial factors in the onset of diabetic neuropathy. Substances derived from algae have also shown protective benefits for the kidneys, such as decreasing albuminuria, minimizing renal fibrosis, and modifying inflammatory pathways. For those with diabetes, these substances could serve as a natural option to prevent or reverse kidney damage. Algae hold significant potential as an adjunct therapy for diabetes and DN, providing a natural alternative to conventional medications with possibly fewer side effects as research in this area advances. Emerging therapeutic strategies from this evolving field may help alleviate the burden of diabetic nephropathy and improve patient outcomes.

2 Diabetic Nephropathy

Diabetes affects millions of individuals globally; by 2045, there probably will be 700 million cases worldwide. The main cause of diabetic nephropathy (DN), a key contributor to end-stage renal failure, is diabetes mellitus (DM). It involves interactions between genetic and non-genetic factors and results in structural kidney changes such as tubulointerstitial fibrosis, mesangial matrix expansion, and GBM thickening. One of the main side effects of diabetes that raises morbidity and mortality is diabetic nephropathy (DN). With more than 50,000 new instances of end-stage renal disease (ESRD), the patients will require dialysis, and a few may be eligible for a kidney transplant. Timely renal replacement therapy is necessary to prevent death. If preventive measures are not improved, the rising prevalence of diabetes, especially in emerging nations, suggests a corresponding rise in DN (Varghese and Jialal 2024, Qi et al. 2017).

Microalbuminuria, the presence of trace levels of albumin in urine is frequently used to assess DN. However, because not all diabetes patients with advanced renal disease have considerable albuminuria, it has limits and cannot accurately reflect the severity or prognosis of the disease. Since some patients with low albuminuria still have advanced kidney disease, relying only on proteinuria may not be enough to assess the severity of DN. There is ongoing discussion on DN screening, including issues on recognizing patients at risk and knowing when to step in. Even though microalbuminuria frequently denotes extensive glomerular impairment, some individuals eventually regain normal albumin levels. Furthermore, diagnosis is made more difficult by nonproteinuric DN and DN without retinopathy, which is frequently observed in type 2 diabetes and may avoid early detection (Qi et al. 2017; Samsu 2021).

The Diabetes Complications and Control Trial (DCCT) showed that strict blood glucose control dramatically lowers microvascular consequences. Research revealed that the risk of problems is not entirely reduced by beginning strict glucose control. Despite glucose normalization, oxidative stress, AGEs (advanced glycation end products), and activation of NF-κB and MAPKs are attributed to persistent consequences. However, DN is a common and serious microvascular consequence that is greatly influenced by AGEs and TGF-β new therapeutic strategies could come from better understanding of the inflammatory processes behind DN. Recent research suggests that targets for the development of creative anti-inflammatory drugs and other treatments are inflammation, angiotensin II, and oxidative stress. (Li et al. 2021; Samsu 2021).

3 Pathogenesis

Numerous routes and mediators are involved in the intricate and multifaceted pathophysiology of DN development and progression. The developmental mechanism of DN is traditionally believed to be caused by aberrant homeostasis, which includes metabolic problems, hemodynamic abnormalities, and hormone production such as Ang-II. The renin-angiotensin-aldosterone system (RAAS), the production of advanced glycation end products (AGE), the activation of connective tissue growth factor (CTGF), transforming growth factor-β1 (TGF-β1), protein kinase C (PKC), mitogen-activated protein kinase (MAPKs), and reactive oxygen species (ROS) are important pathways to the development and progression of DN. To cause damage, each pathway engages in interactions with other pathways or makes use of a range of mediators.

Many pathways and mediators overlap; for example, RAAS causes injury through oxidative stress, while oxidative stress causes injury through Ang-II. Nicotinamide adenine phosphate dehydrogenase (NADPH) oxidase increases TGF-β, and TGF-β increases ROS through activation of NADPH oxidase. As a result, it is still unknown what the exact pathogenic mechanism and molecular incidence of DN are, as well as the relative contributions of each route to the formation of DN (Kopel et al. 2019; Tavafi 2013).

3.1 Role of Oxidative Stress

Oxidative stress, a major contributor to DN, is an imbalance between oxidants and antioxidants, leading to tissue damage. Hyperglycemia-induced ROS, driven mainly by advanced glycation end-products (AGEs), NADPH oxidase particularly isoform Nox 4, and the polyol pathway, contribute significantly to DN pathology. Also produced by mitochondrial dysfunction, xanthine oxidase, uncoupled nitric oxide synthase, and lipoxygenase. This oxidative stress activates macrophages to release proinflammatory cytokines, directly damaging podocytes, mesangial cells, and endothelial cells, resulting in proteinuria and fibrosis. Oxidative damage to DNA, lipids, and proteins exacerbates the condition, linking ROS generation with DN progression. Furthermore, oxidative stress can both result from and promote other pathogenic pathways, intensifying kidney injury in diabetes (Brownlee 2005; Krishan and Chakkarwar 2011).

3.2 Role of Inflammation

Recent research highlights the critical role of inflammation in the onset and progression of diabetic nephropathy (DN). Inflammatory mediators, including interleukins such as IL-1, IL-6, and IL-18, are known to significantly contribute to the pathophysiology of DN. Additionally, markers of inflammation are associated with an increased risk of developing DN, with immune cells like leukocytes, monocytes, and macrophages playing an active role. Persistent inflammation is one of the key underlying factors in the development of DN within the circulatory system and renal tissue. Elevated levels of inflammatory substances, including IL-6, TNF-α, TGF-β1, and IL-18, which are associated with albumin excretion in urine, promote the progression of DN. Consequently, the extent of inflammatory cell accumulation in the kidneys is closely linked to DN (Lim and Tesch 2012; Kiritoshi et al. 2003). The proinflammatory and fibrogenic cytokines present in the renal microenvironment can directly damage kidney tissues, initiating the epithelial-to-mesenchymal transition (EMT) and leading to the accumulation of extracellular matrix (ECM). Increased renal fibrosis is directly associated with this process. Furthermore, by drawing circulating leukocytes and promoting their infiltration into renal tissues, the

aging and studying disease progression, emphasizing the importance of early intervention, especially through blood glucose and blood pressure control, to slow or prevent the advancement of renal complications in diabetes (Mogensen et al. 1983).

5 Management and Treatment

In diabetic nephropathy management, standard treatment continues to prioritize controlling blood glucose and blood pressure to slow disease progression and reduce albuminuria. This focus on albuminuria is based on studies showing that lowering albuminuria correlates with improved renal and cardiovascular outcomes in people with diabetes. However, while this approach slows DN's advancement, it cannot entirely prevent or reverse the condition, so the prevalence of DN remains on the rise. Albuminuria persists as a significant predictor of eGFR decline and is central to renoprotective strategies, especially for patients with moderate-to-severe kidney impairment. Additionally, proteinuria is a critical risk factor in DN progression, reinforcing the need for therapies that reduce protein levels in urine. Alongside glucose and blood pressure management, lifestyle changes like weight loss, protein and lipid control, and smoking cessation are recommended. Weight loss, particularly in obese patients, has shown benefits in reducing albuminuria by mitigating hyperfiltration and hormonal imbalances associated with adipokines, which can worsen DN (de Zeeuw et al. 2004; Quiroga et al. 2015).

5.1 Glycemic Control

Diabetic nephropathy affects roughly 20% of diabetic patients and can develop even with effective glycemic control showing that stricter blood glucose control can delay DN onset and slow its progression, reducing renal complications in diabetics over both short and long terms. However, hypoglycemia led to updated guidelines advocating for personalized glycemic targets based on individual patient factors. Various medications assist in controlling, and some have nephroprotective benefits beyond lowering blood glucose. Incretin-based drugs like DPP-4 inhibitors and GLP-1 receptor agonists may protect kidney function independently, though more studies are needed to confirm these effects (Stratton et al. 2000). DPP-4 inhibitors lower blood glucose by increasing GLP-1 levels, enhancing insulin secretion, and suppressing glucagon. They are safe for renal impairment, with linagliptin requiring no dose adjustment. Beyond glycemic control, DPP-4 inhibitors offer renoprotective benefits, such as antioxidant, anti-inflammatory, and antifibrotic effects by inhibiting TGF-β pathways, particularly benefiting the kidneys where DPP-4 activity is high in proximal tubular cells (Drucker and Nauck 2006). SGLT2 inhibitors lower blood glucose by promoting urinary glucose excretion and offer additional kidney protection beyond glycemic control. Empagliflozin decreases kidney disease progression by 39%, while mechanisms include lowering blood pressure, reducing glomerular pressure, and minimizing tubular injury. Canagliflozin also confirmed reduced albuminuria progression and a slower decline in GFR, helping to delay the need for renal replacement therapy (Kim 2017).

5.2 *Blood Pressure Control*

Lowering systolic blood pressure by 10 mmHg significantly reduces diabetic microvascular complications like nephropathy. A target BP of <140/90 mmHg for diabetic patients and adjusted to ≤130/80 mmHg for those with albuminuria is recommended. Due to strong evidence supporting their efficacy in slowing DN progression, RAAS inhibitors, including ARB and ACE inhibitors, are the primary recommendation for blood pressure control. Major trials like IDNT, RENAAL, and IRMA-2 demonstrate the benefits of RAAS inhibition while combining ARBs with ACE inhibitors is discouraged due to the increased risk of side effects like hyperkalemia. Additionally, nondihydropyridine calcium channel blockers, such as diltiazem, may slow DN progression, while dihydropyridines have variable effects on albumin excretion (Samsu 2021).

5.3 *Lipid Control*

Rosuvastatin administration in patients with DN, controlled hypertension led to a reduction in urinary albumin excretion, oxidative stress markers (8-OHdG), and L-FABP. These benefits occurred independent of blood pressure and lipid changes, suggesting a lipid-lowering effect beyond just cholesterol management. Rosuvastatin also decreased cystatin C, indicating potential glomerular filtration rate (GFR) preservation. Rosuvastatin may slow tubulointerstitial lesions and reduce cardiovascular risks in diabetic nephropathy patients (Abe et al. 2011).

5.4 *Weight Loss*

Overweight and obesity are common comorbidities in diabetes and significantly contribute to chronic kidney disease progression. These conditions may lead to increased hyperfiltration and hormonal imbalances associated with adipokines. Weight loss in obese diabetic patients has been shown to reduce albuminuria and may also decrease serum creatinine, although this could be due to muscle mass loss. Additionally, bariatric surgery has demonstrated benefits in improving diabetes control, renal function, and albuminuria in morbidly obese patients. However, no specific trials have yet examined its direct effects on diabetic kidney disease (Quiroga et al. 2015).

5.5 *Protein Intake*

Dietary management for diabetic kidney disease is challenging due to the need to balance various comorbidities. In addition to regulating carbohydrate intake, patients often require low-salt diets for hypertension, fat-free diets for dyslipidemia, and hypocaloric diets for obesity. It has been demonstrated that patients, especially those with decreased GFR, benefit from moderate protein restriction (up to 0.8 g/kg/day) (Quiroga et al. 2015). To control diabetes and its related effects, a multifaceted approach involving targeted therapy, glucose-lowering medications, and lifestyle changes is required. Despite the advancements in established diabetes medications, interest in alternative therapies is growing since they may be more effective and have fewer side effects. Algal natural products are attracting interest as potential diabetes treatment tools. Bioactive substances found in these sources can control blood sugar and lessen or prevent the development and effects of diabetes. Research into these natural products can lead to the development of potential new therapies addressing more general management issues and regulating blood sugar (Pereira and Valado 2023).

6 Mitigation by Algae

Bioactive compounds found in algae show promise for diabetes management due to their various pharmacological effects, such as enhancing insulin sensitivity, reducing diabetes-related complications, and improving glucose metabolism. Further research into these substances could lead to innovative treatments that enhance patient outcomes and diabetes care (Davies et al. 2022; Abo-Shady et al. 2023). Certain algae species, rich in polyphenols, polysaccharides, and peptides, have demonstrated significant antidiabetic potential by effectively lowering blood glucose levels, improving insulin sensitivity, and reducing diabetes complications. However, to confirm their benefits in humans, extracts from these algae need to undergo rigorous scientific validation, particularly through clinical trials. This research approach, which leverages the bioactive components of algae, may lead to personalized medications that improve diabetes management and patient outcomes (Pradhan et al. 2020). Additionally, compounds derived from algae might help protect the kidneys of diabetic patients by reducing albuminuria, preventing renal fibrosis, and modulating inflammatory pathways that contribute to kidney damage. These effects provide targeted renal benefits crucial for preventing diabetic nephropathy, alongside aiding in blood glucose control. Algae-based therapies could offer the dual benefits of kidney protection and glucose regulation, potentially improving the prognosis for diabetics and slowing the progression of chronic kidney disease (Abo-Shady et al. 2023).

6.1 *Arthrospira Platensis*

Spirulina, scientifically known as *Arthrospira platensis*, is a blue-green microalga that has become increasingly popular as a dietary supplement in recent years due to its rich content of macro- and micronutrients (Gentscheva et al. 2023). Its impressive levels of proteins, polyunsaturated fatty acids (PUFAs), polysaccharides, vitamins, minerals, and other beneficial compounds such as polyphenols, carotenoids, and chlorophylls make it particularly appealing in the nutraceutical industry. The valuable molecular makeup of Spirulina biomass is commonly utilized in the production of nutritional supplements, fortified foods, and cosmetics, especially after the FDA confirmed its safety for human consumption (Bortolini et al. 2022) (Fig. 8.1).

In addition to the commercial applications of Spirulina biomass, the nutraceutical sector might focus on particular Spirulina compounds or groups of compounds.

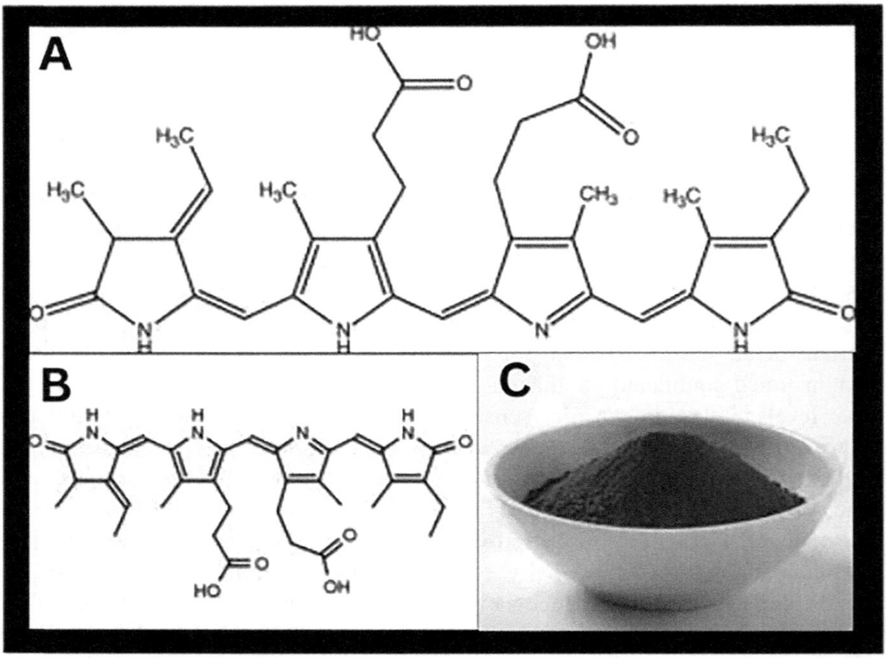

Fig. 8.1 (**a**) Phycocyanin, (**b**) Phycocyanobilin, (**c**) Spirulina extract

One notable example is the blue pigment C-phycocyanin (C-PC). It belonged to the phycobiliprotein family. C-phycocyanin is a water-soluble, non-toxic blue pigment used in a variety of goods. C-PC and its chromophore, Phycocyanobilin B (PCB-B), have been proven to have biological activity because their chromophores are chemically comparable to bile pigments. These chromophores are open-chain tetrapyrroles linked to cysteine residues via thioester bonds. As a result, phycocyanin and phycocyanobilin exhibit antioxidant properties comparable to those of biliverdin and bilirubin. It has been demonstrated in multiple studies that phycocyanin possesses anti-inflammatory, antioxidant, and radical scavenging qualities (Papadopoulos et al. 2022; Citi et al. 2024).

Both PCB-B and C-PC have a role in modulating DN, by inhibiting renal oxidative stress and preventing albuminuria and renal mesangial enlargement. It also restored normal levels of fibronectin and transforming growth factor beta (TGF-β) expression. PCB-B was demonstrated to inhibit the activity of Nox4, which causes renal dysfunction, and NOX, the primary cause of cellular ROS formation, linked to vascular complications in diabetic subjects (Citi et al. 2024). In addition to normalizing albuminuria, treatment with phycocyanobilin derived from phycocyanin (89% purity) also restored renal and urine oxidative stress indicators. The antioxidant effect of bilirubin is due to its strong radical scavenging activity which directly inhibits NAD(P)H oxidase activity. Research has indicated that nonphagocytic NAD(P) H oxidases might be the primary producers of ROS in the vascular tissues of diabetic patients and animals. Thus, vascular NAD(P)H oxidase inhibition may provide a powerful treatment strategy to avoid diabetic vascular consequences, such as nephropathy (Zheng et al. 2013). Research also indicates that oral administration of phycocyanin and phycocyanobilin inhibited increases in the expression of the NAD(P)H oxidase component Nox4, which is believed to be the primary cause of oxidative stress in diabetic kidneys. Antisense oligonucleotides that downregulated Nox4 completely suppressed oxidative stress, preventing renal hypertrophy and fibronectin expression in diabetic rats. Thus, phycocyanin and phycocyanobilin's ability to normalize Nox4 expression may be crucial in lowering oxidative stress and averting renal failure in diabetic kidney disease (Zheng et al. 2013).

6.2 *Saccharina japonica*

The brown algae *Saccharina japonica* is significant both environmentally and economically. The high concentration of carotenoid fucoxanthin and other phyeophycean tannins in their chloroplasts gives them their distinctive brown color. Primarily found in marine environments, they appear as an intertidal component. Certain marine organisms may survive in brackish settings and contribute significantly to the biodiversity of salt marshes. The highest species and physical diversity of brown algae can be found in temperate to subpolar environments (Davis et al. 2003) (Fig. 8.2).

Fig. 8.2 *Saccharina japonica*

Saccharina japonica is the source of fucoidan, a sulfated polymer made up mostly of fucose, galactose, and sulfate but also including trace amounts of mannoses, uronic acid, glucose, rhamnose, arabinose, and xylose. According to reports, fucoidans may provide a wide range of therapeutic properties, including anti-inflammatory, reno-protective, antioxidant, antiuropathy, and antirenopathy (Wang et al. 2019). Fucoidan effectively slows diabetic kidney disease progression and corrects metabolic imbalances in DN by targeting oxidative stress and inflammation. It reduces albumin, beta-N-acetyl-D-glucosaminidase, and the albumin-to-creatinine ratio, helping to protect renal function. Fucoidan's renoprotective effects stem from its inhibition of TGF-β1 signaling, lowering oxidative stress, inflammation, hyperglycemia, and obesity. Low molecular weight fucoidan (LMWF) preserves the glomerular structure, blocks P-selectin to prevent GBM damage, and outperforms Captopril (an ACE inhibitor used for the treatment of nephropathy) in reducing P-selectin expression. By regulating PKC and TGF-β pathways, fucoidan helps prevent glucose-induced DN (Xu et al. 2016).

LMWF reverses early signs of diabetic renal impairment by restoring M-TP, uAlb, and urine volumes to baseline. It addresses tubular and glomerular damage markers, such as proteinuria and hematuria, which are linked to inflammation, GBM rupture, and sclerosis in DN. LMWF reduces mesangial growth, glomerular area, and inflammatory infiltration, preserving renal structure. By downregulating key inflammatory mediators like ICAM-1, JNK, MAPK, TNF-α, IL-6, and TGF-β, LMWF minimizes leukocyte and macrophage buildup, protecting kidney tissue from DN-associated damage (Xu et al. 2016). Studies reveal that the blood biochemical of kidneys starts to exhibit pathological alterations, leading to noticeably elevated BUN (blood urea nitrogen) and Scr levels. Thus, SPS (sulfated polysaccharide—fucoidan) can shield the kidneys of diabetic rats. It was also observed that the urine output of SPS rats was significantly lesser than that of diabetic rats. This implies that SPS has a gradual rather than an immediate effect on renal protection. Studies showed that the development of diabetes significantly changed the organization of kidney skeletal muscle, which results in the formation of podocyte fusion. Additionally, diabetes aggravates the kidney by resulting in inflammation and fibrosis, two common signs of diabetic kidney disease. Immunohistochemical analyses of the kidneys reveal that PKC and p65 are highly expressed; however, the effects of SPS substantially decrease these protein expressions. Inhibiting PKC expression in kidneys reduces DN symptoms such as inflammation, according to pathological tests. PKC inhibition also reduced NF-κB and p65 expression when SPS was given. Thus, the PKC/NF-κB signaling pathway may have a role in the development of DN. This suggests that PKC and NF-κB could be potential therapeutic targets for treating or managing DN (Xu et al. 2021).

6.3 *Phyaeodactylum tricornutum*

High-value metabolites such as fucoxanthin, eicosapentaenoic acid (EPA), and chrysolaminarin are produced by the microalgae *Phyaeodactylum tricornutum* and are used in medicines, nutraceuticals, and cosmetics. Mostly derived from diatoms and brown algae, fucoxanthin (Fx) is a major marine carotenoid that accounts for over 10% of global carotenoid output. Along with β-carotene and chlorophylls a and c, this orange pigment is frequently found in the Chromophyta, which includes diatoms and brown seaweeds. Brown seaweeds, commonly found in Asian and certain European diets, and diatoms, which are golden-brown microalgae, use fucoxanthin in their light-harvesting complexes to support photosynthetic efficiency (Celi et al. 2022; Peng et al. 2011).

Fx differs from other carotenoids like β-carotene due to its distinct structure, which has an allenic link and oxygen-rich functional groups that increase its antioxidant properties. Fucoxanthin has garnered interest from the pharmaceutical, cosmetic, and nutraceutical industries because of its demonstrated anti-inflammatory, anti-cancer, antioxidant, and neuroprotective qualities. Because it is more economical and environmentally friendly than chemical synthesis, natural extraction is still the method of choice (Peng et al. 2011; Lourenço-Lopes et al. 2020). Research suggests that Fx possesses encouraging antioxidant and anti-inflammatory qualities that are pertinent to the treatment of diabetes. By downregulating inflammasome components such as NLRP3 and ASC and upregulating antioxidant genes Nrf2 and HO-1 in HaCaT keratinocytes, the combination of Fx and rosmarinic acid intensifies these effects.

Fx has demonstrated a high therapeutic index, with toxicity studies showing it is safe even at high single and repeated doses. Additionally, Fx reduces extracellular matrix (ECM) secretion and reactive oxygen species (ROS) generation, which helps combat oxidative stress—a significant factor in the progression of diabetic nephropathy (DN). In DN, oxidative stress and inflammation, driven by NF-κB signaling and impaired Nrf2/ARE signaling, worsen the disease progression (Xie et al. 2013) (Fig. 8.3).

Fig. 8.3 Fucoxanthin

Fx effectively lowers ROS generation and ECM secretion triggered by elevated glucose levels, indicating its potential in managing DN. Fx, at a concentration of just 2 μM, demonstrated nearly comparable efficacy to a much higher dose of resveratrol in reducing the expression of fibronectin and collagen IV, key markers of ECM buildup. This highlights Fx's potential as a potent renoprotective agent. By lowering ECM accumulation and ROS production, Fx may help mitigate DN progression FoxO3α, a key transcription factor in oxidative stress defense, drives the expression of stress-resistance genes, enhancing hydrogen peroxide scavenging and oxidative resilience. In high-glucose conditions, FoxO3α levels drop, leading to increased ROS and oxidative damage. Fx and resveratrol restore Sirt1, decreasing FoxO3α acetylation and boosting its activity. High glucose also activates Akt, reducing nuclear FoxO3α and raising ROS. Fx counteracts this by inhibiting Akt, which helps restore FoxO3α and reduce ROS levels. As a result, Fx effectively alleviates oxidative stress caused by high glucose in diabetic nephropathy (DN). Sirt1, an important regulator of FoxO3α, plays a vital role in slowing the progression of DN and enhancing antioxidant defenses. When activated by resveratrol or Fx, Sirt1 reduces markers of fibrosis and ROS while increasing the activity of FoxO3α and the expression of MnSOD. Under high-glucose conditions, Fx promotes the upregulation of Sirt1, thereby restoring the function of FoxO3α to effectively combat oxidative stress (Huang et al. 2015; Yang et al. 2019). Mitochondria play a crucial role in diabetic kidney disease, where the production of ROS due to chronic hyperglycemia activates harmful pathways. Mitochondrial dysfunction, characterized by an excess of ROS production and a decline in antioxidant enzyme activity, accelerates the progression of DN. Research has indicated that in diabetes models, tubular cells exhibit mitochondrial abnormalities such as fragmentation, increased ROS levels, and reduced mitophagy. In renal mesangial cells exposed to high glucose, there is an increase in ROS, a decrease in MnSOD expression, and a reduction in mitochondrial membrane potential. However, Fx therapy has been shown to reverse these mitochondrial dysfunctions and lower oxidative stress by inhibiting Akt and activating Sirt1, which restores FoxO3α levels (Forbes et al. 2008; Xiao et al. 2016). Fx treatment effectively diminishes ROS production and extracellular matrix accumulation in high-glucose-induced glomerular mesangial cells while enhancing the nuclear translocation and transcriptional activity of FoxO3α by reversing the inhibition of Sirt1 and the activation of Akt. This restoration underscores the potential of Fx in halting the progression of DN by increasing MnSOD expression, reducing oxidative stress, and inhibiting the expression of FN and collagen induced by high glucose (Yang et al. 2019).

6.4 Ecklonia Cava

The southern coast of Korea is known for the edible brown algae called Ecklonia cava, which is rich in phlorotannins like eckol, dieckol, and phlorofucofuroeckolA. These compounds are recognized for their antidiabetic, immunomodulatory, neuroprotective, anti-inflammatory, and anti-cancer effects. Due to its high content of bioactive substances, including carotenoids, peptides, polysaccharides, phlorotannins, fucoidans, and minerals, Ecklonia cava, which flourishes along the shores of Jeju Island in Korea, is commonly utilized in food, cosmetics, and medicinal applications. Notably, phlorotannins with phloroglucinol structures, such as triphlorethol-A, dieckol, and eckol found in E. cava, offer various health benefits, including anti-inflammatory, antioxidant, anticancer, antibacterial, and skin-whitening effects.

Dieckol is known for its anti-inflammatory properties, triphlorethol-A exhibits antioxidant activity, and eckol provides radioprotective benefits. Consequently, E. cava plays a vital role in the formulation of medicines, cosmetics, and functional foods (Wijesekara et al. 2010) (Fig. 8.4)

Glycation of proteins leads to the formation of advanced glycation end-products (AGEs), which accumulate in tissues, especially in people with diabetes, resulting in complications such as blindness, cardiovascular disease, and nephropathy. A highly reactive metabolic byproduct, methylglyoxal (MGO), plays a significant role in the formation of AGEs and worsens kidney injury in diabetic kidney disease

Fig. 8.4 Ecklonia cava and its extract

(DN) by thickening the glomerular basement membrane (GBM). To mitigate AGE-related kidney damage, it is essential to prevent the accumulation of MGO and AGEs (Cho et al. 2022). Research suggests that Ecklonia cava extract (ECE) shows great potential as a natural anti-glycation agent. It works by capturing MGO and reactive oxygen species (ROS), chelating metal ions, and regulating the expression of RAGE, a receptor involved in AGE-induced inflammation, thereby breaking the cross-links between AGE and collagen and reducing AGE production. This indicates that ECE could be highly effective as a preventive strategy against kidney damage caused by oxidative stress, potentially delaying diabetes-related complications. In mesangial cells, pretreatment with ECE significantly reduced oxidative stress and apoptosis induced by MGO. MGO enhances RAGE expression through AGE-RAGE interactions, leading to inflammation, oxidative stress, and cell death. By effectively blocking these pathways, ECE protected the cells from dying. Furthermore, ECE activates the Nrf2 pathway, increasing Nrf2 expression and antioxidant enzyme levels, which helps lower ROS production and mitigate damage caused by MGO.

ECE reduced oxidative stress and apoptosis by inhibiting the activation of ERK, JNK, and p38 MAPKs. This suggests that by regulating key stress pathways and minimizing oxidative damage, ECE could serve as a promising natural treatment for preventing kidney damage in diabetic nephropathy (DN) (Lee et al. 2013; Cho et al. 2022) (Fig. 8.5).

Research indicates that dieckol, a phlorotannin found in E. cava, possesses anti-inflammatory properties. In studies involving mouse renal stromal cells, dieckol demonstrated antiapoptotic effects, reduced interactions between AGE and RAGE, regulated the production of RAGE protein, and exhibited strong antiglycation effects. Overall, E. cava influences multiple pathways associated with AGE production, oxidative damage, and inflammation, positioning it as a promising natural approach to mitigate oxidative stress and kidney damage in diabetic nephropathy (Cho et al. 2023).

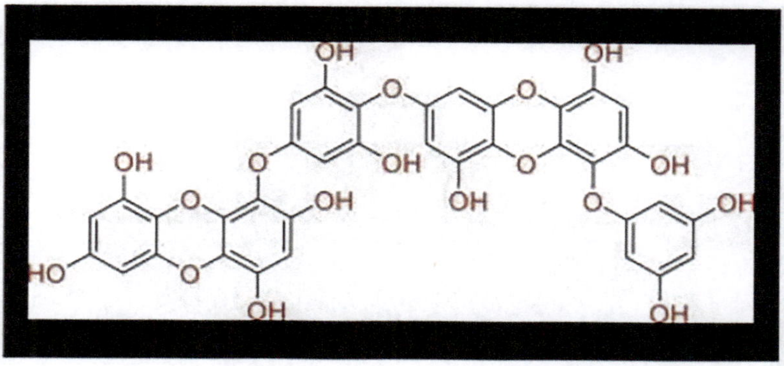

Fig. 8.5 Dieckol

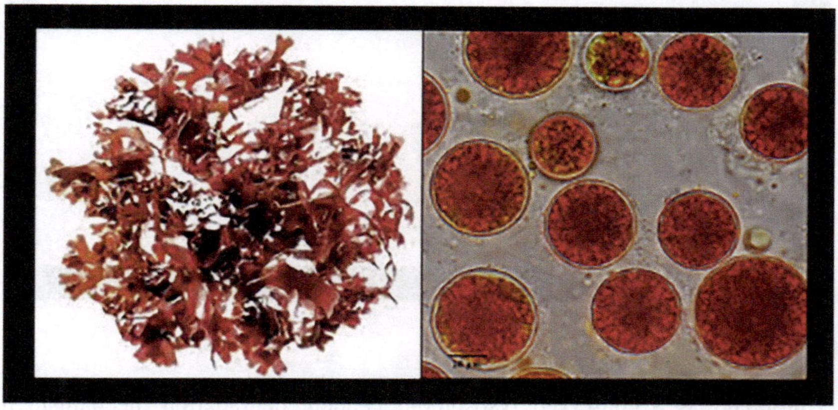

Fig. 8.6 *Haematococcus pluvialis*

6.5 Haematococcus pluvialis

The green microalga *Haematococcus pluvialis* is rich in various beneficial bioactive compounds, such as lutein, proteins, fatty acids, carotenoids, and astaxanthin. Notably, astaxanthin offers numerous health benefits and is utilized in both the pharmaceutical and nutraceutical industries (Oslan et al. 2021) (Fig. 8.6).

One of the most potent marine carotenoids is astaxanthin (ATX), which has a 38-fold greater ability to destroy free radicals than β-carotene and 500 times the antioxidant activity of vitamin E. Astaxanthin is a powerful antioxidant that also has anti-inflammatory and therapeutic benefits for a number of illnesses, such as oxidative stress caused by UV light, inflammation, cancer, liver, heart, joints, and skin, aging, prostate health, stomach ulcers, hypertension, and diabetes mellitus. Astaxanthin-rich oleoresin from *Haematococcus pluvialis* can be safely taken in dietary supplements at levels of up to 40–80 mg/day, as per the European Commission Implementing Regulation (Zou et al. 2013; Yamashita 2021) (Fig. 8.7).

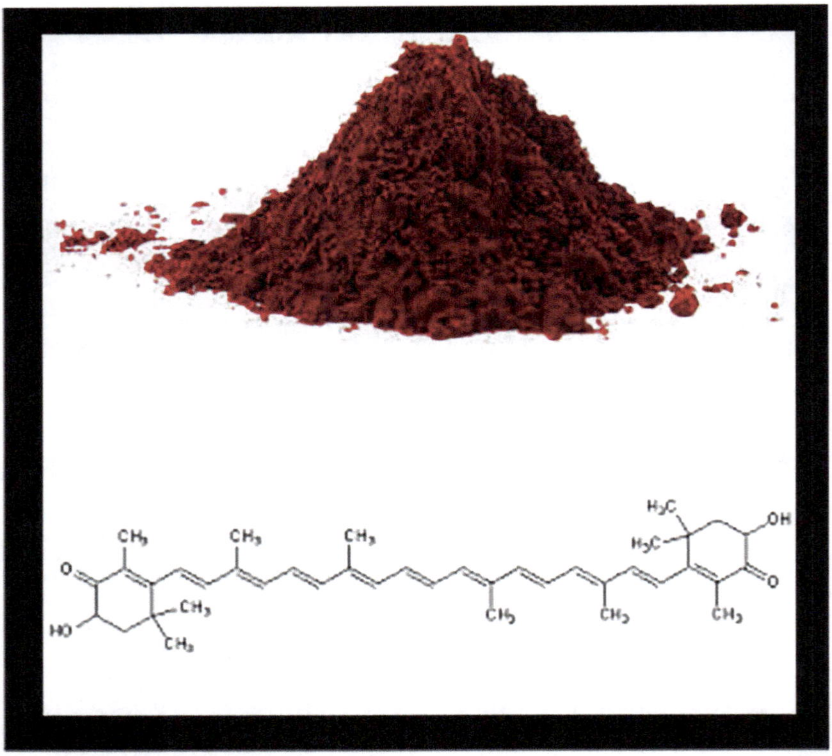

Fig. 8.7 Astaxanthin

Autophagy is a conserved lysosome-dependent degradation mechanism that eliminates pathogens, damaged organelles, and aggregated proteins. It is crucial for maintaining cellular homeostasis. It has been hypothesized that the autophagic stress response is impaired by modifying nutrient-sensing pathways in diabetic circumstances, which may aggravate organelle malfunction and aid in the formation of DN. Autophagy plays a central role in numerous diseases and metabolic disorders, and it is crucial for myofibroblast differentiation, tissue remodeling, and fibrogenesis. LC3 serves as a biomarker for autophagy and p62 a crucial autophagy receptor shows an inverse correlation with autophagic lysosome count and is key in intracellular signaling regulation. Impaired autophagic flux and the accumulation of p62 contribute to podocyte damage and dysfunction which plays a significant role in the progression of DN. An increase in fibrosis markers α-SMA and collagen IV was observed in high-glucose cultured podocytes and diabetic rats. ATX treatment significantly upregulated autophagy in diabetic conditions by enhanced LC3-II levels and reduced p62 accumulation in podocytes. Therefore, resulted in a decrease in the number of autophagosomes, suggesting an improvement in autophagic flux. ATX treatment also suggests that the upregulation of autophagy may play a role in reducing fibrosis in DN and protecting podocytes. Enhancing autophagy in kidney cells could improve DN. This indicates that activating autophagy may help mitigate fibrosis and slow the progression of diabetic nephropathy (DN) (Zhu et al. 2018; Su et al. 2022; Hong et al. 2024). DN arises from issues with mesangial cells and podocytes, and the progression of the disease is influenced by the interactions between these cell types. Hyperglycemia triggers intracellular signaling pathways that disrupt the regulation of VEGF and TGF-β, leading to abnormal angiogenesis and albuminuria. When glucose levels are elevated, podocytes produce increased amounts of VEGF-B and TGF-β. Treatment with astaxanthin (ATX) enhances autophagy in human mesangial cells (HMC), reduces fibrosis, and inhibits the release of harmful signaling factors (Fu et al. 2014; Hong et al. 2024). ATX shows potential in treating DN by boosting antioxidant enzyme activity, minimizing structural damage to the kidneys, and protecting against oxidative stress, inflammation, and apoptosis. Furthermore, studies suggest that ATX can prevent damage to renal cells subjected to high glucose levels, reduce proteinuria, and lower oxidative stress markers. The protective effects of ATX on the kidneys are likely attributed to its ability to modulate oxidative stress (Naito et al. 2004; Hong et al. 2024). Excessive accumulation of extracellular matrix (ECM) is a significant factor in diabetic kidney disease (DN), with high-glucose cultured mesangial cells and the presence of ICAM-1 and TGF-β linked to inflammation and glomerular hypertrophy. Renal fibrosis is driven by the overproduction of reactive oxygen species (ROS), which not only increases ECM but also hampers its degradation. Treatment with astaxanthin (AST) has been shown to reduce ECM buildup in DN rats by decreasing inflammation, oxidative stress, and the production of actin stress fibers, a marker of hypertrophy in mesangial cells treated with high glucose. Due to its unique double-bond structure, AST can effectively scavenge free radicals and neutralize singlet oxygen, thereby restoring the activity of antioxidant enzymes. Research involving Nrf2 siRNA indicates that AST promotes the nuclear translocation of Nrf2, enhances the levels of antioxidant enzymes, and lowers the levels of fibronectin (FN) and TGF-β1 through Nrf2 signaling. AST regulates oxidative stress and ECM accumulation through Nrf2 signaling (Chen et al. 2018).

6.6 Chlorella sp.

Supplementing with chlorella shows potential in treating type 2 diabetes and its complications, including diabetic nephropathy (DN). Due to its antioxidant properties, chlorella has been found to reduce glycated hemoglobin, boost antioxidant levels, and delay cataract formation in diabetic animal studies. Additionally, it exhibited a temporary hypoglycemic effect and improved glucose uptake in muscles and the liver, aiding in glycemic control (Bito et al. 2020). Human studies support these findings, revealing that individuals with borderline diabetes who took chlorella supplements for 12 weeks experienced reduced insulin resistance markers such as resistin, as well as lower levels of inflammation-related factors like TNF-α and

Fig. 8.8 Chlorella

IL-6. These results suggest that chlorella may help prevent diabetic nephropathy (DN) and other diabetes-related complications by enhancing glycemic control and reducing inflammation (de Groot et al. 2001) (Fig. 8.8).

Since high cholesterol and triglyceride levels can exacerbate kidney issues, chlorella may help improve kidney health in patients with diabetic nephropathy (DN) by lowering these lipid levels and reducing oxidative stress. Incorporating chlorella into the diet could assist in managing cholesterol levels and related complications in DN. Its antioxidant properties may also benefit DN patients by decreasing oxidative stress, a factor that contributes to kidney damage in this condition (Senthilkumar and Ashokkumar 2012; Vijayavel et al. 2007). Chlorella is rich in protein, dietary fiber, polyunsaturated fatty acids, and essential vitamins such as D2 and B12, offering numerous health benefits. Its nutritional profile is further bolstered by high levels of iron and folate. Research suggests that chlorella has lipid-lowering, anti-diabetic, anti-hypertensive, immunomodulatory, and antioxidant properties, likely due to the combined effects of its nutritional and antioxidant components. While current studies highlight its health benefits, further research is needed to discover additional bioactive compounds that may enhance these therapeutic effects, solidifying chlorella's role as a valuable natural supplement (Bito et al. 2020). These algae and their bioactive compounds show significant potential in treating diabetic nephropathy and diabetes. They contain vital vitamins, polyunsaturated fatty acids, and antioxidants that offer a strategy for managing DN. Their antioxidant, anti-inflammatory, and cholesterol-lowering properties are crucial in addressing the primary factors contributing to DN, such as oxidative stress, inflammation, and disrupted lipid metabolism. By reducing inflammation and oxidative damage, these compounds help mitigate kidney injury caused by chronic hyperglycemia. Additionally, they regulate lipid metabolism and enhance insulin sensitivity, which is beneficial in managing diabetes-related complications, including DN.

7 Conclusion

Algal-based substances are emerging as a promising supplementary therapy as research progresses, offering a natural alternative to conventional diabetic treatments that may improve kidney function and patient outcomes. The bioactive compounds found in algae have the potential to target critical pathophysiological mechanisms, making them a viable treatment option for diabetes and diabetic nephropathy. These substances influence several key pathways that play a significant role in the development and progression of diabetic nephropathy (DN), including NF-κB, PKC, Nrf2, and AGE. The NF-κB pathway is crucial for managing oxidative stress and inflammation, both of which contribute to kidney damage in DN. Algal compounds have shown the ability to modulate NF-κB activity, potentially reducing oxidative damage and inflammation, thereby slowing the progression of kidney impairment.

Additionally, algal extracts impact the PKC pathway, which is associated with kidney fibrosis and vascular dysfunction, presenting a possible approach to mitigate renal damage caused by diabetes. Moreover, algae enhance the body's defense against oxidative stress by activating Nrf2, an important antioxidant regulator, which offers further protection to the kidneys. Finally, algal substances may help reduce damage related to advanced glycation end-products (AGE) and support kidney function by influencing the AGE pathway, which is vital in the accumulation of harmful substances that lead to renal fibrosis. These findings suggest that algae can serve as a safe and effective complement to traditional diabetes and DN therapies by targeting these essential pathways. Continued research into algae's role in preventing diabetic complications could pave the way for innovative, personalized treatments with fewer side effects, ultimately improving both kidney function and overall patient outcomes.

References

Abe, M., Maruyama, N., Okada, K., Matsumoto, S., Matsumoto, K., & Soma, M. (2011). Effects of Lipid-Lowering Therapy with Rosuvastatin on Kidney Function and Oxidative Stress in Patients with Diabetic Nephropathy. *Journal of Atherosclerosis and Thrombosis*, *18*(11), 1018–1028. https://doi.org/10.5551/jat.9084

Abo-Shady, A. M., Gheda, S. F., Ismail, G. A., Cotas, J., Pereira, L., & Abdel-Karim, O. H. (2023). Antioxidant and Antidiabetic Activity of Algae. *Life*, *13*(2), 460. https://doi.org/10.3390/life13020460

Bito, T., Okumura, E., Fujishima, M., & Watanabe, F. (2020). Potential of Chlorella as a Dietary Supplement to Promote Human Health. *Nutrients*, *12*(9), 2524. https://doi.org/10.3390/nu12092524

Bortolini, D. G., Maciel, G. M., de Fernandes, I. A. A., Pedro, A. C., Rubio, F. T. V., Branco, I. G., & Haminiuk, C. W. I. (2022). Functional properties of bioactive compounds from Spirulina spp.: Current status and future trends. *Food Chemistry. Molecular Sciences*, *5*, 100134. https://doi.org/10.1016/j.fochms.2022.100134

Brownlee, M. (2005). The pathobiology of diabetic complications: A unifying mechanism. *Diabetes*, *54*(6), 1615–1625. https://doi.org/10.2337/diabetes.54.6.1615

Celi, C., Fino, D., & Savorani, F. (2022). *Phaeodactylum tricornutum* as a source of value-added products: A review on recent developments in cultivation and extraction technologies. *Bioresource Technology Reports*, *19*, 101122. https://doi.org/10.1016/j.biteb.2022.101122

Chen, Q., Tao, J., & Xie, X. (2018). Astaxanthin Promotes Nrf2/ARE Signaling to Inhibit HG-Induced Renal Fibrosis in GMCs. *Marine Drugs*, *16*(4), 117. https://doi.org/10.3390/md16040117

Cho, C.-H., Lee, C.-J., Kim, M.-G., Ryu, B., Je, J.-G., Kim, Y., & Lee, S.-H. (2022). Therapeutic Potential of Phlorotannin-Rich Ecklonia cava Extract on Methylglyoxal-Induced Diabetic Nephropathy in In Vitro Model. *Marine Drugs*, *20*(6), Article 6. https://doi.org/10.3390/md20060355

Cho, C.-H., Yoo, G., Kim, M., Kurniawati, U. D., Choi, I.-W., & Lee, S.-H. (2023). Dieckol, Derived from the Edible Brown Algae Ecklonia cava, Attenuates Methylglyoxal-Associated Diabetic Nephropathy by Suppressing AGE-RAGE Interaction. *Antioxidants (Basel, Switzerland)*, *12*(3), 593. https://doi.org/10.3390/antiox12030593

Citi, V., Torre, S., Flori, L., Usai, L., Aktay, N., Dunford, N. T., Lutzu, G. A., & Nieri, P. (2024). Nutraceutical Features of the Phycobiliprotein C-Phycocyanin: Evidence from Arthrospira platensis (Spirulina). *Nutrients*, *16*(11), Article 11. https://doi.org/10.3390/nu16111752

Davies, M. J., Aroda, V. R., Collins, B. S., Gabbay, R. A., Green, J., Maruthur, N. M., Rosas, S. E., Prato, S. D., Mathieu, C., Mingrone, G., Rossing, P., Tankova, T., Tsapas, A., & Buse, J. B. (2022). Management of Hyperglycemia in Type 2 Diabetes, 2022. A Consensus Report by the American Diabetes Association (ADA) and the European Association for the Study of Diabetes (EASD). *Diabetes Care*, *45*(11), 2753. https://doi.org/10.2337/dci22-0034

Davis, T. A., Volesky, B., & Mucci, A. (2003). A review of the biochemistry of heavy metal biosorption by brown algae. *Water Research*, *37*(18), 4311–4330. https://doi.org/10.1016/S0043-1354(03)00293-8

de Groot, M., Anderson, R., Freedland, K. E., Clouse, R. E., & Lustman, P. J. (2001). Association of depression and diabetes complications: A meta-analysis. *Psychosomatic Medicine*, *63*(4), 619–630. https://doi.org/10.1097/00006842-200107000-00015

de Zeeuw, D., Remuzzi, G., Parving, H.-H., Keane, W. F., Zhang, Z., Shahinfar, S., Snapinn, S., Cooper, M. E., Mitch, W. E., & Brenner, B. M. (2004). Albuminuria, a therapeutic target for cardiovascular protection in type 2 diabetic patients with nephropathy. *Circulation*, *110*(8), 921–927. https://doi.org/10.1161/01.CIR.0000139860.33974.28

Drucker, D. J., & Nauck, M. A. (2006). The incretin system: Glucagon-like peptide-1 receptor agonists and dipeptidyl peptidase-4 inhibitors in type 2 diabetes. *Lancet (London, England)*, *368*(9548), 1696–1705. https://doi.org/10.1016/S0140-6736(06)69705-5

Forbes, J. M., Coughlan, M. T., & Cooper, M. E. (2008). Oxidative stress as a major culprit in kidney disease in diabetes. *Diabetes*, *57*(6), 1446–1454. https://doi.org/10.2337/db08-0057

Fu, J., Lee, K., Chuang, P. Y., Liu, Z., & He, J. C. (2014). Glomerular endothelial cell injury and cross talk in diabetic kidney disease. *American Journal of Physiology - Renal Physiology*, *308*(4), F287. https://doi.org/10.1152/ajprenal.00533.2014

Gentscheva, G., Nikolova, K., Panayotova, V., Peycheva, K., Makedonski, L., Slavov, P., Radusheva, P., Petrova, P., & Yotkovska, I. (2023). Application of Arthrospira platensis for Medicinal Purposes and the Food Industry: A Review of the Literature. *Life (Basel, Switzerland)*, *13*(3), 845. https://doi.org/10.3390/life13030845

Hong, M., Nie, Z., Chen, Z., & Bao, B. (2024). Astaxanthin attenuates diabetic kidney injury through upregulation of autophagy in podocytes and pathological crosstalk with mesangial cells. *Renal Failure*, *46*(2), 2378999. https://doi.org/10.1080/0886022X.2024.2378999

Huang, K., Chen, C., Hao, J., Huang, J., Wang, S., Liu, P., & Huang, H. (2015). Polydatin promotes Nrf2-ARE anti-oxidative pathway through activating Sirt1 to resist AGEs-induced upregulation of fibronetin and transforming growth factor-β1 in rat glomerular messangial cells. *Molecular and Cellular Endocrinology*, *399*, 178–189. https://doi.org/10.1016/j.mce.2014.08.014

Jiang, T., Huang, Z., Lin, Y., Zhang, Z., Fang, D., & Zhang, D. D. (2010). The Protective Role of Nrf2 in Streptozotocin-Induced Diabetic Nephropathy. *Diabetes*, *59*(4), 850. https://doi.org/10.2337/db09-1342

Kim, M.-K. (2017). Treatment of diabetic kidney disease: Current and future targets. *The Korean Journal of Internal Medicine*, *32*(4), 622. https://doi.org/10.3904/kjim.2016.219

Kiritoshi, S., Nishikawa, T., Sonoda, K., Kukidome, D., Senokuchi, T., Matsuo, T., Matsumura, T., Tokunaga, H., Brownlee, M., & Araki, E. (2003). Reactive oxygen species from mitochondria induce cyclooxygenase-2 gene expression in human mesangial cells: Potential role in diabetic nephropathy. *Diabetes*, *52*(10), 2570–2577. https://doi.org/10.2337/diabetes.52.10.2570

Kopel, J., Pena-Hernandez, C., & Nugent, K. (2019). Evolving spectrum of diabetic nephropathy. *World Journal of Diabetes*, *10*(5), 269. https://doi.org/10.4239/wjd.v10.i5.269

Krishan, P., & Chakkarwar, V. A. (2011, June 1). *Diabetic nephropathy: Aggressive involvement of oxidative stress. | EBSCOhost*. https://openurl.ebsco.com/contentitem/gcd:61821860?sid=ebsco:plink:crawler&id=ebsco:gcd:61821860

Lee, B.-H., Hsu, W.-H., Hsu, Y.-W., & Pan, T.-M. (2013). Dimerumic acid attenuates receptor for advanced glycation endproducts signal to inhibit inflammation and diabetes mediated by Nrf2 activation and promotes methylglyoxal metabolism into d-lactic acid. *Free Radical Biology and Medicine*, *60*, 7–16. https://doi.org/10.1016/j.freeradbiomed.2013.01.030

Li, X., Lu, L., Hou, W., Huang, T., Chen, X., Qi, J., Zhao, Y., & Zhu, M. (2021). Epigenetics in the pathogenesis of diabetic nephropathy: Epigenetics in the pathogenesis of diabetic nephropathy. *Acta Biochimica et Biophysica Sinica*, *54*(2), 163. https://doi.org/10.3724/abbs.2021016

Lim, A. K. H., & Tesch, G. H. (2012). Inflammation in Diabetic Nephropathy. *Mediators of Inflammation*, *2012*, 146154. https://doi.org/10.1155/2012/146154

Lourenço-Lopes, C., Garcia-Oliveira, P., Carpena, M., Fraga-Corral, M., Jimenez-Lopez, C., Pereira, A. G., Prieto, M. A., & Simal-Gandara, J. (2020). Scientific Approaches on Extraction, Purification and Stability for the Commercialization of Fucoxanthin Recovered from Brown Algae. *Foods*, *9*(8), Article 8. https://doi.org/10.3390/foods9081113

Menne, J., Park, J.-K., Boehne, M., Elger, M., Lindschau, C., Kirsch, T., Meier, M., Gueler, F., Fiebeler, A., Bahlmann, F. H., Leitges, M., & Haller, H. (2004). Diminished loss of proteoglycans and lack of albuminuria in protein kinase C-alpha-deficient diabetic mice. *Diabetes*, *53*(8), 2101–2109. https://doi.org/10.2337/diabetes.53.8.2101

Mogensen, C. E., Christensen, C. K., & Vittinghus, E. (1983). The Stages in Diabetic Renal Disease: With Emphasis on the Stage of Incipient Diabetic Nephropathy. *Diabetes*, *32*(Supplement_2), 64–78. https://doi.org/10.2337/diab.32.2.S64

Naito, Y., Uchiyama, K., Aoi, W., Hasegawa, G., Nakamura, N., Yoshida, N., Maoka, T., Takahashi, J., & Yoshikawa, T. (2004). Prevention of diabetic nephropathy by treatment with astaxanthin in diabetic db/db mice. *BioFactors*, *20*(1), 49–59. https://doi.org/10.1002/biof.5520200105

Oslan, S. N. H., Tan, J. S., Oslan, S. N., Matanjun, P., Mokhtar, R. A. M., Shapawi, R., & Huda, N. (2021). *Haematococcus pluvialis* as a Potential Source of Astaxanthin with Diverse Applications in Industrial Sectors: Current Research and Future Directions. *Molecules*, *26*(21), 6470. https://doi.org/10.3390/molecules26216470

Papadopoulos, K. P., Economou, C. N., Markou, G., Nicodemou, A., Koutinas, M., Tekerlekopoulou, A. G., & Vayenas, D. V. (2022). Cultivation of Arthrospira platensis in Brewery Wastewater. *Water*, *14*(10), Article 10. https://doi.org/10.3390/w14101547

Peng, J., Yuan, J.-P., Wu, C.-F., & Wang, J.-H. (2011). Fucoxanthin, a Marine Carotenoid Present in Brown Seaweeds and Diatoms: Metabolism and Bioactivities Relevant to Human Health. *Marine Drugs*, *9*(10), 1806. https://doi.org/10.3390/md9101806

Pereira, L., & Valado, A. (2023). Algae-Derived Natural Products in Diabetes and Its Complications-Current Advances and Future Prospects. *Life*, *13*(9), 1831. https://doi.org/10.3390/life13091831

Pradhan, B., Nayak, R., Patra, S., Jit, B. P., Ragusa, A., & Jena, M. (2020). Bioactive Metabolites from Marine Algae as Potent Pharmacophores against Oxidative Stress-Associated Human

Diseases: A Comprehensive Review. *Molecules*, *26*(1), 37. https://doi.org/10.3390/molecules26010037

Qi, C., Mao, X., Zhang, Z., & Wu, H. (2017). Classification and Differential Diagnosis of Diabetic Nephropathy. *Journal of Diabetes Research*, *2017*, 8637138. https://doi.org/10.1155/2017/8637138

Quiroga, B., Arroyo, D., & de Arriba, G. (2015). Present and Future in the Treatment of Diabetic Kidney Disease. *Journal of Diabetes Research*, *2015*, 801348. https://doi.org/10.1155/2015/801348

Samsu, N. (2021). Diabetic Nephropathy: Challenges in Pathogenesis, Diagnosis, and Treatment. *BioMed Research International*, *2021*, 1497449. https://doi.org/10.1155/2021/1497449

Senthilkumar, T., & Ashokkumar, N. (2012). Impact of *Chlorella pyrenoidosa* on the attenuation of hyperglycemia-mediated oxidative stress and protection of kidney tissue in streptozotocin-cadmium induced diabetic nephropathic rats. *Biomedicine & Preventive Nutrition*, *2*(2), 125–131. https://doi.org/10.1016/j.bionut.2012.01.006

Stratton, I. M., Adler, A. I., Neil, H. A. W., Matthews, D. R., Manley, S. E., Cull, C. A., Hadden, D., Turner, R. C., & Holman, R. R. (2000). Association of glycaemia with macrovascular and microvascular complications of type 2 diabetes (UKPDS 35): Prospective observational study. *BMJ : British Medical Journal*, *321*(7258), 405. https://doi.org/10.1136/bmj.321.7258.405

Takebayashi, K., Matsumoto, S., Aso, Y., & Inukai, T. (2006). Aldosterone blockade attenuates urinary monocyte chemoattractant protein-1 and oxidative stress in patients with type 2 diabetes complicated by diabetic nephropathy. *The Journal of Clinical Endocrinology and Metabolism*, *91*(6), 2214–2217. https://doi.org/10.1210/jc.2005-1718

Tan, S. M., & de Haan, J. B. (2014). Combating oxidative stress in diabetic complications with Nrf2 activators: How much is too much? *Redox Report : Communications in Free Radical Research*, *19*(3), 107. https://doi.org/10.1179/1351000214Y.0000000087

Tavafi, M. (2013). Diabetic nephropathy and antioxidants. *Journal of Nephropathology*, *2*(1), 20. https://doi.org/10.5812/nephropathol.9093

Varghese, R. T., & Jialal, I. (2024). Diabetic Nephropathy. In *StatPearls*. StatPearls Publishing. http://www.ncbi.nlm.nih.gov/books/NBK534200/

Vijayavel, K., Anbuselvam, C., & Balasubramanian, M. P. (2007). Antioxidant effect of the marine algae *Chlorella vulgaris* against naphthalene-induced oxidative stress in the albino rats. *Molecular and Cellular Biochemistry*, *303*(1–2), 39–44. https://doi.org/10.1007/s11010-007-9453-2

Wang, J., Geng, L., Yue, Y., & Zhang, Q. (2019). Use of fucoidan to treat renal diseases: A review of 15 years of clinic studies. In *Progress in Molecular Biology and Translational Science* (Vol. 163, pp. 95–111). Elsevier. https://doi.org/10.1016/bs.pmbts.2019.03.011

Wijesekara, I., Yoon, N. Y., & Kim, S.-K. (2010). Phlorotannins from *Ecklonia cava* (Phaeophyceae): Biological activities and potential health benefits. *BioFactors (Oxford, England)*, *36*(6), 408–414. https://doi.org/10.1002/biof.114

Xiao, L., Xu, X., Zhang, F., Wang, M., Xu, Y., Tang, D., Wang, J., Qin, Y., Liu, Y., Tang, C., He, L., Greka, A., Zhou, Z., Liu, F., Dong, Z., & Sun, L. (2016). The mitochondria-targeted antioxidant MitoQ ameliorated tubular injury mediated by mitophagy in diabetic kidney disease *via* Nrf2/PINK1. *Redox Biology*, *11*, 297. https://doi.org/10.1016/j.redox.2016.12.022

Xie, X., Peng, J., Chang, X., Huang, K., Huang, J., Wang, S., Shen, X., Liu, P., & Huang, H. (2013). Activation of RhoA/ROCK regulates NF-κB signaling pathway in experimental diabetic nephropathy. *Molecular and Cellular Endocrinology*, *369*(1–2), 86–97. https://doi.org/10.1016/j.mce.2013.01.007

Xu, J., Wang, Y., Wang, Z., Guo, L., & Li, X. (2021). Fucoidan mitigated diabetic nephropathy through the downregulation of PKC and modulation of NF-κB signaling pathway: In vitro and in vivo investigations. *Phytotherapy Research*, *35*(4), 2133–2144. https://doi.org/10.1002/ptr.6966

Xu, Y., Zhang, Q., Luo, D., Wang, J., & Duan, D. (2016). Low molecular weight fucoidan modulates P-selectin and alleviates diabetic nephropathy. *International Journal of Biological Macromolecules*, *91*, 233–240. https://doi.org/10.1016/j.ijbiomac.2016.05.081

Yang, G., Jin, L., Zheng, D., Tang, X., Yang, J., Fan, L., & Xie, X. (2019). Fucoxanthin Alleviates Oxidative Stress through Akt/Sirt1/FoxO3α Signaling to Inhibit HG-Induced Renal Fibrosis in GMCs. *Marine Drugs*, *17*(12), 702. https://doi.org/10.3390/md17120702

Zheng, J., Inoguchi, T., Sasaki, S., Maeda, Y., McCarty, M. F., Fujii, M., Ikeda, N., Kobayashi, K., Sonoda, N., & Takayanagi, R. (2013). Phycocyanin and phycocyanobilin from *Spirulina platensis* protect against diabetic nephropathy by inhibiting oxidative stress. *American Journal of Physiology-Regulatory, Integrative and Comparative Physiology*, *304*(2), R110–R120. https://doi.org/10.1152/ajpregu.00648.2011

Zou, T.-B., Jia, Q., Li, H.-W., Wang, C.-X., & Wu, H.-F. (2013). Response Surface Methodology for Ultrasound-Assisted Extraction of Astaxanthin from *Haematococcus pluvialis*. *Marine Drugs*, *11*(5), 1644. https://doi.org/10.3390/md11051644

Chapter 9
Nutraceutical Potential of Algae in Reducing Obesity Related to Diabetes

Ankita Jain and Samiya Pawne

Abstract The marine ecosystem is considered a potentially mysterious reservoir of biodiversity consisting of a broad spectrum of bioactive constituents, whose potential is still under research for pharmaceuticals, nutraceuticals, and cosmeceuticals. The prevalence of obesity has increased and is a health concern worldwide. Due to the concerns regarding synthetic anti-obesity treatments, nowadays natural products have become a trend. Marine algae, especially seaweeds, are a promising source of anti-obesity agents. Marine algae, particularly species like *Laminaria*, *Undaria pinnatifida*, and *Ecklonia cava*, are rich in bioactive compounds such as fucoxanthin, alginates, and polyphenols. These compounds exhibit significant anti-obesity and anti-diabetic properties, primarily through mechanisms like enhanced lipid metabolism, improved insulin sensitivity, and appetite suppression. Fucoxanthin, a carotenoid found in brown algae, has been shown to stimulate thermogenesis, thereby promoting fat loss and reducing body weight. Additionally, alginates have been demonstrated to decrease fat absorption and improve satiety, leading to reduced calorie intake. Polyphenols from marine algae possess strong antioxidant properties, which help mitigate oxidative stress and inflammation, both of which are critical in the pathogenesis of obesity-related diabetes. Clinical and preclinical studies indicate that marine algae can improve lipid profiles, reduce body fat, and enhance glucose metabolism, making them a potent natural intervention for managing obesity and preventing type 2 diabetes. These findings underscore the potential of marine algae as a functional food ingredient, although further research is needed to establish standardized dosages and long-term safety. Studies demonstrate that incorporating algal supplements into the diet can suppress appetite, lower body fat, and improve lipid profiles, offering a multifaceted approach to tackling obesity and diabetes. The findings underscore algae's role as a promising, natural intervention for metabolic health, although further research is necessary to establish standardized dosages and long-term effects.

A. Jain (✉) · S. Pawne
Tilak College of Science and Commerce Affiliated to University of Mumbai, Mumbai, India

R. Raja et al. (eds.), *Algae in Diabetes Management*,
https://doi.org/10.1007/978-981-95-0571-5_9

Keywords Marine ecosystem · Biodiversity · Bioactive compounds · Marine algae · Seaweeds

1 Introduction

- Algae have garnered attention as a sustainable and potent source of bioactive compounds with wide-ranging health benefits. These photosynthetic organisms are rich in secondary metabolites, including carotenoids, phenolics, polysaccharides, and unsaturated fatty acids, which exhibit antioxidant, anti-inflammatory, and hypolipidemic properties. Such bioactive components play a critical role in preventing oxidative damage and managing chronic diseases, including diabetes. Algal extracts have been shown to support cardiovascular health, modulate lipid profiles, and reduce inflammation, making them a promising nutraceutical option. Furthermore, their versatility in cultivation and ability to produce targeted bioactive compounds on a large scale highlight their potential as reliable and eco-friendly resources for health interventions (Shalaby 2011).
- In addition to their biological activities, algae have been an integral part of traditional diets in various cultures, offering essential nutrients and therapeutic benefits. Among the tribes of northeast India, including the Monpa community in Arunachal Pradesh, algae have been consumed for centuries as a vital dietary component. Freshwater species like *Ulva intestinalis* are valued for their high levels of amino acids, dietary fiber, and bioactive compounds that promote metabolic health. These algae are commonly used in traditional recipes, such as soups and meat dishes, particularly during cultural festivals. Such practices not only reflect the nutritional significance of algae but also underscore their potential in promoting health and preventing conditions like obesity and diabetes through natural, culturally embedded dietary habits (Das 2016).
- One promising area for utilizing bioactive compounds derived from microalgae is in addressing obesity, a significant health challenge with a high prevalence worldwide. Obesity is not only a concern in itself but also a major contributor to the development of chronic conditions such as diabetes, cardiovascular diseases, and certain types of cancer. Although modern anti-obesity medications provide a safer and more effective complement to lifestyle changes for most individuals, their success is often hindered by challenges such as side effects, poor adherence, or discontinuation by patients. These limitations can significantly diminish the overall effectiveness of treatment outcomes (Gómez-Zorita et al. 2019).
- Not all algae are fit for consumption, as some species produce harmful toxins that pose significant risks to human and animal health. Toxic algae, like *Alexandrium catenella* and *Alexandrium excavatum*, are marine species known for producing some of the most potent and fast-acting toxins. These substances can lead to severe poisoning, with no specific antidote available, leaving only supportive care as a recourse (Safiyanu et al. 2018). On the other hand, many non-toxic algae are highly nutritious and have been safely incorporated into

human diets for centuries (Shalaby 2011; Das 2016). Understanding the differences between toxic and non-toxic algae is crucial for safely harnessing their nutritional potential while mitigating health risks associated with toxic varieties.

2 Key Algal Species with Anti-obesity and Anti-diabetic Properties

1. Brown Algae:

It is a class of algae called Phaeophyceae. There are approximately 1500 to 2000 species of brown algae found across the globe. *Undaria pinnatifida*, *Saccharina japonica* formerly known as *Laminaria japonica*, and *Saccharina latissima* are some of the most commonly consumed species of brown algae. These algal species are rich in bioactive compounds such as polyphenols, polysaccharides, peptides, phytosterols, alginate, laminarin, fucoxanthin, and flavonoids, which exert antioxidant, anti-inflammatory, and anti-diabetic properties. These bioactive compounds have been found to show antioxidant, anti-inflammatory, anti-diabetic, and immunoregulatory effects that can help in reducing obesity and metabolic diseases (Jang et al. 2024) (Image 9.1).

Image 9.1 Brown Algae–Giant kelp. (Source: https://en.wikipedia.org)

2. Red Algae:

   Red algae, belonging to the phylum Rhodophyta, are a diverse group of marine organisms. They are known for their unique bioactive compounds such as polysaccharides, polyphenols, and carotenoids. These compounds have demonstrated significant potential in addressing obesity and diabetes due to their anti-inflammatory, antioxidant, and metabolism-regulating properties. The bioactive constituents in red algae can help mitigate insulin resistance, improve lipid profiles, and manage hyperglycemia, making them valuable in tackling obesity-related diabetes. Gigartina sp., Chondrus sp., Eucheuma sp., and Iridaea sp. belong to the class of red algae out of which edible red algae Gloiopeltis furcata have been studied for the effects on obesity, diabetes, and related metabolic diseases, and the result showed positive impact (Murakami et al. 2022) (Image 9.2).

3. Green Algae:

   The green algae belong to the phylum Chlorophyta. There are around 4548 species. Even with such heterogeneity it is considered to be one of the least exploited algae (Qudus et al. 2020). The green algae belonging to *Caulerpa* sp. and *Ulva* sp. are considered to be the richest source of bioactive compounds and can help in the reduction of obesity and diabetes (Kurniawan et al. 2023; Chen et al. 2022) (Image 9.3).

Image 9.2 *Gracillaria* sp. (Source: https://en.wikipedia.org)

Image 9.3 *Caulerpa* sp. (Source: https://en.wikipedia.org)

Image 9.4 Microalgae. (Source: https://en.wikipedia.org)

4. Microalgae:

   Microalgae are unicellular photosynthetic algae that are of micrometer dimension (Tamel et al. 2023). Microalgae from the genus *Chlorella*, *Spirulina*, and *Haematococus* are the most well-studied. Microalgae are rich in bioactive compounds: carotenoids, chlorophylls, sterols, polyunsaturated fatty acids, vitamins, minerals, fiber, polysaccharides (Gómez-Zorita et al. 2019). These compounds have been seen to have anti-obesity and anti-diabetic property (Tamel et al. 2023) (Image 9.4).

3 What Is Diabetes and Its Types?

In order to understand diabetes, we need to first understand the following:

- Glucose: Our body's main source of energy is glucose which is also referred to as "sugar". Our body has potential to produce glucose on its own. We also receive glucose from the food we eat (https://www.niddk.nih.gov/health-information/diabetes/overview/what-is-diabetes). Glucose is a 6-carbon structure with molecular formula $C_6H_{12}O_6$. Glucose often enters our body in isomeric forms such as galactose and fructose (monosaccharides), lactose and sucrose (disaccharides), or starch (polysaccharides). Excess glucose is stored in the body as glycogen, a glucose polymer, utilized during fasting. In addition, glucose can be produced through gluconeogenesis, a process involving the breakdown of fats and proteins. Given the paramount importance of carbohydrates in maintaining homeostasis, numerous sources contribute to glucose production (Hantzidiamantis et al. 2024).
- Insulin: Insulin is a hormone composed of 51 amino acids, produced by the pancreas of our body. Insulin helps to digest and convert glucose into energy.
- Diabetes: Diabetes is a condition of our body where the body fails to produce enough insulin to digest glucose. Due to lack of insulin content of glucose/sugar increases in the body. This excessive sugar stays in the blood. High sugar in the

blood causes many serious health complications. This condition of the body is called diabetes (https://www.who.int/news-room/fact-sheets/detail/diabetes).

- Types of Diabetes:

  (a) Type 1 diabetes is a condition where the immune system of body mistakenly attacks and destroys β-cell. This destruction of β-cell leads to absolute insulin deficiency.
  (b) Type 2 diabetes is caused when the pancreas of our body is unable to produce enough insulin or the body becomes unable to use insulin properly to digest and convert sugar/glucose into energy. This causes high sugar levels in the blood.
  (c) There are other specific types of diabetes due to other causes, for instance, monogenic diabetes syndromes (such as neonatal diabetes and maturity-onset diabetes of the young), diseases of the exocrine pancreas (such as cystic fibrosis and pancreatitis), and drug- or chemical-induced diabetes (such as with glucocorticoid use, in the treatment of HIV/AIDS, or after organ transplantation) (American Diabetes Association Professional Practice Committee 2022).

4 Obesity and Its Role in Diabetes

Obesity is a significant risk factor for the development of Type 2 diabetes (T2DM). The risk of diabetes increases dramatically with higher body mass index (BMI), a measure of body weight relative to height. For instance, the lifetime risk of diabetes rises sharply from 7% to 70% in men, and from 12% to 74% in women, as BMI increases from less than 18.5 kg/m^2 to more than 35 kg/m^2. This highlights the importance of screening for diabetes in individuals with obesity. In fact, addressing obesity is considered the cornerstone of preventing and managing T2DM. Weight loss has been shown to significantly reduce the incidence of diabetes in high-risk individuals. Studies have demonstrated that lifestyle modifications, including a modest weight loss of 5-10% of body weight combined with regular physical activity, can reduce the risk of diabetes by more than 50%. Similarly, bariatric surgery has been found to drastically reduce the incidence of diabetes over time, with one study showing a five-fold reduction in diabetes risk over a seven-year period. Weight reduction also plays a key role in the treatment of existing T2DM, with glycemic control improving as weight decreases, and in some cases, leading to remission. Consequently, the management of T2DM begins with lifestyle changes, followed by pharmacological treatments and, if necessary, surgical interventions (Yashi and Daley 2024) (Fig. 9.1).

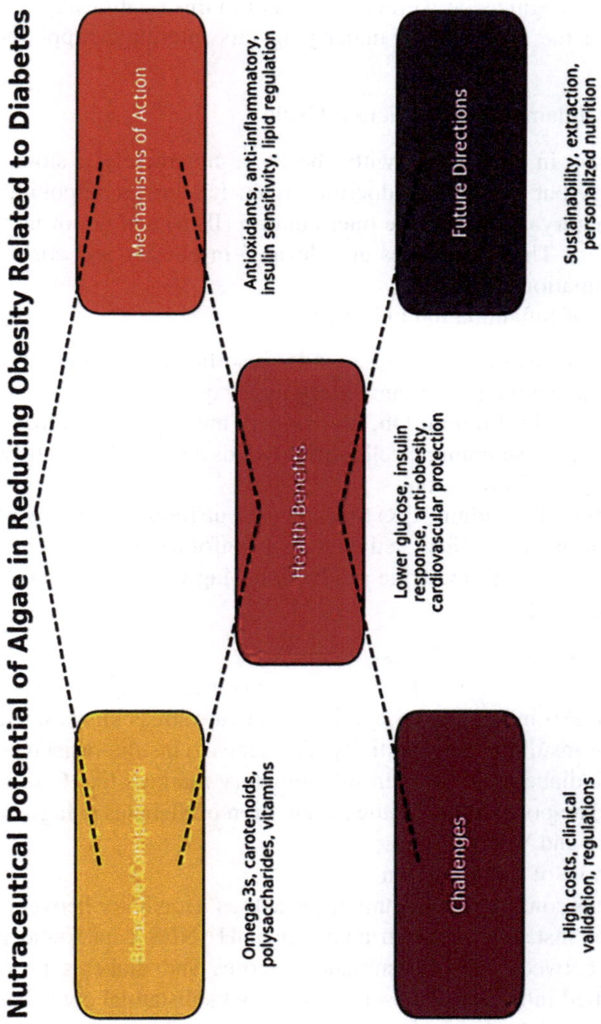

Fig. 9.1 Nutraceutical potential of algae in reducing obesity related diabetes

5 Interconnection Between Obesity, Diabetes, and Inflammation

The global prevalence of obesity and type 2 diabetes has surged, prompting a deeper examination of the biological factors linking these conditions. Chronic low-grade inflammation is recognized as a crucial mediator in this relationship. The following sections explore the mechanisms underlying this interplay, supported by recent research.

- Obesity and Inflammation: A Vicious Cycle

- Adipose tissue in individuals with obesity is not merely a storage depot for excess energy but an active endocrine organ. It secretes adipokines, including pro-inflammatory cytokines like interleukin-6 (IL-6) and tumor necrosis factor-alpha (TNF-α). These cytokines are elevated in obesity and contribute to systemic inflammation.
- Mechanisms of Inflammation in Obesity

  1. Hypoxia: As adipose tissue expands, insufficient vascularization leads to hypoxia, activating pro-inflammatory pathways.
  2. Immune Cell Infiltration: Obesity triggers macrophage infiltration into adipose tissue. These immune cells shift towards a pro-inflammatory phenotype, exacerbating inflammation.
  3. Endoplasmic Reticulum (ER) Stress: Overnutrition causes misfolded protein accumulation in the ER, activating the unfolded protein response, which drives inflammatory cytokine production(Magwaza and Islam 2023; Pereira and Valado 2023).

- Inflammation's Role in Diabetes
  Chronic inflammation disrupts insulin signaling, a key process for glucose homeostasis. Pro-inflammatory cytokines activate stress kinases, such as JNK1, which impair insulin receptor activity. This leads to insulin resistance—a precursor to type 2 diabetes. Moreover, inflammatory markers like C-reactive protein (CRP) correlate positively with the progression of diabetes (Magwaza and Islam 2023; Pereira and Valado 2023).
- Mediating Role of Inflammation
  Recent studies confirm that inflammation acts as a mediator between obesity and diabetes. For instance, research analyzing NHANES data found a significant association between high inflammatory scores and diabetes prevalence. The study identified inflammation as an indirect but substantial contributor to diabetes development in individuals with elevated BMI (Authors from the University of Coimbra 2024).
- Broader Health Implications
  Beyond diabetes, inflammation related to obesity is linked to cardiovascular diseases (CVD). Adipose tissue inflammation promotes atherogenesis by releasing cytokines that impair endothelial function and facilitate vascular damage(Pereira and Valado 2023; Authors from the University of Coimbra 2024).
- Therapeutic Approaches

1. Lifestyle Changes: Regular physical activity and dietary interventions can reduce inflammation and improve metabolic health.
2. Pharmacological Interventions: Anti-inflammatory drugs and therapies targeting cytokine pathways (e.g., IL-6 or TNF-α inhibitors) are being explored for their potential to manage diabetes and related complications.
3. Surgical Interventions: Bariatric surgery not only reduces adiposity but also alleviates systemic inflammation, improving insulin sensitivity (Magwaza and Islam 2023; Pereira and Valado 2023).

6 Algae in Obesity and Diabetes

Algae have demonstrated significant potential in managing both obesity and diabetes due to their multifaceted mechanisms in regulating metabolic processes (Gómez-Zorita et al. 2019). In obesity, algae such as *Spirulina maxima* and *Euglena gracilis* exhibit effects that inhibit the differentiation of pre-adipocytes into mature adipocytes, which helps reduce triglyceride accumulation. For instance, *Euglena gracilis* extracts can reduce triglyceride content in adipocytes by up to 74%, while also downregulating key adipogenic markers like PPARγ and C/EBPα, which are involved in fat cell differentiation. Similarly, *Spirulina maxima* enhances thermogenesis in white adipose tissue by upregulating uncoupling protein 1 (UCP1), boosting energy expenditure through lipolysis and fatty acid oxidation (Gómez-Zorita et al. 2019). Brown algae, including *Saccharina japonica* and *Undaria pinnatifida,* further contribute to obesity management by containing bioactive compounds like fucoidan and alginate, which inhibit lipid absorption and promote the growth of beneficial gut microbiota. These microbiota increase the production of short-chain fatty acids (SCFAs) that aid lipid metabolism and prevent fat storage (Jang et al. 2024; Qudus et al. 2020). Additionally, fucoxanthin, a carotenoid found in brown algae, stimulates energy expenditure by inducing browning in white adipose tissue, thus enhancing metabolic activity and reducing visceral fat accumulation. The combined effects of these algae in regulating fat accumulation, boosting energy utilization, and enhancing lipid metabolism make them valuable candidates for obesity management (Jang et al. 2024). In the context of diabetes, algae offer compelling anti-diabetic properties by improving glucose metabolism and insulin sensitivity. Green algae, such as *Caulerpa racemosa*, have been shown to lower fasting glucose levels and protect against diabetic complications (Gómez-Zorita et al. 2019). Fucoidan, a sulfated polysaccharide from brown algae, helps manage glucose levels by inhibiting sodium-glucose cotransporter 1 (SGLT1), thus reducing postprandial glucose spikes (Jang et al. 2024). Furthermore, polyphenols like phlorotannins, found in *Ecklonia cava*, inhibit digestive enzymes such as alpha-amylase and alpha-glucosidase, which slow carbohydrate breakdown and absorption, leading to better blood sugar control (Jang et al. 2024). Carotenoids like fucoxanthin enhance insulin signaling, improving glucose uptake and reducing insulin resistance, as shown in diabetic animal models (Gómez-Zorita et al. 2019; Jang et al. 2024). Additionally, algae's antioxidant compounds protect pancreatic beta cells from oxidative stress, a major contributor to diabetes progression (Qudus et al. 2020; Tamel et al. 2023). By targeting multiple pathways, from enzyme inhibition to antioxidant defense, algae offer a promising strategy for the prevention and management of diabetes. Their ability to modulate both glucose metabolism and insulin sensitivity, along with their anti-inflammatory effects, underscores their potential as nutraceutical agents in managing obesity and diabetes simultaneously (Gómez-Zorita et al. 2019; Jang et al. 2024; Qudus et al. 2020).

7 Algal Bioactive Compounds with Anti-obesity and Anti-diabetic Effects

Various compounds such as alginates, fucoidans, fucoxanthin, fucosterol, and phlorotannins from marine algae have been proven to have potential for the control of obesity (Wan-Loy and Siew-Moi 2016; Asmita Kumari and Bharadvaja 2023). These bioactive compounds and their derivatives have been proven to play role in digestion and are considered as potential agents for obesity treatment. The mechanisms by which these compounds produce anti-obesity effects are generally associated with their role in modulating the lipoprotein breakdown, suppressing adipogenesis, and delaying stomach emptying (Asmita Kumari and Bharadvaja 2023).

1. Fucoxanthin: Fucoxanthin is a marine xanthophyll carotenoid (Wan-Loy and Siew-Moi 2016; Ye et al. 2022) abundantly found in brown algae like *Undaria pinnatifida and* microalga like *tricornutum and Isochrysis galbana* (Wan-Loy and Siew-Moi 2016; Asmita Kumari and Bharadvaja 2023). Out of all carotenoids fucoxanthin is the most abundant antioxidant with a unique structure. Moreover, it has been shown to have anti-obesity, anti-diabetic, anti-cancerous, anti-angiogenic, and anti-inflammatory properties (Hayato Maeda 2015). Animal models and human clinical trials have shown that fucoxanthin supplementation can lead to significant reductions in body fat, particularly visceral fat, and improvements in metabolic parameters like insulin sensitivity, blood glucose, and cholesterol levels (Hitoe and Shimoda 2017). One of the key mechanisms through which fucoxanthin exerts its anti-obesity effects is by stimulating the expression of uncoupling protein 1 in white adipose tissue (WAT). UCP-1 is a protein in the inner mitochondrial membrane that dissipates energy through heat production and fatty acid oxidation. This process is called thermogenesis, and it helps limit weight gain and favors weight loss (Hayato Maeda 2015; Grasa-López et al. 2016; Kozak and Anunciado-Koza 2008). UCP1 plays a crucial role in thermogenesis, by increasing energy expenditure and promoting the oxidation of stored fat. By activating UCP1, fucoxanthin helps to accelerate fat breakdown and reduce fat accumulation (Grasa-López et al. 2016; Kozak and Anunciado-Koza 2008). This fat is a significant contributor to metabolic syndrome as it releases pro-inflammatory cytokines that exacerbate insulin resistance (Wondmkun 2020). Hence, fucoxanthins may help in reducing obesity related to diabetes. In addition to enhancing thermogenesis, fucoxanthin has been shown

to inhibit adipogenesis, the process by which pre-adipocytes differentiate into mature fat cells. Studies in 3 T3-L1 adipocytes have demonstrated that fucoxanthin downregulates key transcription factors and enzymes involved in adipogenesis, such as peroxisome proliferator-activated receptor gamma (PPAR-γ) and fatty acid synthase, thereby preventing fat cell formation and lipid storage (Hayato Maeda 2015). Furthermore, fucoxanthin has been found to regulate lipid metabolism by reducing the activity of enzymes involved in fat storage, such as acetyl-CoA carboxylase (ACC) and fatty acid synthase (FAS), while also increasing the activity of enzymes that promote fat breakdown, like lipoprotein lipase (LPL) (Ha and Kim 2013). Fucoxanthin has shown promising effects on insulin signaling. By activating AMP-activated protein kinase (AMPK), it enhances glucose uptake in muscle cells and reduces hepatic glucose production. This dual action improves insulin sensitivity and aids in glucose homeostasis (Ye et al. 2022; Steinberg and Carling 2019). Oxidative stress exacerbates insulin resistance and β-cell dysfunction in diabetes. Algal antioxidants such as carotenoids like fucoxanthin can combat free radicals, protecting pancreatic β-cells and maintaining glucose regulation by proper secretion of insulin (Wan-Loy and Siew-Moi 2016; Tangvarasittichai 2015). The combined effects of enhancing fat oxidation, inhibiting fat accumulation, free radical scavenging activity, and improving metabolic health make fucoxanthin a promising natural compound for obesity and obesity-related diabetes prevention and management.

2. Alginates: Alginates are polysaccharides that also act as dietary fibers. They are well known for their anti-obesity potential. They are most commonly found in brown algae. The brown algae of the genus *Laminaria* and *Lessonia* are major sources of alginates (Wan-Loy and Siew-Moi 2016). Alginates exert their anti-obesity effects through multiple mechanisms, primarily related to their ability to interact with the digestive system, modulate fat absorption, influence gut microbiota, and regulate appetite. One of the primary mechanisms by which alginates may help reduce obesity is by decreasing the absorption of dietary fats. Alginates can bind to lipids in the gastrointestinal tract, preventing their absorption and facilitating their excretion. They have also been found to have inhibitory effect on lipase enzyme involved in the breakdown of triacylglycerol in the gastrointestinal tract, resulting in reduced uptake of triacylglycerol which can contribute to obesity (Wan-Loy and Siew-Moi 2016; Kumar and Brown 2018). The gel-forming property of alginates allows them to create a viscous matrix in the stomach and small intestine, which can physically block the absorption of fats and sugar. This has been shown to reduce postprandial lipidemia and glycemia; and help in preventing sugar spikes and decrease the overall caloric intake, thereby supporting weight management (Wan-Loy and Siew-Moi 2016; Kumar and Brown 2018; Zhou et al. 2023). Another potential anti-obesity mechanism of alginates is their ability to enhance feelings of satiety (Hoad et al. 2004; Pelkman et al. 2007). When alginates are ingested, they form a gel in the stomach, which increases gastric volume and slows gastric emptying. This delayed gastric emptying results in prolonged feelings of fullness, which can reduce overall food intake (Wan-Loy and Siew-Moi 2016; Kumar and Brown 2018). In clinical studies, alginate-based products have been shown to reduce hunger levels and lead to

lower caloric consumption, making them a promising option for weight loss diets (Hoad et al. 2004; Pelkman et al. 2007). Emerging evidence suggests that gut microbiota plays a crucial role in the regulation of body weight and metabolism (Davis 2016). Alginates have been shown to influence the composition of the gut microbiome, promoting the growth of beneficial bacteria while suppressing the growth of harmful species (Kumar and Brown 2018). By modulating the microbiota, alginates may enhance metabolic processes, improve insulin sensitivity, and further contribute to weight loss and fat reduction (Zhou et al. 2023; Magwaza and Islam 2023). This mechanism is still under active investigation, but it offers a promising avenue for future anti-obesity strategies. Alginates have demonstrated anti-inflammatory effects by reducing the production of pro-inflammatory cytokines and improving the overall immune response (Lukova et al. 2024). By decreasing systemic inflammation, alginates may help mitigate the complications associated with obesity, including cardiovascular diseases and type 2 diabetes (Magwaza and Islam 2023). Several clinical trials have investigated the effects of alginates on weight loss and obesity-related conditions. Some studies have shown promising results, with alginate-based products reducing body weight and fat mass in overweight and obese individuals. However, more large-scale, long-term clinical trials are needed to fully understand the extent of the anti-obesity effects of alginates and their potential for integration into mainstream weight management strategies.

3. Fucoidans: Fucoidans, a group of sulfated polysaccharides predominantly found in various species of brown seaweed, have gained significant attention for their potential anti-obesity effects (Wan-Loy and Siew-Moi 2016; Magwaza and Islam 2023; Wang et al. 2019). Research has shown that fucoidans may have anti-obesity effects, as demonstrated in animal studies. For example, in a study it was found that mice on a high-fat diet who were also given fucoidans experienced less weight gain, lower food efficiency, and reduced liver and epididymal fat mass compared to those on a high-fat diet alone. Additionally, the mice treated with fucoidans had lower levels of plasma triacylglycerols, total cholesterol, and LDL (low-density lipoprotein). Further gene analysis revealed a down-regulation of adipose tissue genes such as aP2, ACC, and PPAR in the epididymal fat. Another study showed that fucoidans derived from *Sargassum henslowianum* helped reduce blood cholesterol, triacylglycerol, and LDL-cholesterol levels in obese mice (Wan-Loy and Siew-Moi 2016). These findings suggest that fucoidans may play a role in mitigating obesity-related metabolic disturbances. One of the key mechanisms by which fucoidans may help combat obesity is through their ability to regulate lipid metabolism. Studies suggest that fucoidans can inhibit the accumulation of fat in adipocytes (fat cells) by modulating the expression of genes involved in lipogenesis, such as those responsible for fatty acid synthesis and storage (Wan-Loy and Siew-Moi 2016; Magwaza and Islam 2023). Additionally, fucoidans have been shown to increase the activity of lipolytic enzymes, which break down stored fat and promote its utilization as an energy source (Wan-Loy and Siew-Moi 2016). Beyond lipid metabolism, fucoidans also appear to influence insulin sensitivity and glucose metabolism, which

are critical factors in obesity development and its associated complications, such as type 2 diabetes. By improving insulin sensitivity, fucoidans may help prevent the hyperglycemia that often accompanies obesity (Wang et al. 2020). Moreover, fucoidans possess anti-inflammatory properties, which are essential in addressing the chronic low-grade inflammation often observed in obese individuals (Sanjeewa et al. 2021). Fucoidans also exert antioxidant effects, further contributing to their potential in mitigating the oxidative stress commonly linked to obesity-related diseases (Wan-Loy and Siew-Moi 2016; Magwaza and Islam 2023). Some studies suggest that fucoidans might also affect gut microbiota composition, which plays a significant role in weight regulation and metabolic health (Wang et al. 2019). This multi-faceted approach makes fucoidans a promising natural candidate for antiobesity therapies, as they target several aspects of obesity pathology, including fat accumulation, insulin resistance, inflammation, and oxidative stress. However, while preclinical studies and early clinical trials show promising results, further research is needed to fully understand the therapeutic potential of fucoidans in obesity management.

4. Phlorotannins: Phlorotannins, a group of polyphenolic compounds derived from phloroglucinol, are abundant in various brown seaweeds such as *Eisenia bicyclis*, *Ecklonia cava*, and *Undaria pinnatifida*, and have shown considerable promise in addressing obesity and obesity-related disorders (Wan-Loy and Siew-Moi 2016; Magwaza and Islam 2023). These compounds are known to inhibit adipogenesis, the process by which preadipocytes mature into fat cells, thereby preventing excessive fat accumulation (Magwaza and Islam 2023). Studies have demonstrated that phlorotannins, particularly from *E. cava* and *E. stolonifera*, can suppress lipid buildup in adipocytes without adversely affecting cell viability (Wan-Loy and Siew-Moi 2016). Mechanistically, phlorotannins such as eckol and dieckol inhibit key transcription factors involved in fat cell differentiation, including PPARγ and C/EBPα, and activate AMP-activated protein kinase (AMPK), which helps regulate energy balance and reduce adipogenesis (Wan-Loy and Siew-Moi 2016; Magwaza and Islam 2023). In addition to these effects on fat cell formation, phlorotannins also act on other metabolic pathways. For example, they inhibit pancreatic lipase, an enzyme responsible for breaking down dietary fats, thereby limiting fat absorption (Wan-Loy and Siew-Moi 2016). Moreover, phlorotannins like eckol have been shown to block protein tyrosine phosphatase 1B (PTP1B), an enzyme that negatively regulates insulin signaling, making them a potential therapeutic target for improving insulin sensitivity and combating obesity-related metabolic issues (Wan-Loy and Siew-Moi 2016; Medoro et al. 2024). Compounds like phlorotannins inhibit enzymes such as α-amylase and α-glucosidase, slowing carbohydrate digestion and absorption. This results in a lower glycemic response, making algae-derived products suitable for diabetic individuals (Medoro et al. 2024). Although the majority of studies have been conducted in vitro, the evidence suggests that phlorotannins could be a valuable component in the development of anti-obesity treatments, especially for individuals with concomitant type 2 diabetes. However, further research in animal models and clinical trials is needed to fully assess their therapeutic potential in humans.

5. Astaxanthin: Astaxanthin is a red-orange xanthophyll carotenoid renowned for its potent antioxidant and anti-inflammatory properties. It is primarily derived from microalgae such as Haematococcus pluvialis, which is considered the richest natural source, as well as other species like Chlorella zofingiensis and Coelastrella striolata (Medoro et al. 2024; Ambati et al. 2014). Structurally, astaxanthin is unique due to its conjugated double bonds and hydroxyl and keto functional groups, which enhance its bioactivity compared to other carotenoids (Ambati et al. 2014). Astaxanthin exhibits a wide range of biological activities, making it a promising candidate for managing obesity-related diabetes. As an antioxidant, it scavenges free radicals and reduces oxidative stress, a key contributor to the development of insulin resistance and β-cell dysfunction in type 2 diabetes (Landon et al. 2020; Ambati et al. 2014). Its antioxidant potential is reportedly up to 65 times stronger than vitamin C and 54 times more potent than β-carotene (Medoro et al. 2024). By neutralizing reactive oxygen species (ROS), astaxanthin protects pancreatic β-cells, preserving their insulin-secreting function (Landon et al. 2020). Additionally, astaxanthin modulates inflammation by downregulating pro-inflammatory cytokines such as tumor necrosis factor-alpha (TNF-α) and interleukins (IL-1β and IL-6) (Landon et al. 2020). This anti-inflammatory effect is crucial for reducing chronic, low-grade inflammation commonly observed in obesity and metabolic syndrome. Astaxanthin also enhances lipid metabolism by promoting the breakdown of fatty acids and reducing triglyceride and LDL cholesterol levels. This lipid-lowering effect is mediated through the regulation of peroxisome proliferator-activated receptor alpha (PPARα), which facilitates fatty acid oxidation (Nawaz et al. 2021; Landon et al. 2020). A unique aspect of astaxanthin's activity lies in its ability to improve mitochondrial function (Landon et al. 2020). Improved mitochondrial function leads to better fatty acid oxidation and energy utilization, which are critical for weight management and glycemic control. By enhancing mitochondrial efficiency and reducing oxidative damage, it helps maintain energy homeostasis and prevents excessive fat accumulation (Nishida et al. 2021). Furthermore, studies have demonstrated that astaxanthin improves glucose metabolism by enhancing insulin sensitivity and reducing hepatic glucose production. It activates AMP-activated protein kinase (AMPK), a key regulator of energy balance, leading to improved glucose uptake in skeletal muscle and reduced gluconeogenesis in the liver (Nishida et al. 2020). Astaxanthin's comprehensive effects on oxidative stress, inflammation, lipid metabolism, and insulin sensitivity make it a promising natural compound for the prevention and management of obesity-related diabetes. With ongoing research, its therapeutic potential continues to be explored, paving the way for its inclusion in nutraceuticals aimed at combating metabolic disorders.

8 Challenges in Utilizing Algae for Nutraceutical Applications in Obesity and Diabetes

1. High Production Costs: Advanced technology is needed for the cultivation, harvesting, and processing of algae, which raises expenses. Systems like open ponds and photobioreactors require a large amount of energy and money (Lee and Lee 2020; Ramos-Romero et al. 2021).
2. Scalability Issues: Because consistent quality and productivity are required, scaling up from laboratory to industrial levels is difficult. At bigger sizes, problems including contamination and preserving the intended bioactive profiles are frequent.
3. Regulatory Hurdles: To approve nutraceuticals made from algae, regulatory bodies such as the FDA and EFSA need substantial clinical data. This procedure can be costly and time-consuming (Lee and Lee 2020; Ramos-Romero et al. 2021).
4. Bioavailability Concerns: The restricted bioavailability of several bioactive substances found in algae, such as carotenoids and omega-3 fatty acids, lessens their therapeutic effect. To increase absorption, novel delivery methods including nanoencapsulation are being investigated (Ramos-Romero et al. 2021).
5. Variability in Composition: Variability in Composition Environmental factors including light, temperature, and nutrient availability has a significant impact on the nutritional and bioactive profiles of algae. This fluctuation may result in uneven product effectiveness (Ramos-Romero et al. 2021).
6. Consumer Acceptance: Algae-based products often have distinct sensory characteristics (e.g., taste, smell, texture) that may not appeal to consumers, requiring masking agents or reformulation (Lee and Lee 2020).
7. Environmental and Contamination Risks: Algae cultivation can be susceptible to contamination from bacteria, fungi, or heavy metals in the water, which compromises product safety and increases production costs for purification (Lee and Lee 2020; Ramos-Romero et al. 2021).
8. Limited Long-term Studies
   - While there is growing evidence of the health benefits of algae-derived compounds, more robust clinical trials are needed to confirm their long-term effects on obesity and diabetes management (Ramos-Romero et al. 2021).

9 Future Prospects

1. Algae, both microalgae and macroalgae, hold significant promise in combating obesity-related diabetes due to their diverse bioactive compounds and environmental benefits. Here are the detailed future prospects for their application in nutraceuticals:
2. Advancements in Algal Biomass Cultivation: Developments in Algal Biomass Cultivation • Improved cultivation techniques, such as stress-induced synthesis

of valuable components like polysaccharides and omega-3 fatty acids, may boost algae's potential as a medicine. The production of bioactive compounds specifically designed to manage diabetes and obesity may be maximized by precision farming and the use of controlled environments (Lee and Lee 2020; Lomartire and Gonçalves 2022).

3. Development of Functional Foods: To control blood sugar and fight insulin resistance, functional meals can include algae bioactives like phycocyanin, carotenoids, and polyunsaturated fatty acids (PUFAs). Future developments might concentrate on adding these bioactives to staple meals to make them widely accessible (Ramos-Romero et al. 2021).
4. Advanced Bioactive Extraction Techniques: Advanced Bioactive Extraction Techniques: New technologies such as supercritical fluid extraction, ultrasound-assisted extraction, and green solvents provide effective and environmentally friendly ways to separate active ingredients from algae. These developments are intended to improve the environmental friendliness and scalability of algal nutraceutical production (Lee and Lee 2020; Lomartire and Gonçalves 2022).
5. Therapeutic Targeting of Metabolic Disorders: Algal chemicals have demonstrated promise in treating many components of metabolic syndrome, such as improving lipid metabolism, lowering inflammation, and increasing insulin sensitivity. The development of evidence-based nutraceuticals and the validation of these effects in humans depend on clinical trials (Ramos-Romero et al. 2021).
6. Integration with Sustainability Goals: Algal farming's ability to capture CO2 and reduce land use positions it as an eco-friendly option for nutraceutical production. Future developments could align with global sustainability goals, making algae a key component in climate-conscious health interventions(Ramos-Romero et al. 2021).
7. Personalized Nutrition Applications: Nutraceutical therapies for obesity-related diabetes may be more effective thanks to developments in nutrigenomics, which could allow the creation of customized algal-based supplements that target particular genetic profiles or metabolic conditions (Lee and Lee 2020).

References

Shalaby EA (2011). Algae as promising organisms for environment and health. Plant Signal Behaviour, 6(9), 1338–50.

Das, Sudipta Kumar (2016). Dietary use of algae among tribal of North-east India: Special reference to the Monpa tribe of Arunachal Prades. Indian Journal of Traditional Knowledge, Vol. 15(3), pp. 509–513.

Gómez-Zorita S, Trepiana J, González-Arceo M, Aguirre L, Milton-Laskibar I, González M, Eseberri I, Fernández-Quintela A, Portillo MP (2019). Anti-obesity effects of microalgae. *International Journal of Molecular Sciences*, 21(1), 41.

Safiyanu I, Hassan MI, Sale AI, Sadiq IZ and Mardiyya, A.Y (2018). General review on toxic algae. *International Journal of Engineering & Scientific Research*, Vol.6 Issue 9, 2347–6532.

Hyungryun Jang, Jaeeun Lee, Young-Ki Park, Ji-Young Lee (2024). Exploring the health benefits and concerns of brown seaweed consumption: A comprehensive review of bioactive compounds in brown seaweed and its potential therapeutic effects. *Journal ofAgriculture and Food Research*, Volume 17, 101215, 2666–1543.

Murakami S, Hirazawa C, Mizutani T, Yoshikawa R, Ohya T, Ma N, Owaki Y, Owaki T, Ito T, Matsuzaki C (2022). The anti-obesity and anti-diabetic effects of the edible seaweed *Gloiopeltis furcata* (Postels et Ruprecht) J. Agardh in mice fed a high-fat diet. *Food Science and Nutrition*, 11(1), 599–610.

Lomartire S and Gonçalves AMM (2022). An Overview of potential seaweed-derived bioactive compounds for pharmaceutical applications. *Marine Drugs*, 20(2), 141.

Kurniawan R, Nurkolis F, Taslim NA, Subali D, Surya R, Gunawan WB, Alisaputra D, Mayulu N, Salindeho N, Kim B (2023). Carotenoids composition of green algae*Caulerpa racemosa* and their antidiabetic, anti-obesity, antioxidant, and anti-inflammatory properties. *Molecules*, 28(7), 3267.

Chen Y, Wu W, Ni X, Farag MA, Capanoglu E, Zhao C (2022). Regulatory mechanisms of the green alga *Ulva lactuca* oligosaccharide via the metabolomics and gut microbiome in diabetic mice. *Current Research in Food Science*, 5, 1127–1139.

Qudus B Aroyehun A, Abdul Razak S, Palaniveloo K, Nagappan T, Suraiza Nabila Rahmah N, Wee Jin G, Chellappan DK, Chellian J, Kunnath AP (2020). Bioprospecting cultivated tropical green algae, *Caulerpa racemosa* (Forsskal) J. Agardh: A perspective on nutritional properties, antioxidative capacity and anti-diabetic potential. *Foods*, 9(9),1313.

Tamel Selvan K, Goon JA, Makpol S, & Tan, JK (2023). Therapeutic potentials of microalgae and their bioactive compounds on diabetes mellitus. *MarineDrugs*, 21(9), 462

Hantzidiamantis PJ, Awosika AO, Lappin SL. Physiology, Glucose. [Updated 2024 Apr 30]. In: StatPearls [Internet]. Treasure Island (FL): StatPearls Publishing; 2024 Jan-. Available from: https://www.ncbi.nlm.nih.gov/books/NBK545201/

American Diabetes Association Professional Practice Committee; 2. Classification and Diagnosis of Diabetes: Standards of Medical Care in Diabetes—2022. Diabetes Care 1 January 2022; 45 (Supplement_1): S17–S38.

Yashi K, Daley SF. Obesity and type 2 diabetes. [Updated 2023 Jun 19]. In: StatPearls [Internet]. Treasure Island (FL): StatPearls Publishing; 2024 Jan-. Available from: https://www.ncbi.nlm.nih.gov/books/NBK592412/

Magwaza S.N., & Islam M.S. (2023). Roles of marine macroalgae or seaweeds and their bioactive compounds in combating overweight, obesity, and diabetes: A comprehensive review. *Marine Drugs, 21*(4), 258.

Pereira L,& Valado A(2023). Algae-derived natural products in diabetes and its complications—current advances and future prospects. *Life, 13*(9), 1831.

Authors from the University of Coimbra (2024). Seaweeds as nutraceutical elements and drugs for diabetes mellitus:Future perspectives. *Marine Drugs,* 22(4), 168.

Chu Wan-Loy and Phang Siew-Moi (2016). Marine algae as a potential source for anti-obesity agent. *MarineDrugs*, 14, 222.

Asmita Kumari, Garima, Navneeta Bharadvaja (2023). A comprehensive review on algal nutraceuticals as prospective therapeutic agent for different diseases. *Springer*, 3 Biotech 13:44.

Ye J, Zheng J, Tian X, Xu B, Yuan F, Wang B, Yang Z, Huang F(2022). Fucoxanthin attenuates free fatty acid-induced nonalcoholic fatty liver disease by regulating lipid metabolism/oxidative stress/inflammation via the AMPK/Nrf2/TLR4 signaling pathway. *MarineDrugs*, 20(4), 225.

Hayato Maeda (2015). Nutraceitical effects of fucoxanthin for obesity and diabetes therapy: A review. *Journal of oleo science*, 64(2), 125–132.

Hitoe S, Shimoda H(2017). Seaweed fucoxanthin supplementation improves obesity parameters in mildly obese japanese subjects. *Functional Foods in Health and Disease*, 7(4), 246–262.

Grasa-López A, Miliar-García Á, Quevedo-Corona L, Paniagua-Castro N, Escalona-Cardoso G, Reyes-Maldonado E, Jaramillo-Flores ME (2016). Undaria pinnatifida and fucoxanthin ameliorate lipogenesis and markers of both inflammation and cardiovascular dysfunction in an animal model of diet-induced obesity. *Marine Drugs*, 14(8), 148.

Kozak LP, Anunciado-Koza R(2008). UCP1: its involvement and utility in obesity. *International Journal of Obesity*(Lond), 32 Suppl 7, S32–8.

Wondmkun YT (2020). Obesity, insulin resistance, and type 2 diabetes: Associations and therapeutic implications. *Diabetes Metabolic Syndrome and Obesity*, 13, 3611–3616.
Ha AW, Kim WK (2013). The effect of fucoxanthin rich power on the lipid metabolism in rats with a high fat diet. *Nutrition Research and Practice*, 7(4), 287–293.
Steinberg GR, Carling D (2019). AMP-activated protein kinase: the current landscape for drug development. *Nature Reviews Drug Discovery*, 18(7), 527–551.
Tangvarasittichai S (2015). Oxidative stress, insulin resistance, dyslipidemia and type 2 diabetes mellitus. *World Journal of Diabetes*, 6(3), 456–480.
Senthil Arun Kumar and Lindsay Brown (2018). Alginates in metabolic syndrome. *Springernature*, Alginates and Their Biomedical Applications, Springer Series in Biomaterials Science and Engineering 11, 223–235.
Xiaolei Zhou, Zhao Peng, Yuxiao Liao, Dan Li, Shiyin Xu, Yu Wen, Junya Gao, Xinxin Qi, Xinyu Zhang, Liyuan Feng, Hong Zhang, Xingjie Hao, Qi Wang, Liegang Liu, Wei Yang (2023). Weight reduction effect of alginate associated with gut microbiota and bile acids: A double-blind and randomized trial. *Journal of Functional Foods*, Volume 108, 105774.
Hoad CL, Rayment P, Spiller RC, Marciani L, Alonso Bde C, Traynor C, Mela DJ, Peters HP, Gowland PA (2004). In vivo imaging of intragastric gelation and its effect on satiety in humans. *JournalofNutrition*, 134(9), 2293–300.
Pelkman CL, Navia JL, Miller AE, Pohle RJ (2007). Novel calcium-gelled, alginate-pectin beverage reduced energy intake in nondieting overweight and obese women: interactions with dietary restraint status. *The American Journal of Clinical Nutrition*, 86(6), 1595–1602
Davis CD (2016). The gut microbiome and its role in obesity. *Nutr*ition *Today*, 51(4), 167–174.
Lukova, P, Kokova, V, Baldzhieva, A,Murdjeva, M,Katsarov, P, Delattre, C, andApostolova, E (2024). Alginate from ericariacrinita possesses antioxidant activity and attenuates systemic inflammation via downregulation of pro-inflammatory cytokines. *MarineDrugs*, 22(11), 482.
Wang Y, Xing M, Cao Q, Ji A, Liang H, Song S (2019). Biological activities of fucoidan and the factors mediating its therapeutic effects: A review of recent studies. *MarineDrugs*, 17(3),183.
Wang L, Zhang K, Ding X, Wang Y, Bai H, Yang Q, Ben J, Zhang H, Li X, Chen Q, Zhu X 2020. Fucoidan antagonizes diet-induced obesity and inflammation in mice. The Journal of Biomedical Research, 35(3): 197–205.
Sanjeewa KKA, Herath KHINM, Yang HW, Choi CS, Jeon YJ (2021). Anti-inflammatory mechanisms of fucoidans to treat inflammatory diseases: A review. *MarineDrugs*, 19(12), 678.
Alessandro Medoro, Mariano Intrieri, Daniela Passarella, Donald Craig Willcox, Sergio Davinelli, Giovanni Scapagnini 2024. Astaxanthin as a metabolic regulator of glucose and lipid homeostasis. Journal of Functional Foods, Volume 112, 105937, ISSN 1756-4646.
Ambati RR, Phang SM, Ravi S, Aswathanarayana RG (2014). Astaxanthin: sources, extraction, stability, biological activities and its commercial applications—a review. *MarineDrugs*, 12(1), 128–52.
Landon R, Gueguen V, Petite H, Letourneur D, Pavon-Djavid G, Anagnostou F (2020). Impact of astaxanthin on diabetes pathogenesis and chronic complications. *MarineDrugs*, 18(7), 357.
Nawaz A, Nishida Y, Takikawa A, Fujisaka S, Kado T, Aminuddin A, Bilal M, Jeelani I, Aslam MR, Nishimura A, Kuwano T, Watanabe Y, Igarashi Y, Okabe K, Ahmed S, Manzoor A, Usui I, Yagi K, Nakagawa T, Tobe K (2021). Astaxanthin, a marine carotenoid, maintains the tolerance and integrity of adipose tissue and contributes to its healthy functions. *Nutrients*, 13(12), 4374.
Nishida Y, Nawaz A, Hecht K, Tobe K (2021). Astaxanthin as a novel mitochondrial regulator: a new aspect of carotenoids, beyond antioxidants. Nutrients, 14(1), 107.
Nishida Y, Nawaz A, Kado T, Takikawa A, Igarashi Y, Onogi Y, Wada T, Sasaoka T, Yamamoto S, Sasahara M, Imura J, Tokuyama K, Usui I, Nakagawa T, Fujisaka S, Kunimasa Y, Tobe K (2020). Astaxanthin stimulates mitochondrial biogenesis in insulin resistant muscle via activation of AMPK pathway. *Journal of Cachexia Sarcopenia and Muscle*, 11(1), 241–258.
Lee, B. Y., & Lee, J. Y. (2020). Nutraceuticals and functional foods for obesity and diabetes. *Journal of Nutritional Biochemistry*.
Ramos-Romero, S. et al. (2021). Edible microalgae and their bioactive compounds in metabolic alterations. *Nutrients*.

Chapter 10
Mitigation of Diabetes and Cardiovascular Complications by Algae

K. M. G. M. M. Kariyawasam, G. Rathnayake, K. M. G. R. M. Kariyawasam, and Indira Wickramasinghe

Abstract Diabetes mellitus is a metabolic disorder that is linked with many etiologies. It is characterized by chronic hyperglycemia, which interrupts the fat, carbohydrate, and protein metabolism. Lack of insulin production, and impaired insulin sensitivity or insulin resistance are the main factors contributing to dysmetabolism and its complications. Furthermore, uncontrolled diabetes mellitus can cause cardiovascular diseases (CVD). A proper diet can be the management tool for hyperglycemia and its complications as well as for CVD. Several studies have demonstrated the effect of algal bioactive compounds in alleviating different diseases, including diabetes, cardiovascular diseases, and hyperglycemia. In this context, carotenoids, polyphenols, and antioxidant compounds in algae could play a vital role in lowering oxidative stress, leading to diabetes and CVD. Additionally, algal polysaccharides contribute to hypoglycemia by modifying insulin activity and glucose metabolism. Moreover, omega-3 fatty acids, which are necessary for lowering inflammation, improving lipid profiles, and increasing insulin sensitivity, are abundant in algae. Therefore, this chapter discusses how algae-based dietary sources may help to mitigate diabetes and cardiovascular complications.

K. M. G. M. M. Kariyawasam
Department of Biosystems Technology, Faculty of Technological Studies, Uva Wellassa University of Sri Lanka, Badulla, Sri Lanka

Department of Food Science and Biotechnology of Animal Resources, Konkuk University, Seoul, South Korea

G. Rathnayake
Department of Biochemistry, Faculty of Medicine, Uva Wellassa University of Sri Lanka, Badulla, Sri Lanka

K. M. G. R. M. Kariyawasam
Department of Food Science, Obihiro University of Agriculture and Veterinary Medicine, Hokkaidao, Japan

I. Wickramasinghe (✉)
Department of Food Science & Technology, Faculty of Applied Sciences, University of Sri Jayewardenepura, Nugegoda, Sri Lanka
e-mail: indiraw@sjp.ac.lk

R. Raja et al. (eds.), *Algae in Diabetes Management*,
https://doi.org/10.1007/978-981-95-0571-5_10

Keywords Mitigation · Cardiovascular complications · Algal polysaccharides · Chronic hyperglycemia

1 Introduction

Algae accounted for more than 70% of total living mass and produce a large amount of biomass compared to plants (Kumari 2013). They are photosynthetic organisms that consist with various range of organisms (Bishop and Zubeck 2012). Thus far around 300,000 algal species have been discovered (Kumari 2013). As they are photosynthetic organisms, they are able to synthesize a large range of complex molecules and biomass (Dolganyuk et al. 2020). Algae grow in various aquatic habitats such as lakes, oceans, and rivers (Pereira and Valado 2023). Algae can be broadly classified as microscopic cyanobacteria and macroscopic-complex macroalgae such as kelps (Kumari 2013). Macroalgae, also called seaweeds, are commonly found in marine ecosystems, that are multicellular organisms (Kumari 2013). In contrast, microalgae are single-cell organisms found in both marine and freshwater habitats. Additionally, macroalgae are subdivided into various groups based on their morphology, especially considering the color of the pigments they produce, including brown algae (belong to phylum Ochrophyta and class Phaeophyceae), green algae (belong to phylum Chlorophyta, and classes: Ulvophyceae, Bryopsidophyceae, Dasycladophyceae, Prasinophyceae, and Chlorophyceae), and red algae (belong to phylum Rhodophyta and class Rhodophyceae) (Ramanan et al. 2016; Bocanegra et al. 2021; Pereira and Valado 2023).

Algae have been a staple food for many countries since ancient times, according to the various archeological reports. Particularly, they are renowned in countries including the Republic of Korea, China, and Japan. Many varieties of seaweed dishes can be found in these nations' daily diets named Nori or Gim, Konbu (dried seaweed derived from Laminaria species), hizikia (*Hizikia fusiforme*), and wakame (*Undaria pinnatifida*). Seaweeds have also been eaten in some parts of Europe, particularly in coastal part, including Wales, Ireland, Norway, and France (Fernando et al. 2020; Bocanegra et al. 2021). Nowadays, algae are gaining high popularity in non-Asian countries, which makes them available in various restaurant menus as well as in supermarkets. The high consumer acceptance of the algae can be primarily attributed to their abundant nutritional profile, particularly they are rich in protein and minerals. Moreover, algae are low in lipids, rich in polysaccharides, polyunsaturated fatty acids (PUFA), minerals, vitamins, and bioactive compounds (Murugan et al. 2015). For example, brown algae have been consumed by many people and have been a valuable food ingredient in the food industry for many years (Afonso et al. 2019; Barbier et al. 2019). The global annual seaweed production is around 5.5–6 US dollars (Phang 2010). In 1960, Japan released the first commercially cultivated microalgae (Chlorella) in nutraceutical industry. As algae are rich in bioactive compounds, they have become a highly desirable ingredient in nutraceutical industry. In particular, components such as proteins, carbohydrates, and lipids are extracted from algae for pharmacological and nutraceutical applications (Nilesh Hemantkumar and Ilza Rahimbhai 2020).

2 Bioactive Compounds in Algae

There is high interest toward healthy natural products for promoting human health due to their high potential in preventing non-communicable diseases. Algae have chemical diversity and unique components, making them important in the medical, nutraceutical, and cosmetics industries. Moreover, algae are gaining popularity as they are rich in a wide range of bioactive compounds. Bioactive compounds in algae are in two categories: primary metabolites and secondary metabolites. Primary metabolites include polysaccharides, lipids, and protein; and secondary metabolites are sterols, halogenated compounds, terpenes, phenolic compounds, and small peptides (Cho et al. 2022). However, bioactive compounds levels present in algae are influenced by various factors including temperature variations, reproductive development, salinity levels, and nutrient levels in the habitats (Stengel et al. 2011). Many studies have highlighted the role of algae-derived bioactive compounds in treating various non-communicable diseases, including type II diabetes mellitus, cardiovascular diseases, cancers, dementia, and inflammation (Rengasamy et al. 2020).

3 The Primary Bioactive Compound in Seaweeds

3.1 Proteins and Peptides

Algae are a key source of protein and essential amino acids. The protein content of algae varies with the variety (Wells et al. 2017). These peptides contribute to various therapeutic properties, including antioxidant, anti-cancer, immunomodulatory, opioid receptor modulation, anti-thrombotic, anti-atherosclerotic, and antimicrobial effects, depending on their amino acid sequence. A peptide (VECYGPNRPQF) derived from *Chlorella vulgaris* protein waste has shown resistance to gastrointestinal enzyme and radical oxygen activity. Moreover, Y2 polypeptide was isolated from *Spirulina platensis* by trypsin digestion and was shown to have antitumor activity. Additionally, phycobiliprotein, which is an enzymatic hydrolysate of *Porphyra columbina,* demonstrated immunosuppressive activity on mouse splenocytes by inducing IL-10 production and suppression of IFN-γ and TNF-α. Dipeptidyl peptidase-4 (DPP-4) inhibition is a well-known glucose-lowering treatment in type 2 diabetes. There are some bioactive peptides of algae that have been discovered to inhibit DPP-4, thus, these peptides extend the incretin activity, boosting insulin secretion and regulating glucose metabolism (Cian et al. 2012).

3.2 Polysaccharides

Algal polysaccharides are readily isolated and valuable as potential bioactive substances. They are categorized into structural polysaccharides, mucopolysaccharides, and storage polysaccharides (Rengasamy et al. 2020). The major component of algae biomass is polysaccharides, which accounts for 33–47% of dry matter. These polysaccharides serve as energy storage and provide structural support within the algae and have several health benefits (Bocanegra et al. 2021). Algal polysaccharides are distinct from those found in plants. Algal cell wall has structural polysaccharides that comprise of high-molecular-weight, water-insoluble compounds, including mannan, cellulose and xylan, and also, water-soluble polysaccharides including carrageenan, alginate, agar, fucoidan ulvan, and pyropian. These water-soluble types make up the majority of the dietary fiber found in algae (Jiménez-Escrig and Sánchez-Muniz 2000).

The sequence of monomer units usually decides the properties of polysaccharides (Usov and Zelinsky 2013). Algal polysaccharides have low glycemic index (GI), thereby reducing the glycemic load of diets (Goñi et al. 2002). Additionally, algae are a rich source of fiber. These factors make algae a desirable diet for treating T2DM. Moreover, red algae are rich in sulfated polysaccharides, such as agar and carrageenan, which has the ability to inhibit α-glucosidase and α-amylase enzymes, responsible for carbohydrate breakdown. This action helps lower blood glucose levels after meals. Additionally, these polysaccharides show radical oxygen scavenging activity and enhance insulin sensitivity (Lu and Chen 2022; Pereira and Valado 2023).

Algal polysaccharides also have beneficial effects on cardiovascular health. For example, sulfated polysaccharides from *Padina tetrastromatica* demonstrated therapeutic properties for CVD by activating the PI3K/Akt/Nrf2 signaling pathway in mouse models (Lekshmi et al. 2019). Moreover, *Laminaria japonica* (Fucoidan), L-fucose-enriched sulfate (MW = 7000 Da) upregulated eNOS expression in brain microvascular endothelial cells and efficiently enhanced local blood flow and blood pressure in rats (Cui et al. 2014).

3.3 Polyphenols

Polyphenols are predominant compound of algae that are well-known for therapeutic properties including ACE inhibitory activity, cardioprotective effect, anti-inflammatory activity, cancer prevention, diabetes, antioxidant activity, and neurodegenerative disorders (Farhat et al. 2017). Polyphenols are made up of four primary compounds: bromophenols, phlorotannins, flavonoids, and phenolic acids. Brown algae typically contain phlorotannins, which are monomers of phloroglucinol (1,3,5-tryhydroxybenzene). Laminariacea is the marine algae with the highest concentration of phlorotannins (Rengasamy et al. 2020). Many have reported the beneficial effect of phlorotannins, which are common in brown seaweeds including *Ecklonia* and *Ishige* species. They have been found to have therapeutic activities including antioxidant, anticancer, antibacterial, antiallergic, antimutagenic, antidiabetic, anti-inflammatory, antiproliferative, antihypertensive, and antiobesity effects (Wikesekara et al. 2010; Meng et al. 2021; Shrestha et al. 2021). Therefore, it is well documented the role of phlorotannins as a beneficial bioactive compound for treating various disease conditions.

The major cause of cardiovascular disease (CVD) is hypertension. There are many studies, which demonstrated the ACE inhibitory activity of various phlorotannins including 6,6′-bieckol, phlorofucofuroeckol A, eckol, phloroglucinol, dieckol, triphlorethol-A, eckstolonol, and octaphlorethol A, fucosterol isolated from *Ecklonia cava*, *Ishige foliacea*, and *Ecklonia stolonifera* (Jung et al. 2006; Wijesinghe et al. 2011). Additionally, genus Phaeophyta contains higher amounts of polyphenol and fucosterol (Sugawara et al. 2014). Hyperglycemia is a key indicator of all forms of diabetes and is linked to complications like cardiovascular disease. The enzymes α-amylase and α-glucosidase in the small intestine are crucial for breaking down carbohydrates into glucose. Thus, managing postprandial hyperglycemia by slowing down glucose absorption through the inhibition of these enzymes is a vital strategy for treating T2DM (Asgar 2013; Ghani 2015).

3.4 Pigments

There are various pigments in plants such as anthocyanins, betalains, carotenoids, and chlorophylls which are renowned for their therapeutic activities. Seaweeds include useful pigments, such as fucoxanthin, phycoerythrin, and phycocyanin (Rengasamy et al. 2020). Fucoxanthin, astaxanthin, carotene, lycopene, and siphonaxanthin are the primary carotenoid pigments in seaweeds. Bhagavathy and Sumathi (2012) reported that green seaweed extracts rich in carotenoids exhibited antigenotoxic activity. Carotenoids have shown therapeutic properties against type II diabetes through increasing insulin sensitivity (Sluijs et al. 2015; She et al. 2017). For example, fucoxanthin is mainly found in brown algae and has been shown to have an effect on diabetes through improving insulin sensitivity and stimulating glucose uptake in adipocytes and skeletal muscle cells (Ito et al. 2018). It has also been reported that some algal-derived pigments have a potential role in mitigating diabetes related complications via anti-inflammatory and oxygen radical scavenging activities (Figueiredo et al. 2016; Bocanegra et al. 2021).

In addition, in vitro studies suggested the anti-inflammatory and prebiotic activity of fucoxanthin in human epithelial cells (Hwang et al. 2016). Ganesan et al. (2013) reported that fucoxanthin and siphonaxanthin inhibited angiogenesis in vascular endothelial cells by suppressing FGF-2-mediated intracellular signaling (Ganesan et al. 2013). Moreover, Sugawara et al. (2014) reported the anti-angiogenic and anti-inflammatory effects of siphonaxanthin. Ganesan et al. (2011) reported the ability of siphonaxanthin to induce apoptosis in human leukemia (HL-60) cells through activating caspase-3 and reducing expression of Bcl-2.

3.5 Lipids

Normally, algae have higher amount of fat than the regularly consumed vegetables, however, it has low fat percentage of 1.5% to 4%. Algae contain lipids, including glycolipids, phospholipids, and glycerolipids (Wells et al. 2017) as well as wide range of fatty acids, including sterols, terpenoids, tocopherols, and carotenoids (Bocanegra et al. 2009; Kumari 2013). In addition, approximately 10% to 50% of the total fatty acid in algae are rich in omega −3 fatty acids of which docosahexaenoic acid (DHA) and eicosapentaenoic acid (EPA) have shown anti diabetic activity, antihyperglycemic activity, and anti-inflammatory properties (Sinha et al. 2023). Potential for reducing triglyceride levels among diabetic patients is also evident by lipids extracted from *Chlorella* and *Spirulina* (Conde et al. 2021).

Sterols, which have various therapeutic properties, are mostly extracted from seaweeds. For instance, *Pelvetia siliguosa* has fucosterol that enhances antioxidant activity, with glutathione peroxidase (GSH-Px), superoxide dismutase (SOD), and catalase (Lee et al. 2003). Similarly, the sterols extracted from *Sargassum carpophyllum* (3β,28ξ-dihydroxy-24-ethylcholesta-5,23Z-dien and 24ξ-hydroperoxy-24-vinylcholesterol) demonstrated cytotoxicity in human promyelocytic leukemia cells (Tang et al. 2002). Fucosterol and β-sitosterol are sterols that have a number of health benefits. Fucosterol can primarily be found in brown algae, while β-sitosterol is mainly found in red algae (Wells et al. 2017). Carotenoid pigments, β-carotene and fucoxanthin, are nutritionally valuable and comprised in algal lipids (Wells et al. 2017). Algae contain fatty acids, phytosterols (i.e., β-sitosterol), antioxidants (i.e., α-tocopherol), and polyphenols that can prevent CVD and T2DM by reducing inflammation and oxidative stress and improving endothelial function (Murray et al. 2018).

4 Mechanisms of Algal Compounds in Diabetes Management

Impaired insulin secretion or insulin resistance causes blood glucose level to become imbalanced and lipid profile to be altered (Pereira and Valado 2023), and to arise type 2 diabetes mellitus (T2DM), which is totally a dysmetabolic syndrome (Bocanegra et al. 2021). High sugar and high fat diet, lifestyle, genetic factors, gut microbiota, and obesity are found to be significant contributing factors for T2DM (Kolb and Martin 2017). While T2DM leads to a number of complications, hugely affecting quality of life, leading to a higher mortality and morbidity rates of those who were affected, it also significantly impacts millions of people to drain financially, leading to economical constraints and to pressuring healthcare systems (Tomic et al. 2022). An increasing interest in searching for natural products as potential solution for controlling diabetes in recent times, algae have been found to be an important source of curing agents for diabetes with their bioactive compounds, showing potential.

Bioactive compounds, derived from algae, are necessary ingredients for functional foods, which are believed to offer health benefits. It has been found out a positive link between prevention and management of type 2 diabetes and regular consumption of these functional foods by enhanced antioxidant enzyme activity, improved insulin sensitivity, reduced production of pro-inflammatory cytokines, and cholesterol-lowering effects (Macho-González et al. 2021a, b; Mirmiran et al. 2014). Accordingly, algae could serve as an antidiabetic food ingredient. Algae are a key source of bioactive polysaccharides, polyphenols, peptides, and various other compounds. These bioactive compounds possess various therapeutic pathways and mechanisms for the prevention and management of type 2diabetes. These include regulating glucose homeostasis, increasing insulin sensitivity, and showing inhibitory activities against various enzymes involved in carbohydrate digestion and hepatic glucose production (Sharifuddin et al. 2015; Bocanegra et al. 2021).

4.1 Inhibitory Activity of α-Amylase and α-Glucosidase Enzymes

The enzymes α-glucosidase and α-amylase are responsible for carbohydrate digestion and absorption. Inhibition of these enzymes may impede carbohydrate digestion and thus the release of glucose into blood bloodstream. Phlorotannins found in marine algae such as dieckol, eckol, 7-phloroeckol, 6,6′-bieckol, and phlorofucofuroeckol-A can inhibit α-glucosidase and α-amylase activity. Inhibition of these enzymes causes slower release of glucose into blood bloodstream and rises in blood glucose level after a meal (Attjioui et al. 2021 and Popovic-Djordjevic et al. 2021). Algal phlorotannins can boost antidiabetic activity when compared to standard antidiabetic drug such as acarbose (Lee and Jeon 2013).

4.2 Stimulation of Glucose Uptake

Glucose uptake in cells, adipose tissues, and skeletal muscles has been shown to be influenced by algal-based natural products. These natural algal compounds can influence on insulin-mediated glucose uptake through activating PI3K/Akt pathway. Increasing glucose uptake can lower the blood glucose levels and improve the insulin sensitivity (Popovic-Djordjevic et al. 2021 and Nyakundi and Yang 2023).

4.3 Modulation of Insulin Secretion

The main cause of T2DM is loss of adequate β-cell insulin secretion or impaired insulin secretion. However, algal bioactive compounds have been shown to have influence on insulin secretion from pancreatic β-cells. In addition, there are other various mechanisms these bioactive compounds can influence on insulin secretion, including the influx of calcium ions through closure of ATP-sensitive potassium channels following secretion of insulin vesicles (Abo-Shady et al. 2023; Bocanegra et al. 2021).

4.4 Modulation of Insulin Sensitivity

Insulin resistance or impaired response of target tissues to insulin is one of the central aspects causing the development of T2DM. The protein tyrosine phosphatase 1B (PTP1B) can modulate the insulin sensitivity and blood glucose levels (Li et al. 2016), consequently, inhibiting PTP1B, and can lower blood glucose level by reducing insulin sensitivity. Phlorotannins from brown algae such as 7-phloroeckol, phlorofucofuroeckol-A, and ekol have been shown to suppress PTP1B activity (Lopes et al. 2017). Furthermore, some algal extracts have the ability to activate PI3K/Akt pathway and regulate insulin-driven glucose uptake and regulate blood glucose levels (Melloni et al. 2023; Naz et al. 2023).

4.5 Modulation of Hepatic Glucose Production

The hepatic glucose production (HGP) is controlled by liver through regulating the gluconeogenesis and glycogenolysis. Therefore, liver is responsible for glucose level control in the whole body. However, in diabetic patients, hepatic glucose production becomes altered. Shehadeh et al. (2021) reported that some algal-based compounds have been shown to regulate HGP through controlling the enzymes responsible for gluconeogenesis and glycogenolysis. Moreover, Bocanegra et al. (2021) also reported that glucose production by liver is inhibited by algae-derived compounds, which help with blood sugar regulation. In vivo studies suggested that administration of polyphenol-rich extracts (*Ishige okamurae*) continuously for 6 weeks has diminished the phosphoenolpyruvate carboxykinase (PEPCK) and glucose-6-phosphatase (G6Pase) activity in mice, which increased the hepatic glycogen production and consequently, reduced the fasting blood glucose level (Min et al. 2011).

4.6 Modulation of Adipokine Secretion

Adipose tissues secrete various signaling proteins, such as adipokines, which are responsible for the metabolism of glucose and fat. Moreover, it secretes various signaling molecules and hormones essential for maintaining hemostasis, including leptin, adiponectin, and resistin. Dysregulation of these compounds can increase the risk of obesity and T2DM (Fernando et al. 2020). However, natural products derived from algae have been found to influence the secretion of leptin and adiponectin. Adiponectin enhances insulin sensitivity and promotes glucose utilization. Similarly, algae-based compound can enhance adiponectin levels, thereby improving glucose metabolism. In addition, leptin controls appetite and body weight, which are key factors in diabetes management (Selvan et al. 2023).

4.7 Inhibitory Activity of Angiotensin-Converting Enzymes (ACE)

Angiotensin-converting enzyme (ACE) can increase blood pressure level in diabetic patients which can lead to various complications. However, phlorotannins present in algae have inhibitory activity against ACE (Paiva et al. 2017). For example, *Ecklonia stolonifera* is a rich source of phlorotannins and has been examined for its anti-diabetic properties. There are a number of phlorotannins shown to have ACE inhibitory activity including eckol, dieckol, and phlorofucofuroeckol. Jung et al. (2006) demonstrated the ACE inhibitory activity of *Sargassum horneri*, *Sargassum fusiforme,* and *Ecklonia cava.*

4.8 Anti-inflammatory and Antioxidant Effects

Oxidation can cause tissue damage and consequently, contribute to the pathogenesis of diabetes, hypertension, and cardiovascular damages. Moreover, impaired glucose metabolism and insulin resistance are mainly associated with oxidative stress and chronic inflammation. There are some algae-derived polyphenolic compounds that possess radical oxygen scavenging activity and anti-inflammatory activities. Therefore, by alleviating inflammation and oxidative stress, these compounds can ensure better control of glucose levels and overall metabolic health (Sahakyan et al. 2022). Murray et al. (2018) reported that fatty acids, including antioxidants (α-tocopherol), phytosterols (β-sitosterol), and ω-3 PUFAs and polyphenols have anti-inflammatory and antioxidant properties. They observed that such effects could improve endothelial microvascular and macrovascular function, playing preventive roles in T2DM and CVD (Fig. 10.1).

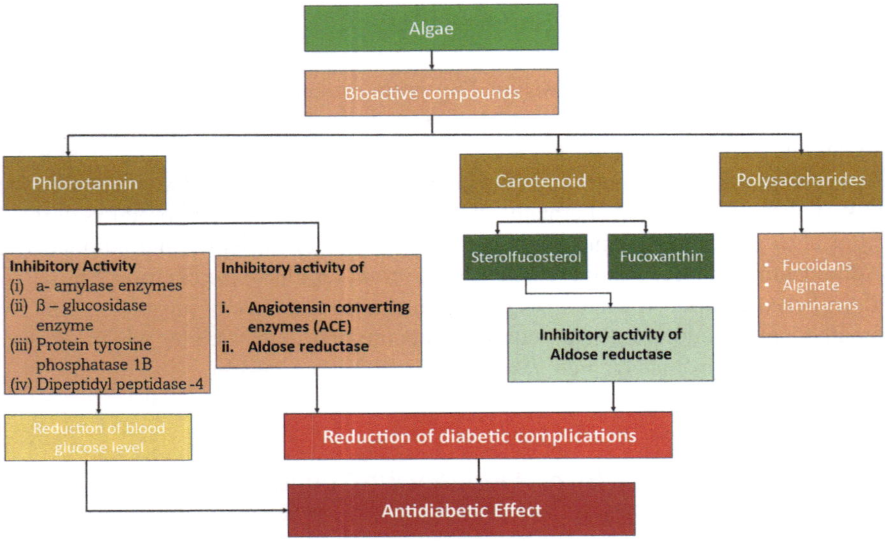

Fig. 10.1 Management of diabetes mellitus by algal bioactive compounds – mechanism. (Adapted from Kumari 2013)

5 Mechanism of Algal Compounds in Cardiovascular Health

Cardiovascular diseases are considered as leading causes of morbidity both in developing and developed world (Nabel 2003, Cheong et al. 2022, Omachi et al. 2024). Diseases of cardiac muscle and vascular system supplying vital organs including heart and brain are included in CVDs. Ischemic heart disease (IHD), stroke, and congestive heart failure (CHF) are considered as predominant cardiovascular diseases (Cheong et al. 2022; Gaziano et al. 2006). Evidences provided by epidemiological studies and randomized controlled trials indicate that these CVDs are largely preventable (Cornish et al. 2015; Nabel 2003). However, it also highlighted that the heritability of these CVDs makes it impossible to achieve 100% prevention (Nabel 2003).

5.1 Risk Factors of CVDs

When it comes to risk factors of CVDs, age, gender, family history, and genetics are considered as non-modifiable risk factors. On the other hand, hypertension, dyslipidemia, smoking, diabetes mellitus, obesity, physical inactivity, dietary habits, excessive alcohol consumption, and stress have been identified as modifiable risk factors (Cheong et al. 2022; Omachi et al. 2024). Oxidative stress and inflammation are also considered as risk factors for CVDs (Omachi et al. 2024). Among these risk factors, obesity, high blood pressure, high blood glucose, smoking, abnormalities in lipid profile, and physical inactivity are considered as major risk factors for CVDs (Gaziano et al. 2006; Omachi et al. 2024; Petrie et al. 2018).

When it comes to pharmacological interventions focused on long-term management of existing vascular diseases, the interventions have been focused on prevention of thrombosis, use of antihypertensives and statins targeting cholesterol (Gaziano et al. 2006). Modifiable risk factors are targeted for nonpharmacological interventions including stopping smoking and dietary modifications (Gaziano et al. 2006). When it comes to marine algae, a variety of nutrients like proteins, peptides, amino acids, polysaccharides, minerals, lipids, and polyphenols have been identified as having potential in improving cardiovascular health (Cheong et al. 2022).

5.2 Cardioprotective Compounds Available in Algae

5.2.1 Phenolic Compounds

Phenolic compounds are considered as secondary metabolites and made up of benzene ring structures that bind to hydroxyl groups. Examples are phlorotannins, bromophenols, flavonoids, and amino acids similar to mycosporine. These phenolic compounds show antidiabetic, anti-inflammatory, and antioxidant activities (Tavares et al. 2023). For example, flavonoids are capable of antidiabetic treatment by decreasing fasting blood glucose and modulation of gut-microbiome (Tavares et al. 2023). Bromophenols show antidiabetic and anti-obesity properties (Tavares et al. 2023).

5.2.2 Lipids

Functional lipids derived from dietary substances have been considered as having effect on human health by reducing risk of diseases including CVDs. Docosahexaenoic acid (DHA), α-linolenic acid (ALA), eicosapentaenoic acid (EPA), oleic acid (OA), and linoleic acid (LA) can be given as several examples for functional lipids which provides positive effects on cardiometabolic health (Omachi et al. 2024). Microalgae are considered as one dietary source of functional lipids ω-3 fatty acid (ALA, EPA, and DHA).

5.2.3 Polysaccharides

Polysaccharides like fucoidan, alginate, carrageenan, porphyrin, agar, ulvans, and sulfated rhamnan derived from brown, red, and green marine algae types have protection against cardiovascular diseases like atherosclerosis, heart diseases, and stroke (Cheong et al. 2022). These marine algal polysaccharides have demonstrated antioxidant, immunomodulatory, and antidiabetic properties in studies under different experimental models (Cheong et al. 2022).

The protective effects of polysaccharides derived from different marine algae are listed below.

1. Fucoidan: increased endothelial protective activity (Cheong et al. 2022).
2. Alginate: reduction of fasting blood glucose, reduction of total cholesterol, reduction of total body fat (Cheong et al. 2022).
3. Agar: antidiabetic effect (Cheong et al. 2022), reducing blood glucose levels (Tavares et al. 2023).
4. Porphyran: antihyperlipidemic activity, antioxidant capacities (Cheong et al. 2022).
5. Carrageenan: reducing total cholesterol, reducing low-density lipoprotein cholesterol (Cheong et al. 2022), reducing blood glucose levels (Tavares et al. 2023).
6. Ulvan: antihyperlipidemic activity (Cheong et al. 2022).
7. Sulfated rhamnan: thrombolytic activity, antithrombotic activity (Cheong et al. 2022).

5.3 Mechanisms of Algae for Prevention/Controlling CVDs

Oxidate stress, inflammatory mechanisms, hyperlipidemias, hypertension, hyperglycemia, and many other physiological conditions are associated with mechanisms involved in development and progression of cardiovascular diseases (Cho et al. 2022; Gaziano et al. 2006; Griendling et al. 2021). Here, let's discuss how primary and secondary algal compounds are involved in cardiovascular health by modifying the risk factors and physiological mechanisms involved. The mechanisms are common that of diabetes mellitus, since diabetes is one of the risk factors of CVDs.

5.3.1 Antioxidant Properties

Several antioxidant mechanisms can be described under the antioxidant properties of algal products that linked with cardiovascular health. For example, fucoxanthin has demonstrated the capacity of removal of reactive oxygen species (ROS) and removal of radical scavengers. This prevents DNA damage and apoptosis (Yamagata 2021). Further, neutralization of ROS leads to reduction of lipid peroxidation (Tavares et al. 2023; Yamagata 2021). Astaxanthin can be given as an example. It has been demonstrated that astaxanthin has the capacity to increase antioxidant enzymes in rat heart tissue, thereby blocking ROS generation, oxidative damage, and lipid peroxidation (Yamagata 2021).

5.3.2 Anti-inflammatory Effects

Inflammatory mechanisms are involved in the pathogenesis of CVDs (Yamagata 2021). It has been found that fucoidan intake leads to reduced interleukin-6 (IL-6) levels, thereby reducing the inflammatory response (Yamagata 2021). Apart from that it has been identified that seaweed specialized compounds that are capable of downregulation of pro-inflammatory cytokines (TNF-α, IL-6) and inhibition of NF-KB signaling, reducing endothelial inflammation and vascular dysfunction (Conde et al. 2021; Rocha et al. 2022).

5.3.3 Lipid Regulation

Several mechanisms have been described in the literature that regulate lipids related to cardiovascular health. Reduction of LDL cholesterol (Holdt and Kraan 2011) and increased HDL cholesterol via upregulation of hepatic LDL receptors is one mechanism (Yamagata 2021). For example, it has been shown that intake of fucoidan increased HDL cholesterol levels by blocking the gene expression of liver enzymes fatty acid synthase and acetyl-CoA carboxylase. Further, fucoidan has been shown to reduce expression of 3-hydroxy-3-methyl-glytaryl-CoA reductase which is the rate-limiting step of the cholesterol synthesis pathway (Yamagata 2021). At the gut level, hypocholesterolemic and hypolipidemic response due to reduced cholesterol gut absorption has been identified as another mechanism (Holdt and Kraan 2011).

5.3.4 Antihypertensive Activity

Hypertension or high blood pressure is one of the risk factor of CVDs. In these cases, hypertension is associated with vasodilation dysfunctions associated with endothelial NO synthase. It has been shown that algal products including fucoidan have the capacity of enhancing nitric oxide (NO) production for vasodilation and improved endothelial function (Yamagata 2021). Inhibition of angiotensin-converting enzyme (ACE), reducing angiotensin II formation, is another mechanism by which algal peptides have demonstrated antihypertensive activity (Cho et al. 2022; Jiang et al. 2021). For example, previous studies have demonstrated that seaweed-derived phlorotannins have shown non-competitive ACE inhibition (Seca and Pinto 2018).

5.3.5 Platelet Aggregation Inhibition

Platelet activation, aggregation, and later development of thrombus lead to acute ischemic conditions as a result of exposed atherosclerotic plaque components, coagulation factors, and platelet receptors. This mechanism leads to IHD and stroke (Badimon and Vilahur 2014). Anticoagulant potential of algal products plays an important role here in the prevention of such platelet aggregation and subsequent events (Holdt and Kraan 2011). As an example, induction of release of tissue plasminogen activator by fucoidan can be given (Yamagata 2021). Inhibition of platelet aggregation by modulating thromboxane pathways (Carvalhal et al. 2019; Ren et al. 2013) and enhancement of fibrinolysis to prevent clot formation are other two mechanisms by which algal products are capable of modulating cardiovascular effects (Carvalhal et al. 2019).

6 Future Directions and Challenges in Algae Research for Diabetes and CVD

The future research and developments focused on the role of algae's primary and secondary products in managing diabetes and cardiovascular diseases (CVDs) are concerned, there are several concerns and prospects available. One approach is the research and development of enhancement of clinical applications of algal bioactive compounds. Since current research is limited to in vitro and animal studies, it is important to move to human studies with robust clinical trials maintaining consistency and efficacy (Tamel Selvan et al. 2023). Further, elucidation of the dosage, dietary intake, effectiveness, and mechanisms of actions of metabolites is important (Omachi et al. 2024). Another challenge is the insufficiency and reliability of information available in the literature regarding effectiveness and associated mechanism of action of these products (Omachi et al. 2024). Exploration of development of algal products as a potential prevention strategy for CVDs via targeting gut microbiota is another future avenue (Cheong et al. 2022). Development of efficient green methods for compound extraction and purification (Baghel et al. 2023) and the effect of toxicity and antinutrients due to concerns of radioactive and microplastic contaminants, chemical hazards, and biological hazards will be challenges that need to be addressed (Baghel et al. 2023).

References

Abo-Shady, A.M.; Gheda, S.F.; Ismail, G.A.; Cotas, J.; Pereira, L.; Abdel-Karim, O.H. 2023 Antioxidant and Antidiabetic Activity of Algae. *Life*, 13, 460

Afonso, N.C.; Catarino, M.D.; Silva, A.M.S.; Cardoso, 2019 S.M. Brown Macroalgae as Valuable Food Ingredients. *Antioxidants*, 8, 365.

Asgar, A. 2013 Anti-diabetic potential of phenolic compounds: A review. *International Journal of FoodProperties*, 16, 91–103.

Attjioui, M.; Ryan, S.; Ristic, A.K.; Higgins, T.; Goñi, O.; Gibney, E.R.; Tierney, J.; O'Connell, S. 2021 Comparison of edible brown algae extracts for the inhibition of intestinal carbohydrate digestive enzymes involved in glucose release from the diet. *Journal of Nutritional Science*., 10, e5.

Badimon, L. and Vilahur, G. 2014 Thrombosis formation on atherosclerotic lesions and plaque rupture. *Journal of internal medicine* 276, 618–632.

Baghel, R. S., Choudhary, B., Pandey, S., Pathak, P. K., Patel, M. K., and Mishra, A. 2023. Rehashing Our Insight of Seaweeds as a Potential Source of Foods, Nutraceuticals, and Pharmaceuticals. *Foods*, 12(19), 3642.

Barbier, M.; Charrier, B.; Araujo, R.; Holdt, S.L.; Jacquemin, B.; Rebours, C. 2019 Pegasus—Phycomorph European Guidelines for a Sustainable Aquaculture ofSeaweeds; Barbier, M., Charrier, B., Eds.; COST Action FA1406: Roscoff, France.

Bhagavathy, S.; Sumathi, P. 2012 Evaluation of antigenotoxic effects of carotenoids from green algae Chlorococcum humicola using human lymphocytes. *Asian Pacific Journal of Tropical Biomedicine*2, 109–117.

Bishop W. M, Zubeck HM 2012 Evaluation of Microalgae for use as Nutraceuticals and Nutritional Supplements. *Journal of Nutrition Food Science and Technology*2:147. doi:https://doi.org/10.4172/2155-9600.1000147

Bocanegra, A.; Bastida, S.; Benedí, J.; Ródenas, S.; Sánchez-Muniz, F.J 2009 Characteristics and nutritional and cardiovascular-health properties of seaweeds. *Journal of Medicinal Food*, 12, 236–258.

Bocanegra, A.; Macho-González, A.; Garcimartín, A.; Benedí, J.; Sánchez-Muniz, F.J 2021 Whole Alga, Algal Extracts, and Compounds as Ingredients of Functional Foods: Composition and Action Mechanism Relationships in the Prevention and Treatment of Type-2 Diabetes Mellitus. *International Journal of Molecular Sciences*, 22, 3816.

Carvalhal F, Cristelo RR, Resende DISP, Pinto MMM, Sousa E, Correia-da-Silva M. 2019 Antithrombotics from the Sea: Polysaccharides and Beyond. *Marine Drugs*17(3):170. doi: https://doi.org/10.3390/md17030170.

Cheong, K.; Yu, B.; Chen, J.; Zhong, S. A. 2022 Comprehensive Review of the Cardioprotective Effect of Marine Algae Polysaccharide on the Gut Microbiota. *Foods*, 11, 3550.

Cho, C.H.; Lu, Y.A.; Kim, M.Y.; Jeon, Y.J.; Lee; S.H. 2022 Therapeutic potential of seaweed-derived bioactive compounds for cardiovascular disease treatment. *Applied Sciences* 12, 1025

Cian, R.E.; Martínez-Augustin, O.; Drago, S.R. 2012 Bioactive properties of peptides obtained by enzymatic hydrolysis from protein byproducts of Porphyra columbina. Food Ressearch International, 49, 364–372.

Conde, T.A.; Zabetakis, I.; Tsoupras, A.; Medina, I.; Costa, M.; Silva, J.; Neves, B.; Domingues, P.; Domingues, M.R. 2021 Microalgal Lipid Extracts Have Potential to Modulate the Inflammatory Response: A Critical Review. *International Journal of Molecular Sciences*, 22, 9825.

Cornish, M. L.; Critchley, A. T.; Mouritsen, O. G. 2015 A role for dietary macroalgae in the amelioration of certain risk factors associated with cardiovascular disease. *Phycologia* 54, 649–666.

Cui, W.; Zheng, Y.; Zhang, Q.; Wang, J.; Wang, L.; Yang, W.; Guo, C.; Gao, W.; Wang, X.; Luo, D. 2014 Low-molecular-weight fucoidan protects endothelial function and ameliorates basal hypertension in diabetic Goto-Kakizaki rats. *Labotary Investigation*.94, 382–393

Dolganyuk, V.; Belova, D.; Babich, O.; Prosekov, A.; Ivanova, S.; Katserov, D.; Patyukov, N.; Sukhikh, S. 2020 Microalgae: A Promising Source of Valuable Bioproducts. *Biomolecules*, 10, 1153.

Farhat, G., Drummond, S., and Al-Dujaili, E. A. S. 2017. Polyphenols and their role in obesity management: A systematic review of randomized clinical trials. *Phytotherapy Research*, 31, 1005–1018.

Fernando, I.P.S.; Ryu, B.; Ahn, G.; Yeo, I.-K.; Jeon, Y.-J. 2020 Therapeutic potential of algal natural products against metabolic syndrome: A review of recent developments. *Trends in Food Science Technology*, 97, 286–299.

Figueiredo, F.; Encarnação, T.; Campos, M. 2016 Algae as Functional Foods for the Elderly. Food Nutritional Science, 7, 1122–1148.

Ganesan, P.; Matsubara, K.; Sugawara, T.; Hirata, T. 2013 Marine algal carotenoids inhibit angiogenesis by down-regulating FGF-2- mediated intracellular signals in vascular endothelial cells. *Molecular and Cellular Biochemistry*, 380, 1–9.

Ganesan, P.; Noda, K.; Manabe, Y.; Ohkubo, T.; Tanaka, Y.; Maoka, T.; Sugawara, T.; Hirata, T. 2011 Siphonaxanthin, a marine carotenoid from green algae, effectively induces apoptosis in human leukemia (HL-60) cells. Biochimica et Biophysica Acta. Biochimica et Biophysica Acta (BBA)1810, 497–503.

Gaziano, T.; Reddy, K. S.; Paccaud, F.; Horton, S.; Chaturvedi, V. 2006 Cardiovascular disease. Disease Control Priorities in Developing Countries. 2nd edition.

Ghani, U. 2015 Re-exploring promising α-glucosidase inhibitors for potential development into oral anti-diabetic drugs: Finding needle in the haystack. *European journal of medicinal chemistry*, 103, 133–162.

Goñi, I.; Valdivieso, L.; Gudiel-Urbano, M. 2002 Capacity of edible seaweeds to modify in vitro starch digestibility of wheat bread. *Food/Nahrung*, 46, 18–20.

Griendling K. K, Camargo LL, Rios FJ, Alves-Lopes R, Montezano AC, Touyz RM. 2021 Oxidative Stress and Hypertension. *Circulation Research*. 2;128(7):993–1020.

Holdt, S. L. and Kraan, S. 2011 Bioactive compounds in seaweed: functional food applications and legislation. *Journal of applied phycology* 23, 543–597.

Hwang, P.-A.; Phan, N.N.; Lu, W.-J.; Hieu, B.T.N.; Lin, Y.-C. 2016 Low-molecular-weight fucoidan and high-stability fucoxanthin from brown seaweed exert prebiotics and anti-inflammatory activities in Caco-2 cells. *Food Nutrition Research*, 60, 32033.

Ito, M.; Koba, K.; Hikihara, R.; Ishimaru, M.; Shibata, T.; Hatate, H.; Tanaka, R 2018 Analysis of functional components and radical scavenging activity of 21 algae species collected from the Japanese coast. *Food Chemistry*, 255, 147–156.

Jiang Q, Chen Q, Zhang T, Liu M, Duan S, Sun X. 2021 The Antihypertensive Effects and Potential Molecular Mechanism of Microalgal Angiotensin I-Converting Enzyme Inhibitor-Like Peptides: A Mini Review. *International Journal of Molecular Sciences*, 22(8):4068. doi: https://doi.org/10.3390/ijms22084068.

Jiménez-Escrig, A.; Sánchez-Muniz, F. 2000 Dietary fibre from edible seaweeds: Chemical structure, physicochemical properties and effects on cholesterol metabolism. Nutritional Research, 20, 585–598.

Jung, H.A.; Hyun, S.K.; Kim, H.R.; Choi, J.S. 2006 Angiotensin-converting enzyme I inhibitory activity of phlorotannins from *Ecklonia stolonifera*. *Fisheries Science*, 72, 1292–1299.

Kolb, H.; Martin, S. 2017 Environmental/lifestyle factors in the pathogenesis and prevention of type 2 diabetes. *BMC Medicine*, 15, 1–11.

Kumari, P. 2013 Algal lipids, fatty acids and sterols. In Functional Ingredients from Algae for Foods and Nutraceuticals, 1st ed.; Domínguez, H., Ed.; Woodhead Publishing Limited: Cambridge, UK,pp. 87–132

Lee, S.; Lee, Y.S.; Kang, S.S.; Shin, K.H. 2003 Anti-oxidant activities of fucosterol from the marine algae *Pelvetia siliquosa*. *Archives of Pharmacal Research*, 26, 719–722.

Lekshmi, V.S.; Arun, A.; Rauf, G.; Muraleedhara, K. 2019 Sulfated polysaccharides from the edible marine algae *Padina tetrastromatica* attenuates isoproterenol-induced oxidative damage via activation of PI3K/Akt/Nrf2 signaling pathway-An in vitro and in vivo approach. *Chemico-Biological Interactions*, 308, 258–268

Lu, L.W.; Chen, J.-H. 2022 Seaweeds as Ingredients to Lower Glycemic Potency of Cereal Foods Synergistically—A Perspective. *Foods*, 11, 714.

Macho-González, A.; Bastida, S.; Garcimartín, A.; López-Oliva, M.E.; González, P.; Benedí, J.; González-Muñoz, M.J.; Sánchez- Muniz, F.J. 2021a Functional meat products as oxidative stress modulators: A review. *Advances in Nutrition*182.

Macho-González, A.; Garcimartín, A.; Benedí, J.; Sánchez-Muniz, F.J. 2021b Whole Alga, Algal Extracts, and Compounds as Ingredients of Functional Foods: Composition and Action Mechanism Relationships in the Prevention and Treatment of Type-2 Diabetes Mellitus. *International journal of molecular sciences*, 22, 3816. https://doi.org/10.3390/ijms22083816

Melloni, M.; Sergi, D.; Simioni, C.; Passaro, A.; Neri, L.M. 2023 Microalgae as a Nutraceutical Tool to Antagonize the Impairment of Redox Status Induced by SNPs: *Implications on Insulin Resistance. Biology*, 12, 449.

Meng, W.; Sun, T.M.H.; Garcia-Vaquero, M. 2021 Phlorotannins: A review of extraction methods, structural characteristics, bioactivities, bioavalability, and future trends. Algal Research, 60, 102484.

Min, K.H., Kim H.J, Jeon, Y.J. and Han, J.S. (2011). *Ishige okamurae* ameliorates hyperglycemia and insulin resistance in C57BL/KsJ-db/db mice. Diabetes Research and Clinical Practice 93: 70–76.

Mirmiran, P.; Bahadoran, Z.; Azizi, F. 2014 Functional foods-based diet as a novel dietary approach for management of type 2 diabetes and its complications: A review. *World journal of diabetes*, 5, 267–281.

Murray, M.; Dordevic, A.L.; Ryan, L.; Bonham, M.P. 2018 An emerging trend in functional foods for the prevention of cardiovascular disease and diabetes: Marine algal polyphenols. *Critical Review in Food Science and Nutrition*, 58, 1342–1358.

Murugan, A. C., Karim, M. R., Yusoff, M. B. M., Tan, S. H., Asras, M. F. B. F., and Rashid, S. S. 2015, New insights into seaweed polyphenols on glucose homeostasis. *Pharmaceutical Biology*, 53, 1087–1097.

Naz, R.; Saqib, F.; Awadallah, S.; Wahid, M.; Latif, M.F.; Iqbal, I.; Mubarak, M.S. 2023 Food Polyphenols and Type II Diabetes Mellitus: Pharmacology and Mechanisms. *Molecules*, 28, 3996

Nyakundi, B.B.; Yang, J. 2023 Uses of Papaya Leaf and Seaweed Supplementations for Controlling Glucose Homeostasis in Diabetes. *International Journal of Molecular Sciences*, 24, 6846.

Omachi, D. O.; Aryee, A. N.; Onuh, J. O. 2024 Functional Lipids and Cardiovascular Disease Reduction: A Concise Review. *Nutrients* 16, 2453.

Pereira, L.; Valado, A. 2023 Algae-Derived Natural Products in Diabetes and Its Complications—Current Advances and Future Prospects. *Life*, 13, 1831. https://doi.org/10.3390/life13091831

Petrie, J. R.; Guzik, T. J.; Touyz, R. M. 2018 Diabetes, Hypertension, and cardiovascular disease: Clinical Insights and Vascular Mechanisms. *Canadian Journal of Cardiology* 34, 575–584.

Phang, S.-M.2010. Potential products from tropical algae and seaweeds, especially with reference to Malaysia. *Malaysian Journal of Science*, 29, 160–166.

Popovic-Djordjevic, B.J.; Katanic Stankovic, S.J.; Mihailovic, V.; Pereira, G.A.; Garcia-Oliveira, P.; Prieto, A.M.; Simal-Gandara, J 2021 Algae as a Source of Bioactive Compounds to Prevent the Development of Type 2 Diabetes Mellitus. *Current Medicinal Chemistry*, 28, 4592–4615.

Ramanan, R.; Kim, B.-H.; Cho, D.-H.; Oh, H.-M.; Kim, H.-S 2016 Algae–bacteria interactions: Evolution, ecology and emerging applications. *Biotechnology Advances*, 34, 14–29

Ren, R, Azuma Y, Ojima T, Hashimoto T, Mizuno M, Nishitani Y, Yoshida M, Azuma T, Kanazawa K 2013 Modulation of platelet aggregation-related eicosanoid production by dietary F-fucoidan from brown alga Laminaria japonica in human subjects. *British Journal of Nutrition*, 110(5):880–90. doi: https://doi.org/10.1017/S000711451200606X.

Rengasamy, K.R.; Mahomoodally, M.F.; Aumeeruddy, M.Z.; Zengin, G.; Xiao, J.; Kim, D.H. 2020 Bioactive compounds in seaweeds: An overview of their biological properties and safety. *Food and Chemical Toxicology*, 135, 111013.

Rocha, D. H.; Pinto, D. C.; Silva, A. M. 2022 Macroalgae specialized metabolites: evidence for their anti-inflammatory health benefits. *Marine Drugs* 20, 789.

Sahakyan, G.; Vejux, A.; Sahakyan, N. 2022 The Role of Oxidative Stress-Mediated Inflammation in the Development of T2DM-Induced Diabetic Nephropathy: Possible Preventive Action of Tannins and Other Oligomeric Polyphenols. *Molecules*, 27, 9035

Seca, A. M. & Pinto, D. C. 2018 Overview on the antihypertensive and anti-obesity effects of secondary metabolites from seaweeds. *Marine drugs* 16, 237.

Sharifuddin, Y.; Chin, Y.-X.; Lim, P.-E.; Phang, S.-M. 2015 Potential Bioactive Compounds from Seaweed for Diabetes Management. *Marine Drugs*, 13, 5447–5491.

She, C.; Shang, F.; Zhou, K.; Liu, N. 2017 Serum carotenoids and risks of diabetes and diabetic retinopathy in a Chinese population sample. *Current molecular medicine*, 17, 287–297.

Shehadeh, M.B.; Suaifan, G.A.R.Y.; Abu-Odeh, A.M. 2021 Plants Secondary Metabolites as Blood Glucose-Lowering Molecules. *Molecules*, 26, 4333.

Sinha, S.; Haque, M.; Lugova, H.; Kumar, S. 2023 The Effect of Omega-3 Fatty Acids on Insulin Resistance. *Life*, 13, 1322.

Sluijs, I.; Cadier, E.; Beulens, J.; van der A, D.L.; Spijkerman, A.; van der Schouw, Y. 2015 Dietary intake of carotenoids and risk of type 2 diabetes. *Nutrition, Metabolism and Cardiovascular Diseases*, 25, 376–381.

Stengel, D.B; Connan, S.; Popper, Z.A. 2011 Algal chemodiversity and bioactivity: Sources of natural variability and implications for commercial application. Biotechnology Advances, 29, 483–501.

Sugawara, T, Ganesan, P, Li, Z.; Manabe, Y.; Hirata, T. 2014 Siphonaxanthin, a green algal carotenoid, as a novel functional compound. Marine Drugs, 12, 3660–3668.

Tamel Selvan, K.; Goon, J.A.; Makpol, S.; Tan, J.K 2023 Effects of Microalgae on Metabolic Syndrome. *Antioxidants*, 12, 449.

Tang, HF.; Yi, YH.; Yao, XS.; Xu, QZ, Zhang, SY, Lin, HW. 2002 Bioactive steroids from the brown alga *Sargassum carpophyllum*. *Journal of Asian Natural Products Research*, 4, 95–101.

Tavares, J. O, Cotas, J, Valado, A, Pereira, L. 2023 Algae Food Products as a Healthcare Solution. *Marine Drugs* 21, 578

Tomic, D, Shaw, JE.; Magliano, DJ 2022 The burden and risks of emerging complications of diabetes mellitus. *Nature Reviews Endocrinology*, 18, 525–539

Usov, A.I.; Zelinsky, N.D 2013 Chemical structures of algal polysaccharides. In Functional Ingredients from Algae for Foods and Nutraceuticals, 1st ed.; Domínguez, H., Ed.; Woodhead Publishing Limited: Cambridge, UK, pp. 23–86

Wells, M.L.; Potin, P.; Craigie, J.S.; Raven, J.A.; Merchant, S.S.; Helliwell, K.E.; Smith, A.G.; Camire, M.E.; Brawley, S.H. 2017 Algae as nutritional and functional food sources: Revisiting our understanding. *Journal of Applied Phycology*29, 949–982.

Wijesinghe, W.A.J.P.; Koo, SC.; Jeon, YJ. 2011 Effect of phlorotannins isolated from Ecklonia cava on angiotensin I-convering enzyme (ACE) inhibitory activity. *Nutrition Research and Practice*.5, 93–100.

Yamagata, K. 2021 Prevention of cardiovascular disease through modulation of endothelial cell function by dietary seaweed intake. *Phytomedicine Plus 1*, 100026.

Chapter 11
Cardio-protective and Anti-obesity Properties of Polyunsaturated Fatty Acids from Algae

K. M. G. R. M. Kariyawasam, K. M. G. M. M. Kariyawasam, and Indira Wickramasinghe

Abstract Polyunsaturated fatty acids (PUFAs) have been shown to exert a wide range of health benefits. In particular, n-3 long chain polyunsaturated fatty acids (n-3 LC-PUFAs), abundant in marine dietary products, play a major role in the prevention and treatment of obesity and obesity-related disorders, particularly type 2 diabetes, hypertension, and cardiovascular diseases. The etiology of obesity and its associated disorders involves multiple biochemical and physiological pathways; thus, prevention and treatment require action on multiple targets/organs. Recent research showed that n-3 PUFAs limit fatty acid availability in white adipose tissues, therefore preventing and/or controlling overweight and obesity. In addition, PUFAs are also involved in other processes including gene and cytokine expression, eicosanoid production, and gut microbial regulation, all of which can significantly improve obesity and obesity related diseases. Macroalgae (seaweeds) and microalgae are capable of de novo PUFAs synthesis and hence represent a vast and untapped PUFAs source, which may be more resourceful and sustainable than fish.

Keywords PUFAs · Macroalgae · Microalgae · Cardiovascular disease · Obesity · Inflammation

---

K. M. G. R. M. Kariyawasam
Department of Food Science, Obihiro University of Agriculture and Veterinary Medicine, Hokkaidao, Japan

K. M. G. M. M. Kariyawasam
Department of Biosystems Technology, Faculty of Technological Studies, Uva Wellassa University of Sri Lanka, Badulla, Sri Lanka

Department of Food Science and Biotechnology of Animal Resources, Konkuk University, Seoul, South Korea

I. Wickramasinghe (✉)
Department of Food Science & Technology, Faculty of Applied Sciences, University of Sri Jayewardenepura, Nugegoda, Sri Lanka
e-mail: indiraw@sjp.ac.lk

R. Raja et al. (eds.), *Algae in Diabetes Management*,
https://doi.org/10.1007/978-981-95-0571-5_11

1 Introduction

Cardiovascular diseases (CVD) and obesity are among the leading causes of morbidity and mortality worldwide. According to the World Health Organization (WHO), cardiovascular diseases account for approximately 32% of global deaths, while obesity has reached epidemic proportions, affecting over 650 million adults worldwide (WHO 2016). These conditions not only decrease life expectancy but also significantly impair the quality of life, making their prevention and management a global health priority. Polyunsaturated fatty acids (PUFAs), particularly omega-3 (n-3) and omega-6 (n-6) fatty acids, play a pivotal role in regulating numerous physiological functions. They have been associated with reduced inflammation, improved lipid profiles, enhanced cardiovascular health, and the regulation of body weight (Adkins and Kelley 2010). Among these, n-3 fatty acids, eicosapentaenoic acid (EPA) and docosahexaenoic acid (DHA), are particularly noteworthy for their cardio-protective and anti-obesity properties.

Traditionally, n-3 PUFAs have been sourced from fish, but concerns regarding sustainability, mercury contamination, and overfishing have raised significant environmental and health concerns. As a result, the focus has shifted toward finding alternative sources of these essential fatty acids. Algae, specifically microalgae, have emerged as a promising and sustainable source of n-3 PUFAs, particularly EPA and DHA, without the risks associated with fish-derived oils. This chapter explores the cardio-protective and anti-obesity effects of polyunsaturated fatty acids derived from algae, emphasizing the underlying mechanisms, clinical evidence, and the advantages of using algae-based omega-3 oils.

2 Overview of Polyunsaturated Fatty Acids (PUFAs)

2.1 PUFAs Structure and Metabolism

Polyunsaturated fatty acids (PUFAs) are hydrocarbon chains that contain two or more double bonds. These chains have a methyl group ($-CH_3$) at one end and a carboxyl group (-COOH) at the other end. They are categorized according to the position of the first double bond from the methyl end of the molecule, which separates them into two main groups: omega-3 (n-3 PUFAs) and omega-6 (n-6 PUFAs) fatty acids. n-3 PUFAs are synthesized from the essential fatty acid α-linolenic acid (ALA, 18:3) while n-6 PUFAs are synthesized from the essential fatty acid precursor linoleic acid (LA, 18:2). ALA and LA are termed "essential" as they cannot be synthesized de novo in mammals, due to a predisposed genetic lack of the delta-15 desaturase enzyme. Thus, ALA and LA must be obtained extrinsically from the diet.

Even though these two fatty acids are primarily found in plant sources such as vegetable oil, nuts, and seeds, land plants do not have enzymatic capacity to further metabolize LA and ALA, and hence do not contain very long-chain PUFAs

(VLCPUFAs). Nevertheless, in mammalian tissue, LA and ALA are subjected to further metabolization that includes series of elongation and desaturation to yield VLCPUFAs (Fig. 1.1). The major product of n-6 (ALA) pathway is arachidonic acid (AA, 20:4) while eicosapentaenoic acid (EPA, 20:5) is the major product of n-3 (LA) pathway. EPA can be further converted into docosahexaenoic acid (DHA, 22:6) through a series of elongation and desaturation steps (Harwood 2019). Once 20- or 22-carbon PUFAs are produced, they can undergo oxidation to produce wide array of bioactive molecules called eicosanoids, which include prostaglandins (PGs), leukotrienes (LTs), and thromboxanes (TXs) (Fig. 1.2). These eicosanoids are key regulators of inflammation, vascular function, and other physiological processes. As a generalization n-6 eicosanoids (e.g., those derived from AA) tend to promote inflammation and vasoconstriction, while n-3 eicosanoids (e.g., those derived from DHA and EPA) have anti-inflammatory effects and are associated with beneficial cardiovascular outcomes. n-3 eicosanoids also promote the resolution of inflammation, which is important for tissue repair (c).

Various enzymes, genetic factors, and dietary intake regulate PUFAs metabolism. The key enzymes involved in PUFAs metabolism are delta-6 desaturase, delta-5 desaturase, and elongases, which determine the extent to which fatty acids are converted into VLCPUFAs and their derivatives (Sprecher et al. 1999) Genetic variations in the enzymes involved in PUFAs metabolism, such as FADS1 and FADS2 (genes encoding for delta-6 and delta-5 desaturase enzymes), can affect an individual's ability to convert ALA to EPA and DHA. This variability can influence how dietary n-3 fatty acids are processed and their availability for incorporation into tissues, potentially affecting health outcomes related to cardiovascular disease, inflammation, and cognitive function. Furthermore, the ratio of n-3 to n-6 fatty acids in the diet plays a crucial role in modulating PUFAs metabolism and the subsequent production of eicosanoids. Modern Western diets, which tend to be high in n-6 fatty acids (from vegetable oils like sunflower, corn, and soybean oils) and low in n-3 fatty acids (from sources like fatty fish, algae and flaxseed), can lead to an imbalance that favors the production of pro-inflammatory eicosanoids and upregulation of inflammation (Calder 2001, 2006; Harwood and Caterson 2006). Therefore, a balanced n-3/n-6 ratio (ideally 1:4) is essential for maintaining healthy inflammatory responses and reducing the risk of chronic diseases.

2.2 *Algae as a Source of PUFAs*

While fish are generally the most convenient source of essential fatty acids, including VLCPUFAs, many commercially available fish such as cod contain only small amounts of oil. Thus, to achieve human nutritional requirements, fish rich in fatty acids are necessary as it is evident that n-3 VLCPUFAs make up 14–31% of the total fatty acids, with significant variation in the EPA:DHA ratio (Table 1.1). Even though humans can convert EPA and DHA into each other, due to the high levels of DHA found in humans most focus has been placed on ensuring sufficient DHA intake.

Although fish remain the main source of n-3 VLCPUFAs (Tocher et al. 2019), there is increasing concern about whether they can meet the rising global demand. According to the FAO reports, a notable rise in consumption of fish and marine product was observed in the past few decades (FAO 2018). Therefore, access to fish and seafood plays a key role in determining whether the supply can meet the growing demand for n-3 PUFAs. On the other hand, fish and seafood may contain harmful contaminants such as dioxins and mercury. Though these toxins can be removed, the process is costly. Given the rising demand for VLCPUFAs and the challenges in sustaining fish consumption and supplies, it is evident that this is a critical issue that requires urgent attention.

Algae are the primary producers in ocean ecosystems, forming the foundation of the oceanic food chain. Specifically, algae synthesize n-3 fatty acids, which are subsequently consumed by other marine organisms, aAlgae are usually enriched in n-3 VLCPUFAs than that of fish (Lenihan-Geels et al. 2013) because fish are often inefficient at converting ALA (or LA) into VLCPUFAs (Tocher et al. 2003). In addition, algae are absence of toxic compounds that may be present in fish (Cohen and Ratledge 2005). Furthermore, algae or algal-derived oils are not only vegetarian-friendly but also scalable for large-scale production due to their smaller size and efficient growth rates. Algae can be mainly derived into two groups, microalgae and macroalgae. Microalgae are single-celled organisms, whereas macroalgae are multicellular. There are extensive varieties of microalgae and macroalgae, and they show variability in the synthesis of EPA and DHA (Scott et al. 2007; Kyle 2001).

For an instant, green algae (Chlorophyta) have higher LA and ALA content (7.40 ± 6.04% and 5.06 ± 4.90%, respectively) than brown algae (Ochrophyta) (4.41 ± 2.70% and 5.97 ± 3.47%) and red algae (Rhodophyta) (1.93 ± 3.26% and 1.93 ± 1.23%) (Jaworowska and Murtaza 2022). Furthermore, Rhodophyta and Ochrophyta are rich in AA and EPA while Chlorophyta predominantly contains hexadecatetraenoic, oleic, and palmitic acids (Biancarosa et al. 2018; Peng et al. 2015). The average DHA content in macroalgae is lower than that of EPA, ranging from 2.16 ± 4.19% in brown seaweed, 3.56 ± 7.77% in green seaweed, and 6.91 ± 16.76% in red seaweed. Rhodophyta *Gracilaria changii* has been found to have the highest concentration of DHA, with 48.36 ± 6.76%. (Jaworowska and Murtaza 2022). Further, the ratio of n-6 to n-3 fatty acids, as well as the ratio of PUFAs to saturated fatty acids (SFAs), found in Rhodophyta and Ochrophyta are more beneficial for humans compared to those found in Chlorophyta (Table 1.2) (Jayasinghe et al. 2018).

Dinophytes, Haptophytes, some Cryptophytes, Thraustochytrids, or Euglenoids are known as the higher DHA producers among microalgae (Fan et al. 2007, Leblond et al. 2003, Pernet et al. 2003, Taipale et al. 2020). Dinophytes are capable of producing significant amounts of DHA, with certain species containing up to 40% DHA of their total fatty acids. (Mendes et al. 2009; Leblond and Chapman 2000; Usup et al. 2008). Diatoms, Eustigmatophyte, and some Haptophytes are known as EPA synthesizing taxa (Arao et al. 1994; Gillan et al. 1981; Liang et al. 2014; Moreno et al. 1979; Taipale et al. 2020; Tonon et al. 2002; Yao et al. 2015). Some diatoms are known to synthesize around 20% of EPA while some

Eustigmatophyte produces between 15 and 30% of EPA (Liang et al. 2014; Yao et al. 2015). In addition to inter-species variability, growth conditions, nutrition (nitrogen (N), phosphorus (P), and sulfur (S)), seasonal variations (e.g., light intensity, seawater salinity, temperature), and developmental stages also affect the fatty acid composition of algae (Guschina and Harwood 2006; Harwood 1989; Kendel et al. 2013; Pohl and Zurheide 1979).

Taipale et al. (2020) examined how the growth stage affects the production of n-3 and n-6 PUFAs in 16 species across six major groups of phytoplankton namely: Cryptophytes, Dinophytes, Chrysophytes, Diatoms, Chlorophytes, and Cyanobacteria. Their study found that higher temperatures and nitrogen concentrations were associated with increased production of EPA, DHA, and DPA-6 (22:5n-6) in Cryptophytes, Chrysophytes, and Dinophytes. They also observed that PUFA production in some species was more pronounced during the stationary phase (e.g., Cryptophytes and Chrysophytes), while in others, it occurred during the exponential growth phase (e.g., Diatoms and Dinophytes) (Taipale et al. 2020).

Recent developments in the use of algal oil in the food industry highlight its potential as a sustainable source of PUFAs (Biris-Dorhoi et al. 2020). For example, OmegaTech (USA) uses *Schizochytrium* to produce DHA Gold (Spolaore et al. 2006; Doughman et al. 2007) which has currently been used as a dietary supplement for adults (Spolaore et al. 2006). In another trial, Babuskin et al. (2014) incorporated *Nannochloropsis* to produce pasta. This, incorporating algae oils enriched with n-3 PUFAs into various food products appears to be a promising idea for new food development, particularly for vegetarians. Nevertheless, algal PUFAs functionality can be used wide variety of industries including aquaculture and food and feed industries.

3 Cardio-Protective Effects of Algal PUFAs

Numerous extensive studies, including both primary and secondary prevention clinical trials and cohort meta-analyses, have determined that the intake of n-3 PUFAs is an effective dietary approach to reduce CVD-related morbidity and mortality, as well as various traditional and emerging risk factors (Holub 2009). These risk factors include elevated serum triglycerides (TGs), low high-density lipoprotein (HDL), high low-density lipoprotein (LDL), high blood pressure, increased postprandial lipaemia, cardiac arrhythmia, abnormal heart rate, endothelial dysfunction, and elevated inflammation (Bernstein et al. 2012; MacArtain et al. 2007).

Unlike heart medications, n-3 PUFAs (derived from fish or algae) tend to have fewer side effects and are classified as generally recognized as safe (GRAS) by the U.S. Food and Drug Administration. The current dietary recommendations from the American Heart Association and the American College of Cardiology are (1) 1 g of n-3 PUFAs (EPA and DHA) daily for secondary prevention in individuals with existing coronary and other vascular diseases; (2) 2 to 4 g of n-3 PUFAs daily for managing hypertriglyceridemia; and (3) one serving of oily fish two times per week (around 0.5 g of n-3 PUFAs daily) for those without documented coronary heart disease (Kris-Etherton et al. 2002; Lichtenstein et al. 2006; Mozaffarian and Rimm 2006). Therefore, the PUFAs found in seaweed may have beneficial properties in relation to cardiovascular disease (Guil-Guerrero 2007; Miyashita et al. 2011; Pereira et al. 2012; Rodrigues et al. 2015; Santos et al. 2015). The cardio-protective effects of algal n-3 PUFAs also encompass the prevention of arrhythmias, reduction of plasma triacylglycerol levels, improvement of vascular relaxation, anti-inflammatory responses, inhibition of platelet aggregation, enhancement of plaque stability, and anti-atherosclerotic effects (Bernstein et al. 2012; Ryan et al. 2009).

3.1 Reduction of Blood TGs

Elevated TGs are a well-known risk factor for atherosclerosis and cardiovascular disease. n-3 s from algae have been shown to lower triglyceride levels by increasing the clearance of triglyceride-rich particles from the bloodstream and reducing their synthesis in the liver (Bernstein et al. 2012). A meta-analysis that included 11 randomized control trials, with a total of 485 participants, conducted between 1996 and 2011 showed a dose of 1.68 g/d algal DHA is associated with 0.2 mmol/L (95% CI -0.27 to 0.14) reduction in TGs (Bernstein 2012). Further, the same order of magnitude was observed in meta-analyses that used fish oil (1.95 g EPA and 1.35 g DHA) (Eslick et al. 2009). The increase in HDL-C was also similar to that with fish oil (0.01 mmol/L; 95% CI: 0.00–0.02). In another double-blind trial, 93 healthy adults with hypertriglyceridemia (TGs 150–499 mg/dL) were randomized to receive either marine algal oil (2.4 g/day, DHA to EPA ratio of 2.7:1), or fish oil (2.0 g/day, DHA to EPA ratio of 0.7:1). After 14 weeks, the study demonstrated that algal oil providing 2.4 g/day DHA + EPA lowered TG levels to a degree that was not different from that of a standard fish oil product (Maki et al. 2014). A study conducted using a double-blind randomized placebo-controlled parallel-design trial in 39 men and 40 women to evaluate algal oil on cardiovascular risk factors found that 4 g algal oil/day (1.5 g DHA and 0.6 g DPA) increased Serum total, LDL- and HDL-cholesterol by 0·33 mmol/l (7·3%), 0·26 mmol/l (10·4%) and 0·14 mmol/l (9·0%) respectively, compared with placebo fed with 4 g of refined olive oil (Sanders et al. 2006).

However, another recent study observed a significant decrease in mean VLDL particle size and increases in mean LDL and HDL particle sizes in algal DHA-supplemented group compared to the control group (Neff et al. 2011). They further demonstrated that DHA supplementation also increased the concentrations of large LDL and large HDL particles and decreased the concentrations of small LDL ($P = 0.009$) and medium HDL particles ($P = 0.001$). Nevertheless, two earlier meta-analyses of fish oil supplementation reported an increase in LDL similar to the above studies with algal oil (Friedberg et al. 1998; Harris 1997). Unlike purified or extracted algal DHA, algal oil contained palmitic and oleic acid. Thus, it is also likely that the saturated fat content of the algal oil contributed to the lipid concentrations (Bernstein et al. 2012).

3.2 Anti-inflammatory Effect

Evidence from observational studies, prospective cohort studies, and randomized clinical intervention trials suggests that moderate doses of n-3 PUFAs significantly reduce the risk of fatal coronary heart disease by modulating atherosclerotic lesion development through decreasing the synthesis of pro-inflammatory agents (Allaire et al. 2016, Mori et al. 2003, Robertson et al. 2015). Based on the limited comparative studies available, algal DHA seems to be more effective than algal EPA in reducing a variety of pro-inflammatory biomarkers in individuals with subclinical inflammation (Allaire et al. 2016; Vors et al. 2017). Both EPA and DHA, however, effectively reduced oxidative stress biomarkers (F2 isoprostanes) (Mas et al. 2010; Mori et al. 2000; Mori et al. 2003). Robertson et al. (2015) showed that PUFAs extracted from several macroalgae downregulate LPS-induced pro-inflammatory responses in human macrophages through inhibition of IL-6 and IL-8 production (Robertson et al. 2013; Robertson et al. 2015). Similarly, both Ishihara et al. (1998) and Khan et al. (2007) isolated PUFAs from macroalgae to demonstrate their anti-inflammatory properties. In another clinical study, 77 hypertensive patients were randomized to receive 1.2 g EPA + 0.8 g DHA or corn oil for 90 days. Authors observed a significant reduction in TNF-α (-1.87 ± 2.71 vs -0.64 ± 2.62; $p = 0.02$) and CRP (-0.85 ± 2.49 vs 0.56 ± 2.14; $p = 0.01$) in the DHA + EPA group compared to placebo. The reduction in TNF-α was positively correlated with decreased cardiometabolic risk in subjects taking EPA and DHA ($p = 0.02$), but not in the control subjects. ($p = 0.54$) (Yang et al. 2020). The study by Robertson et al. (2015) assessed the anti-inflammatory potential of lipid extracts from the macroalgae *Porphyra dioica*, *Palmaria palmata*, and *Chondrus crispus*, as well as the microalga *Pavlova lutheri*, in lipopolysaccharide (LPS)-stimulated human THP-1 macrophages. The researchers observed that the lipid extracts inhibited the production of pro-inflammatory cytokines IL-6 and IL-8, downregulated the expression of 14 pro-inflammatory genes, and blocked the LPS-induced pro-inflammatory signaling pathways, which are mediated through toll-like receptors, chemokines, and NF-κB signaling molecules. Additionally, the study found that the lipid extract from *Chondrus crispus* exhibited the lowest anti-inflammatory potential, while the extracts from *Pavlova lutheri* and *Palmaria palmata* were the most effective at inhibiting inflammatory cytokine production. The extract from *Porphyra dioica* inhibited the expression of more inflammatory genes than any of the other species tested, indicating significant variation in functional properties between species.

3.3 Improvement of Vascular Endothelial Function

The n-3 PUFAs offer protective effects on the function of vascular endothelial cells by reducing endothelial activation. Endothelial cells play a central role in inflammation and thrombosis, which help attract immune cells and platelets. These molecules also contribute to the early stages of atherosclerosis. When endothelial cells are activated by inflammatory cytokines, the expression of these adhesion molecules increases. However, n-3 PUFAs have been shown to suppress the production of inflammatory cytokines that normally trigger this endothelial activation (Adkins and Kelley 2010). Vascular tone is regulated by the balance between vasoconstrictors (such as TXA2, PGH2, and endothelin-1) and vasodilators (such as NO, endothelium-derived hyperpolarizing factor, and PGI) produced by the endothelium. The vasorelaxant effects of DHA are thought to result from a reduction in Ca2+ influx into vascular smooth muscle cells (VSMCs) (Hirafuji et al. 2001). As previously mentioned, n-3 PUFAs can influence eicosanoid production, promoting vasodilation and antithrombotic effects. It has also been proposed that n-3 PUFAs enhance endothelium-dependent relaxation by increasing NO release (Abeywardena and Head 2001). NO plays a crucial role in inhibiting platelet aggregation, leukocyte adhesion, and smooth muscle cell proliferation. Hirafuji et al. (2002) demonstrated that DHA (and, to a lesser extent, EPA, but not AA) increased IL-1β-induced NO production and upregulated iNOS expression in rat VSMCs, via the p44/42 MAPK signaling pathway. Additionally, DHA was shown to reduce ischemia-induced arrhythmias in hypertensive rats by inhibiting TXA2-like vasoconstrictor responses in the aorta (McLennan et al. 1996). Similar study by De Caterina et al. (1994) revealed that culturing endothelial cells with DHA, following inflammatory challenges (such as IL-1, IL-4, TNF-α, or LPS) led to a reduction in the expression of VCAM-1, ICAM-1, and E-selectin, as well as a decrease in the secretion of cytokines like IL-6 and IL-8 (De Caterina et al. 1994). Additionally, DHA and EPA reduced the mRNA levels of these adhesion molecules in human umbilical vein endothelial cells (HUVECs) stimulated with IL-1β. Another study that used 30 hypercholesterolemic subjects with no other known cause for endothelial dysfunction was able to demonstrate that marine n-3 fatty acids improve large artery endothelium-dependent dilation in subjects with hypercholesterolemia without affecting endothelium-independent dilation (Goodfellow et al. 2000).

4 Anti-obesity Effects of Algal PUFAs

Obesity is increasing at an epidemic rate globally, affecting both developed and developing countries (Dunstan et al. 2002; Flegal et al. 1998). One of the major issues with obesity is that it is both a disease in itself and a risk factor for several other conditions, such as cardiovascular disease (CVD), type 2 diabetes mellitus, respiratory diseases, musculoskeletal disorders, and certain cancers, all of which significantly affect quality of life and longevity. (Bhaskaran et al. 2014, De Mello et al. 2018, Knight 2011, WHO 2016). Current guidelines for combating obesity emphasize increasing physical activity and adopting a healthy, balanced diet. However, many individuals find it challenging to adhere to these lifestyle modifications, especially over the long term. For instance, a recent meta-analysis revealed that almost half of the initial weight loss achieved through diet or exercise is typically regained within a year (Curioni and Lourenço 2005). Hence, despite widespread recommendations to change lifestyle, the prevalence of obesity continues to rise. The failure of these measures, mainly with regard to the maintenance of the results, has emphasized the need for adjuvant therapeutic resources (Pucci and Finer 2015). On the other hand, currently available pharmacological therapies for weight loss have been shown to be associated with serious adverse effects. Given the rising obesity rates worldwide and the current challenges in treatment, further studies are needed to expand treatment options. In this context, research has shown that n-3 PUFAs have the potential to help prevent and treat obesity (Campbell and Bello 2012, DeFina et al. 2011; Golub et al. 2011).

4.1 Modulation of Fat Metabolism and Insulin Sensitivity

A growing number of studies have shown that n-3 PUFAs have the potential to induce several beneficial effects for the treatment of obesity, such as reduced body fat gain, decreased fat accumulation in the visceral region, improved lipid profile, enhanced insulin sensitivity, reduced glucose intolerance, alleviated hepatic steatosis, and a reduction in inflammation in peripheral tissues (Bargut et al. 2015; Kalupahana et al. 2010; Liu et al. 2013; Rokling-Andersen et al. 2009; Rossmeisl et al. 2014). Further, several animal studies revealed that n-3 PUFAs have the ability to modulate inflammation and hypothalamic function in rodents (Cintra et al. 2012; Pimentel et al. 2013; Viggiano et al. 2016) and to reverse the negative effects of obesity on brain-derived neurotrophic factor (BDNF) gene expression in the hypothalamus of rats (Abdel-Maksoud et al. 2017).

In an animal study, Wistar rats were fed a low-fat control diet (10% corn oil, 70% starch and 20% casein) or one of three high-fat diets: 1. high-lard diet (48% lard, 2% corn oil), 2. lard–corn oil diet (35% lard, 15% corn oil), or 3. lard-fish oil diet (33% lard, 2% corn oil, and 15% fish oil), along with 30% starch and 20% casein ad libitum for 16–20 days. Exceptionally, rats that consumed lard–fish oil diet showed a 30% reduction in visceral (retroperitoneal and epididymal) and 20% reduction in subcutaneous (inguinal) fat by the end of the study. Yook et al. (2015) investigated the effect of microalgal oil on body weight management in obesity-induced C57BL/6 J mice. The animals were orally supplemented with one of the following for 8 weeks: beef tallow (BT), corn oil, fish oil (FO), microalgal oil (MO), or none, as a high-fat diet control group (HD). After the experimental period, FO and MO groups showed significant decreases in body weight gain, epididymal fat pad weights, serum triglycerides, and total cholesterol levels compared to the HD and BT groups. Further, they observed significantly higher expression of PPAR-α, a gene involved in lipid catabolism and acyl-CoA oxidase (ACO), a gene involved in β-oxidation in peroxisomes, and lower expression in lipid anabolic gene, stearoyl-CoA desaturase enzyme (SCD)-1 which is involved in lipogenesis in the MO and FO groups compared to the HD group. A remarkable difference was observed in the carnitine palmitoyltransferase (CPT)-1, a gene involved in β-oxidation in mitochondria, in which the expression was significantly higher in the MO group than that of the FO group. Additionally, serum insulin and leptin were also notably lower in the MO group compared to the FO group suggesting n-3 PUFAs in algal oil are not only effective in lowering TGs and cholesterol synthesis and accumulation in the liver but also in reducing hyperinsulinemia and hyperleptinemia, which could be attributed to the protective effects against obesity. Consistent with above findings, most recent study that used marine algal oil from *Phaeodactylum tricornutum* (PO) and *Laminaria japonica* (LO) (PO had 19.18% EPA and 1.24% DHA, while LO had 15.23% EPA and 1.16% DHA) on high fat diet fed mice also revealed the involvement of algal n-3 PUFAs in inhibiting lipid biosynthesis occurs through the reduction of SREBP-1 expression and its targeted genes, including ACC, FAS, and SCD1, as well as PPARγ. Additionally, n-3 PUFAs enhance lipid degradation by increasing the expression of PPARα and its targeted genes, including CPT-1α and CPT-2 (Lü et al. 2024). Belzung et al. (1993) demonstrated that the protective effect of n-3 PUFAs depends on dose. Further, they showed plasma triglyceride concentrations were positively correlated with the size of retroperitoneal and epididymal adipocytes. Parrish et al. (1990) also confirmed the reduction in adipocyte hypertrophy in an earlier study.

Micallef et al. (2009) measured plasma n-3 PUFAs concentrations in 124 healthy adults with BMI 20–40 kg/m^2. They observed that the percentage of n-3 PUFA was inversely related to BMI and obesity (Micallef et al. 2009). A systematic review with meta-analysis conducted by Bender et al. (2014) showed that the consumption of fish or encapsulated fish oil (rich in n-3 PUFAs) is associated with slight reductions in body weight and waist circumference. However, another meta-analysis conducted by Zhang et al. (2017) found no significant difference in weight loss between n-3 PUFAs and placebo, whereas n-3 PUFAs might effectively reduce waist circumference and TG levels in overweight and obese adults.

4.2 *Modulation of Gut Microbiota*

Gastrointestinal microbial community plays a vital role in host metabolism, digestion, immune function, and the regulation of inflammatory responses. Disruptions in the microbial balance, referred to as dysbiosis, have been implicated in the development of a range of diseases, including obesity, metabolic syndrome, and inflammatory bowel disease. Numerous studies have reported that long-term consumption of high-fat diet results in an increase in Firmicutes and a decrease in Bacteroidetes (Turnbaugh et al. 2006). A study conducted to evaluate the effect of algal oil on high-fat diet-fed mice found that algal oil supplementation significantly reduced the HFD-induced Firmicute to Bacteroidetes ratio (Lü et al. 2024). These results align with the theory that body fat percentage is positively correlated with the abundance of Firmicutes in the gut microbiota of both mice and humans. Further, they noticed the increased abundance of short-chain fatty acid (SCFA) producing microbiota including *Bifidobacteriaceae*, *Lachnospiraceae*, and *Lactobacillaceae* in algal oil supplemented groups. Evidences suggest that SCFAs-producing microbiota, *Bifidobacteriaceae*, *Lachnospiraceae*, and *Lactobacillaceae* have negative correlation with obesity and insulin resistance (Morrison and Preston 2016). Another animal study by Li et al. (2018) used PUFAs from microalgae *Spirulina platensis* (SP) to evaluate the effect on gut microbiota in high-fat diet rats. In their study, they revealed a decrease in Firmiutes to Bacteroidetes ratio in cecal contents. Further, they observed increased proportions of *Porphyromonadaceae*, which was previously associated with atherosclerosis (Henao-Mejia et al. 2012). Further evidence by Li et al. (2019) also suggests that PUFAs from *Spirulina platensis* decrease the abundances of *Turicibacter*, *Clostridium_XlVa*, and *Romboutsia*, which were positively correlated with lipid metabolism indicators.

Similarly, another study discovered that PUFAs from microalga *Chlorella pyrenoidosa* can also improve the abundance of *Prevotella,* and *Ruminococcus* which are correlated with lipid metabolism (Wan et al. 2018).

5 Conclusion

Algal-derived PUFAs, particularly EPA and DHA, offer significant potential for improving cardiovascular and metabolic health. Through their effects on lipid profiles, inflammation, adipogenesis, and fat oxidation, algal n-3 s play a critical role in preventing and managing cardiovascular diseases and obesity. Furthermore, algae offer a sustainable, ethical, and contaminant-free source of n-3 s, making them an attractive alternative to fish oils. As research continues to uncover the full range of benefits associated with algal n-3 s, these compounds hold promise not only for individual health but also for global health solutions, given their sustainability and bioavailability. Further clinical studies and advancements in algae biotechnology will be critical in expanding the use of these n-3-rich oils for therapeutic purposes.

References

Abdel-Maksoud, S. M., Hassanein, S. I., Gohar, N. A., Attia, S. M. M., & Gad, M. Z. (2017). Investigation of brain-derived neurotrophic factor (BDNF) gene expression in hypothalamus of obese rats: Modulation by omega-3 fatty acids. Nutritional Neuroscience, 20(8), 443–448.

Abeywardena, M. Y., & Head, R. J. (2001). Long chain n− 3 polyunsaturated fatty acids and blood vessel function. Cardiovascular Research, 52(3), 361–371.

Adkins, Y., & Kelley, D. S. (2010). Mechanisms underlying the cardioprotective effects of omega-3 polyunsaturated fatty acids. The Journal of Nutritional Biochemistry, 21(9), 781–792.

Allaire, J., Couture, P., Leclerc, M., Charest, A., Marin, J., Lépine, M. C., Talbot, D., Tchernof, A., & Lamarche, B. (2016). A randomized, crossover, head-to-head comparison of eicosapentaenoic acid and docosahexaenoic acid supplementation to reduce inflammation markers in men and women: The Comparing EPA to DHA (ComparED) Study. American Journal of Clinical Nutrition, 104(2), 280–287.

Arao, T., Sakaki, T., & Yamada, M. (1994). Biosynthesis of polyunsaturated lipids in the diatom, *Phaeodactylum tricornutum*. Phytochemistry, 36(3), 629–635.

Babuskin, S., Krishnan, K. R., Saravana Babu, P. A., Sivarajan, M., & Sukumar, M. (2014). Functional foods enriched with marine microalga *Nannochloropsis oculata* as a source of ω-3 fatty acids. Food Technology and Biotechnology, 52(3), 292–299.

Bargut, T. C. L., Mandarim-de-Lacerda, C. A., & Aguila, M. B. (2015). A high-fish-oil diet prevents adiposity and modulates white adipose tissue inflammation pathways in mice. The Journal of Nutritional Biochemistry, 26(9), 960–969.

Belzung, F., Raclot, T., & Groscolas, R. (1993). Fish oil n-3 fatty acids selectively limit the hypertrophy of abdominal fat depots in growing rats fed high-fat diets. American Journal of Physiology - Regulatory Integrative and Comparative Physiology, 264(6) 33–6.

Bender, N., Portmann, M., Heg, Z., Hofmann, K., Zwahlen, M., & Egger, M. (2014). Fish or n3-PUFA intake and body composition: A systematic review and meta-analysis. Obesity Reviews, 15(8), 657–665.

Bernstein, A. M., Ding, E. L., Willett, W. C., & Rimm, E. B. (2012). A Meta-Analysis Shows That Docosahexaenoic Acid from Algal Oil Reduces Serum Triglycerides and Increases HDL-Cholesterol and LDL-Cholesterol in Persons without Coronary Heart Disease3. The Journal of Nutrition, 142(1), 99–104.

Bhaskaran, K., Douglas, I., Forbes, H., dos-Santos-Silva, I., Leon, D. A., & Smeeth, L. (2014). Body-mass index and risk of 22 specific cancers: a population-based cohort study of 5· 24 million UK adults. The lancet, 384(9945), 755–765.

Biancarosa, I., Belghit, I., Bruckner, C. G., Liland, N. S., Waagbø, R., Amlund, H., Heesch, S., & Lock, E. J. (2018). Chemical characterization of 21 species of marine macroalgae common in Norwegian waters: benefits of and limitations to their potential use in food and feed. Journal of the Science of Food and Agriculture, 98(5), 2035–2042.

Biris-Dorhoi, E. S., Michiu, D., Pop, C. R., Rotar, A. M., Tofana, M., Pop, O. L., Socaci, S. A., & Farcas, A. C. (2020). Macroalgae—A Sustainable Source of Chemical Compounds with Biological Activities. Nutrients 2020, Vol. 12, Page 3085, 12(10), 3085.

Calder, P. C. (2001). Polyunsaturated fatty acids, inflammation, and immunity. Lipids, 36(9), 1007–1024.

Calder, P. C. (2006). Polyunsaturated fatty acids and inflammation. Prostaglandins Leukotrienes and Essential Fatty Acids, 75(3), 197–202.

Campbell, Sara & Bello, Nicholas. (2012). Omega-3 Fatty Acids and Obesity. Journal of Food & Nutritional Disorders. 01. 10.

Cintra, D. E., Ropelle, E. R., Moraes, J. C., Pauli, J. R., Morari, J., de Souza, C. T., Grimaldi, R., Stahl, M., Carvalheira, J. B., Saad, M. J., & Velloso, L. A. (2012). Unsaturated fatty acids revert diet-induced hypothalamic inflammation in obesity. PLoS ONE, 7(1).

Cohen, Z., & Ratledge, C. (2005). Single cell oils: microbial and algal oils. AOCS Publishing.

Curioni, C. C., & Lourenço, P. M. (2005). Long-term weight loss after diet and exercise: a systematic review. International Journal of Obesity, 29(10), 1168–1174.

De Caterina, R., Cybulsky, M. I., Clinton, S. K., Gimbrone Jr, M. A., & Libby, P. (1994). The omega-3 fatty acid docosahexaenoate reduces cytokine-induced expression of proatherogenic and proinflammatory proteins in human endothelial cells. Arteriosclerosis and thrombosis: a journal of vascular biology, 14(11), 1829-1836.

DeFina, L. F., Marcoux, L. G., Devers, S. M., Cleaver, J. P., & Willis, B. L. (2011). Effects of omega-3 supplementation in combination with diet and exercise on weight loss and body composition. The American journal of clinical nutrition, 93(2), 455-462.

De Mello, A. H., Uberti, M. F., De Farias, B. X., De Souza, N. A. R., & Rezin, G. T. (2018). n-3 PUFA and obesity: from peripheral tissues to the central nervous system. British Journal of Nutrition, 119(11), 1312–1323.

Dunstan, D. W., Zimmet, P. Z., Welborn, T. A., De Courten, M. P., Cameron, A. J., Sicree, R. A., & AusDiab Steering Committee. (2002). The rising prevalence of diabetes and impaired glucose tolerance: the Australian Diabetes, Obesity and Lifestyle Study. Diabetes Care, 25(5), 829–834

Eslick, G. D., Howe, P. R., Smith, C., Priest, R., & Bensoussan, A. (2009). Benefits of fish oil supplementation in hyperlipidemia: a systematic review and meta-analysis. International Journal of Cardiology, 136(1), 4–16.

Fan, K. W., Jiang, Y., Faan, Y. W., & Chen, F. (2007). Lipid characterization of mangrove thraustochytrid - Schizochytrium mangrovei. Journal of Agricultural and Food Chemistry, 55(8), 2906–2910.

FAO, F. (2018). Agriculture Organization of the United Nations 2018 The state of world fisheries and aquaculture 2018—Meeting the sustainable development goals. CC BYNC-SA, 3.

Flegal, K. M., Carroll, M. D., Kuczmarski, R. J., & Johnson, C. L. (1998). Overweight and obesity in the United States: prevalence and trends, 1960–1994. International Journal of Obesity, 22(1), 39–47.

Friedberg, C. E., Janssen, M. J., Heine, R. J., & Grobbee, D. E. (1998). Fish oil and glycemic control in diabetes: a meta-analysis. Diabetes Care, 21(4), 494–500.

Gillan, F. T., Mcfadden, G. I., Wetherbee, R., & Johns, R. B. (1981). Sterols and fatty acids of an Antarctic Sea ice diatom, *Stauroneis amphioxys*. Phytochemistry, 20(8), 1935–1937.

Golub, N., Geba, D., Mousa, S. A., Williams, G., & Block, R. C. (2011). Greasing the wheels of managing overweight and obesity with omega-3 fatty acids. Medical Hypotheses, 77(6), 1114–1120.

Goodfellow, J., Bellamy, M. F., Ramsey, M. W., Jones, C. J. H., & Lewis, M. J. (2000). Dietary supplementation with marine omega-3 fatty acids improves systemic large artery endothelial function in subjects with hypercholesterolemia. Journal of the American College of Cardiology, 35(2), 265–270.

Guil-Guerrero, J. L. (2007). Stearidonic acid (18:4n-3): Metabolism, nutritional importance, medical uses and natural sources. European Journal of Lipid Science and Technology, 109(12), 1226–1236.

Guschina, I. A., & Harwood, J. L. (2006). Lipids and lipid metabolism in eukaryotic algae. Progress in Lipid Research, 45(2), 160–186.

Harwood, J. L., & Caterson, B. (2006). Dietary omega-3 polyunsaturated fatty acids and inflammation. Lipid Technology, 18, 7–10.

Harwood, J. (1989). Lipid metabolism in plants. Critical Reviews in Plant Sciences, 8(1), 1–43.

Harwood, J. L. (2019). Algae: Critical Sources of Very Long-Chain Polyunsaturated Fatty Acids. Biomolecules, 9(11), 708.

Harris, W. S. (1997). n-3 fatty acids and serum lipoproteins: human studies. The American Journal of Clinical Nutrition, 65(5), 1645S-1654S.

Henao-Mejia, J., Elinav, E., Jin, C., Hao, L., Mehal, W. Z., Strowig, T., & Flavell, R. A. (2012). Inflammasome-mediated dysbiosis regulates progression of NAFLD and obesity. Nature, 482(7384), 179–185.

Hirafuji, M., Ebihara, T., Kawahara, F., Hamaue, N., Endo, T., & Minami, M. (2001). Inhibition by docosahexaenoic acid of receptor-mediated Ca2+ influx in rat vascular smooth muscle cells stimulated with 5-hydroxytryptamine. European Journal of Pharmacology, 427(3), 195–201.

Hirafuji, M., Machida, T., Tsunoda, M., Miyamoto, A., & Minami, M. (2002). Docosahexaenoic acid potentiates interleukin-1β induction of nitric oxide synthase through mechanism involving p44/42 MAPK activation in rat vascular smooth muscle cells. British Journal of Pharmacology, 136(4), 613–619.

Holub, B. J. (2009). Docosahexaenoic acid (DHA) and cardiovascular disease risk factors. Prostaglandins, Leukotrienes and Essential Fatty Acids, 81(2–3), 199–204.

Ishihara, K., Murata, M., Kaneniwa, M., Saito, H., Shinohara, K., & Maeda-Yamamoto, M. (1998). Inhibition of icosanoid production in MC/9 mouse mast cells by n-3 polyunsaturated fatty acids isolated from edible marine algae. Bioscience, Biotechnology and Biochemistry, 62(7), 1412–1415.

Jaworowska, A., & Murtaza, A. (2022). Seaweed derived lipids are a potential anti-inflammatory agent: a review. International Journal of Environmental Research and Public Health, 20(1), 730.

Jayasinghe, G. D. T. M., Jinadasa, B. K. K. K., & Chinthaka, S. D. M. (2018). Study on Lipid Content and Fatty Acid Profile of Four Marine Macro Algae (Seaweeds) Collected from South East Coast of Sri Lanka. Asian Journal of Chemistry and Pharmaceutical Sciences, 3(1), 1–6.

Kalupahana, N. S., Claycombe, K., Newman, S. J., Stewart, T., Siriwardhana, N., Matthan, N., & Moustaid-Moussa, N. (2010). Eicosapentaenoic acid prevents and reverses insulin resistance in high-fat diet-induced obese mice via modulation of adipose tissue inflammation. The Journal of Nutrition, 140(11), 1915–1922.

Kendel, M., Barnathan, G., Fleurence, J., Rabesaotra, V., & Wielgosz-Collin, G. (2013). Non-methylene interrupted and hydroxy fatty acids in polar lipids of the alga *Grateloupia turuturu* over the four seasons. Lipids, 48(5), 535–545.

Khan, M. N. A., Cho, J. Y., Lee, M. C., Kang, J. Y., Nam, G. P., Fujii, H., & Hong, Y. K. (2007). Isolation of two anti-inflammatory and one pro-inflammatory polyunsaturated fatty acids from the brown seaweed *Undaria pinnatifida*. Journal of Agricultural and Food Chemistry, 55(17), 6984–6988.

Knight, J. A. (2011). Diseases and disorders associated with excess body weight. Annals of Clinical & Laboratory Science, 41(2), 107–121.

Kris-Etherton, P. M., Harris, W. S., & Appel, L. J. (2002). Fish consumption, fish oil, omega-3 fatty acids, and cardiovascular disease. Circulation, 106(21), 2747–2757.

Kyle, D. J. (2001). The large-scale production and use of a single-cell oil highly enriched in docosahexaenoic acid. ACS Symposium Series, 788, 92–107.

Leblond, J. D., & Chapman, P. J. (2000). Lipid class distribution of highly unsaturated long chain fatty acids in marine dinoflagellates. Journal of Phycology, 36(6), 1103–1108.

Leblond, J. D., Evans, T. J., & Chapman, P. J. (2003). The biochemistry of dinoflagellate lipids, with particular reference to the fatty acid and sterol composition of a *Karenia brevis* bloom. Phycologia, 42(4), 324–331.

Lenihan-Geels, G., Bishop, K. S., & Ferguson, L. R. (2013). Alternative Sources of Omega-3 Fats: Can We Find a Sustainable Substitute for Fish? Nutrients, 5(4), 1301–1315.

Li, T. T., Liu, Y. Y., Wan, X. Z., Huang, Z. R., Liu, B., & Zhao, C. (2018). Regulatory efficacy of the polyunsaturated fatty acids from microalgae Spirulina platensis on lipid metabolism and gut microbiota in high-fat diet rats. International Journal of Molecular Sciences, 19(10), 3075.

Li, T. T., Tong, A. J., Liu, Y. Y., Huang, Z. R., Wan, X. Z., Pan, Y. Y., Jia, R. B., Liu, B., Chen, X. H., & Zhao, C. (2019). Polyunsaturated fatty acids from microalgae Spirulina platensis modulates lipid metabolism disorders and gut microbiota in high-fat diet rats. Food and Chemical Toxicology, (131) 110558.

Liang, Y., Maeda, Y., Yoshino, T., Matsumoto, M., & Tanaka, T. (2014). Profiling of fatty acid methyl esters from the oleaginous diatom *Fistulifera* sp. strain JPCC DA0580 under nutrition-sufficient and -deficient conditions. Journal of Applied Phycology, 26(6), 2295–2302.

Lichtenstein, A. H., Appel, L. J., Brands, M., Carnethon, M., Daniels, S., Franch, H. A., Franklin, B., Kris-Etherton, P., Harris, W. S., Howard, B., Karanja, N., Lefevre, M., Rudel, L., Sacks, F., Horn, L. van, Winston, M., & Wylie-Rosett, J. (2006). Summary of American heart association diet and lifestyle recommendations revision 2006. Arteriosclerosis, Thrombosis, and Vascular Biology, 26(10), 2186–2191.

Liu, X., Xue, Y., Liu, C., Lou, Q., Wang, J., Yanagita, T., Xue, C., & Wang, Y. (2013). Eicosapentaenoic acid-enriched phospholipid ameliorates insulin resistance and lipid metabolism in diet-induced-obese mice. Lipids in Health and Disease, 12(1).

Lü, L., Li, S. X., Li, X. Y., Ma, J. R., He, J. X., Xing, H. Z., Chen, C., Miyashita, K., Yang, Y. H., & Du, L. (2024). Marine algae oils from *Phaeodactylum tricornutum* and *Laminaria japonica* alleviate obesity, insulin resistance, and gut microbiota dysbiosis in high-fat diet-fed mice. Journal of Functional Foods, (117) 106234.

MacArtain, P., Gill, C. I., Brooks, M., Campbell, R., & Rowland, I. R. (2007). Nutritional value of edible seaweeds. Nutrition reviews, 65(12), 535–543.

Maki, K. C., Yurko-Mauro, K., Dicklin, M. R., Schild, A. L., & Geohas, J. G. (2014). A new, microalgal DHA- and EPA-containing oil lowers triacylglycerols in adults with mild-to-moderate hypertriglyceridemia. Prostaglandins, Leukotrienes, and Essential Fatty Acids, 91(4), 141–148.

Mas, E., Woodman, R. J., Burke, V., Puddey, I. B., Beilin, L. J., Durand, T., & Mori, T. A. (2010). The omega-3 fatty acids EPA and DHA decrease plasma F2- isoprostanes: Results from two placebo-controlled interventions. Free Radical Research, 44(9), 983–990.

McLennan, P., Howe, P., Abeywardena, M., Muggli, R., Raederstorff, D., Mano, M., ... & Head, R. (1996). The cardiovascular protective role of docosahexaenoic acid. European journal of pharmacology, 300(1–2), 83–89.

Mendes, A., Reis, A., Vasconcelos, R., Guerra, P., & Lopes Da Silva, T. (2009). *Crypthecodinium cohnii* with emphasis on DHA production: A review. Journal of Applied Phycology, 21(2), 199–214.

Micallef, M., Munro, I., Phang, M., & Garg, M. (2009). Plasma n-3 polyunsaturated fatty acids are negatively associated with obesity. British Journal of Nutrition, 102(9), 1370–1374.

Miyashita, K., Narayan, B., Tsukui, T., Kamogawa, H., Abe, M., & Hosokawa, M. (2011). Brown seaweed lipids as potential source of omega-3 PUFA in biological systems. Handbook of Marine Macroalgae: Biotechnology and Applied Phycology, 329–339.

Moreno, V. J., de Moreno, J. E. A., & Brenner, R. R. (1979). Biosynthesis of unsaturated fatty acids in the diatom Phaeodactylum tricornutum. Lipids, 14(1), 15–19.

Mori, T. A., Puddey, I. B., Burke, V., Croft, K. D., Dunstan, D. W., Rivera, J. H., & Beilin, L. J. (2000). Effect of ω3 fatty acids on oxidative stress in humans: GC-MS measurement of urinary F2-isoprostane excretion. Redox Report, 5(1), 45–46.

Mori, T. A., Woodman, R. J., Burke, V., Puddey, I. B., Croft, K. D., & Beilin, L. J. (2003). Effect of eicosapentaenoic acid and docosahexaenoic acid on oxidative stress and inflammatory markers in treated-hypertensive type 2 diabetic subjects. Free Radical Biology and Medicine, 35(7), 772–781.

Morrison, D. J., & Preston, T. (2016). Formation of short chain fatty acids by the gut microbiota and their impact on human metabolism. Gut Microbes, 7(3), 189–200.

Mozaffarian, D., & Rimm, E. B. (2006). Fish intake, contaminants, and human health: evaluating the risks and the benefits. Jama, 296(15), 1885–1899.

Neff, L. M., Culiner, J., Cunningham-Rundles, S., Seidman, C., Meehan, D., Maturi, J. & Breslow, J. L. (2011). Algal docosahexaenoic acid affects plasma lipoprotein particle size distribution in overweight and obese adults. The Journal of nutrition, 141(2), 207–213.

Parrish, C. C., Pathy, D. A., & Angel, A. (1990). Dietary fish oils limit adipose tissue hypertrophy in rats. Metabolism, 39(3), 217–219.

Peng, Y., Hu, J., Yang, B., Lin, X. P., Zhou, X. F., Yang, X. W., & Liu, Y. (2015). Chemical composition of seaweeds. In Seaweed sustainability Academic Press. pp. 79–124.

Pereira, H., Barreira, L., Figueiredo, F., Custódio, L., Vizetto-Duarte, C., Polo, C., Rešek, E., Aschwin, E., & Varela, J. (2012). Polyunsaturated fatty acids of marine macroalgae: Potential for nutritional and pharmaceutical applications. Marine Drugs, 10(9), 1920–1935.

Pernet, F., Tremblay, R., Demers, E., & Roussy, M. (2003). Variation of lipid class and fatty acid composition of *Chaetoceros muelleri* and *Isochrysis* sp. grown in a semicontinuous system. Aquaculture, 221(1–4), 393–406.

Pimentel, G. D., Lira, F. S., Rosa, J. C., Oller do Nascimento, C. M., Oyama, L. M., Harumi Watanabe, R. L., & Ribeiro, E. B. (2013). High-Fat Fish Oil Diet Prevents Hypothalamic Inflammatory Profile in Rats. ISRN Inflammation, 2013, 1–7.

Pohl, P., & Zurheide, F. (1979). Fatty acids and lipids of marine algae and the control of their biosynthesis by environmental factors. Marine Algae in Pharmaceutical Science, 2, 473523.

Pucci, A., & Finer, N. (2015). New medications for treatment of obesity: metabolic and cardiovascular effects. Canadian Journal of Cardiology, 31(2), 142–152.

Robertson, R. C., Guihéneuf, F., Bahar, B., Schmid, M., Stengel, D. B., Fitzgerald, G. F., Ross, R. P., & Stanton, C. (2015). The Anti-Inflammatory Effect of Algae-Derived Lipid Extracts on Lipopolysaccharide (LPS)-Stimulated Human THP-1 Macrophages. Marine Drugs, 13(8), 5402–5424.

Robertson, R., Guihéneuf, F., Schmid, M., Stengel, D. B., Fitzgerald, G., Ross, P., & Stanton, C. (2013). Algae-derived polyunsaturated fatty acids: implications for human health. Polyunsaturated fatty acids: sources, antioxidant properties and health benefits, 45–99.

Rodrigues, D., Freitas, A. C., Pereira, L., Rocha-Santos, T. A. P., Vasconcelos, M. W., Roriz, M., Rodríguez-Alcalá, L. M., Gomes, A. M. P., & Duarte, A. C. (2015). Chemical composition of red, brown and green macroalgae from Buarcos bay in Central West Coast of Portugal. Food Chemistry, 183, 197–207.

Rokling-Andersen, M. H., Rustan, A. C., Wensaas, A. J., Kaalhus, O., Wergedahl, H., Røst, T. H., & Drevon, C. A. (2009). Marine n-3 fatty acids promote size reduction of visceral adipose depots, without altering body weight and composition, in male Wistar rats fed a high-fat diet. British Journal of Nutrition, 102(7), 995–1006.

Rossmeisl, M., Medrikova, D., Van Schothorst, E. M., Pavlisova, J., Kuda, O., Hensler, M., & Kopecky, J. (2014). Omega-3 phospholipids from fish suppress hepatic steatosis by integrated inhibition of biosynthetic pathways in dietary obese mice. Biochimica et Biophysica Acta (BBA)-Molecular and Cell Biology of Lipids, 1841(2), 267–278.

Ryan, A. S., Keske, M. A., Hoffman, J. P., & Nelson, E. B. (2009). Clinical overview of algal-docosahexaenoic acid: effects on triglyceride levels and other cardiovascular risk factors. American Journal of Therapeutics, 16(2), 183–192.

Sanders, T. A., Gleason, K., Griffin, B., & Miller, G. J. (2006). Influence of an algal triacylglycerol containing docosahexaenoic acid (22: 6n-3) and docosapentaenoic acid (22: 5n-6) on cardiovascular risk factors in healthy men and women. British Journal of Nutrition, 95(3), 525–531.

Santos, S. A. O., Vilela, C., Freire, C. S. R., Abreu, M. H., Rocha, S. M., & Silvestre, A. J. D. (2015). Chlorophyta and Rhodophyta macroalgae: A source of health promoting phytochemicals. Food Chemistry, 183, 122–128.

Scott D. Doughman, Srirama Krupanidhi, & Carani B. Sanjeevi. (2007). Omega-3 Fatty Acids for Nutrition and Medicine: Considering Microalgae Oil as a Vegetarian Source of EPA and DHA. Current Diabetes Reviews, 3(3), 198–203.

Spolaore, P., Joannis-Cassan, C., Duran, E., & Isambert, A. (2006). Commercial applications of microalgae. Journal of Bioscience and Bioengineering, 101(2), 87–96.

Sprecher, H., Chen, Q., & Yin, F. Q. (1999). Regulation of the biosynthesis of 22:5n-6 and 22:6n-3: A complex intracellular process. Lipids, 34(1), S153–S156.

Taipale, S., Peltomaa, E., & Salmi, P. (2020). Variation in ω-3 and ω-6 polyunsaturated fatty acids produced by different phytoplankton taxa at early and late growth phase. Biomolecules, 10(4).

Tocher, D. R., Bell, J. G., Dick, J. R., & Crampton, V. O. (2003). Effects of Dietary Vegetable Oil on Atlantic Salmon Hepatocyte Fatty Acid Desaturation and Liver Fatty Acid Compositions. Lipids, 38(7), 723–732.

Tocher, D. R., Betancor, M. B., Sprague, M., Olsen, R. E., & Napier, J. A. (2019). Omega-3 long-chain polyunsaturated fatty acids, EPA and DHA: Bridging the gap between supply and demand. Nutrients, 11(1).

Tonon, T., Harvey, D., Larson, T. R., & Graham, I. A. (2002). Long chain polyunsaturated fatty acid production and partitioning to triacylglycerols in four microalgae. Phytochemistry, 61(1), 15–24.

Turnbaugh, P. J., Ley, R. E., Mahowald, M. A., Magrini, V., Mardis, E. R., & Gordon, J. I. (2006). An obesity-associated gut microbiome with increased capacity for energy harvest. Nature, 444(7122), 1027–1031.

Usup, G., Hamid, S. Z., Chiet, P. K., Wah, C. K., & Ahmad, A. (2008). Marked differences in fatty acid profiles of some planktonic and benthic marine dinoflagellates from Malaysian waters. Phycologia, 47(1), 105–111.

Viggiano, E., Mollica, M. P., Lionetti, L., Cavaliere, G., Trinchese, G., De Filippo, C., & Monda, M. (2016). Effects of a high-fat diet enriched in lard or in fish oil on the hypothalamic amp-activated protein kinase and inflammatory mediators. Frontiers in Cellular Neuroscience, 10, 150.

Vors, C., Allaire, J., Marin, J., Lépine, M. C., Charest, A., Tchernof, A., Couture, P., & Lamarche, B. (2017). Inflammatory gene expression in whole blood cells after EPA vs. DHA supplementation: Results from the ComparED study. Atherosclerosis, 257, 116–122.

Wan, X., Li, T., Liu, D., Chen, Y., Liu, Y., Liu, B., Zhang, H., & Zhao, C. (2018). Effect of Marine Microalga *Chlorella pyrenoidosa* Ethanol Extract on Lipid Metabolism and Gut Microbiota Composition in High-Fat Diet-Fed Rats. Marine Drugs 2018, Vol. 16, Page 498, 16(12), 498.

World Health Organization (2016) Obesity and overweight fact sheet no. 311.

Yang, B., Ren, X. L., Li, Z. H., Shi, M. Q., Ding, F., Su, K. P., & Li, D. (2020). Lowering effects of fish oil supplementation on proinflammatory markers in hypertension: results from a randomized controlled trial. Food & function, 11(2), 1779–1789.

Yao, L., Gerde, J. A., Lee, S. L., Wang, T., & Harrata, K. A. (2015). Microalgae lipid characterization. Journal of Agricultural and Food Chemistry, 63(6), 1773–1787.

Yook, J. S., Kim, K. A., Park, J. E., Lee, S. H., & Cha, Y. S. (2015). Microalgal oil supplementation has an anti-obesity effect in C57BL/6J mice fed a high fat diet. Preventive nutrition and food science, 20(4), 230.

Zhang, Y. Y., Liu, W., Zhao, T. Y., & Tian, H. M. (2017). Efficacy of omega-3 polyunsaturated fatty acids supplementation in managing overweight and obesity: a meta-analysis of randomized clinical trials. The Journal of nutrition, health and aging, 21(2), 187-192.

Chapter 12
Algal Pigments and Diabetes

Jayada Gallege, Saummi Uduwana, Udari Menaka, and Indira Wickramasinghe

Abstract The health condition of elevated levels of glucose in the bloodstream, or, in simple terms, diabetes, has disastrously affected millions of people worldwide, causing a significant strain on the healthcare system. It is a chronic metabolic condition mainly due to decreased insulin synthesis or utilization. Algae are known for their extraordinary chemical diversity and have been discovered to be rich in several pigments. These compounds contribute to the functional properties of algae, one of which is their antidiabetic nature. These pigments are bioactive compounds that can be used as natural compounds in developing novel therapeutic approaches to treating diabetes. Chlorophylls, carotenoids, and phycobiliproteins are some of the examples of algal pigments that exert various biological functions. For instance, a carotenoid called fucoxanthin found in brown algae can accelerate glucose uptake into the skeletal muscle cells and adipocytes while creating an environment to enhance insulin sensitivity. Thus, it can be used as a promising treatment for diabetes. Moreover, several algal-derived pigments have antioxidant and anti-inflammatory characteristics, which can prevent diabetes-related problems. This chapter is dedicated to algal pigments and their potential to control diabetes, with a comprehensive discussion uncovering multiple curative approaches.

Keywords Algal pigments · Chlorophylls · Carotenoids · Phycobiliproteins Diabetes Antidiabetic properties

---

J. Gallege
School of Health Sciences, Queen Margaret University, Edinburgh, UK

S. Uduwana · I. Wickramasinghe (✉)
Department of Food Science and Technology, Faculty of Applied Sciences, University of Sri Jayewardenepura, Nugegoda, Sri Lanka
e-mail: indiraw@sjp.ac.lk

U. Menaka
Department of Food Science and Technology, Faculty of Applied Sciences, University of Sri Jayewardenepura, Nugegoda, Sri Lanka

Faculty of Graduate Studies, University of Sri Jayewardenepura, Nugegoda, Sri Lanka

R. Raja et al. (eds.), *Algae in Diabetes Management*,
https://doi.org/10.1007/978-981-95-0571-5_12

1 Introduction

Diabetes mellitus occurs due to increased blood glucose and lack of insulin. There were 536.6 million individuals living with diabetes in 2021, which will increase to 783.2 million by 2045 according to the forecast (Kaushik et al. 2023). Due to diabetes mellitus, other diseases such as kidney disease, heart attack, blindness, and stroke can occur (Bae et al. 2020). Diabetes is categorized into two primary forms: type 1 and type 2. Type 1 diabetes is due to genetic problems that prevent the pancreas from producing insulin, and Type 2 diabetes is defined by elevated blood sugar levels that are not directly influenced by insulin (Kaushik et al. 2023).

Macroalgae or else seaweeds are mainly categorized as green algae (*Chlorophyta*), red algae (*Rhodophyta*), and brown algae (*Phaeophyta*). As well as microalgae are very small cells that can be visible under microscopic conditions and are also very popular now due to their potential for applying as nutraceuticals and pharmaceuticals (Imchen and Singh 2023). There are several advantages of using microalgae as a food ingredient such as their rapid growth, short harvesting cycle, and high productivity. Moreover, microalgae can be grown using non-potable water such as wastewater and seawater, even in deserts or wetlands (Baek et al. 2018). Even a thousand years ago, algae were a food item that existed in the human diet. Seaweeds have been eaten by several Asian countries for a long time. However, this practice is now becoming popular in other countries (Bocanegra et al. 2021).

Chlorophylls, phycobiliproteins, and carotenoids are the basic pigment categories that can be found in algae. Chlorophylls are found in all autotrophic algae, and they are non-polar, green colored pigments. The structure of the chlorophyll pigment has porphyrin or hydroporphyrin rings bound to a magnesium atom. Carotenoids show yellow, red, orange, or purple color. The structure of carotenoids belongs to the terpenoid class and is considered non-polar. There are two main groups of carotenoids which are carotenes and xanthophylls. Carotenes contain unsaturated hydrocarbons and xanthophylls identical to carotenes but have at least one functional group with oxygen. Phycobiliproteins are fluorescent and water-soluble pigment groups. Purple-colored phycoerythrin and phycoerythrocyanin, blue-colored phycocyanin, and bluish-green-colored allophycocyanin are the four main categories of phycobiliproteins (Osório et al. 2020). In photosynthesis, pigments in algae are used to grab the sunlight as the primary energy source. These pigments have antioxidants, anti-carcinogenic, anti-obesity, anti-inflammatory, and antitumor activities other than anti-diabetes properties (Patel et al. 2022). This chapter will discuss how algal pigments are related to diabetes and the usage and importance of algal pigments in antidiabetic properties.

2 Phycobiliproteins

Phycobiliproteins have gained attention as algal-derived bioactive compounds due to their anti-carcinogenic, anti-inflammatory, antioxidant, anti-diabetic, immunomodulatory, anti-tumor, hepatoprotective properties, etc. (Dagnino-Leone et al. 2022; Imchen and Singh 2023; Chen et al. 2022).

Due to the vibrant colors and the functional properties exerted by phycobiliproteins, the applications of phycobiliproteins are widespread in numerous fields of biotechnology including food, nutraceuticals, clinical diagnostics, cosmetics, etc. The trend toward the usage of natural colorants over synthetic ones has led to research preferences toward advanced applications of phycobiliproteins.

Phycobiliproteins are natural, hydrophilic, non-toxic, and highly fluorescent pigment proteins present in algae, mainly found in Rhodophyta, Cryptophyta, and Glaucophyta (Chen et al. 2022; Aslam et al. 2020). Phycobiliproteins are considered auxiliary photosynthetic complexes capable of light harvesting. Although chlorophyll is the predominant photosynthetic pigment in aquatic environments, auxiliary photosynthetic complexes or accessory pigments assist in photosynthesis where chlorophylls do not get sufficient sunlight due to circumstances like reflection, scattering, and absorption. Under such conditions, accessory pigments like phycobiliproteins can harvest light in the spectral range of 500–600 nm and transfer the energy to chlorophyll A through an energy transfer chain (Dagnino-Leone et al. 2022).

As stated by Derbel et al. (2022), phycobiliproteins are categorized into four types based on their spectral properties, namely phycoerythrin (λ_{max} = 540–570 nm), phycocyanin (λ_{max} = 610–620 nm), allophycocyanin (λ_{max} = 650–655 nm), and phycoerythrocyanin (λ_{max} = 560–600 nm). The bright color of red algae is referred to as phycoerythrin and phycocyanin content which are prevalent over others. Phycocyanin and phycoerythrin are the most studied for their anti-diabetic properties.

3 Phycocyanin

Phycocyanin is a dark blue pigment-protein complex owing to a spectrum of bioactivity (Ziyaei et al. 2023). It possesses many beneficial functionalities such as antioxidant, anti-inflammatory, immune boosting, anticarcinogenic, and hepatoprotective properties among which its commendable impact on anti-diabetic properties is the primary objective of this chapter. Several studies have been conducted on the anti-diabetic properties of phycocyanin as an algal pigment.

According to the study conducted by Hao et al. (2022), phycocyanin led to a downregulation of fasting blood glucose and serum indicators while improving glucose tolerance and the histological changes in the liver and pancreas, by AKT and AMPK signaling pathways in high-insulin-induced insulin-resistant human hepatocellular carcinoma (HCC) cell line models and mice with type 2 diabetes mellitus induced by a high-glucose and high-fat diet. This research lays a scientific theoretical background for the potential use of natural marine products as a functional food. Ren et al. (2018) stated phycocyanin as a promising hypoglycemic agent based on the research conducted on human hepatoma HepG2 cells. The study revealed that phycocyanin promoted glucose homeostasis by hindering hepatic gluconeogenesis and increasing glycogen synthesis through activating specific signaling pathways, namely, IRS/PI3 K/AKT and SIRT1/LKB1/AMPK in insulin-resistant liver cells.

Phycocyanin has been observed to exert anti-diabetic and anti-glycation properties on streptozotocin (STZ)-induced diabetic Wistar rats (Husain et al. 2022; Hao et al. 2022). Persistence of hyperglycemia results in the rise of advanced glycation end products (AGEs), which leads to severe complications of diabetes over time. Furthermore, El-Sayed et al. (2018) reported that phycocyanin has resulted in an increase of insulin response hence a decline in blood glucose levels in mice with type 2 diabetes.

Phycocyanin possesses inhibitory action against α-amylase, α-glucosidase, and β-glucosidase (Munawaroh et al. 2020; Prabakaran et al. 2020), which are involved in carbohydrate metabolism and hence has been claimed for its potential as an anti-diabetic or therapeutic agent.

The anti-diabetic nature can likely be aided by the radical scavenging activity exerted by phycocyanin. Oxidative stress is causative for many ailments ranging from inflammation to severe ends like cancers, where diabetes can potentially lie amidst. Hence, antioxidants can stand out in preventing or reducing the risk of diabetes.

4 Phycoerythrin

Phycoerythrin is an algal pigment that appears red or intense pink (Chen et al. 2023). Phycoerythrin is predominantly produced by red algae (Imchen and Singh 2023); however, *Gelidium amansii* and *Porphyra haitanensis* are the species extensively used in the extraction of phycoerythrin as a natural colorant in the food processing industry (Chen et al. 2023). As far as cryptophytes are concerned, *Cryptomonas curvata* and *C. lundii* yield high phycoerythrin contents (Mercier et al. 2022). The applications of phycoerythrin extend to cosmetics, and the pharmaceutical industry as well.

It acts as a bioactive compound owing to numerous functional properties, among which anti-diabetic properties are notable. Phycoerythrin exerts antioxidant and anti-inflammatory properties protecting the body against reactive oxygen species (ROS) and thereby minimizing the damages caused by oxidative stress (Tan et al. 2022).

Soni et al. (2009) investigated the relationship of C-phycoerythrin dose-dependent administration on streptozotocin–nicotinamide treatment-induced type 2 diabetic male albino rats with diabetic complications. This study evidenced that phycoerythrin demonstrated a significant impact on diabetes complications by diminishing oxidative stress and oxidized low-density lipoprotein (LDL)-triggered atherogenesis.

Sonani et al. (2014) reported the antioxidant capacity of phycoerythrin to be the highest of that of phycobiliproteins, followed by phycocyanin and allophycocyanin. As per the study conducted by Ghosh et al. (2016), purified C-phycoerythrin tends to show a higher total antioxidant capacity than purified C-phycocyanin. Moreover, the authors stated that C-phycoerythrin demonstrated an inhibitory action on α-glucosidase and α-amylase enzymes which negatively impacts the cleavage of carbohydrates, thereby reducing the release of glucose molecules into the bloodstream.

Studies have been conducted on the hypoglycemic properties of bioactive compounds of the extracts of red seaweeds. Inhibitory actions on α-glucosidase and α-amylase activity in extracts of *Gracillaria edulis*, *Gracilaria corticate,* and *Acanthophora spicifera* (Reka et al. 2017) and development of advanced glycation end products (AGEs) in the extracts of *Gracillaria edulis* (Gotama et al. 2018) have been reported.

As far as red algae are concerned, phycoerythrin and phycocyanin stand to be dominant proteinic pigments attributed to its bright color. Hence, they are considered as vital bioactive compounds in the extracts of red algae; it can be assumed that phycoerythrin and phycocyanin can have an impact on the anti-diabetic properties.

5 Fucoxanthin

Fucoxanthins are categorized under carotenoids that are contained in algae (Patel et al. 2022). It gives a brown color to Phaeophyta (brown seaweed). For instance, macroalgae that contain fucoxanthin are *Alaria crassifolia, Fucus serratus, Cystoseira hakodatensis, Eisenia bicyclis, Hijikia fusiformis, Sargassum fulvellum, Laminaria japonica, S. heterophyllum, S. cinerium, Undaria pinnatifida,* and some microalgae are *Chaetoceros* sp., *Odontella aurita, Cylindrotheca closterium,* and *Phaeodactylum tricornutum* (Imchen and Singh 2023). Moreover, Ganesan et al. (2019) mentioned that fucoxanthin in wakame, laminaria, and sargassum species fluctuated from 1.3 to 2.4 mg/g on a dry weight basis. The Fucoxanthin content of brown algae depends on many factors such as type of species, salinity, season, location, temperature, light intensity, etc. Conventional solvent extraction is the basic method for the extraction of fucoxanthin. However, disadvantages include high time consumption, requirement of organic solvents in large volumes, etc. So, there are some advanced methods such as enzyme-assisted extraction, pressurized liquid extraction, and so on (Miyashita et al. 2020).

Fucoxanthin is classified under primary carotenoids based on its function and metabolism (Imchen and Singh 2023). Specially, fucoxanthin is a carotenoid that is contained only within algae. Fucoxanthin cannot be found in terrestrial plants (Miyashita et al. 2020). According to animal studies, supplementation of fucoxanthin has no harmful effects (Bae et al. 2020). Fucoxanthin has antioxidant properties because these pigments have long conjugated double bonds in polyene chains (Patel et al. 2022). Fucoxanthin molecule has allenic bonds, hydroxyl, carboxyl, epoxy, and carbonyl groups. Allenic bonds are known to exhibit free radical scavenging activity (Bae et al. 2020). There is a unique ability of fucoxanthin and other carotenoids that protect tissue in oxygen-deficient situations (Imchen and Singh 2023).

Nevertheless, fucoxanthin is unstable for heating, aerial exposure, or lighting (Bae et al. 2020). Fucoxanthinol and amarouciaxanthin A are the primary metabolites of fucoxanthin. In the gastrointestinal tract, fucoxanthin is converted into fucoxanthinol with the help of cholesterol esterase and lipase. In the liver, it is converted into amarouciaxanthin A (Bae et al. 2020).

Diabetes is one of the largest health issues facing the world today. The main reason for diabetes is the accumulation of acylation products (AGE—Advance Glycation End-product) developed during the nonenzymatic glycation of the Millard reaction. The main solution for diabetes is preventing the formation of AGE. Therefore, many algae pigments including the carotenoids group are used to develop some drugs for diabetes (Patel et al. 2022). Fucoxanthin does not affect the levels of fasting blood glucose and glycated hemoglobin. However, it can decrease triglycerides, LDL cholesterol, and total cholesterol. Fucoxanthin plays a significant role in safeguarding the retina by promoting catalase activation and thereby reducing oxidative stress induced by elevated glucose levels. Fucoxanthin reduces the levels of inflammatory cytokines that are linked to insulin resistance. Also, it improves diabetic kidney function and oxidative stress-mediated diabetic nephropathy. Moreover, fucoxanthin could prevent Alzheimer's pyemia, lipid metabolic disease, etc. Fucoxanthin inhibits carbohydrate-digesting enzyme α-amylase. An example of notable findings in this area is the inhibitory activity of fucoxanthin, a compound derived from *Undaria pinnatifida*, against α-amylase. It has been measured with an IC50 value of 0.047 ± 0.001 mg/ml, highlighting its potential significance. Additionally, fucoxanthin extracted from *Phaeodactylum tricornutum* plays a key role in the inhibitory activity against AGEs. Fucoxanthin from *Undaria pinnatifida* exhibited the greatest α-glucosidase activity. According to clinical trials, 2.4 mg of fucoxanthin taken from brown seaweeds can reduce the fat content and total body weight. (Kaushik et al. 2023). In another study, they proved that 3 mg of fucoxanthin can somewhat reduce blood glucose levels while reducing body weight (Kaushik et al. 2023). Fucoxanthin, a compound derived from the microalga *Phaeodactylum tricornutum*, exhibited significant inhibitory effects on α-amylase in a concentration-dependent manner, with an IC50 value of 0.68 mmol/L. Additionally, it displayed notable inhibitory activity against α-glucosidase, as indicated by an IC50 value of 4.75 mmol/L(Abo-Shady et al. 2023; Kaushik et al. 2023).

Fucoxanthin extracted from *Sargassum hemiphyllum* prevents α-amylase, maltase, and sucrase absorption. Moreover, fucoxanthin extracted from *Sirophysalis trinodis* inhibits intestinal α-amylase. Fucoxanthin extracted from *U. pinnatifida* remarkably reduced adipocyte size, insulin level, and fasting blood glucose as compared to overweight mice (Ganesan et al. 2019). The microalga *P. tricornutum* contains 17.5–18 mg of fucoxanthin per 1 g of microalga dry weight (Kawee-Ai et al. 2019). Kawee-Ai et al. (2019) declared that, when increasing the fucoxanthin concentration from 0.1 to 1.0 mmol/L, the α-amylase inhibitory activity was increased from 28.38 to 59.92%. As the same way, when increasing fucoxanthin concentration α-glucosidase was increased. Moreover, they stated that the inhibitory activity of fucoxanthin toward α-amylase was stronger than α-glucosidase. According to the inhibitory kinetics of fucoxanthin, when increasing the concentration of fucoxanthin, the velocity reaction catalyzed by α-glucosidase decreased. Fucoxanthin from microalgae may help manage diabetes by inhibiting carbohydrate-digesting enzymes. This action delays the absorption of dietary carbohydrates in the digestive system and assists in suppressing the increase in blood glucose levels after meals. (Kawee-Ai et al. 2019).

Oliyaei et al. (2021)experimented with the in vivo antidiabetic action of fucoxanthin extracted from *Sargassum angustifolium.* They introduced encapsulated fucoxanthin with porous starch and free fucoxanthin to type 2 diabetic mice induced by streptozotocin (STZ) and nicotinamide (NA) orally for 3 weeks. There were five groups of mice such as normal control, diabetic mice as negative control, diabetic mice ingested with metformin (50 mg/Kg in 0.3 ml normal saline) as positive control, diabetic mice ingested with fucoxanthin (400 mg/Kg BW), and diabetic mice ingested with encapsulated fucoxanthin (400 mg/Kg BW). After every 24 hours from the dosage, blood samples were collected, and they were subjected to analysis. According to Oliyaei et al. (2021), the weight increment of normal mice was lower than diabetic mice as negative control (without treatment). However, no significant differences existed between normal control mice and diabetic mice with treatments. When considering the fasting blood glucose levels (FBS) of mice, there was a more significant increase in FBS in diabetic mice without treatment when compared with the standard control group. Moreover, diabetic mice with treatments had lower FBS levels than diabetic mice without treatments. However, mice treated with fucoxanthin had lower FBS values than mice treated with encapsulated fucoxanthin. Additionally, plasma insulin levels were increased in treated mice than in control diabetic mice. Finally, this study confirmed considerable antidiabetic properties in fucoxanthin (Oliyaei et al. 2021).

Zhang et al. (2018)extracted fucoxanthin from *Laminaria japonica* and inserted 0.2 and 0.4% for C57BL/KsJ-db/db mice (db/db) and their nondiabetic heterozygote littermates (db/+). Also, there were standard, control, and metformin (0.02%) groups. After 6 weeks, the body weight and epididymal fat of control mice increased while the body weight and epididymal of fucoxanthin and metformin-treated mice were decreased. Therefore, they have proven that dietary fucoxanthin reduces body weight and epididymal fat. According to the fasting blood glucose level, metformin-treated mice, 0.4% fucoxanthin-treated mice, and 0.2% fucoxanthin-treated mice displayed lower levels within 2 weeks, 4 weeks, and 6 weeks respectively by regulating phosphoenolpyruvate carboxykinase and glucokinase mRNA expressions in the liver. It revealed that fucoxanthin regulates glucose metabolism. Furthermore, the plasma insulin and homeostasis model assessment of insulin resistance (HOMA-IR) levels decreased in fucoxanthin-treated mice compared to the control. So, it showed that fucoxanthin can ameliorate insulin resistance in diabetic mice. In the same way, fucoxanthin can regulate plasma lipid levels by reducing total cholesterol and triglyceride levels.

Fucoxanthin can modulate a nuclear receptor (PPARγ) for improving insulin sensitivity, controlling lipid metabolism, and helping adipocyte differentiation (Pereira and Valado 2023). Some factors are responsible for the absorption of fucoxanthin in our body, but they are not verified completely. Some of them are the number and probably the type of dietary lipids that we consume, stability of binding matrix, dietary fiber, etc. (Pangestuti and Kim 2011).

6 Zeaxanthin

It is a primary carotenoid (Imchen and Singh 2023). It is a yellow color phytochemical concentrated in the macula of the retina (Lin et al. 2024). Zeaxanthin from *Chlorella ellipsoidea* showed the best α-glucosidase activity (Kaushik et al. 2023). *G. asiatica* and *G. livida* contain zeaxanthin as a significant carotenoid around 80.2μg/g (Ganesan et al. 2019).

Qi and Kim (2018) studied the α-glucosidase inhibitory activity of zeaxanthin extracted from *Chlorella ellipsoidea* (green algae). An effective approach for managing type 2 diabetes involves inhibiting or delaying the digestion and absorption of carbohydrates within the digestive system. This strategy can assist individuals in maintaining healthier blood sugar levels and promoting overall metabolic health. α-Glucosidase is an essential enzyme found in the epithelium of the small intestine, which catalyzes the breakdown of starch and disaccharides into glucose. So, α-glucosidase inhibitors can be used as a treatment for type 2 diabetes because (*9-Z*)-zeaxanthin displays the α-glucosidase inhibitory activity with IC_{50} values of 53.5 $\mu molL^{-1}$ against *Saccharomyces cerevisiae* α-glucosidase. Furthermore, IC_{50} values of (9-Z)-zeaxanthin against *Bacillus stearothermophilus* and rat-intestinal α-glucosidase were 805.1 and 671.2$\mu molL^{-1}$ respectively. Therefore, they proved that *Chlorella ellipsoidea* can be used as a solution for type-2 diabetes (Qi and Kim 2018).

7 Lutein

Lutein is a fat-soluble carotenoid belonging to the xanthophyll family, which is a bright yellow color (Anusree et al. 2023; Hu et al. 2018). The maximum absorption of lutein can be observed around 445 nm (Arunkumar et al. 2020). It is present in various genera of microalgae, some of which include *Scenedesmus*, *Chlorella*, *Coccomyxa*, *Parachlorella*, *Tetraselmis, Haematococcus*, and *Arthrospira* (Xie et al. 2021; Tamel Selvan et al. 2023) whereas *Monostroma*, *Cladophora, Grateloupia, Chaetomorpha,* and *Grateloupia* are a few of macroalgae containing lutein available in the literature (Bhat et al. 2021*;* Kanda et al. 2020).

Lutein exerts numerous health benefits attributed to its bioactivity. The free-radical scavenging activity and anti-inflammatory activity of lutein are mentioned to be promising factors against many diseases (Montuori et al. 2024). Damages caused by oxidative stress are stated to be suppressed due to the antioxidant and anti-inflammatory activity of lutein and thereby plays a protective role against neurological problems, cardiovascular diseases, cancer, gastrointestinal impairments, eye diseases, liver dysfunction, diabetes, and diabetes retinopathy (Ahn and Kim 2021).

Lutein is referred to as a macular pigment hence strongly affecting eye health. It absorbs blue light and scavenges ROS to protect the retinal pigment epithelial cells from light-induced oxidative stress (Arunkumar et al. 2020). Several studies focused on the impact of lutein on diabetes and related complications have been included in this chapter.

Diabetic retinopathy is a key complication among diabetes patients threatening the vision caused due to damaged retina which eventually leads to blindness (Wang et al. 2020). The study reported that the long-term administration of lutein led to the preservation of retinal function by hindering retinal inflammation and damage to retinal blood vessels in Ins2Akita/+ mice, a hereditary model that develops diabetes due to a genetic mutation.

A decline in diabetes-related complications upon administration of lutein has been reported in certain studies. Administration of lutein has resulted in reduced oxidative damage and protective functions against inflammation in the testicular cells (Fatani et al. 2015) and cerebral cortex (Fatani et al. 2016) of streptozotocin-induced diabetic rats. Shivarudrappa and Ponesakki (2020) stated that lutein decreases the risk of diabetes-related retinal disorders. The study was conducted on ARPE-19 human retinal pigment epithelial cells and evidenced a protective function of lutein against diabetes-mediated damages.

Sharavana et al. (2017) observed an improvement in glucose tolerance in streptozotocin-induced diabetic rats administered with lutein. Another study outlined improvements in visual dysfunction of patients aged between 40 and 60 years with type 2 diabetes upon supplementation of lutein (10 mg), zeaxanthin (2 mg), and meso-zeaxanthin (10 mg) once a day for 2 years (Moschos et al. 2017). Pan et al. (2022) discovered that serum lutein level acts as a biomarker of type 2 diabetes and diabetic kidney dysfunction in individuals over 60 years.

8 Macular Pigments

Among more than 750 carotenoids available in nature, due to the accumulation in the human macula in the retina only zeaxanthin, meso-zeaxanthin, lutein, and their oxidative metabolites are collectively considered as macular pigments (MP) (Arunkumar et al. 2018; Baek et al. 2018). MP has the ability to protect the eyes through its capacity to filter phototoxic blue light radiation and its antioxidant properties (Arunkumar et al. 2018). Lin et al. (2024)proved that these macular pigments could reduce the signs of type 2 diabetes mellitus by experimenting with mice exhibiting typical features of diabetes-related phenotypes observed in humans. Macular pigments (lutein: zeaxanthin = 5:1) extracted and produced from *Chlorella* sp. AT1 microalgae significantly lowered the glucose tolerance by 25% and insulin resistance by 41% than the diabetic mice group. Microalgal MP had an ability to reduce the diabetic apolipoprotein E-deficient mice's type 2 diabetes symptoms. Therefore, MP has the potential to be a useful component of nutritious food, enhancing biological efficiency by scavenging free radicals and regulating blood sugar levels (Lin et al. 2024). However, using wild-type microalgae for the production of macular pigments is not recommended (Baek et al. 2018).

9 Astaxanthin

Another member of the carotenoid group is astaxanthin which gives it a reddish hue. *Haematococcu spluvialis* is a major astaxanthin-producing alga that is used in the commercial production of the pigment. *Chlorella zofingiensis* and *Chlorococcum* sp. are also identified as natural astaxanthin producing algae with potential for commercial extraction (Yuan et al. 2011). The bioactivity of astaxanthin extracted naturally from algal biomass tends to have an increased bioactivity over the synthetically produced astaxanthin (Villaró et al. 2021).

Astaxanthin is a potential therapeutic agent for immune-impaired diseases including type 2 diabetes which is the main concern of this chapter. Astaxanthin is reported to be one of the most robust antioxidants among all known carotenoids as well as vitamins E and C (Yuan et al. 2011) which is attributed to its alternative double bonds that can donate electrons to stabilize ROS and free radicles and thereby, terminating the chain reactions (Ambati et al. 2014).

The onset, progression, and complications of diabetes mellitus are strongly associated with oxidative stress (Landon et al. 2020). The anti-diabetic property of astaxanthin is strongly associated with its antioxidant activity. The oxidative damages induced by hyperglycemia cause impaired functioning of pancreatic β-cells. Astaxanthin has demonstrated promising results in glucose tolerance, particularly in improving serum insulin levels. The same study demonstrated a significant decrease in non-fasting blood glucose levels (Uchiyama et al. 2002).

Hirakida et al. (2022) suggested the anti-diabetic effect of astaxanthin through its potential to increase insulin secretion, stimulated by glucose. Moreover, this study proved that astaxanthin's synergistic action with biosynthetic intermediates of β-carotene can produce an anti-diabetic effect.

A comparative in vivo study was done by Gao et al. (2020) on the effectiveness of astaxanthin *n*-octanoic acid diester (AOD), a stable form of astaxanthin bound to medium-chain fatty acids and unbound or free form of astaxanthin on glucose tolerance, insulin sensitivity, and modulation of gut microflora, which generated superior results for AOD on alleviating the above factors. Moreover, it revealed that the modulated gut microbiota exerted an increased carbohydrate metabolism in AOD fed mice compared to unbound astaxanthin.

A significant increase in insulin sensitivity has been observed in high-fat-high-fructose diet-induced diabetic mice which were chronically administered with astaxanthin (Arunkumar et al. 2012; Bhuvaneswari and Anuradha 2012). As reported, the reduced oxidative stress caused by astaxanthin is a key factor that has resulted in increased insulin sensitivity. However, based on the in vivo studies by Nishida et al. (2020), the improvement in insulin sensitivity is independent of the antioxidant activity of astaxanthin.

Zhu et al. (2018) described the protective effect of astaxanthin on diabetic nephropathy with an in vivo study conducted on streptozotocin-induced diabetic rats. The study revealed that the diabetic rats treated with astaxanthin delayed the pathological process of diabetic nephropathy and ameliorated renal morphological injuries.

Park et al. (2015) discovered the protective effect of astaxanthin on hepatic dysfunction in type 1 diabetic rats and suggested that the blockage of AGE formation and anti-inflammatory effect of astaxanthin to be the causative. Administration of astaxanthin has ameliorated hepatic insulin resistance and non-alcoholic fatty liver conditions in genetically and high-fat diet-induced obese mice and turned out to be more effective than vitamin E (Ni et al. 2015).

Moreover, astaxanthin is claimed for its protective function against diabetic retinopathy which is attributed to its strong antioxidant potential (Yeh et al. 2016).

10 β-Carotene

β-carotene is a dark red to brown in color pigment produced in algae. This is considered a secondary metabolite and the quantity of double bond cis-trans isomerization contributes for the brightness of the color of this pigment due to exposure to high temperatures (Anand et al. 2022). *Dunaliella, Haematococcus, Chlorella*, and *Spirulina* are some algae that produce β-Carotene in large amounts. *Dunaliella salina* contains a high amount of β-carotene which produces 13% of its biomass as β-carotene. β-carotene is considered as the most common pro-vitamin A which can be converted into micronutrient retinol (Wang et al. 2022). Lycopene acts as a precursor in the production of β-carotene and α-carotene. Lycopene belongs to the carotenoids group, which is a class of unsaturated hydrocarbons soluble in organic solvents but not in water (Chen et al. 2023). Natural β-carotene easily dissolves in bile and can mix with chylomicrons. However, synthesized β-carotene shows side effects and is not easy to absorb by the body due to the slightly oil-soluble single crystal structure of the all-trans configuration (Ma et al. 2016).

The β-carotene extract from *Spirulina platensis* has shown antidiabetic properties. Following 10 days of administration of β-carotene extract [100mgkg^{-1} body weight (BW)] the blood glucose level of streptozotocin-induced diabetic mice was significantly decreased from 15.81 ± 1.71 mmolL^{-1} to 8.1 ± 0.88 mmolL^{-1} (Ma et al. 2016). Kurniawan et al. (2023) identified seven carotenoids from the carotenoids extract of *Caulerpa racemosa* green alga and that extract was dominant with β-carotene and canthaxanthin. This carotenoid extract was good at inhibiting α-glucosidase, α amylase and it had a safe cytotoxicity value indicating the potential of using this to develop pharmaceuticals and nutraceuticals (Kurniawan et al. 2023).

Canthaxanthin is a secondary carotenoid (Imchen and Singh 2023). It is a bioactive compound that is useful in feed, pharmaceuticals, cosmeceuticals, nutraceuticals, and so on (Lin et al. 2024).

11 Chlorophylls

There are several types of chlorophylls: chlorophyll *a*, chlorophyll *b*, chlorophyll *c*, and chlorophyll *d*. These pigments are overall visible in green color due to their ability to reflect green light. Chlorophyll plays a major role in photosynthesis and consists of an unstable Mg porphyrin compound which is easily soluble in organic solvents but less soluble in water. In algae chlorophyll can be identified as the most common pigment and they have mainly chlorophyll *a, b ,c, d,* and *f* types. The two most important ones for food are chlorophyll *a* and *b*, which are both present in and extracted from algae, particularly green algae (Chen et al. 2023). Several environmental growth factors can affect for the content of chlorophylls in algae. For instance, when increasing the concentration of salinity, total chlorophyll content is significantly decreased in a green alga *Chlorella vulgaris. Spirulina platensis* produces the highest amount of chlorophyll *a* at pH 8.5 conditions (Patel et al. 2022).

Pheophorbide *a* is a compound generated by the breaking down of chlorophyll, which forms in higher plants and algae. It is a dephytylation and demetallation product of chlorophyll *a* and Mg dechatalase, and chlorophyllase enzymes are responsible for this transformation. Pheophorbide *a* shows a similar UV-vis spectrum of chlorophyll *a* including Q-bands in between 500-700 nm and Soret band at 390 nm (Saide et al. 2020). A study done by Kim et al. (2019) showed that a diabetic mice group who were treated with pheophorbide *a*, extracted from *Gelidium amansii* alga has significantly decreased the increment of blood glucose level than the control group. In addition, pheophorbide *a* treated normal mice group had shown a significant decrease in postprandial blood glucose levels compared to the control group. Furthermore, pheophorbide *a* inhibits the activity of bothα-glucosidase and α-amylase. Inhibiting these enzymes helps for an effective delay of carbohydrate digestion and glucose absorption that helps to control the postprandial hyperglycemia. Postprandial hyperglycemia is considered as a risk factor of having diabetic complications in type 2 diabetes patients (Kim et al. 2019).

Pheophytin *a* is a pigment that contributes to photosynthesis and has a similar structure to chlorophyll *a* but does not contain the chelated Mg^{2+} ion (Lin and Charng 2021). Pradhan et al. (2021) reported that pheophorbide *a* and pheophytin *a* display aldose reductase inhibition. The aldose reductase catalyzed polyol pathway, which reduces glucose, has been associated with secondary diabetes problems (Srivastava et al. 2005). The porphyrin derivatives pheophorbide *a* and pheophytin *a* present in the dichloromethane fraction extracted from a brown seaweed variety *Laminaria japonica* showed rat lens aldose reductase inhibition (RLAR) activity and inhibition of advanced glycation end products (AGE) formation. The results reflected that pheophorbide *a* has a potential of inhibiting both RLAR and AGE than pheophytin *a*. Moreover, having a carboxyl group without a phytyl group at C-17^2 position of the porphyrin is considered as a reason for this inhibitory task (Son et al. 2011).

12 Conclusion

Algal pigments such as phycocyanin, phycoerythrin, fucoxanthin, lutein, astaxanthin, zeaxanthin, β-carotene, pheophorbide *a,* and pheophytin *a* show antidiabetic properties and these pigments can be used as natural sources for treating diabetic patients. These pigments have shown a potential to include inhibiting α-glucosidase and α-amylase enzymatic activities, reducing blood glucose levels, regulating phosphoenolpyruvate carboxykinase, and glucokinase mRNA expressions in the liver, and regulating glucose metabolism, inhibiting the formation of advanced glycation end products (AGEs). Thus, algal pigments can be used as a promising treatment for diabetes.

References

Abo-Shady AM, Gheda SF, Ismail GA, Cotas J, Pereira L and Abdel-Karim OH (2023) Antioxidant and Antidiabetic Activity of Algae. Life 13(2): 460. https://doi.org/10.3390/life13020460

Ahn YJ and Kim H (2021) Lutein as a Modulator of Oxidative Stress-Mediated Inflammatory Diseases. *Antioxidants*, 10(9): 1448. https://doi.org/10.3390/antiox10091448

Ambati RR, Phang SM, Ravi S and Aswathanarayana RG (2014) Astaxanthin: sources, extraction, stability, biological activities and its commercial applications: a review. *Marine Drugs*,12(1): 128–152. https://doi.org/10.3390/md12010128

Anand R, Mohan L and Bharadvaja N (2022) Disease prevention and treatment using β-Carotene: the ultimate provitamin A. *RevistaBrasileira de Farmacognosia*, 32(4): 491–501. https://doi.org/10.1007/s43450-022-00262-w

Anusree MK, Leela KM, Sreehari M, Raj S, Sreenikethanam A and Bajhaiya AK (2023) Marine microalgae: an emerging source of pharmaceuticals and bioactive compounds. In Meena SN, Nandre V, Kodam K and Meena RS (Eds.), *New Horizons in Natural Compound Research*, 251–265. Academic Press. https://doi.org/10.1016/B978-0-443-15232-0.00025-4

Arunkumar E, Bhuvaneswari S and Anuradha CV (2012) An intervention study in obese mice with astaxanthin, a marine carotenoid–effects on insulin signaling and pro-inflammatory cytokines. *Food & Function*, 3(2): 120–126. https://doi.org/10.1039/c1fo10161g

Arunkumar R, Calvo CM, Conrady CD and Bernstein PS (2018) What do we know about the macular pigment in AMD: the past, the present, and the future. *Eye*, 32(5): 992–1004. https://doi.org/10.1038/s41433-018-0044-0

Arunkumar R, Gorusupudi A and Bernstein PS (2020) The macular carotenoids: A biochemical overview. *Biochimica et BiophysicaActa (BBA) - Molecular and Cell Biology of Lipids*, 1865(11): 158617. https://doi.org/10.1016/j.bbalip.2020.158617

Aslam A, Fazal T, Uz Zaman Q, Shan A, Rehman F, Iqbal J, Rashid N and Rehman MS (2020) Biorefinery of microalgae for nonfuel products. In Yousuf A (Ed.), *Microalgae Cultivation for Biofuels Production*, 197–209, Academic Press. https://doi.org/10.1016/B978-0-12-817536-1.00013-8

Bae M, Kim MB, Park YK and Lee JY (2020) Health benefits of fucoxanthin in the prevention of chronic diseases. *Biochimica et BiophysicaActa - Molecular and Cell Biology of Lipids,* 1865(11): 158618. https://doi.org/10.1016/j.bbalip.2020.158618

Baek K, Yu J, Jeong J, Sim, SJ, Bae S and Jin E (2018) Photoautotrophic production of macular pigment in a *Chlamydomonasreinhardtii* strain generated by using DNA-free CRISPR-Cas9 RNP-mediated mutagenesis. *Biotechnology and Bioengineering*, 115(3): 719–728. https://doi.org/10.1002/bit.26499

Bhat I, Haripriya G, Jogi N and Mamatha BS (2021) Carotenoid composition of locally found seaweeds of Dakshina Kannada district in India. *Algal Research*, 53: 102154. https://doi.org/10.1016/j.algal.2020.102154

Bhuvaneswari S and Anuradha CV (2012) Astaxanthin prevents loss of insulin signaling and improves glucose metabolism in liver of insulin resistant mice. *Canadian Journal of Physiology and Pharmacology*, 90(11): 1544–1552. https://doi.org/10.1139/y2012-119

Bocanegra A, Macho-González A, Garcimartín A, Benedí J and Sánchez-Muniz FJ (2021) Whole alga, algal extracts, and compounds as ingredients of functional foods: composition and action mechanism relationships in the prevention and treatment of type-2 diabetes mellitus. *International Journal of Molecular Sciences*, 22(8): 3816. https://doi.org/10.3390/ijms22083816

Chen H, Qi H and Xiong P (2022) Phycobiliproteins - A Family of Algae-Derived Biliproteins: Productions, Characterization and Pharmaceutical Potentials. *Marine Drugs*, 20(7): 450. https://doi.org/10.3390/md20070450

Chen Z, Wu W, Wen Y, Zhang L, Wu Y, Farid MS, El-Seedi HR, Capanoglu E and Zhao C (2023) Recent advances of natural pigments from algae. *Food Production, Processing and Nutrition*, 5(1): 39. https://doi.org/10.1186/s43014-023-00155-y

Dagnino-Leone J, Figueroa CP, Castañeda ML, Youlton AD, Vallejos-Almirall A, Agurto-Muñoz A, Pavón Pérez J and Agurto-Muñoz C (2022) Phycobiliproteins: Structural aspects, functional characteristics, and biotechnological perspectives. *Computational and Structural BiotechnologyJournal,* 20: 1506–1527. https://doi.org/10.1016/j.csbj.2022.02.016

Derbel H, Elleuch J, Tounsi L, Nicolo MS, Rizzo MG, Michaud P, Fendri I and Abdelkafi S (2022) Improvement of biomass and phycoerythrin production by a strain of *Rhodomonassp.* isolated from the Tunisian coast of Sidi Mansour. *Biomolecules*, 12(7): 885. https://doi.org/10.3390/biom12070885

El-Sayed ES, Hikal MS, Abo El-Khair BE, El-Ghobashy RE and El-Assar AM (2018) Hypoglycemic and hypolipidemic effects of spirulina platensis, phycocyanin, phycocyanopeptide and phycocyanobilin on male diabetic rats. *Arab Universities Journal of Agricultural Sciences*, 26:1121–1134. https://doi.org/10.21608/ajs.2018.28365

Fatani AJ, Al-Rejaie SS, Abuohashish HM, Al-Assaf A, Parmar MY and Ahmed MM (2015) Lutein dietary supplementation attenuates streptozotocin-induced testicular damage and oxidative stress in diabetic rats. *BMC Complementary and Alternative Medicine*, 15: 1–10. https://doi.org/10.1186/s12906-015-0693-5

Fatani AJ, Parmar MY, Abuohashish HM, Ahmed MM and Al-Rejaie SS (2016) Protective effect of lutein supplementation on oxidative stress and inflammatory progression in cerebral cortex of streptozotocin-induced diabetes in rats. *Neurochemical Journal*, 10: 69–76. https://doi.org/10.1134/S1819712416010074

Ganesan AR, Tiwari U and Rajauria G (2019) Seaweed nutraceuticals and their therapeutic role in disease prevention. *Food Science and Human Wellness,* 8(3): 252–263. https://doi.org/10.1016/j.fshw.2019.08.001

Gao Y, Yang L, Chin Y, Liu F, Li RW, Yuan S, Xue C, Xu J and Tang Q (2020) Astaxanthin *n*-Octanoic Acid Diester Ameliorates Insulin Resistance and Modulates Gut Microbiota in High-Fat and High-Sucrose Diet-Fed Mice. *International Journal of Molecular Sciences*, 21(6): 2149. https://doi.org/10.3390/ijms21062149

Ghosh T, Bhayani K, Paliwal C, Maurya R, Chokshi K, Pancha I and Mishra S (2016) Cyanobacterial pigments as natural anti-hyperglycemic agents: an in vitro study. *Frontiers in Marine Science*, 3: 146. https://doi.org/10.3389/fmars.2016.00146

Gotama TL, Husni A and Ustadi (2018) Antidiabetic Activity of *Sargassumhystrix* Extracts in Streptozotocin-Induced Diabetic Rats. *Preventive Nutrition and Food Science*, 23(3): 189–195. https://doi.org/10.3746/pnf.2018.23.3.189

Hao S, Li F, Li Q, Yang Q and Zhang W (2022) Phycocyanin Protects against High Glucose High Fat Diet Induced Diabetes in Mice and Participates in AKT and AMPK Signaling. *Foods*, 11(20): 3183. https://doi.org/10.3390/foods11203183

Hirakida H, Nakamura S, Inagaki S, Tsuji S, Hayashi M, Shimazawa M and Hara H (2022) Anti-diabetic effects of astaxanthin-rich extract derived from *Paracoccuscarotinifaciens* on pancreatic β cells. *Journal of Functional Foods*, 97: 105252. https://doi.org/10.1016/j.jff.2022.105252

Hu J, Nagarajan D, Zhang Q, Chang JS and Lee DJ (2018) Heterotrophic cultivation of microalgae for pigment production: A review. *Biotechnology Advances*, 36(1): 54–67.https://doi.org/10.1016/j.biotechadv.2017.09.009

Husain A, Alouffi S, Khanam A, Akasha R, Farooqui A and Ahmad S (2022) Therapeutic Efficacy of Natural Product 'C-Phycocyanin' in Alleviating Streptozotocin-Induced Diabetes via the Inhibition of Glycation Reaction in Rats. *International Journal of Molecular Sciences*, 23(22): 14235. https://doi.org/10.3390/ijms232214235

Imchen T and Singh KS (2023) Marine algae colorants: Antioxidant, anti-diabetic properties and applications in food industry. *Algal Research,* 69: 102898. https://doi.org/10.1016/j.algal.2022.102898

Kanda H, Wahyudiono, Machmudah S and Goto M (2020) Direct Extraction of Lutein from Wet Macroalgae by Liquefied Dimethyl Ether without Any Pretreatment. *ACS Omega*, 5(37): 24005–24010. https://doi.org/10.1021/acsomega.0c03358

Kaushik A, Sangtani R, Parmar HS and Bala K (2023) Algal metabolites: Paving the way towards new generation antidiabetic therapeutics. *Algal Research,* 69: 102904. https://doi.org/10.1016/j.algal.2022.102904

Kawee-Ai A, Kim AT and Kim SM (2019) Inhibitory activities of microalgal fucoxanthin against α-amylase, α-glucosidase, and glucose oxidase in 3T3-L1 cells linked to type 2 diabetes. *Journal of Oceanology and Limnology*, 37(3): 928–937. https://doi.org/10.1007/s00343-019-8098-9

Kim MJ, Kim HJ and Han JS (2019)Pheophorbide A from *Gelidiumamansii* improves postprandial hyperglycemia in diabetic mice through α-glucosidase inhibition. *Phytotherapy Research*, 33: 702–707. DOI: https://doi.org/10.1002/ptr.6260

Kurniawan R, Nurkolis F, Taslim NA, Subali D, Surya R, Gunawan WB, Alisaputra D, Mayulu N, Salindeho N and Kim B (2023) Carotenoids composition of green algae *Caulerpa racemosa* and their antidiabetic, anti-obesity, antioxidant, and anti-inflammatory properties. *Molecules,* 28(7): 3267. https://doi.org/10.3390/molecules28073267

Landon R, Gueguen V, Petite H, Letourneur D, Pavon-Djavid G and Anagnostou F (2020) Impact of Astaxanthin on Diabetes Pathogenesis and Chronic Complications. *Marine Drugs*, 18(7): 357. https://doi.org/10.3390/md18070357

Lin MW, Chiu WH, Lin CH, Liu DH., Wu PC and Lin CS (2024)Macular pigments produced from microalga *Chlorella* sp. and applied to alleviate the pathogenic process in diabetic mice.*Algal Research, 78*: 103414. https://doi.org/10.1016/j.algal.2024.103414

Lin YP and Charng YY (2021) Chlorophyll dephytylation in chlorophyll metabolism: a simple reaction catalyzed by various enzymes. *Plant Science,* 302: 110682, https://doi.org/10.1016/j.plantsci.2020.110682

Ma QY, Fang M, Zheng JH, Ren DF and Lu J (2016) Optimised extraction of β-carotene from *Spirulina platensis* and hypoglycaemic effect in streptozotocin-induced diabetic mice, *Journal of the Science of Food and Agriculture*, 96(5): 1783–1789. https://doi.org/10.1002/jsfa.7286

Mercier L, Peltomaa E and Ojala A (2022) Comparative analysis of phycoerythrin production in cryptophytes. *Journal of Applied Phycology*, 34(2): 789–797. https://doi.org/10.1007/s10811-021-02657-z

Miyashita K, Beppu F, Hosokawa M, Liu X and Wang S (2020) Nutraceutical characteristics of the brown seaweed carotenoid fucoxanthin. *Archives of Biochemistry and Biophysics,* 686: 108364. https://doi.org/10.1016/j.abb.2020.108364

Montuori E, Lima S, Marchese A, Scargiali F and Lauritano C (2024) Lutein Production and Extraction from Microalgae: Recent Insights and Bioactive Potential. *International Journal of Molecular Sciences*, 25(5): 2892. https://doi.org/10.3390/ijms25052892

Moschos MM, Dettoraki M, Tsatsos M, Kitsos G and Kalogeropoulos C (2017) Effect of carotenoids dietary supplementation on macular function in diabetic patients. *Eye and Vision (Lond)*, 4: 23. https://doi.org/10.1186/s40662-017-0088-4

Munawaroh HS, Gumilar GG, Nurjanah F, Yuliani G, Aisyah S, Kurnia D, Wulandari AP, Kurniawan I, Ningrum A, Koyande AK and Show PL (2020) In-vitro molecular docking analysis of microalgae extracted phycocyanin as an anti-diabetic candidate. *Biochemical Engineering Journal*, 161: 107666. https://doi.org/10.3390/foods11203183

Ni Y, Nagashimada M, Zhuge F, Zhan L, Nagata N, Tsutsui A, Nakanuma Y, Kaneko S and Ota T (2015) Astaxanthin prevents and reverses diet-induced insulin resistance and steatohepatitis in mice: A comparison with vitamin E. *Scientific Reports*, 5: 17192. https://doi.org/10.1038/srep17192

Nishida Y, Nawaz A, Kado T, Takikawa A, Igarashi Y, Onogi Y, Wada T, Sasaoka T, Yamamoto S, Sasahara M and Imura J (2020) Astaxanthin stimulates mitochondrial biogenesis in insulin resistant muscle via activation of AMPK pathway. *Journal of Cachexia, Sarcopenia and Muscle*, 11(1): 241–258. https://doi.org/10.1002/jcsm.12530

Oliyaei N, Moosavi-Nasab M, Tamaddon AM andTanideh N (2021) Antidiabetic effect of fucoxanthin extracted from *Sargassum angustifolium* on streptozotocin-nicotinamide-induced type 2 diabetic mice. *Food Science and Nutrition*, 9(7): 3521–3529. https://doi.org/10.1002/fsn3.2301

Osório C, Machado S, Peixoto J, Bessada S, Pimentel FB, C. Alves R and Oliveira MBPP (2020) Pigments Content (Chlorophylls, Fucoxanthin and Phycobiliproteins) of Different Commercial Dried Algae. *Separations*, 7(2): 33. doi:https://doi.org/10.3390/separations7020033

Pan F, Cui W, Gao L, Shi X, Yang H, Hu Y and Li M (2022) Serum lutein is a promising biomarker for type 2 diabetes mellitus and diabetic kidney disease in the elderly. *Journal of Clinical Laboratory Analysis*, 36(4): e24350. https://doi.org/10.1002/jcla.24350

Pangestuti R and Kim SK (2011) Biological activities and health benefit effects of natural pigments derived from marine algae. *Journal of Functional Foods*, 3(4): 255–266. https://doi.org/10.1016/j.jff.2011.07.001

Park CH, Xu FH, Roh SS, Song YO, Uebaba K, Noh JS and Yokozawa T (2015)Astaxanthin and CorniFructus protect against diabetes-induced oxidative stress, inflammation, and advanced glycation end product in livers of streptozotocin-induced diabetic rats. *Journal of Medicinal Food*, 18(3): 337–344. https://doi.org/10.1089/jmf.2014.3174

Patel AK, Albarico FPJB, Perumal PK, Vadrale AP, Ntan CT, Chau HTB, Anwar C, ud din Wani HM, Pal A, Saini R, Ha LH, Senthilkumar B, Tsang YS, Chen CW, Dong CD and Singhania RR (2022) Algae as an emerging source of bioactive pigments. *Bioresource Technology*, 351: 126910. https://doi.org/10.1016/j.biortech.2022.126910

Pereira L and Valado A (2023) Algae-Derived Natural Products in Diabetes and Its Complications—Current Advances and Future Prospects. *Life*, 13(9): 1831. https://doi.org/10.3390/life13091831

Prabakaran G, Sampathkumar P, Kavisri M and Moovendhan M (2020) Extraction and characterization of phycocyanin from *Spirulina platensis* and evaluation of its anticancer, antidiabetic and anti-inflammatory effect. *International Journal of Biological Macromolecules*, 153: 256–263. https://doi.org/10.1016/j.ijbiomac.2020.03.009

Pradhan B, Nayak R, Patra S, Jit BP, Ragusa A and Jena M (2021) Bioactive metabolites from marine algae as potent pharmacophores against oxidative stress-associated human diseases: A Comprehensive Review. *Molecules*, 26(1): 37. https://doi.org/10.3390/molecules26010037

Qi J and Kim SM (2018) α-Glucosidase Inhibitory Activities of Lutein and Zeaxanthin Purified from Green Alga *Chlorella ellipsoidea*. *Journal of Ocean University of China*, 17: 983–989. https://doi.org/10.1007/s11802-018-3465-2

Reka P, Banu T and Seethalakshmi M (2017) Alpha amylase and alpha glucosidase inhibition activity of selected edible seaweeds from south coast area of India. *International Journal of Pharmacy and Pharmaceutical Sciences,* 9(6): 64–68. http://dx.doi.org/10.22159/ijpps.2017v9i6.17684

Ren Z, Xie Z, Cao D, Gong M, Yang L, Zhou Z and Ou Y (2018) C-Phycocyanin inhibits hepatic gluconeogenesis and increases glycogen synthesis via activating Akt and AMPK in insulin resistance hepatocytes. *Food & Function*, 9(5): 2829–2839. https://doi.org/10.1039/c8fo00257f

Saide A, Lauritano C and Ianora A (2020)Pheophorbide *a*: state of the art. *Marine Drugs*, 18(5):257. https://doi.org/10.3390/md18050257

Sharavana G, Joseph GS and Baskaran V (2017) Lutein attenuates oxidative stress markers and ameliorates glucose homeostasis through polyol pathway in heart and kidney of STZ-induced hyperglycemic rat model. *European Journal of Nutrition*, 56(8): 2475–2485.https://doi.org/10.1007/s00394-016-1283-0

Shivarudrappa AH and Ponesakki G (2020) Lutein reverses hyperglycemia-mediated blockage of Nrf2 translocation by modulating the activation of intracellular protein kinases in retinal pigment epithelial (ARPE-19) cells. *Journal of Cell Communication and Signaling*, 14(2): 207–221. https://doi.org/10.1007/s12079-019-00539-1

Son, Y. K., Jin, S. E., Kim, H. R., Woo, H. C., Jung, H. A., and Choi, J. S. (2011) Inhibitory activities of the edible brown alga *Laminaria japonica* on glucose-mediated protein damage and rat lens aldose reductase. *Fisheries Science*, 77: 1069–1079. DOI https://doi.org/10.1007/s12562-011-0406-z

Sonani RR, Singh NK, Kumar J, Thakar D and Madamwar D (2014) Concurrent purification and antioxidant activity of phycobiliproteins from *Lyngbya* sp. A09DM: An antioxidant and anti-aging potential of phycoerythrin in *Caenorhabditis elegans. Process Biochemistry*, 49(10): 1757–1766.https://doi.org/10.1016/j.procbio.2014.06.022

Soni B, Visavadiya NP and Madamwar D (2009) Attenuation of diabetic complications by C-phycoerythrin in rats: antioxidant activity of C-phycoerythrin including copper-induced lipoprotein and serum oxidation. *Brritish Journal of Nutrrition*, 102(1): 102–109. https://doi.org/10.1017/s0007114508162973

Srivastava SK, Ramana KV and Bhatnagar A (2005) Role of aldose reductase and oxidative damage in diabetes and the consequent potential for therapeutic options. *Endocrine Reviews*, 26(3): 380–392. doi: https://doi.org/10.1210/er.2004-0028

Tamel Selvan K, Goon JA, Makpol S and Tan JK (2023) Therapeutic Potentials of Microalgae and Their Bioactive Compounds on Diabetes Mellitus. *Marine Drugs*, 21(9): 462. https://doi.org/10.3390/md21090462

Tan HT, Yusoff FM, Khaw YS, Noor Mazli NAI, Nazarudin MF, Shaharuddin NA, Katayama T and Ahmad SA (2022) A Review on a Hidden Gem: Phycoerythrin from Blue-Green Algae. *Marine Drugs*, 21(1): 28. https://doi.org/10.3390/md21010028

Uchiyama K, Naito Y, Hasegawa G, Nakamura N, Takahashi J and Yoshikawa T (2002) Astaxanthin protects β-cells against glucose toxicity in diabetic db/db mice. *Redox Report*, 7(5): 290–293. https://doi.org/10.1179/135100002125000811

Villaró S, Ciardi M, Morillas-España A, Sánchez-Zurano A, Acién-Fernández G and Lafarga T (2021) Microalgae Derived Astaxanthin: Research and Consumer Trends and Industrial Use as Food. *Foods*, 10(10): 2303. https://doi.org/10.3390/foods10102303

Wang J, Hu X, Chen J, Wang T, Huang X and Chen G (2022) The extraction of β-carotene from microalgae for testing their health benefits. *Foods*, 11(4): 502. https://doi.org/10.3390/foods11040502

Wang W, Tam KC, Ng TC, Goit RK, Chan KLS and Lo ACY (2020) Long-term lutein administration attenuates retinal inflammation and functional deficits in early diabetic retinopathy using the Ins2Akita/+ mice. *BMJ Open Diabetes Research & Care*, 8(1): e001519. https://doi.org/10.1136/bmjdrc-2020-001519

Xie Y, Xiong X and Chen S (2021) Challenges and Potential in Increasing Lutein Content in Microalgae. *Microorganisms*, 9(5): 1068. https://doi.org/10.3390/microorganisms9051068

Yeh PT, Huang HW, Yang CM, Yang WS and Yang CH (2016) Astaxanthin Inhibits Expression of Retinal Oxidative Stress and Inflammatory Mediators in Streptozotocin-Induced Diabetic Rats. *PLoS One*, 11(1): e0146438. https://doi.org/10.1371/journal.pone.0146438

Yuan JP, Peng J, Yin K and Wang JH (2011) Potential health-promoting effects of astaxanthin: a high-value carotenoid mostly from microalgae. *Molecular Nutrition and Food Research*, 55(1): 150–165. https://doi.org/10.1002/mnfr.201000414

Zhang Y, Xu W, Huang X, Zhao Y, Ren Q, Hong Z, Huang M and Xing X (2018) Fucoxanthin ameliorates hyperglycemia, hyperlipidemia and insulin resistance in diabetic mice partially through IRS-1/PI3K/Akt and AMPK pathways. *Journal of Functional Foods,* 48: 515–524. https://doi.org/10.1016/j.jff.2018.07.048

Zhu X, Chen Y, Chen Q, Yang H and Xie X (2018)Astaxanthin Promotes Nrf2/ARE Signaling to Alleviate Renal Fibronectin and Collagen IV Accumulation in Diabetic Rats. *Journal of Diabetes Research*, 2018, 6730315. https://doi.org/10.1155/2018/6730315

Ziyaei K, Abdi F, Mokhtari M, Daneshmehr MA and Ataie Z (2023) Phycocyanin as a nature-inspired antidiabetic agent: A systematic review. *Phytomedicine*, 119: 154964. https://doi.org/10.1016/j.phymed.2023.154964

Chapter 13
Clinical Trials and Patents Related to Anti-diabetic Properties of Algae

R. Esther Shoba, Deepa Ramachandran, and M. Sonia Angeline

Abstract Natural products have gained attention as potential treatments for diabetes, with algae emerging as a rich source of bioactive compounds. Sea algae contain various health-promoting molecules such as carotenoids, polysaccharides, fatty acids, amino acids, polyphenols, tocopherols, pigments, proteins, and omega-3 fatty acids. These compounds offer anti-diabetic effects by improving insulin sensitivity, regulating glucose metabolism, and combating oxidative stress. As a result, algae are widely used in medications, cosmetics, dietary supplements, and functional foods. Clinical studies have examined the glycemic control benefits of both microalgae and macroalgae species like *Spirulina, Chlorella, Dunaliella,* and *Haematococcus pluvialis* in diabetic patients. The increasing preference for natural remedies among consumers emphasizes the potential of algae-based solutions in managing diabetes. Additionally, patents involving algae-derived formulations and innovative extraction methods support their increasing role in pharmaceutical and nutraceutical development. Combining clinical research with patented technologies strengthens the case for algae-based interventions in diabetes management.

Keywords Algae · Anti-diabetic · Bioactive compounds · Clinical trials

1 Introduction

Diabetes mellitus (DM), a metabolic disorder of high blood glucose levels, includes type 1 and type 2 diabetes, maturity-onset diabetes of the young (MODY), gestational diabetes, neonatal diabetes, along with secondary causes from endocrinopathies, steroid use, etc. Primarily, two kinds of diabetes mellitus: Type 1 diabetes mellitus (T1DM) and Type 2 diabetes mellitus (T2DM), are typically caused by

R. E. Shoba (✉) · D. Ramachandran · M. S. Angeline
Department of Life Sciences, School of Biological and Forensic Sciences, Kristu Jayanti (Deemed-to-be University), Bengaluru, India
e-mail: esther@kristujayanti.com

R. Raja et al. (eds.), *Algae in Diabetes Management*,
https://doi.org/10.1007/978-981-95-0571-5_13

impaired insulin secretion (T1DM) and/or action (T2DM). While T1DM typically manifests in children or teenagers, T2DM is believed to effect middle-aged and older adults who have chronic hyperglycemia and insulin sensitivity as a result of metabolic and adipose tissue disorders linked with obesity and lipodystrophy (Makki et al. 2013). Because of the significant differences in their pathophysiologies, T1DM and T2DM have distinct etiologies, manifestations, and therapies (Sapra and Bhandari 2024).

An individual with diabetes mellitus may experience hyperglycemia. Because the disease can frequently be caused by multiple sources, the pathology of DM might be obscure. Hyperglycemia on its own can affect insulin secretion and beta-cell function of the pancreas. The cause of insulin resistance is proinflammatory cytokines and excess fatty acids, which hinder glucose transport and accelerate fat breakdown. In reaction to insufficient insulin synthesis or response, the body raises glucagon improperly, which exacerbates hyperglycemia.

Another effect of chronic hyperglycemia is the nonenzymatic glycation of lipids and proteins. By using the glycation hemoglobin (HbA1c) test, the degree of this can be determined. Glycation damages peripheral nerves, kidney, and retinal tiny blood vessels (Sapra and Bhandari 2024; Unger and Orci 2010).

Further increasing the overall disease burden, diabetes leads to an amplified risk of various illnesses such as dyslipidemia, hypertension, and non-alcoholic fatty liver disease. A multifaceted strategy is needed to manage diabetes and related complications, involving lifestyle changes, glucose-lowering drugs, and focused therapies to address particular issues (Li et al. 2021; Kosmalski et al. 2023). Even while traditional diabetes treatment has advanced, there is increasing interest in investigating alternative therapy approaches that may be able to increase efficacy, reduce side effects, and thereby decrease the development of illness (Davies et al. 2022; Williams et al. 2022).

Natural products and their derivatives have drawn curiosity in this area for treating diabetes. These substances have a wide spectrum of pharmacological effects and different chemical ingredients (Abo-Shady et al. 2023; Quitério et al. 2021). They come from a variety of sources such as plants, algae, fungi, and marine organisms. Utilizing natural products' potential in diabetes research could lead to the creation of innovative treatments and therapies that can successfully regulate blood glucose levels and lessen the disease's effects (Liu et al. 2022; Salehi et al. 2019). An natural antioxidant astaxanthin from marine microalgae in conjugation with s-allyl cysteine has shown to inhibit alpha glucosidase activity of Saccharomyces cerevisiae, mammalian (rat intestinal) α-glucosidases (maltase and sucrase) activity in a dose dependent manner (Sakayanathan et al.2018).

Algae are also a natural source of bioactive molecules. Sea algae, a well-known one, have beneficial raw material in the form of manifold bioactive molecules that include carotenoids, fatty acids, polysaccharides, amino acids, and tocopherols. As a result, they are used as components in medications and cosmetics and nutritional and dietary supplements (Alsenani et al. 2020; Wells et al. 2017). So yet, the chemical composition of freshwater algae is largely unknown. The valuable advantages of algae on humans are the main topic of this article. This chapter investigates how algae and their extracts could be used in the fields of medicine and cosmetics. The pharmaceutical and cosmetic industries can benefit greatly from biologically active chemicals derived from marine macroalgae, although their use has been limited thus far (Bhattacharyya et al. 2016).

As natural sources of bioactive chemicals, algae hold great promise during algal culture enabling the production of structurally complex molecules that are challenging or impossible to produce by chemical synthesis. (Ślusarczyk et al. 2021). These algae-derived natural substances provide fresh potential for researching the pathogenesis of diabetes and developing revolutionary diabetes care strategies (Magwaza and Islam 2023). Figure 13.1 represents different types of algae.

Additionally, natural chemicals derived from algae have the potential to cure issues unique to diabetes. The microvascular disorder known as diabetic retinopathy is defined by inflammation and oxidative stress in the retina. Chemicals generated from algae that include anti-inflammatory and antioxidant properties have demonstrated efficacy in the protection of retinal cells and delaying their progression. Also compounds produced from algae have demonstrated reno-protective effects by reducing renal fibrosis and albuminuria (Jin and Arroo 2023; Zhou and Chen 2023).

Natural substances obtained from algae may be able to fill therapeutic gaps and enhance existing diabetes treatments. They can be added to conventional anti-diabetic medications as adjuvants, boosting their effectiveness or providing

Fig. 13.1 Types of algae

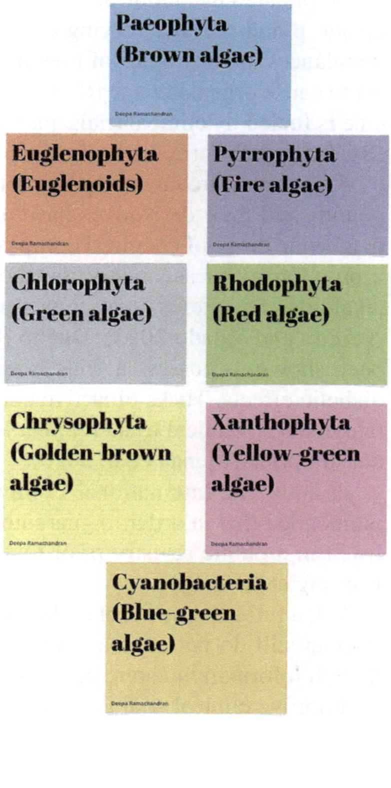

complementary advantages (Alam et al. 2022; Lankatillake et al. 2019). Additionally, individuals who show intolerance or resistance to traditional medications may benefit from alternative treatment choices offered by chemical compounds generated from algae (Menaa et al. 2021).

In diabetes research, escalating our knowledge of significance and impending uses of natural chemicals generated from algae may help develop new treatment approaches and enhance patient results. With an emphasis on their significance and possible future directions in the area, this review aims to provide a thorough examination of role of naturally occurring chemicals produced from algae in diabetes research (Khaligh and Asoodeh 2022; Zhuang et al. 2022).

Preclinical and clinical studies, including human, in vivo, and in vitro participants, provide a more thorough understanding of both the benefits and drawbacks of using algae and algae-derived products to treat diabetes mellitus. The anti-diabetic and antilipidemic properties of certain algae have been extensively researched. This chapter discusses the study's specifics as well as the patents granted for goods created to help manage and treat diabetes.

2 Clinical Trials

A complicated metabolic and endocrine condition is type 2 diabetes. Defective pancreatic β and α cells, varying degrees of insulin intolerance, and other hormonal imbalances are examples of how hereditary and environmental variables can interact to cause organ damage (Stumvoll et al. 2005, p. 2). The growth of clinical medicine is fueled by clinical trials, particularly well-designed randomized clinical trials (RCTs), which form the basis of evidence-based medicine.

As possible treatment agents for diabetes and its consequences, natural compounds and their derivatives have attracted a lot of attention lately. Natural products' wide range of chemical composition is one of their main benefits. For example, a broad array of secondary metabolites, including compounds that are phenolic, alkaloids, flavonoids, and terpenoids, are produced by plants, including algae (Pereira and Valado 2023). Due to their extraordinary chemical variety, algae have been shown to possess a wide range of bioactive substances that may have anti-diabetic effects. These bioactive compounds are extracted and formulated as drugs that undergo clinical trials for years and approved to be available in the markets for consumption (Menaa et al. 2020).

In 2004, the International Committee of Medical Journal Editors (ICMJE) recommended that in order to guarantee process transparency, clinical trials should be listed in a public registry prior to participant recruitment (De Angelis et al. 2004; Hartung et al. 2014).

Yet, a full assessment of diabetes-related studies has not yet been carried out, and doctors still do not have a good understanding of clinical trials on diabetes, even though information about ongoing and completed clinical trials is readily available. To improve clinical trial designs and uncover understudied research areas of algal

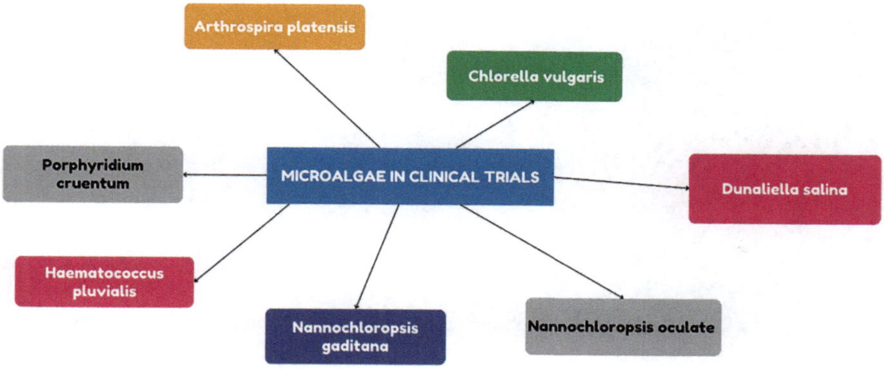

Fig. 13.2 Alga under pre-clinical and clinical trials

anti-diabetic properties, a deeper comprehension of the present characteristics of T2DM-related clinical trials is crucial. This will enhance the potential to translate results into benefits for patients (Long et al. 2021). Figure 13.2 represents examples of few algae used for their evaluation for anti-diabetic characteristics and the clinical trials associated with it.

2.1 *Blue Green Algae—Spirulina—*Arthrospira platensis

A microalga that is a member of the cyanobacteria family, *Arthrospira platensis* or *Spirulina platensis,* commonly known as spirulina (Fig. 13.3), has the most preventative and therapeutic nutritional properties. It has high levels of protein, phycocyanins, polyphenols, carotenoids, vitamins, and vital fatty acids, which contribute to its cardioprotective and antioxidant properties (Ovando et al. 2018; van den Driessche et al. 2020). Studies carried out previously have explained the therapeutic effects of spirulina in a variety of non-communicable disorders (Wells et al. 2017). In people who are healthy, have cardiac disease, or have diabetes, spirulina may lower blood lipid levels (Mani et al. 2000; Ramamoorthy and Premakumari 1996). Studies on healthy older adults have shown that spirulina lowers blood pressure, total and LDL cholesterol, and plasma triglycerides (Kim and Park 2016; Park et al. 2008). Among its various biological properties, spirulina can prevent anemia and herpes simplex infection (Hernández-Corona et al. 2002). Additionally, lipid peroxidation products and liver lipid profiles are reduced (Abd El-Baky et al. 2009). Table 13.1 gives the details on the composition of Spirulina.

Because of its anti-inflammatory and antioxidant qualities, research has demonstrated that *A. platensis* biomass administration lowers hyperglycemia among diabetic induced rats (Bitam and Aissaoui 2020; Bohórquez-Medina et al. 2021). Male Wistar rats with diabetes caused by streptozotocin (STZ) were given *A. platensis* in

Fig. 13.3 Spirulina

Table 13.1 Composition of Spirulina

Component	Amount in gm or mg	Component	Amount in gm or mg
Proteins	50-70 g	**Biotin**	0.01 mg
Glucose	15-25 g	**Beta carotene**	130 mg
Lipids	6 g	**Riboflavin**	4 mg
Calcium	770 mg	**Inositol**	35 mg
Phosphorous	780 mg	**Cobalamin**	0.2 mg
Magnesium	250 mg	**Tocopherol**	20 mg
Sodium	450 mg	**Thiamine**	5 mg
Potassium	1100 mg	**Folic acid**	0.05 mg
Ferrous/iron	120 mg	**Pantothenic acid**	13 mg
Zinc	3 mg	**Pyridoxin**	0.7 mg
Manganese	3 mg		

mg/kg of body weight (BW) in increasing dose of 10, 20, or 30 daily for three weeks (Nasirian et al. 2018).

In diabetic rats, extracts at the concentration of 20 and 30 mg/kg decreased elevated levels of triglycerides, total cholesterol, low-density lipoprotein cholesterol, and glucose. Hyperglycemia and hyperlipidemia have been demonstrated to be improved by trace minerals (TMs) included in *A. platensis* extract. The concentrate includes chromium (Cr), zinc (Zn), copper (Cu), iron (Fe), magnesium (Mg), and selenium (Se) according to a chemical analysis. Research indicates that TMs have insulin-modulatory impacts (Akhuemokhan et al. 2013). Serum malondialdehyde (MDA), an end product of lipid peroxidation levels, were lowered as a result of the doses of *A. platensis* extract that increased quantities of the enzymes catalase, superoxide dismutase, and glutathione peroxidases. The author claims that zinc and selenium affect the activity of antioxidant enzymes via participating in antioxidant processes and serving as cofactors for superoxide dismutase (Battin and Brumaghim 2009; Chen et al. 1998).

TMs plays an important role in insulin signalling, and it has been indicated that Mg defiecency affects glucose metabolim and insulin resistance. It was observed that the highest dosage decreased plasma levels of cytokines namely interleukin-6 (IL-6) and alpha tumor necrosis factor (TNF-α). These results imply that the antioxidant qualities of *A. platensis* extract are responsible for its anti-inflammatory action. As a result, TMs enhance glucose metabolism while reducing oxidative stress and inflammation (Nazem et al. 2023; Viktorínová et al. 2009).

Reactive oxygen species (ROS) from hyperglycemia harm pancreatic β-cells, preventing the generation of insulin and leading to insulin insufficiency (Robertson and Harmon 2006).

In alloxan-induced diabetic Wistar or Swiss rats, supplementation with powder of *A. platensis* for 5–10 days of about 100 mg/kg/day showed enhanced surface area of pancreatic islet and improved β-cell characteristics. This dose also reduced pro-inflammatory TNF-α levels, indicating potential anti-diabetic effects (Joventino et al. 2012). Less dosages of 5, 10, and 25 mg/kg significantly reduced mouse licking time in a formalin test, indicating anti-inflammatory activity. Additionally, in mouse paw tissues, doses of 50 and 100 mg/kg lessened neutrophil infiltration, elevated tissue architecture, and decreased oedema. Glucose levels in serum sharply decreased, especially on day 10 (Tamel Selvan et al. 2023).

In diabetic rats on a high-fat and high-sucrose diet, 150 mg/kg/day of *A. platensis* both ethanol and water extracts for eight weeks altered intestinal microbiota composition and improved glucose tolerance. However, the anti-diabetic effects varied depending on the extract type. The results reveal that diabetic rats fed with water extract of *A. platensis* had considerably lower fasting glucose levels in blood (FBG) levels than non-supplemented diabetic rats. In addition, the oral glucose tolerance test (OGTT) indicated that aqueous as well ethanol extracts enhanced glucose tolerance (Wan et al. 2019). Effectiveness of water extract of *A. platensis* supplementation in lowering hyperglycemia, a feature of diabetes mellitus, was observed. PUFAs in extracts could enhance pancreatic β-cell function and energy metabolism (Zhao et al. 2015).

Decreased serum glucose and HbA1c levels, increased insulin levels, and prevention of pancreatic β-cells from undergoing apoptosis was observed when 500 mg/kg of *A. platensis* powdered extract was administered twice weekly for two months in male Albino rats with Streptozotocin (STZ)-induced diabetes which indicates enhanced glucose metabolism (Sadek et al. 2017). Spirulina at dosages of 100, 200, and 300 mg/kg body weight when administered to alloxan induced diabetic male Sprague-Dawley albino rats showed higher insulin, HDL, and phospholipids levels and significantly lower blood glucose, cholesterol, VLDL, LDL, and triglycerides levels (Shaman et al. 2024).

Studies revealed that *A. platensis's* capacity to decrease pyruvate carboxylase (PC) expression may be responsible for its glucose-lowering effects. Furthermore, supplementation repaired pancreatic islet architecture while increasing glutathione levels. Furthermore, glutathione-s-transferase (GST), superoxide dismutase (SOD), and catalase activities were enhanced that was supported by the analysis of increased expression of SOD, catalase, and GST mRNA in these rats. Enzymes protect pancreatic β-cells by neutralizing ROS (reactive oxygen species) (Hughey and Crawford 2019; Jörns et al. 1999).

2.1.1 Human Studies with Spirulina

Karun and group in their work supplemented spirulina capsules named "SUNOVA spirulina" 2 mg to E1 group and 4 mg to E2 group along with oral glycemic medications for 2 months to the subjects that are male diabetic individuals in the age of 40–60. The treatment significantly reduced fasting and post-prandial blood glucose levels. The mean carbohydrate and protein intake decreased significantly ($p = 0.01$) in group E2 while non-significantly in group E1 and C. However, non-significant change was observed in blood glucose levels of the subjects of group C. A significant decrease was observed in TC, TG, LDL-C, and VLDL-C levels in the E groups.

Gamma linolenic acid (GLA) may be the cause of the decrease in triglycerides and total cholesterol, according to the results of this study.

The decrease in LDL-C in both experimental groups may be attributed to the effects of spirulina supplementation (Kaur et al. 2008).

The benefits of 500 mg of spirulina twice a day for 12 weeks were evaluated in a trial including 160 male NIDDM—noninsulin dependent diabetic male volunteers, aged 45 to 60. The lipid profile, fasting blood glucose, and glycosylated hemoglobin levels all significantly decreased, while HDL cholesterol increased (Anitha and Chandralekha 2010).

Similar work by Mani and group on 15 NIDDM patients administering 2gm of spirulina every day for 2 months reported FBS levels decrease by 27%, and glycated serum proteins decrease by 34%, while uronic acid levels remained unchanged. Triglyceride levels also decreased by 22% and free fatty acids by 34% (Mani et al. 2000).

Single blind randomized standard control study conducted on 40 patients of type 2 DM by Alam and group by administering 7 gm of test drug Tahlab (Spirulina) on 30 subjects of test group and Metformin (500 mg) to 10 subjects of control group for a period over 45 days observed significant decrease in PPBS and blood glucose levels compared to the control group (Alam et al. 2016). The decrease might be because of spirulina's high fiber content that inhibits glucose absorption through the effect of peptides and polypeptides produced by its protein breakdown. Spirulina a protein-rich food also promotes insulin production and this impact may be responsible for a decrease in PPBS (Anwer et al. 2013; Parikh et al. 2001).

Fifty individuals with type II diabetes randomly chosen for study were split into two equal groups, of which one group served as a control. Spirulina tablets (8 gm) given to the patient group for three months showed patient group's fasting plasma glucose dropped from 158.1 mg/dl to 127.8 mg/dl after spirulina supplementation, suggesting short-term effectiveness in lowering fasting blood glucose; however, a longer period is required to impact HbA1c (Beihaghi et al. 2017).

Sixty patients were randomly assigned to *A. platensis* (2 g/day) or placebo groups for 3 months. Supplementing diabetic patients with A. platensis resulted in a significant decrease in both HbA1c and FBS levels. During a 90-day course of treatment, it also reduced LDL-C and total cholesterol (TC) while somewhat raising HDL-C (Park et al. 2008).

The major clinical trials conducted on spirulina supplements lead to a conclusion that Spirulina tablets provide a natural reduction in fasting blood sugar, rather than a temporary decrease. Several ideas have been offered to validate the hypoglycemic impact of spirulina. It was believed that spirulina's fiber content, which lowers glucose absorption, was the cause of its hypoglycemic effect (Hatami et al. 2021a). Another idea proposes that spirulina protein digestion produces peptides and polypeptides that may act. Gamma-linolenic acid-rich spirulina may lower blood cholesterol by inhibiting the production of fat and cholesterol. Those who are obese and lack the enzyme delta 6-desaturase benefit from this important fatty acid. Furthermore, spirulina peptides have the potential to activate the vagus nerve in the sympathetic nervous system and promote insulin secretion (Hatami et al. 2021b; Mani et al. 2000).

Spirulina supplementation reduced triglyceride and VLDL levels via decreasing VLDL triglyceride synthesis and increasing VLDL clearance in the periphery due to its high protein and fiber content. Reduced triglyceride levels may result from increased insulin output, which stimulates lipoprotein lipase activity (Shils et al. 1994). Molecular docking studies has reported that the pigments of this cyanobacteria to have inhibitory function on diabetic enzymes. An experimental approach with molecular docking results will be significant for the discovery of diabetes mellitus drugs and treatments (Munaworah et al. 2022).

2.2 *Chlorella Vulgaris*

Protein, dietary fiber, choline, lipid-soluble vitamins, and vital minerals are all found in good amounts in chlorella. Chlorella's biochemical and physiological properties, including increasing animal development, strengthening the immune system, preventing ulcers brought on by stress, and reducing dyslipidemia are the center of numerous studies. Chlorella's impacts, including fatigue recovery, heavy metal removal, toxic material degradation, and environmental applicability, are investigated (Cherng and Shih 2005; Jeong et al. 2009; Jong-Yuh and Mei-Fen 2005).

In addition to a significant quantity of excellent protein, dietary fibers, and PUFA, such as α-linolenic and linoleic acids, commercially available *chlorella* products provide a number of elements that are necessary for human health (Bito et al. 2020). Table 13.2 enlists the composition of Chlorella.

When streptozotocin (STZ)-induced mice or rats were given C*hlorella vulgaris* (Fig. 13.4) acutely, it resulted in a notable hypoglycemic impact. The hypoglycemic impact of prolonged *chlorella* consumption was hard to explain to this group (Cherng and Shih 2005). Hypoglycemic effects of chlorella were analyzed by

Table 13.2 *Chlorella vulgaris* powder composition

Components	Content (g)
Protein	60.6
Carbohydrate	3.7
Fat	12.8
Dietary fiber	13
Ash	4.5
Moisture	5.4
Chlorophyll	2.4
Carotene	0.106
Vitamin C	0.074
Vitamin E	0.023
Total	100
Calorie (kcal)	372

Fig. 13.4 *Chlorella vulgaris*

acclimatizing 7 days pellet meal of Chlorella to both normal Wistar rats and type 2 diabetic GK—Goto-Kakizaki rats. Insulin secretion capacity was either negligible or unaffected by *chlorella* consumption however showed decreased plasma glucose (Jeong et al. 2009).

However in six-week-old diabetic GK rats, the chlorella-fed group had lower fasting blood glucose levels than the non-fed group, which resulted in a minor drop in plasma insulin levels. The slight variation in insulin concentration may be explained by the reduced insulin resistance in younger diabetic rats (Jeong et al. 2009; Rodríguez-López and López-Quijada 1971).

Diabetes induced in ICR mice by STZ revealed low hypoglycemic effects when administered with Chlorella treatment. The findings imply that the increased absorption of glucose by liver and soleus muscles is the cause of *chlorella's* hypoglycemic effects (Cherng and Shih 2006).

The effects of unicellular green algae Chlorella and Spirulina on type 2 diabetes (T2DM)-induced rats were examined with male Sprague-Dawley albino rats ($n = 6$) in eight groups. The findings indicated that the control group that was injected with saline had significantly higher levels of triglycerides, low-density lipoprotein (LDL), blood sugar, and very low-density lipoprotein (VLDL) and considerably lower levels of insulin, high-density lipoprotein (HDL), and phospholipids. However, rats given Spirulina (groups 3, 4, and 5) showed notably higher insulin, HDL, and phospholipids levels and significantly lower blood glucose, cholesterol, VLDL, LDL, and triglycerides levels. Rats fed with Chlorella (groups 6, 7, and 8) also showed increased effectiveness of feed ratios, greater final weights, and body weight growth percentages compared to the control group. This investigation paves an insight that Spirulina and Chlorella may be used as functional meals to help control and cure diabetes.

Type 2 diabetes (Otsuka Long-Evans Tokushima Fatty [OLETF] rats of 20 week old of 4 different groups ($n = 7$ per group) showed decreased fasting blood glucose, insulin levels, and total glucose which followed combination of aerobic exercise training and chlorella intake as compared to the rats which followed only sedentary life style, physical aerobic exercise and diabetic rats administered with chlorella alone for 8 weeks. Improvement of glycemic control was attributed to the activation of muscle hosphatidylinositol-3 kinase (PI3K) activity, protein kinase B (Akt) phosphorylation, and glucose transporter 4 (GLUT4) translocation levels. (Horii et al. 2019).

With all the above-mentioned preclinical trial information, *chlorella* seems to be a promising therapeutic agent for diabetes. Planned execution and detailed research can prove its efficiency. The metabolism in vivo and in vitro cannot exactly replicate a human body. The clinical trials performed on human volunteers can show the potency of the drugs and their side effects.

Using volunteers with borderline diabetes, a randomized, double-blind, placebo-controlled trial was carried out. A group of 57 participants was randomly assigned to consume either chlorella or a placebo. In order to conduct laboratory tests, blood samples were obtained. Using RNA from peripheral blood cells, gene expression assessments were conducted both before and 12 weeks following the trial. Resistin, an insulin resistance inducer, was one of six genes linked to type 2 diabetes that showed dramatically decreased expression when consumed with chlorella (Itakura et al. 2015). Changes in HbA1c, TNF-α, and IL-6 levels, all of which are closely linked to glucose metabolism and/or inflammation, showed a substantial correlation with resistin mRNA expression (Kaser et al. 2003).

Balb/C mice fed either with standard rodent chow diet or high-fat diet (HFD) received concomitant treatment with *C. vulgaris* for 12 consecutive weeks. High-fat diet-induced insulin resistance in obese mice showed increased glucose and insulin tolerance. Western blot studies showed increased phosphorylation of insulin receptor (IR) β, insulin receptor substrate (IRS)-1 on tyrosine residue, protein kinase B (Akt), and lower phosphorylation levels of IRS-1 on serine residue(ser307). This indicates that *C. vulgaris* played its part in the improvement of the insulin signaling pathway at its main target tissues such as liver, skeletal muscle, and adipose tissue. It was also found that C. *vulgaris* administration prevented high-fat diet-induced dyslipidemia by reducing triglyceride, cholesterol, and free fatty acid levels. Thus, Chlorella can prevent insulin resistance (Vecina et al. 2014). Another species of Chlorella, Chlorella pyrenoidosa functional formulations CFF showed the presence of eicosapentaenoic acid – PEA, docosahexaenoic acid - DHA, adrenic acid, and succinic acid, all of which showed improves glucose tolerance and better hypoglycemic effect on T2DM mice (Huang et al.2023).

In an eight-week study of 70 non-alcoholic fatty liver disease (NAFLD) patients confirmed through ultra-sonographic findings, divided as intervention group who were administered four 300 mg tablets of *C. vulgaris* and the placebo group who were administered four 300 mg tablets of placebos. Non-invasive quantitative measurements showed a significant reduction in both groups ($p < 0.001$). In *C. vulgaris*-treated patients, the average weight loss was noticeably greater. While liver enzyme, FSG (fasting serum glucose) and hs-CRP (high sensitive C-reactive protein)) concentrations likewise sharply declined and serum insulin concentration and homeostatic model assessment (HOMA) score increased significantly. This paves the way to use C. vulgaris supplementation as an adjunctive therapy to improve weight and glycemic status (Ebrahimi-Mameghani et al. 2017).

However similar studies carried out by Hosseini and group on 84 T2DM patients who were divided into two groups and given either a placebo or 1500 mg of *C. vulgaris* daily for eight weeks showed no meaningful change in any of the variables examined in the blood samples for lipid profile, insulin concentration, insulin resistance, FBS, and HbA1c. Might be higher dosage of *C. vulgaris* and treatment for more than 8 weeks could be studied in the future (Hosseini et al. 2021). Investigations revealed that Chlorella protected H2O2-induced oxidative damaged INS-1 (832/13) cells from apoptosis. Chlorella was able enhance glucose stimulated insulin secretion partially and preventing mitochondria disruption and maintaining cellular ATP levels after H2O2 toxicity (Lin et al. 2014).

Studies on both humans and animals have demonstrated the many health benefits of consuming chlorella on a daily basis. Studies on C*hlorella* have documented pharmacological properties such as immunomodulation, antioxidative activity, and benefits against hypertension, diabetes, and hyperlipidemia. *Chlorella* may have positive benefits by promoting the synergy of several antioxidant and nutritional components (Bito et al. 2020). Future research could still uncover new bioactive substances that give it its pharmacological effects.

2.3 Dunaliella salina

Dunaliella salina is a salt-loving single-cell bi-flagellate green that contains high-value biomolecules such as carotenoids, oleic acid (OA), α-linolenic acid (ALA), and linoleic acid (LA), which can reduce diabetes-related oxidative stress through antioxidant effects. (Andriopoulos et al. 2022a; Barbosa et al. 2023). ALA, EPA, DHA, pigments, and polyphenols are among the compounds in algae that may have anti-diabetic effects (Tamel Selvan et al. 2023). It's important to remember that pigments like β-carotene might possibly be involved in this function. According to published research, β-carotene from *Dunaliella salina* contains several natural characteristics, such as anti-inflammatory, hepatoprotective, neuroprotective, and antiproliferative properties, along with a high potential for antioxidants and could be studied in great length for novel functional components that could serve for non-pharmacological treatment of diabetes, obesity, metabolic syndrome and additional diseases (Wang et al. 2022; Pais et al. 2024).

Apart from treating diabetes, algae can be used to treat diabetic related complications as well. Damage to peripheral nerves is one of the many incapacitating symptoms of diabetic neuropathy (DN). Other than glycemic management, there is no treatment for DN (Figueroa-Romero et al. 2008). One of the main factors in DN pathogenesis is thought to be hyperglycemia, which causes reactive oxygen species (ROS) to develop and harm nerves (Grisold et al. 2017; Kaur et al. 2011). A natural source of carotenoids, *Dunaliella salina* (*D. salina*), is thought to be an antioxidant treatment that can help diabetes linked to oxidative stress.

In a study with fifty adult male albino Wister rats that were induced for diabetes with STZ, treatment with glimepiride to group 3 of the diabetic rats and *D. salina* powder (100 & 200 mg/kg to groups 4 and 5 diabetic rats for 30 days, *D. salina* treatment decreased glucose blood level and increased insulin level with normal islets of Langerhans when compared with rats treated with STZ alone that were grouped as group 2.

In the STZ-treated rats DN progressed due to the stimulation of oxidative stress, which can be evidenced by decreased serum levels of TAC, SOD, and catalase. However, *D. saline* high content of carotenoids bettered oxidative stress indices through an elevation in antioxidant contents of TAC, SOD, and catalase. Thereby, natural products as carotenoids from *D. saline* have benefits on diabetes and protect from oxidative damage caused by ROS and can protect the neurons of the brain from oxidative stress and inflammation induced by STZ in rats (El-Baz et al. 2020).

Antioxidant thiol-reductase systems glutathione and thioredoxin (Trx) scavenge ROS and shield cells from oxidative damage (Holmgren et al. 2005). Trx is controlled by a protein called Trxip, which decreases Trx activity and causes oxidative stress, which hinders cell growth (Morrison et al. 2014; Nagaraj et al. 2018).

In STZ-induced hyperglycemic rats, thioredoxin (Trx) antioxidative activity in the brain was reduced. Treatment with *D. salina* powder (100 mg/kg and 200 mg/kg) increased brain Trx levels by 24-fold and 28-fold, respectively, with similar effects observed with glimepiride (El-Baz et al. 2020).

Diabetes mellitus is a complex metabolic disorder marked by chronic hyperglycemia due to reduced insulin secretion or action, leading to abnormalities in carbohydrate, lipid, and protein metabolism (Mahfouz et al. 2009). Paraoxonase (PON), an enzyme associated with high-density lipoprotein (HDL), has an antioxidant and anti-inflammatory effect in HDL. It may also protect against problems (Namitha et al. 2015). PON1, an antioxidant enzyme linked to high-density lipoprotein (HDL), was reported to decrease in T2DM. Reduced PON1 activity may predict CVD in individuals with T2DM (Poh and Muniandy 2010). *D. salina* treatment improved PON1 activity in diabetic rats by 11.56% relative to the normal group. Algae containing β-carotene can reduce LDL-oxidation in diabetes and protect against oxidative stress (El-Baz et al. 2020; Levy et al. 2000).

STZ-treated rats are considered an ideal model for studying systemic oxidative stress in diabetic patients and evaluating the effectiveness of antioxidant therapies (Rupérez et al. 2008). STZ-diabetic rats showed a positive response to a moderate dose of 750 mg of *D. salina* extract over days. This effect was successfully measured using pattern recognition methods, including urine fingerprinting by capillary electrophoresis and target metabolite analysis (Ruperez et al. 2009). Another significant work from El-Baz and group observed that the treatment of diabetic rats with D. salina ethanolic extract, demonstrated an increase in PON1 level as increase in antioxidant potential , while reduction in the elevated TGF-β level contributing to therapeutic potential in hyperglycemia condition (El-Baz et al. 2016).

In another study, although supplementation did not improve blood glucose levels, it significantly reduced triglyceride (TG) levels, indicating an antihyperlipidemic effect that may help alleviate complications associated with diabetes mellitus (Tamel Selvan et al. 2023).

D. salina has been shown to be safe in preclinical investigations. However, nothing is known about the safety of higher dosages or prolonged use in humans, therefore the ideal dosage is still uncertain (El-Baz et al. 2019). Moreover, contamination brought on by environmental heavy metal buildup or cyanobacteria-produced toxins in the culture can be harmful (Leong and Chang 2020). As a result, harmful effects from consuming some microalgae have been recorded (ElFar et al. 2022). Therefore, each microalgae's safety is dependent on its source, growing environment, and quality control (Fig. 13.5).

Fig. 13.5 *Dunaliella salina*

2.4 Nannochloropsis oculate

A marine unicellular microalga from the Eustigmatophyceae family is noted for its nutritional value. *Nannochloropsis* microalgae (NOM) contain several components such as soluble and insoluble polysaccharides, protein, and eicosapentaenoic acid (Pangestuti and Kim 2011). Studies have also revealed that eicosapentaenoic acid (EPA) and docosahexaenoic acid (DHA) affect blood lipid levels (Brown et al. 1997).

In male Wistar rats with STZ-induced diabetes, supplementing with *N. oculate* (10 or 20 mg/kg/day) for three weeks reduced hyperglycemia and hyperlipidemia. While serum glucose, triglycerides (TG), total cholesterol (TC), and LDL-C concentrations decreased, body weight, serum insulin, and HDL-C levels increased (Brown et al. 1997; Tamel Selvan et al. 2023).

Similar to insulin therapy in STZ-induced diabetic rats, *N. oculate*'s antihyperglycemic properties may be linked to its protective effects against oxidative damage on pancreatic β-cell islets or enhanced glucose uptake by peripheral tissues (Jubaidi et al. 2021; Ramos-Romero et al. 2021). The improved lipid profile points to improved glucose and lipid metabolism, mostly as a result of improved insulin action and secretion. Furthermore, the lipid profile may have improved as a result of the EPA and DHA in *N. oculate* (Kosmalski et al. 2022).

Through the modulation of pro-inflammatory cytokine levels and antioxidant enzymes, another study using a similar study methodology and dosages of *N. oculate* powder has shown its antihyperglycemic effect (Fereidouni et al. 2022) Furthermore, there was a sharp decline in the levels of pro-inflammatory markers such as TNF-α, NF-κB, IL-1β, and IL-6 in the serum. An improvement in hyperglycemia in diabetic rats may therefore be the cause of *N. oculate's* anti-inflammatory effect. Similar to *D. salina*, these bioactivities may have been induced by carotenoids found in *N. oculate*, according to the authors (Andriopoulos et al. 2022b) (Fig. 13.6).

Fig. 13.6 *Nannochloropis gaditana* and *Nannochloropis occulate*

2.5 Nannochloropsis gaditana

Administering male Wistar rats with STZ-induced diabetes, 10% *N. gaditana* powder daily for two months increased their body weight and serum glucose levels, just like *N. oculate* did (Nacer et al. 2020). The trial demonstrated *N. gaditana's* promise in treating diabetes-related problems by lowering HbA1c levels and reducing symptoms including glycosuria and polyphagia (Kreilmeier-Berger et al. 2021). Prior studies have demonstrated that *N. gaditana* reduces lipid peroxidation and scavenges free radicals.

2.6 Porphyridium cruentum

P. cruentum, a red algae of the family Porphyridiophyceae, has therapeutic effects on diabetes by promoting the growth and protection of pancreatic β-cell islets. In STZ-induced male Sprague-Dawley rats, powdered algae in doses of 1200 and 1800 mg/kg for 14 days improved β-cell granulation and increased the pancreatic islet area, which typically reduced diabetes. Pancreatic β-cells are crucial for insulin production, and their impairment leads to hyperglycemia (Perego et al. 2019). Bioactive compounds found in *P. cruentum*, such as omega-3 fatty acids, carotenoids, B-phycoerythrin, and phycobiliproteins, have anti-inflammatory and antioxidant qualities that may help lower hyperglycemia. Nevertheless, *P. cruentum* supplementation did not result in a reduction of blood glucose levels in diabetic rats, indicating that the dosages were not ideal for a hypoglycemic effect (S. H. Kim et al. 2021; Setyaningsih et al. 2020) (Fig. 13.7).

2.7 Haematococcus pluvialis

Astaxanthin is abundant in *Haematococcus pluvialis*. Microalgal photosynthetic pigments, such as chlorophylls, carotenoids (β-carotene and astaxanthin), and phycobilins (phycocyanin, phycoerythrin, and phycoerythrocyanin), are widely used as dyes in the food, cosmetic, and pharmaceutical industries (Ambati et al. 2014; D'Alessandro and Antoniosi Filho 2016).

In genetically diabetic female mice, 1.0 mg/mouse/day of marine microalgae-derived astaxanthin for 12 weeks reduced non-fasting blood glucose, improved intraperitoneal glucose tolerance (IPGT), and increased serum insulin levels (Uchiyama et al. 2002). Reactive oxygen species (ROS) generate oxidative stress in the pancreas, which is susceptible to damage from hyperglycemia. This is thought to be the reason for the protective effect. Together with ROS, advanced glycosylation end products (AGEs) produced by hyperglycemia prevent the transcription of the insulin gene and cause the death of pancreatic β-cells (Nowotny et al. 2015). Astaxanthin affects the catalytic site of S. cerevisiae α-glucosidase by way of

Fig. 13.7 *Porphyridium cruentum* and its formulation

Fig. 13.8 *Haematococcus pluvialis*

hydrophobic and hydrogen bonding interactions with its active site. Additionally, it interacts with human α-glucosidases and can be mixed with other substances to improve its enzyme interactions and antioxidant activity (Fig. 13.8).

2.8 Polysaccharides of Various Microalga

The antihyperglycemic impact of microalgal polysaccharides has been studied; it inhibits stomach emptying and small intestine glucose absorption while shielding pancreatic β-cells (Lunn and Buttriss 2007; Setyaningsih et al. 2020). For four weeks, male Kunming diabetic mice were given a diet heavy in fat, sugar, and STZ along with 150 or 300 mg/kg of *C. pyrenoidosa* polysaccharides (Qiu et al. 2022). Through the modulation of the IL-6R/FOXO-1 and GLP-1R/FOXO-1 pathways, CPP supplementation increased body weight and insulin levels; lower levels of phenylpyruvic acid were associated with improved glucose absorption. For 14 days, *P. cruentum* extracellular polysaccharides (EPs) were administered at 150–450 mg/kg per day. In rats with STZ-induced diabetes, it decreased blood glucose and food intake, having effects similar to those of glibenclamide. Additionally, EPs may affect insulin secretion and reduce intestinal glucose absorption to assist control blood sugar (Setyaningsih et al. 2020). Microalgae-derived EPs may be effective in diabetes management, according to the study's findings (Tamel Selvan et al. 2023).

3 Patents

A patent is a legal document issued by a government authority that grants the patent holder exclusive rights to a new invention or technique for a set length of time, usually 20 years from the date of filing. This privilege is granted in exchange for public disclosure of the invention, which allows others to learn from it and maybe generate new ideas based on it. Patents serve an important role in stimulating innovation and investment by granting inventors a temporary monopoly on their ideas. This article examines the notion of patents, its various varieties, the patent process, and their importance in modern society. A patent is a type of intellectual property (IP) that allows the owner to prevent others from creating, using, selling, or importing an innovation without authorization. The innovation can be a product, process, or improvement that provides a novel and valuable solution to a problem. To be patentable, an invention often meets the criteria of novelty (it is new), non-obviousness (it is not something that anyone skilled in the field might easily deduce), and utility.

Drug/medicinal patents come under formulation patenting. The microalgal products, used in the treatment of diabetes, are formulated with less toxicity for consumption. Such formulations are patented and rights are reserved. There are a huge number of rights and patents claimed on algae and algal products for the treatment of diabetes. Below are listed a few of important, innovative, and recent patents registered and issued by the World Intellectual Property Organization (WIPO) (Table 13.3).

4 Conclusion

By highlighting the anti-diabetic qualities of bioactive algal components, this review shows how beneficial they are both in vitro and in vivo. Because of their abundance of active ingredients, algae extracts have shown promise in the treatment of diabetes and may find new uses in sectors like pharmaceuticals and functional foods. Algal extracts can promote adequate nutrition and health by reversing histological damage

Table 13.3 Patents obtained on microalgae for DM

Patent obtained	Year of publication	Country
FUNCTIONAL PET FEED COMPOSITION USING MICROALGAE HAVING ANTI-OBESITY AND ANTI-DIABETIC FUNCTIONALITY	2019	Republic of Korea
MICROALGAE/BFGF.COF HYDROGEL MICRONEEDLE PATCH AND APPLICATION THEREOF IN PREPARATION OF DIABETIC WOUND TREATMENT SYSTEM	2024	China
COMPANION ANIMAL FEED COMPOSITION FOR PREVENTION OF OBESITY, DIABETES, AND EYE DISEASE COMPRISING MICROALGAE AS ACTIVE COMPONENT	2023	Republic of Korea
COMPOSITION CONTAINING FUCOXANTHIN FOR PREVENTING OR TREATING OBESITY OR DIABETES	2011	Republic of Korea
COMPOSITION INCLUDING SILICON-ENRICHED MICROALGAE FOR THERAPEUTIC USE	2016	France
METHOD OF MICROALGAE SELF-OXYGEN SUPPLY CONDUCTIVE ANTIBACTERIAL MULTIFUNCTIONAL HYDROGEL	2023	China
INHIBITOR OF RISE IN BLOOD GLUCOSE LEVEL, DIABETES INHIBITOR, AND FOOD COMPOSITION	2018	Japan
COMPOSITION COMPRISING FRACTION OF TETRASELMIS SUECICA FOR PREVENTING OR TREATING OBESITY OR DIABETES	2014	Republic of Korea
MARINE ORGANISM TYPE ENTERAL NUTRITION PREPARATION FOR DIABETES PATIENTS, PREPARATION METHOD OF MARINE ORGANISM TYPE ENTERAL NUTRITION PREPARATION AND APPLICATION	2012	China
A COMPOSITION FOR TREATMENT OF DIABETES MELLITUS AND A PREPARATION AND AN USE THEREOF	2009	India
A COMPOSITION FOR TREATMENT OF DIABETES MELLITUS AND A PREPARATION AND USE	2007	Australia
SPIRULINA BEVERAGE	1999	China
PREPARATION OF FREEZE-DRIED SPIRULINA POWDER CONTAINING DIABETES RESISTING FACTOR	2000	China
HEALTH CARE FOOD COMPOSITION COMPRISING SPIRULINA POWDER AND CAIAPO POWDER FOR PROTECTING ADULT DISEASES SUCH AS DIABETES MELLITUS	2003	Republic of Korea
SELF-HELP FLOUR CONTAINING SPIRULINA	2015	China
SPIRULINA FRUIT JAM	1997	China
SELF-HELP RICE CONTAINING SPIRULINA	2015	China
CULTIVATION METHOD OF HYPERGLYCEMIC SPIRULINA	2014	China
SPIRULINA WINE AND PREPARING PROCESS THEREOF	1996	China
GINSENG AND SPIRULINA BREAD FOR REDUCING BLOOD SUGAR	2017	China
LOW SUGAR HIGH PROTEIN SPIRULINA RESIDUE AS WELL AS PREPARATION METHOD AND APPLICATION THEREOF	2023	China

(continued)

Table 13.3 (continued)

Patent obtained	Year of publication	Country
CAPSULE MADE OF ANT EXTRACT, SPIRULINA PLANTESIS DRY POWDER AND CHINESE CATERPILLAR FUNGUS EXTRACT	2006	China
MEDICINE FOR TREATING DIABETES AND ITS PREPARING METHOD.	2005	China
HEALTH FUNCTIONAL FOOD COMPOSITION FOR PREVENTING DIABETES	2018	Republic of Korea
DIABETES PREVENTIVE FOOD	2006	Japan
COMPOSITION FOR TREATING DIABETES AND PREPARATION METHOD OF COMPOSITION	2020	China
A COMPOSITION FOR TREATMENT OF DIABETES MELLITUS AND A PREPARATION AND AN USE THEREOF	2008	China
SOLID DRINK FOR DIABETIC PATIENTS	2016	China
MEDICAMENT FOR TREATING DIABETES AND PREPARATION METHOD OF MEDICAMENT.	2016	China
TRADITIONAL CHINESE MEDICINAL COMPOSITION FOR TREATING DIABETES AND COMPLICATION	2007	China
TRADITIONAL CHINESE MEDICINAL COMPOSITION FOR TREATING HYPERGLYCEMIA AND DIABETES AND PREPARATION METHOD THEREOF	2018	China
METHOD FOR MAKING BLOOD SUGAR ADJUSTING HEALTH FOOD	2016	China
MARINE ORGANISM TYPE ENTERAL NUTRITION PREPARATION FOR DIABETIC PATIENTS' PREPARATION METHOD FOR MARINE ORGANISM TYPE ENTERAL NUTRITION PREPARATION AND APPLICATION	2012	China
CHINESE MEDICINAL PREPARATION FOR TREATING DIABETES	2013	China
COMPOSITION OF HEALTHY BEVERAGE USING COMBINED ORGANIC ACID FERMENTATION AND FAST ABSORBING DISTILLATION METHOD TO HELP ALLEVIATE TYPE 2 DIABETES AND IMPROVE ADRENAL FUNCTION	2019	Republic of Korea
CHITOSAN OLIGOSACCHARIDE AND *DUNALIELLA SALINA* POWDER-CONTAINING DIETARY FORMULA FOR DIABETIC PATIENTS	2015	China
ICE CREAM CONE HAVING HEALTHCARE FUNCTION	2003	China
FORTIFIED FOOD FOR PREVENTING AND TREATING DIABETES	2022	China

associated with diabetes and improving physiological indices. They do poorly on the market even though they are promoted as useful meals and medications. However, algae's significance in medications and functional foods may increase due to consumers' increased interest in natural bioactive compounds.

Natural substances derived from algae have the potential to help control diabetes, but further study is required to fully realize this potential. Unknown algae species and their metabolites may be studied in order to find novel bioactive substances with better anti-diabetic properties (Agarwal et al. 2023; Zhao et al. 2018).

To evaluate the safety, effectiveness, and best use of therapies derived from algae, strong preclinical research and carefully planned clinical trials are necessary, guaranteeing that preclinical findings are successfully translated into clinical practice (Ahmed et al. 2023; Wang et al. 2023). Additional research and advancements in this area could revolutionize diabetes treatment and improve patient outcomes.

References

Abd El-Baky, H. H., El-Baz, F. K., & El-Baroty, G. S. (2009). Natural preservative ingredient from marine alga Ulva lactuca L. *International Journal of Food Science & Technology*, *44*(9), 1688–1695. https://doi.org/10.1111/j.1365-2621.2009.01926.x

Abo-Shady, A. M., Gheda, S. F., Ismail, G. A., Cotas, J., Pereira, L., & Abdel-Karim, O. H. (2023). Antioxidant and Antidiabetic Activity of Algae. *Life (Basel, Switzerland)*, *13*(2), 460. https://doi.org/10.3390/life13020460

Agarwal, S., Singh, V., & Chauhan, K. (2023). Antidiabetic potential of seaweed and their bioactive compounds: A review of developments in last decade. *Critical Reviews in Food Science and Nutrition*, *63*(22), 5739–5770. https://doi.org/10.1080/10408398.2021.2024130

Ahmed, M., Semreen, A. M., El-Huneidi, W., Bustanji, Y., Abu-Gharbieh, E., Alqudah, M. A. Y., Alhusban, A., Shara, M., Abuhelwa, A. Y., Soares, N. C., Semreen, M. H., & Alzoubi, K. H. (2023). Preclinical and Clinical Applications of Metabolomics and Proteomics in Glioblastoma Research. *International Journal of Molecular Sciences*, *24*(1), Article 1. https://doi.org/10.3390/ijms24010348

Akhuemokhan, I. K., Eregie, A., & Fasanmade, O. A. (2013). *Diabetes prevention and management: The role of trace minerals*. *21*(2).

Alam, A., Quamri, S., Fatima, S., Roqaiya, M., & Ahmad, Z. (2016). Efficacy of Spirulina (Tahlab) in Patients of Type 2 Diabetes Mellitus (Ziabetus Shakri)—A Randomized Controlled Trial. *Journal of Diabetes & Metabolism*, *7*(10). https://doi.org/10.4172/2155-6156.1000710

Alam, S., Sarker, M. M. R., Sultana, T. N., Chowdhury, M. N. R., Rashid, M. A., Chaity, N. I., Zhao, C., Xiao, J., Hafez, E. E., Khan, S. A., & Mohamed, I. N. (2022). Antidiabetic Phytochemicals From Medicinal Plants: Prospective Candidates for New Drug Discovery and Development. *Frontiers in Endocrinology*, *13*, 800714. https://doi.org/10.3389/fendo.2022.800714

Alsenani, F., Tupally, K. R., Chua, E. T., Eltanahy, E., Alsufyani, H., Parekh, H. S., & Schenk, P. M. (2020). Evaluation of microalgae and cyanobacteria as potential sources of antimicrobial compounds. *Saudi Pharmaceutical Journal: SPJ: The Official Publication of the Saudi Pharmaceutical Society*, *28*(12), 1834–1841. https://doi.org/10.1016/j.jsps.2020.11.010

Ambati, R. R., Phang, S.-M., Ravi, S., & Aswathanarayana, R. G. (2014). Astaxanthin: Sources, Extraction, Stability, Biological Activities and Its Commercial Applications—A Review. *Marine Drugs*, *12*(1), Article 1. https://doi.org/10.3390/md12010128

Andriopoulos, V., Gkioni, M. D., Koutra, E., Mastropetros, S. G., Lamari, F. N., Hatziantoniou, S., & Kornaros, M. (2022a). Total Phenolic Content, Biomass Composition, and Antioxidant Activity of Selected Marine Microalgal Species with Potential as Aquaculture Feed. *Antioxidants*, *11*(7), Article 7. https://doi.org/10.3390/antiox11071320

Andriopoulos, V., Lamari, F. N., Hatziantoniou, S., & Kornaros, M. (2022b). Production of Antioxidants and High Value Biomass from Nannochloropsis oculata: Effects of pH, Temperature and Light Period in Batch Photobioreactors. *Marine Drugs*, *20*(9), Article 9. https://doi.org/10.3390/md20090552

Anitha, L., & Chandralekha, K. (2010). *Effect of Supplementation of Spirulina on Blood Glucose, Glycosylated Hemoglobin and Lipid Profile of Male Non- Insulin Dependent Diabetics*. https://www.semanticscholar.org/paper/Effect-of-Supplementation-of-Spirulina-on-Blood-and-Anitha-Chandralekha/96993689d499b7ecfc09558638dafb2bcc945990

Anwer, R., Alam, A., Khursheed, S., Kashif, S. Mohd., Kabir, H., & Fatma, T. (2013). Spirulina: Possible pharmacological evaluation for insulin-like protein. *Journal of Applied Phycology*, *25*(3), 883–889. https://doi.org/10.1007/s10811-012-9924-z

Barbosa, M., Inácio, L. G., Afonso, C., & Maranhão, P. (2023). The microalga Dunaliella and its applications: A review. *Applied Phycology*, *4*(1), 99–120. https://doi.org/10.1080/26388081.2023.2222318

Battin, E. E., & Brumaghim, J. L. (2009). Antioxidant activity of sulfur and selenium: A review of reactive oxygen species scavenging, glutathione peroxidase, and metal-binding antioxidant mechanisms. *Cell Biochemistry and Biophysics*, *55*(1), 1–23. https://doi.org/10.1007/s12013-009-9054-7

Beihaghi, M., Ghodrati Azadi, H., Taherzadeh, Z., & Bahrami, H. Reza. (2017). THE EFFECTS OF ORAL ADMINISTRATION OF SPIRULINA PLATENSIS (CULTURED IRANIAN) ON BLOOD GLUCOSE AND GLYCOSYLATED HEMOGLOBIN BLOOD IN TYPE II DIABETES MELLITUS PATIENTS. *Iranian Journal of Diabetes and Metabolism*, *16*(3), 183–190.

Bhattacharyya, D. S., Deep, P., Singh, S., & Nayak, B. (2016). Lichen Secondary Metabolites and Its Biological Activity. *American Journal of Pharmtech Research 2249-3387*, *6*, 28–44.

Bitam, A., & Aissaoui, O. (2020). Chapter 32—Spirulina platensis, oxidative stress, and diabetes. In V. R. Preedy (Ed.), *Diabetes (Second Edition)* (pp. 325–331). Academic Press. https://doi.org/10.1016/B978-0-12-815776-3.00033-4

Bito, T., Okumura, E., Fujishima, M., & Watanabe, F. (2020). Potential of Chlorella as a Dietary Supplement to Promote Human Health. *Nutrients*, *12*(9), 2524. https://doi.org/10.3390/nu12092524

Bohórquez-Medina, S. L., Bohórquez-Medina, A. L., Benites Zapata, V. A., Ignacio-Cconchoy, F. L., Toro-Huamanchumo, C. J., Bendezu-Quispe, G., Pacheco-Mendoza, J., & Hernandez, A. V. (2021). Impact of spirulina supplementation on obesity-related metabolic disorders: A systematic review and meta-analysis of randomized controlled trials. *NFS Journal*, *25*, 21–30. https://doi.org/10.1016/j.nfs.2021.09.003

Brown, M. R., Jeffrey, S. W., Volkman, J. K., & Dunstan, G. A. (1997). Nutritional properties of microalgae for mariculture. *Aquaculture*, *151*(1), 315–331. https://doi.org/10.1016/S0044-8486(96)01501-3

Chen, M. D., Liou, S. J., Lin, P. Y., Yang, V. C., Alexander, P. S., & Lin, W. H. (1998). Effects of zinc supplementation on the plasma glucose level and insulin activity in genetically obese (ob/ob) mice. *Biological Trace Element Research*, *61*(3), 303–311. https://doi.org/10.1007/BF02789090

Cherng, J.-Y., & Shih, M.-F. (2005). Preventing dyslipidemia by *Chlorella pyrenoidosa* in rats and hamsters after chronic high fat diet treatment. *Life Sciences*, *76*(26), 3001–3013. https://doi.org/10.1016/j.lfs.2004.10.055

Cherng, J.-Y., & Shih, M.-F. (2006). Improving glycogenesis in Streptozocin (STZ) diabetic mice after administration of green algae *Chlorella*. *Life Sciences*, *78*(11), 1181–1186. https://doi.org/10.1016/j.lfs.2005.06.050

D'Alessandro, E. B., & Antoniosi Filho, N. R. (2016). Concepts and studies on lipid and pigments of microalgae: A review. *Renewable and Sustainable Energy Reviews*, *58*, 832–841. https://doi.org/10.1016/j.rser.2015.12.162

Davies, M. J., Aroda, V. R., Collins, B. S., Gabbay, R. A., Green, J., Maruthur, N. M., Rosas, S. E., Del Prato, S., Mathieu, C., Mingrone, G., Rossing, P., Tankova, T., Tsapas, A., & Buse, J. B. (2022). Management of Hyperglycemia in Type 2 Diabetes, 2022. A Consensus Report by the American Diabetes Association (ADA) and the European Association for the Study of Diabetes (EASD). *Diabetes Care*, *45*(11), 2753–2786. https://doi.org/10.2337/dci22-0034

De Angelis, C., Drazen, J. M., Frizelle, F. A., Haug, C., Hoey, J., Horton, R., Kotzin, S., Laine, C., Marusic, A., Overbeke, A. J. P. M., Schroeder, T. V., Sox, H. C., Van Der Weyden, M. B., & International Committee of Medical Journal Editors. (2004). Clinical trial registration: A statement from the International Committee of Medical Journal Editors. *The New England Journal of Medicine*, *351*(12), 1250–1251. https://doi.org/10.1056/NEJMe048225

Ebrahimi-Mameghani, M., Sadeghi, Z., Abbasalizad Farhangi, M., Vaghef-Mehrabany, E., & Aliashrafi, S. (2017). Glucose homeostasis, insulin resistance and inflammatory biomarkers in patients with non-alcoholic fatty liver disease: Beneficial effects of supplementation with microalgae *Chlorella vulgaris*: A double-blind placebo-controlled randomized clinical trial. *Clinical Nutrition*, *36*(4), 1001–1006. https://doi.org/10.1016/j.clnu.2016.07.004

El-Baz, F. K., Aly, H. F., & Salama, A. A. A. (2019). Toxicity assessment of the green Dunaliella salina microalgae. *Toxicology Reports*, *6*, 850–861. https://doi.org/10.1016/j.toxrep.2019.08.003

El-Baz, F. K., Khalil, W. K. B., Booles, H. F., Aly, H. F., & Ali, G. H. (2016). *Dunaliella salina Suppress Oxidative Stress, Alterations in the Expression of Pro-Apoptosis and Inflammation Related Genes Induced by STZ in Diabetic Rats. 40.*

El-Baz, F. K., Salama, A., & Salama, R. A. A. (2020). Dunaliella salina Attenuates Diabetic Neuropathy Induced by STZ in Rats: Involvement of Thioredoxin. *BioMed Research International*, *2020*, 1295492. https://doi.org/10.1155/2020/1295492

ElFar, O. A., Billa, N., Lim, H. R., Chew, K. W., Cheah, W. Y., Munawaroh, H. S. H., Balakrishnan, D., & Show, P. L. (2022). Advances in delivery methods of Arthrospira platensis (spirulina) for enhanced therapeutic outcomes. *Bioengineered*, *13*(6), 14681–14718. https://doi.org/10.1080/21655979.2022.2100863

Fereidouni, A., Khaleghian, A., Mousavi-Niri, N., & Moradikor, N. (2022). The effects of supplementation of Nannochloropsis oculata microalgae on biochemical, inflammatory and antioxidant responses in diabetic rats. *Biomolecular Concepts*, *13*(1), 314–321. https://doi.org/10.1515/bmc-2022-0025

Figueroa-Romero, C., Sadidi, M., & Feldman, E. L. (2008). Mechanisms of disease: The oxidative stress theory of diabetic neuropathy. *Reviews in Endocrine & Metabolic Disorders*, *9*(4), 301–314. https://doi.org/10.1007/s11154-008-9104-2

Grisold, A., Callaghan, B. C., & Feldman, E. L. (2017). Mediators of diabetic neuropathy: Is hyperglycemia the only culprit? *Current Opinion in Endocrinology, Diabetes, and Obesity*, *24*(2), 103–111. https://doi.org/10.1097/MED.0000000000000320

Hartung, D. M., Zarin, D. A., Guise, J.-M., McDonagh, M., Paynter, R., & Helfand, M. (2014). Reporting discrepancies between the ClinicalTrials.gov results database and peer-reviewed publications. *Annals of Internal Medicine*, *160*(7), 477–483. https://doi.org/10.7326/M13-0480

Hatami, E., Ghalishourani, S.-S., Najafgholizadeh, A., Pourmasoumi, M., Hadi, A., Clark, C. C. T., Assaroudi, M., Salehi-sahlabadi, A., Joukar, F., & Mansour-Ghanaei, F. (2021a). The effect of spirulina on type 2 diabetes: A systematic review and meta-analysis. *Journal of Diabetes and Metabolic Disorders*, *20*(1), 883. https://doi.org/10.1007/s40200-021-00760-z

Hatami, E., Ghalishourani, S.-S., Najafgholizadeh, A., Pourmasoumi, M., Hadi, A., Clark, C. C. T., Assaroudi, M., Salehi-sahlabadi, A., Joukar, F., & Mansour-Ghanaei, F. (2021b). The effect of spirulina on type 2 diabetes: A systematic review and meta-analysis. *Journal of Diabetes and Metabolic Disorders*, *20*(1), 883–892. https://doi.org/10.1007/s40200-021-00760-z

Hernández-Corona, A., Nieves, I., Meckes, M., Chamorro, G., & Barron, B. L. (2002). Antiviral activity of Spirulina maxima against herpes simplex virus type 2. *Antiviral Research*, *56*(3), 279–285. https://doi.org/10.1016/s0166-3542(02)00132-8

Holmgren, A., Johansson, C., Berndt, C., Lönn, M. E., Hudemann, C., & Lillig, C. H. (2005). Thiol redox control via thioredoxin and glutaredoxin systems. *Biochemical Society Transactions*, *33*(Pt 6), 1375–1377. https://doi.org/10.1042/BST0331375

Horii, N., Hasegawa, N., Fujie, S., Uchida, M., Iemitsu, K., Inoue, K., & Iemitsu, M. (2019). Effect of combination of chlorella intake and aerobic exercise training on glycemic control in type 2 diabetic rats. *Nutrition (Burbank, Los Angeles County, Calif.)*, *63–64*, 45–50. https://doi.org/10.1016/j.nut.2019.01.008

Hosseini, A. M., Keshavarz, S. A., Nasli-Esfahani, E., Amiri, F., & Janani, L. (2021). The effects of Chlorella supplementation on glycemic control, lipid profile and anthropometric measures on patients with type 2 diabetes mellitus. *European Journal of Nutrition*, *60*(6), 3131–3141. https://doi.org/10.1007/s00394-021-02492-5

Huang, Z., Chen, J., Wang, C., Xiao, M., Zhu, Y., Li, N., Huang, Z., Liu, B., & Huang, Y. (2023). Antidiabetic potential of *Chlorella pyrenoidosa* functional formulations in streptozocin-induced type 2 diabetic mice. *Journal of Functional Foods*, *103*, 105489. https://doi.org/10.1016/j.jff.2023.105489

Hughey, C. C., & Crawford, P. A. (2019). Pyruvate Carboxylase Wields a Double-Edged Metabolic Sword. *Cell Metabolism*, *29*(6), 1236–1238. https://doi.org/10.1016/j.cmet.2019.05.013

Itakura, H., Kobayashi, M., & Nakamura, S. (2015). Chlorella ingestion suppresses resistin gene expression in peripheral blood cells of borderline diabetics. *Clinical Nutrition ESPEN*, *10*(3), e95–e101. https://doi.org/10.1016/j.clnesp.2015.04.002

Jeong, H., Kwon, H. J., & Kim, M. K. (2009). Hypoglycemic effect of Chlorella vulgaris intake in type 2 diabetic Goto-Kakizaki and normal Wistar rats. *Nutrition Research and Practice*, *3*(1), 23. https://doi.org/10.4162/nrp.2009.3.1.23

Jin, Y., & Arroo, R. (2023). The protective effects of flavonoids and carotenoids against diabetic complications-A review of in vivo evidence. *Frontiers in Nutrition*, *10*, 1020950. https://doi.org/10.3389/fnut.2023.1020950

Jong-Yuh, C., & Mei-Fen, S. (2005). Potential hypoglycemic effects of *Chlorella* in streptozotocin-induced diabetic mice. *Life Sciences*, *77*(9), 980–990. https://doi.org/10.1016/j.lfs.2004.12.036

Jörns, A., Tiedge, M., Lenzen, S., & Munday, R. (1999). Effect of superoxide dismutase, catalase, chelating agents, and free radical scavengers on the toxicity of alloxan to isolated pancreatic islets in vitro. *Free Radical Biology & Medicine*, *26*(9–10), 1300–1304. https://doi.org/10.1016/s0891-5849(98)00325-6

Joventino, I. P., Alves, H. G. R., Neves, L. C., Pinheiro-Joventino, F., Leal, L. K. A. M., Neves, S. A., Ferreira, F. V., Brito, G. A. C., & Viana, G. B. (2012). The Microalga Spirulina platensis Presents Anti-inflammatory Action as well as Hypoglycemic and Hypolipidemic Properties in Diabetic Rats. *Journal of Complementary and Integrative Medicine*, *9*(1). https://doi.org/10.1515/1553-3840.1534

Jubaidi, F. F., Zainalabidin, S., Taib, I. S., Hamid, Z. A., & Budin, S. B. (2021). The Potential Role of Flavonoids in Ameliorating Diabetic Cardiomyopathy via Alleviation of Cardiac 0Oxidative Stress, Inflammation and Apoptosis. *International Journal of Molecular Sciences*, *22*(10), Article 10. https://doi.org/10.3390/ijms22105094

Kaser, S., Kaser, A., Sandhofer, A., Ebenbichler, C. F., Tilg, H., & Patsch, J. R. (2003). Resistin messenger-RNA expression is increased by proinflammatory cytokines in vitro. *Biochemical and Biophysical Research Communications*, *309*(2), 286–290. https://doi.org/10.1016/j.bbrc.2003.07.003

Kaur, K., Sachdeva, R., & Grover, K. (2008). *EFFECT OF SUPPLEMENTATION OF SPIRULINA ON BLOOD GLUCOSE AND LIPID PROFILE OF THE NON-INSULIN DEPENDENT DIABETIC MALE SUBJECTS*. *27*(3).

Kaur, S., Pandhi, P., & Dutta, P. (2011). Painful diabetic neuropathy: An update. *Annals of Neurosciences*, *18*(4), 168–175. https://doi.org/10.5214/ans.0972-7531.1118409

Khaligh, S. F., & Asoodeh, A. (2022). Recent advances in the bio-application of microalgae-derived biochemical metabolites and development trends of photobioreactor-based culture systems. *3 Biotech*, *12*(10), 260. https://doi.org/10.1007/s13205-022-03327-8

Kim, S. H., Lee, U. H., Lee, S. B., Jeong, G.-T., & Kim, S.-K. (2021). *Improvement of Unsaturated Fatty Acid Production from Porphyridium cruentum Using a Two-Phase Culture System in a*

Photobioreactor with Light-Emitting Diodes (LEDs). *31*(3), 456–463. https://doi.org/10.4014/jmb.2011.11004

Kim, W. Y., & Park, J. Y. (2016). The Effect of Spirulina on Lipid Metabolism, Antioxidant Gapacity and Immune Function in Korean Elderlies. *Korean Journal of Nutrition*, *36*(3), 287–297.

Kosmalski, M., Pękala-Wojciechowska, A., Sut, A., Pietras, T., & Luzak, B. (2022). Dietary Intake of Polyphenols or Polyunsaturated Fatty Acids and Its Relationship with Metabolic and Inflammatory State in Patients with Type 2 Diabetes Mellitus. *Nutrients*, *14*(5), Article 5. https://doi.org/10.3390/nu14051083

Kosmalski, M., Śliwińska, A., & Drzewoski, J. (2023). Non-Alcoholic Fatty Liver Disease or Type 2 Diabetes Mellitus—The Chicken or the Egg Dilemma. *Biomedicines*, *11*(4), Article 4. https://doi.org/10.3390/biomedicines11041097

Kreilmeier-Berger, T., Zeugswetter, F. K., Blohm, K.-O., Schwendenwein, I., Baszler, E., Ploderer, B., Burgener, I. A., & Künzel, F. (2021). Successful Insulin Glargine Treatment in Two Pet Guinea Pigs with Suspected Type 1 Diabetes Mellitus. *Animals*, *11*(4), Article 4. https://doi.org/10.3390/ani11041025

Lankatillake, C., Huynh, T., & Dias, D. A. (2019). Understanding glycaemic control and current approaches for screening antidiabetic natural products from evidence-based medicinal plants. *Plant Methods*, *15*, 105. https://doi.org/10.1186/s13007-019-0487-8

Leong, Y. K., & Chang, J.-S. (2020). Bioremediation of heavy metals using microalgae: Recent advances and mechanisms. *Bioresource Technology*, *303*, 122886. https://doi.org/10.1016/j.biortech.2020.122886

Levy, Y., Zaltsberg, H., Ben-Amotz, A., Kanter, Y., & Aviram, M. (2000). Dietary supplementation of a natural isomer mixture of beta-carotene inhibits oxidation of LDL derived from patients with diabetes mellitus. *Annals of Nutrition & Metabolism*, *44*(2), 54–60. https://doi.org/10.1159/000012821

Li, K., Ferguson, T., Embil, J., Rigatto, C., Komenda, P., & Tangri, N. (2021). Risk of Kidney Failure, Death, and Cardiovascular Events After Lower Limb Complications in Patients With CKD. *Kidney International Reports*, *6*(2), 381–388. https://doi.org/10.1016/j.ekir.2020.11.010

Lin, C.-Y., Huang, P.-J., & Chao, C.-Y. (2014). Chlorella protects against hydrogen peroxide-induced pancreatic β-cell damage. *Journal of Medicinal Food*, *17*(12), 1273–1280. https://doi.org/10.1089/jmf.2013.3002

Liu, Y., Zeng, S., Ji, W., Yao, H., Lin, L., Cui, H., Santos, H. A., & Pan, G. (2022). Emerging Theranostic Nanomaterials in Diabetes and Its Complications. *Advanced Science*, *9*(3), 2102466. https://doi.org/10.1002/advs.202102466

Long, J., Liang, R., Zheng, Q., Yuan, G., Xin, Z., Chen, X., Lai, F., & Liu, Y. (2021). Overview of Clinical Trials on Type 2 Diabetes Mellitus: A Comprehensive Analysis of the ClinicalTrials.gov Database. *Diabetes, Metabolic Syndrome and Obesity: Targets and Therapy*, *14*, 367. https://doi.org/10.2147/DMSO.S288065

Lunn, J., & Buttriss, J. L. (2007). Carbohydrates and dietary fibre. *Nutrition Bulletin*, *32*(1), 21–64. https://doi.org/10.1111/j.1467-3010.2007.00616.x

Magwaza, S. N., & Islam, M. S. (2023). Roles of Marine Macroalgae or Seaweeds and Their Bioactive Compounds in Combating Overweight, Obesity and Diabetes: A Comprehensive Review. *Marine Drugs*, *21*(4), 258. https://doi.org/10.3390/md21040258

Mahfouz, M. H., Emara, I. A., Shouman, M. S., & Ezz, M. K. (2009). Asymmetrical dimethylarginine (ADMA) and nitric oxide as potential cardiovascular risk factors in type 2 diabetes mellitus. *African Journal of Biochemistry Research*, *3*(8), 293–301. https://doi.org/10.5897/AJBR.9000040

Makki, K., Froguel, P., & Wolowczuk, I. (2013). Adipose tissue in obesity-related inflammation and insulin resistance: Cells, cytokines, and chemokines. *ISRN Inflammation*, *2013*, 139239. :// doi.org/https://doi.org/10.1155/2013/139239

Mani, U. V., Desai, S., & Iyer, U. (2000). Studies on the Long-Term Effect of Spirulina Supplementation on Serum Lipid Profile and Glycated Proteins in NIDDM Patients. *Journal*

of Nutraceuticals, Functional & Medical Foods, *2*(3), 25–32. https://doi.org/10.1300/J133v02n03_03
Menaa, F., Wijesinghe, P. a. U. I., Thiripuranathar, G., Uzair, B., Iqbal, H., Khan, B. A., & Menaa, B. (2020). Ecological and Industrial Implications of Dynamic Seaweed-Associated Microbiota Interactions. *Marine Drugs*, *18*(12), 641. https://doi.org/10.3390/md18120641
Menaa, F., Wijesinghe, U., Thiripuranathar, G., Althobaiti, N. A., Albalawi, A. E., Khan, B. A., & Menaa, B. (2021). Marine Algae-Derived Bioactive Compounds: A New Wave of Nanodrugs? *Marine Drugs*, *19*(9), Article 9. https://doi.org/10.3390/md19090484
Morrison, J. A., Pike, L. A., Sams, S. B., Sharma, V., Zhou, Q., Severson, J. J., Tan, A.-C., Wood, W. M., & Haugen, B. R. (2014). Thioredoxin interacting protein (TXNIP) is a novel tumor suppressor in thyroid cancer. *Molecular Cancer*, *13*, 62. https://doi.org/10.1186/1476-4598-13-62
Munawaroh, H. S. H., Hazmatulhaq, F., Gumilar, G. G., Pratiwi, R. N., Kurniawan, I., Ningrum, A., Hidayati, N. A., Koyande, A. K., Kumar, P. S., & Show, P.-L. (2022). Microalgae as a potential sustainable solution to environment health. *Chemosphere*, *295*, 133740. https://doi.org/10.1016/j.chemosphere.2022.133740
Nacer, W., Baba Ahmed, F. Z., Merzouk, H., Benyagoub, O., & Bouanane, S. (2020). Evaluation of the anti-inflammatory and antioxidant effects of the microalgae Nannochloropsis gaditana in streptozotocin-induced diabetic rats. *Journal of Diabetes & Metabolic Disorders*, *19*(2), 1483–1490. https://doi.org/10.1007/s40200-020-00681-3
Nagaraj, K., Lapkina-Gendler, L., Sarfstein, R., Gurwitz, D., Pasmanik-Chor, M., Laron, Z., Yakar, S., & Werner, H. (2018). Identification of thioredoxin-interacting protein (TXNIP) as a downstream target for IGF1 action. *Proceedings of the National Academy of Sciences of the United States of America*, *115*(5), 1045–1050. https://doi.org/10.1073/pnas.1715930115
Namitha, D., Nusrath, A., Rajeshwari, A., & Rani, N. A. (2015). Serum Paraoxonase Levels in Type 2 Diabetes Mellitus Patients: A Case-Control Study. *Journal of Medical Sciences and Health*, *1*(2), 14–18. https://doi.org/10.46347/jmsh.2015.v01i02.003
Nasirian, F., Dadkhah, M., Moradi-Kor, N., & Obeidavi, Z. (2018). Effects of Spirulina platensis microalgae on antioxidant and anti-inflammatory factors in diabetic rats. *Diabetes, Metabolic Syndrome and Obesity: Targets and Therapy*, *11*, 375–380. https://doi.org/10.2147/DMSO.S172104
Nazem, M. R., Asadi, M., Adelipour, M., Jabbari, N., & Allameh, A. (2023). Zinc supplementation ameliorates type 2 diabetes markers through the enhancement of total antioxidant capacity in overweight patients. *Postgraduate Medical Journal*, *99*(1174), 862–867. https://doi.org/10.1136/postgradmedj-2021-140878
Nowotny, K., Jung, T., Höhn, A., Weber, D., & Grune, T. (2015). Advanced Glycation End Products and Oxidative Stress in Type 2 Diabetes Mellitus. *Biomolecules*, *5*(1), Article 1. https://doi.org/10.3390/biom5010194
Ovando, C. A., Carvalho, J. C. de, Vinícius de Melo Pereira, G., Jacques, P., Soccol, V. T., & Soccol, C. R. (2018). Functional properties and health benefits of bioactive peptides derived from Spirulina: A review. *Food Reviews International*, *34*(1), 34–51. https://doi.org/10.1080/87559129.2016.1210632
Pais, R., Conde, T., Neves, B. B., Pinho, M., Coelho, M., Pereira, H., Rodrigues, A. M. C., Domingues, P., Gomes, A. M., Urbatzka, R., Domingues, R., & Melo, T. (2024). Bioactive Lipids in Dunaliella salina: Implications for Functional Foods and Health. *Foods*, *13*(20), Article 20. https://doi.org/10.3390/foods13203321
Pangestuti, R., & Kim, S.-K. (2011). Biological activities and health benefit effects of natural pigments derived from marine algae. *Journal of Functional Foods*, *3*(4), 255–266. https://doi.org/10.1016/j.jff.2011.07.001
Parikh, P., Mani, U., & Iyer, U. (2001). Role of Spirulina in the Control of Glycemia and Lipidemia in Type 2 Diabetes Mellitus. *Journal of Medicinal Food*, *4*(4), 193–199. https://doi.org/10.1089/10966200152744463
Park, H. J., Lee, Y. J., Ryu, H. K., Kim, M. H., Chung, H. W., & Kim, W. Y. (2008). A randomized double-blind, placebo-controlled study to establish the effects of spirulina in elderly Koreans. *Annals of Nutrition & Metabolism*, *52*(4), 322–328. https://doi.org/10.1159/000151486

Perego, C., Da Dalt, L., Pirillo, A., Galli, A., Catapano, A. L., & Norata, G. D. (2019). Cholesterol metabolism, pancreatic β-cell function and diabetes. *Biochimica et Biophysica Acta (BBA) - Molecular Basis of Disease*, *1865*(9), 2149–2156. https://doi.org/10.1016/j.bbadis.2019.04.012

Pereira, L., & Valado, A. (2023). Algae-Derived Natural Products in Diabetes and Its Complications—Current Advances and Future Prospects. *Life*, *13*(9), 1831. https://doi.org/10.3390/life13091831

Poh, R., & Muniandy, S. (2010). Paraoxonase 1 activity as a predictor of cardiovascular disease in type 2 diabetes. *The Southeast Asian Journal of Tropical Medicine and Public Health*, *41*(5), 1231–1246.

Qiu, Y., Gao, X., Chen, R., Lu, S., Wan, X., Farag, M. A., & Zhao, C. (2022). Metabolomics and biochemical insights on the regulation of aging-related diabetes by a low-molecular-weight polysaccharide from green microalga *Chlorella pyrenoidosa*. *Food Chemistry: X*, *14*, 100316. https://doi.org/10.1016/j.fochx.2022.100316

Quitério, E., Soares, C., Ferraz, R., Delerue-Matos, C., & Grosso, C. (2021). Marine Health-Promoting Compounds: Recent Trends for Their Characterization and Human Applications. *Foods*, *10*(12), Article 12. https://doi.org/10.3390/foods10123100

Ramamoorthy, A., & Premakumari, S. (1996). Effect of supplementation of Spirulina on hypercholesterolemic patients. *Journal of Food Science and Technology-Mysore*. https://www.semanticscholar.org/paper/Effect-of-supplementation-of-Spirulina-on-patients-Ramamoorthy-Premakumari/2a8602c4713154e25e7ff9ad9a4e7a3f687f0ab3

Ramos-Romero, S., Torrella, J. R., Pagès, T., Viscor, G., & Torres, J. L. (2021). Edible Microalgae and Their Bioactive Compounds in the Prevention and Treatment of Metabolic Alterations. *Nutrients*, *13*(2), Article 2. https://doi.org/10.3390/nu13020563

Robertson, R. P., & Harmon, J. S. (2006). Diabetes, glucose toxicity, and oxidative stress: A case of double jeopardy for the pancreatic islet β cell. *Free Radical Biology and Medicine*, *41*(2), 177–184. https://doi.org/10.1016/j.freeradbiomed.2005.04.030

Rodríguez-López, M., & López-Quijada, C. (1971). Plasma-glucose and plasma-insulin in normal and alloxanized rats treated with chlorella<math><mtext>chlorella</mtext></math>. *Life Sciences*, *10*(1, Part 2), 57–60. https://doi.org/10.1016/0024-3205(71)90225-6

Rupérez, F. J., García-Martínez, D., Baena, B., Maeso, N., Cifuentes, A., Barbas, C., & Herrera, E. (2008). Evolution of oxidative stress parameters and response to oral vitamins E and C in streptozotocin-induced diabetic rats. *Journal of Pharmacy and Pharmacology*, *60*(7), 871–878. https://doi.org/10.1211/jpp.60.7.0008

Ruperez, F. J., Garcia-Martinez, D., Baena, B., Maeso, N., Vallejo, M., Angulo, S., Garcia, A., Ibañez, E., Señorans, F. J., Cifuentes, A., & Barbas, C. (2009). *Dunaliella salina* extract effect on diabetic rats: Metabolic fingerprinting and target metabolite analysis. *Journal of Pharmaceutical and Biomedical Analysis*, *49*(3), 786–792. https://doi.org/10.1016/j.jpba.2008.12.041

Sadek, K. M., Lebda, M. A., Nasr, S. M., & Shoukry, M. (2017). *Spirulina platensis* prevents hyperglycemia in rats by modulating gluconeogenesis and apoptosis via modification of oxidative stress and MAPK-pathways. *Biomedicine & Pharmacotherapy*, *92*, 1085–1094. https://doi.org/10.1016/j.biopha.2017.06.023

Sakayanathan, P., Loganathan, C., Iruthayaraj, A., Periyasamy, P., Poomani, K., Periasamy, V., & Thayumanavan, P. (2018). Biological interaction of newly synthesized astaxanthin-s-allyl cysteine biconjugate with *Saccharomyces cerevisiae* and mammalian α-glucosidase: *In vitro* kinetics and *in silico* docking analysis. *International Journal of Biological Macromolecules*, *118*, 252–262. https://doi.org/10.1016/j.ijbiomac.2018.06.027

Salehi, B., Ata, A., V. Anil Kumar, N., Sharopov, F., Ramírez-Alarcón, K., Ruiz-Ortega, A., Abdulmajid Ayatollahi, S., Valere Tsouh Fokou, P., Kobarfard, F., Amiruddin Zakaria, Z., Iriti, M., Taheri, Y., Martorell, M., Sureda, A., N. Setzer, W., Durazzo, A., Lucarini, M., Santini, A., Capasso, R., … Sharifi-Rad, J. (2019). Antidiabetic Potential of Medicinal Plants and Their Active Components. *Biomolecules*, *9*(10), Article 10. https://doi.org/10.3390/biom9100551

Sapra, A., & Bhandari, P. (2024). Diabetes. In *StatPearls*. StatPearls Publishing. http://www.ncbi.nlm.nih.gov/books/NBK551501/

Setyaningsih, I., Prasetyo, H., Agungpriyono, D. R., & Tarman, K. (2020). Antihyperglycemic activity of *Porphyridium cruentum* biomass and extra-cellular polysaccharide in streptozotocin-induced diabetic rats. *International Journal of Biological Macromolecules*, *156*, 1381–1386. https://doi.org/10.1016/j.ijbiomac.2019.11.178

Shaman, A. A., Zidan, N. S., Alzahrani, S., AlBishi, L. A., Sakran, M. I., & Almutairi, F. M. (2024). Anti-diabetic Activity of Spirulina and Chlorella in In vivo Experimental Rats. *Biomedical and Pharmacology Journal*, *17*(2), 903–913.

Shils, M. E. Maurice E., Olson, J. A., & Shike, M. (with Internet Archive). (1994). *Modern nutrition in health and disease*. Philadelphia : Lea & Febiger. http://archive.org/details/modernnutritioni00shil

Ślusarczyk, J., Adamska, E., & Czerwik-Marcinkowska, J. (2021). Fungi and Algae as Sources of Medicinal and Other Biologically Active Compounds: A Review. *Nutrients*, *13*(9), Article 9. https://doi.org/10.3390/nu13093178

Stumvoll, M., Goldstein, B. J., & van Haeften, T. W. (2005). Type 2 diabetes: Principles of pathogenesis and therapy. *Lancet (London, England)*, *365*(9467), 1333–1346. https://doi.org/10.1016/S0140-6736(05)61032-X

Tamel Selvan, K., Goon, J. A., Makpol, S., & Tan, J. K. (2023). Therapeutic Potentials of Microalgae and Their Bioactive Compounds on Diabetes Mellitus. *Marine Drugs*, *21*(9), Article 9. https://doi.org/10.3390/md21090462

Uchiyama, K., Naito, Y., Hasegawa, G., Nakamura, N., Takahashi, J., & Yoshikawa, T. (2002). Astaxanthin protects β-cells against glucose toxicity in diabetic db/db mice. *Redox Report*, *7*(5), 290–293. https://doi.org/10.1179/135100002125000811

Unger, R. H., & Orci, L. (2010). Paracrinology of islets and the paracrinopathy of diabetes. *Proceedings of the National Academy of Sciences of the United States of America*, *107*(37), 16009–16012. https://doi.org/10.1073/pnas.1006639107

van den Driessche, J. J., Plat, J., Konings, M. C. J. M., & Mensink, R. P. (2020). Effects of spirulina and wakame consumption on intestinal cholesterol absorption and serum lipid concentrations in non-hypercholesterolemic adult men and women. *European Journal of Nutrition*, *59*(5), 2229–2236. https://doi.org/10.1007/s00394-019-02073-7

Vecina, J. F., Oliveira, A. G., Araujo, T. G., Baggio, S. R., Torello, C. O., Saad, M. J. A., & Queiroz, M. L. de S. (2014). Chlorella modulates insulin signaling pathway and prevents high-fat diet-induced insulin resistance in mice. *Life Sciences*, *95*(1), 45–52. https://doi.org/10.1016/j.lfs.2013.11.020

Viktorínová, A., Toserová, E., Krizko, M., & Duracková, Z. (2009). Altered metabolism of copper, zinc, and magnesium is associated with increased levels of glycated hemoglobin in patients with diabetes mellitus. *Metabolism: Clinical and Experimental*, *58*(10), 1477–1482. https://doi.org/10.1016/j.metabol.2009.04.035

Wan, X.-Z., Li, T.-T., Zhong, R.-T., Chen, H.-B., Xia, X., Gao, L.-Y., Gao, X.-X., Liu, B., Zhang, H.-Y., & Zhao, C. (2019). Anti-diabetic activity of PUFAs-rich extracts of Chlorella pyrenoidosa and Spirulina platensis in rats. *Food and Chemical Toxicology: An International Journal Published for the British Industrial Biological Research Association*, *128*, 233–239. https://doi.org/10.1016/j.fct.2019.04.017

Wang, J., Hu, X., Chen, J., Wang, T., Huang, X., & Chen, G. (2022). The Extraction of β-Carotene from Microalgae for Testing Their Health Benefits. *Foods*, *11*(4), Article 4. https://doi.org/10.3390/foods11040502

Wang, X., Chan, Y. S., Wong, K., Yoshitake, R., Sadava, D., Synold, T. W., Frankel, P., Twardowski, P. W., Lau, C., & Chen, S. (2023). Mechanism-Driven and Clinically Focused Development of Botanical Foods as Multitarget Anticancer Medicine: Collective Perspectives and Insights from Preclinical Studies, IND Applications and Early-Phase Clinical Trials. *Cancers*, *15*(3), 701. https://doi.org/10.3390/cancers15030701

Wells, M. L., Potin, P., Craigie, J. S., Raven, J. A., Merchant, S. S., Helliwell, K. E., Smith, A. G., Camire, M. E., & Brawley, S. H. (2017). Algae as nutritional and functional food sources: Revisiting our understanding. *Journal of Applied Phycology*, *29*(2), 949–982. https://doi.org/10.1007/s10811-016-0974-5

Williams, D. M., Jones, H., & Stephens, J. W. (2022). Personalized Type 2 Diabetes Management: An Update on Recent Advances and Recommendations. *Diabetes, Metabolic Syndrome and Obesity: Targets and Therapy*, *15*, 281–295. https://doi.org/10.2147/DMSO.S331654

Zhao, C., Wu, Y., Yang, C., Liu, B., & Huang, Y. (2015). Hypotensive, hypoglycaemic and hypolipidaemic effects of bioactive compounds from microalgae and marine micro-organisms. *International Journal of Food Science & Technology*, *50*(8), 1705–1717. https://doi.org/10.1111/ijfs.12860

Zhao, C., Yang, C., Liu, B., Lin, L., Sarker, S. D., Nahar, L., Yu, H., Cao, H., & Xiao, J. (2018). Bioactive compounds from marine macroalgae and their hypoglycemic benefits. *Trends in Food Science & Technology*, *72*, 1–12. https://doi.org/10.1016/j.tifs.2017.12.001

Zhou, J., & Chen, B. (2023). Retinal Cell Damage in Diabetic Retinopathy. *Cells*, *12*(9), Article 9. https://doi.org/10.3390/cells12091342

Zhuang, D., He, N., Khoo, K. S., Ng, E.-P., Chew, K. W., & Ling, T. C. (2022). Application progress of bioactive compounds in microalgae on pharmaceutical and cosmetics. *Chemosphere*, *291*, 132932. https://doi.org/10.1016/j.chemosphere.2021.132932

Chapter 14
Biomaterial Composites with Algal Polysaccharides for the Treatment of Diabetes Foot Ulcers

Jayachandran Venkatesan, Neema Kumari, and Kranti Kiran Reddy Ealla

Abstract Alginate is an anionic polysaccharide and is extensively used for drug delivery applications due to its biodegradable, biocompatible, and drug-loading capacity. Alginate can be modified into any form, such as hydrogel, scaffold, nanofibers, 3D printed scaffold, microspheres, etc. Due to the tuning properties of alginate, its application is extended to various applications, including diabetic foot ulcer (DFU) treatment. The current book chapter discusses alginate and its role in DFU. Alginate with chitosan makes polyelectrolyte complexes, which help make rigid composite scaffolds. The chapter provides the overview of the studies where various molecules such as drugs growth factors, proteins, peptides, metal ions, and exosomes were loaded into the composite and evaluated for DFU. Further the chapter discusses the alginate dressings in the ongoing clinical trials. With promising results seen in this field, alginate-based dressing materials has potential to treat DFU.

Keywords Alginate · Chitosan · Diabetic Foot Ulcer · Growth Factors · Metal Nanoparticles

---

J. Venkatesan (✉)
School of Dental Sciences and Technology, Department of Biomaterials and Transplant Sciences, Malla Reddy Vishwavidyapeeth, Hyderabad, Telangana, India

N. Kumari
Department of Microbiology, Malla Reddy Institute of Medical Sciences, Malla Reddy Vishwavidyapeeth, Hyderabad, India

K. K. R. Ealla
Department of Oral and Maxillofacial Pathology, Malla Reddy Institute of Dental Sciences, Malla Reddy Vishwavidyapeeth, Hyderabad, India

R. Raja et al. (eds.), *Algae in Diabetes Management*,
https://doi.org/10.1007/978-981-95-0571-5_14

1 Introduction

Chronic wounds are a serious problem for adults and children and significantly affect the quality of life. Proper treatment usually requires diagnosis, identification, and management of wounds (Hoversten et al. 2020). Common types of wounds in elderly patients are ulcers, including vascular ulcers, pressure ulcers, and neuropathic ulcers. Neuropathic ulcers commonly occur from diabetes and other disease conditions. Diabetes patients often have an infection in their wounds even though all the parameters in the bodies are normal condition. Wound healing often follows the coagulation, inflammation, proliferation, and remodeling process to heal the wounds. Various products are often used to treat wounds, such as film, membranes, hydrogels, hydrocolloids, calcium alginate, foam forms, charcoal, and collagenase products for wound care conditions (Martin and Hawkins 2013). Apart from these regular wound care products, other materials, including cotton wool, synthetic bandages, lint, and gauze, are used to treat wounds (Bharambe et al. 2013). Traditional Wound Healing Agents (Topical Pharmaceutical Formulation, traditional dressing); hydrocolloid dressings, alginate dressing, hydrogel dressings, semi-permeable adhesive film dressings, foam dressing, biological dressings, and tissue engineering skin graft substitutes are widely used in wound care management (Thomas 2013). These dressings often combine with polymeric substances, adjuvants, drugs, and other growth factors (Narayana Reddy Karri et al. 2015). A defect or interruption in the epidermis or skin is classified as a wound. Further wounds can be classified, depending on the nature of the wound, acute or chronic wounds (Borda et al. 2016). In addition, antibiotic resistance, inflammation, and allergies cause delayed wounds (Macha and Ben-Nissan 2022).

2 Diabetes and Diabetic Foot Ulcers

Diabetes mellitus is common in the Indian population, and it is observed in various age groups of people. Diabetes is a serious problem when it is combined with co-morbidities. Diabetic foot ulcer (DFU) is one of the complications commonly observed in people, which severely affects the quality of patients' life. These problems are worsening in patients in developing countries and underdeveloped countries. In worst cases, amputation becomes one of the final solutions for the patient (Armstrong et al. 2023; Wang et al. 2022; Chang and Nguyen 2021). Several treatment options are available for DFU, such as debridement, hyperbaric oxygen therapy, offloading, routine check-ups with clinicians, etc. Several researchers are studying the role of growth factors in treating DFU, including platelet-derived growth factor (PDGF), vascular endothelial growth factor (VEGF), fibroblast growth factor (FGF), epidermal growth factor (EGF), keratinocyte growth factor (KGF), and transforming growth factor (TGF) (Rao et al. 2020). Delays in DFU might be due to the presence of microbial community. Different kinds of microbes make polymicrobial film very difficult to treat with a single drug system. In addition, antimicrobial resistance still worsens the treatment procedure. Scientists are trying to use different kinds of materials, including hydrogels, alginate, collagen, and hydrocolloids, which are often tried to deliver drugs simultaneously and also deliver multiple drugs in a single point. Recently, several other modern therapies have been utilized to treat diabetic wounds, including cell therapy, bioengineered skin, scaffolding systems, artificial skin scaffolding systems, 3D printed scaffolds, etc. (Rozman et al. 2020). There are several kinds of wound dressing agents for DFU; among all the dressings, alginate-based dressing systems have gained much attention due to their properties, such as being biocompatible, biodegradable, and having water adsorption capacity(Sahana and Rekha 2018).

3 Alginate

Alginate is an anionic polysaccharide and is commonly isolated from brown seaweed. Various methods are often used to extract alginate from seaweed, but one of the standard methods for isolating alginate is the calcium precipitation technique. The percentage of alginate in seaweed varies depending on the species; around 25 to 30 percent of alginate is commonly available in various types of seaweed. Alginate is frequently made of α-l-guluronic acid (G) and β-d-mannuronic acid (M) with 1,4-glycosidic linkages. Alginate has several advantages, including water solubility, nontoxicity, biocompatibility, and biodegradability (Lee and Mooney 2012; Mohapatra et al. 2023). Alginate in films and wafers was developed using ciprofloxacin and fluconazole to target microbial infection in DFU. Subsequently, the developed film and wafers were characterized by scanning electron microscopy, X-ray diffraction, and Fourier transform microscopy. The results indicated that alginate wafers show better results compared to the films in the simulated wound fluid. Burst release of drugs from the matrix, followed by sustainable release, inhibits *Escherichia coli, Staphylococcus aureus, Pseudomonas aeruginosa,* and *Candida Albicans.* The advantages of this drug delivery system were biocompatiblity with human primary adult keratinocytes (Viability greater than 70%), and the vitro scratch results assay showed 65–68% wound closure within 7 days (Ahmed et al. 2021).

Cellulose, hyaluronic acid, and alginate are commonly studied for diabetic wound healing applications. These polysaccharides are further modified with cross-linking, grafting, quaternate, nanoparticle preparation, and composite development to meet the needs for healing diabetic wounds. These polysaccharides may have the ability to provide sufficient mechanical strength for the scaffolds, re-epithelialization, and collagen synthesis, which can relate to diabetic wounds. In addition, the stimuli-responsive polymer may have additional advantages by delivering drugs whenever needed for diabetes wound healing (Zainuddin et al. 2023).

Calcium alginate sulfate has been used as a wound-healing agent and tested in diabetic mouse models. Different degrees of substitution, such as 0.2, 0.5, and 0.9, were developed using crosslinking with calcium ions and the freeze-dried method. The developed scaffold can be used as a scaffolding system for diabetes wound healing. The results suggest that the developed calcium alginate sulfate group shows better results and heals the wound faster than the non-sulfated alginate group. The histological studies suggest that collagen deposition and tissue regeneration occur similarly to normal skin. (Zare-Gachi et al. 2024).

The mechanism involved in diabetic wound ulceration with calcium alginate was elucidated. A diabetic rat model was used for the experiments. Calcium alginate dressing was applied to the wound created on the rat, and the Vaseline dressing group was kept as control. Collagen deposition and hydroxyproline content were estimated using ELISA assay at different times 3 d, 7 d, and 14 d after wounding skin. Collagen I in Alg-group was enhanced from day 3 (0.66 ± 0.25 vs. 0.42 ± 0.09, $P < 0.05$) to day 14 (1.09 ± 0.14 vs. 0.78 ± 0.16, $P < 0.05$). The study suggested that calcium alginate accelerates the process of wound healing by type 1 collagen synthesis (Wang et al. 2015).

4 Alginate Composites for Diabetes Treatment

In a study, *Moringa oleifera* leaf extract was mixed with 3% sodium alginate-pectin film, prepared the film using the solvent casting method, and evaluated for diabetic wound healing. Various compositions, such as 0.1, 0.5, and 1% of *Moringa oleifera* leaf extract, were mixed with the films. Excision and abrasion wounds were created on the diabetic rat model using STZ/HFD, and healing was studied for days. *Moringa oleifera* leaf extract did not show any cytotoxicity, and the 0.5% group showed better results by $77.67 \pm 7.28\%$ at day 7 compared to the control group (Chin et al. 2019). In another study, natural rubber latex (*Hevea brasiliensis* – the rubber tree) was extensively studied as a drug delivery system for diabetic wounds. Alginate loaded with latex has been developed, which is highly absorptive, permeable, and has a high amount of swelling and is biocompatible (Barros et al. 2021). Due to the low cost and availability of gallic acid, it has been studied for wound care applications. The hydrocolloid film containing gallic acid with alginate composites was prepared using the solvent casting method (Chin and Gan 2021). Sodium alginate with chondroitin sulfate hydrogel was developed using a solvent casting method with the addition of curcumin for collagen deposition, angiogenesis, and epithelialization. The controlled release of curcumin from the hydrogel was evaluated using high-performance liquid chromatography. The mean residence time and area under the curve increased to 16.18 h and 203.64 ± 30.1 µg/mL*h, respectively. In vivo, results suggest that the injectable hydrogel shows collagen deposition and upregulation of fibroblast-like cells for wound healing applications(Shah et al. 2021). The film containing sodium alginate, sodium carboxymethyl cellulose, tween 80, PEG-400, and Glycerol was used using the solvent casting method. Formulation containing three minutes of microwave treatment shows better physicochemical properties with high moisture adsorption, water vapor transmission rate, water vapor permeability, and high tensile strength. The in vivo study suggests that wounds were healed at 16 days with increased epithelialization and enhanced collagen when compared to the untreated group (Mahmood et al. 2023) (Fig. 14.1).

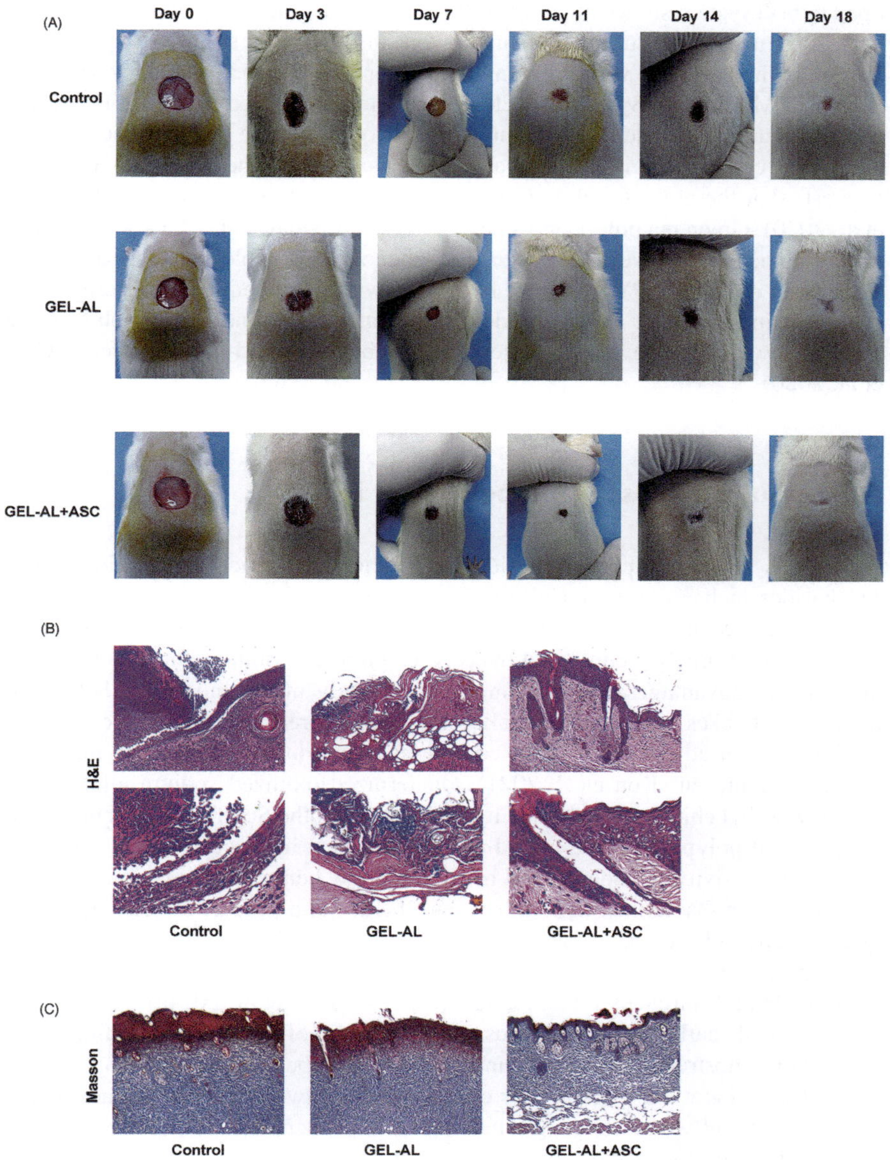

Fig. 14.1 Adding adipose-derived mesenchymal stem cells onto alginate/gelatin hydrogels enhances wound healing. (**a**) Representative images of the wound healing (**b**, **c**). Images of full-thickness wounds on day 14 from hematoxylin and eosin (**b**) and Masson staining (**c**) (Sheng et al. 2023)

Sulfated alginate combined with a gelation hybrid scaffold was developed and used as a dermal substitute for diabetic ulcers in a diabetic mouse model. Different amounts of sulfated alginate, from 10% up to 50%, were prepared using chemical crosslinking by carbodiimide chemistry and freeze-drying techniques. The developed hybrid system shows biocompatibility and biodegradability and increased cell growth by increasing sulfate alginate content in the hybrid system (Sadeghi et al. 2023). Sodium alginate with Polyvinylpyrrolidone-Iodine and Frequency Rhythmic Electrical Modulation System (FREMS) technology was used for DFU. The results show that composite and FREMS show better wound closure (Contardi et al. 2024). Okra fruit combined with sodium alginate was studied. The animal model study suggests that the alginate-Okra group shows better results in wound healing (Ilmi et al. 2020). Green tea polyphenols have potent antioxidants and therapeutic effects on wound healing. However, green tea polyphenols are solubleand easily oxidisible. To prevent oxidation, PVA-alginate hydrogels with polyphenol was developed and evaluated in a diabetic animal model to evaluate the wound healing ability. The results show that the developed materials show better wound-healing effects (Chen et al. 2020).

5 Alginate Chitosan Composites for DFU

Alginate is an anionic polysaccharide and efficiently interacts with cationic polysaccharides such as chitosan. Chitosan is a natural polysaccharide abundantly available in crustacean shells and other resources. Chemical and enzymatic biological methods often produce chitin and hydrolyzed chitosan. Both methods have advantages and disadvantages; chemical methods have advantages in yield and effectiveness, and it takes less time to make chitosan, whereas the enzymatic isolation method has an advantage in mild conditions in producing the known molecular weight of chitosan (Kou et al. 2021). Quaternized oxidized sodium alginate and carboxymethyl chitosan can form a hydrogel through the Schiff base reaction. Deer antler blood polypeptide was added to the composite hydrogel system, which has antioxidant activity. The developed hydrogels show better biocompatibility, biodegradability, antibacterials, antioxidant, and hemostatic properties. The developed quaternised oxidised sodium alginate-carboxymethyl chitosan hydrogel shows better in promoting diabetic wound healing by expression of growth factors at the wound site and mitigates inflammatory responses (Ma et al. 2024a). Alginate can combine with carboxymethyl chitosan in the form of hydrogel, nanofibers, and core-shell nanostructures, which can be used for diabetic wound healing applications. The integration, blending, or cross-linking of two materials has advantages and can meet all wound-healing requirements (Astaneh et al. 2024). The biocompatible dressing consists of chitosan and alginate and can accelerate the wound-healing process through inflammatory phases, stimulating fibroblast proliferation and remodeling. Chitosan-alginate membranes were studied against acute skin wounds of hyperglycemic mice. The results showed that the chitosan-alginate membrane treatment within 10 days shows better organization of collagen fibers. High amounts of interleukin (IL)-1a, IL-1b, granulocyte colony-stimulating factor (G-CSF), and tumor necrosis factor-alpha (TNF-a) were detected in the first and second days of treatment (Breder et al. 2020). Sodium alginate blended with polyethylene glycol with *Satureja cuneifolia* was prepared and checked for diabetic ulcers. Different alginate concentrations, 1, 3, and 5 wt.%, were mixed with 9% polyethylene glycol. The antimicrobial activity of the developed blends was evaluated with *Escherichia coli and Staphylococcus aureus.* At the same time, the developed blends showed biocompatibility with fibroblast L929 cells. The developed blends can potentially help treat diabetic wounds (Ilhan et al. 2020). Polyvinyl alcohol combined with sodium alginate with berberine hydrogel was prepared and evaluated for diabetic wound healing applications. Calcium ions were used to crosslink the hydrogel to get better mechanical strength. Animal experiment results suggested that the berberine from the hydrogels released and promoted wound healing in diabetic rats. In addition, berberine inhibits the expression of NF-κB, TNF-a, and IL-6. Still, it increases the expression of F VEGF, CD 31, and SMA by activating Sirt 1, which was beneficial for wound healing in diabetic rats (Zhang et al. 2020) (Fig. 14.2).

Chitosan-alginate membranes were prepared in the presence of glycerol. Animal groups were divided into normal glycemic groups with saline solution and

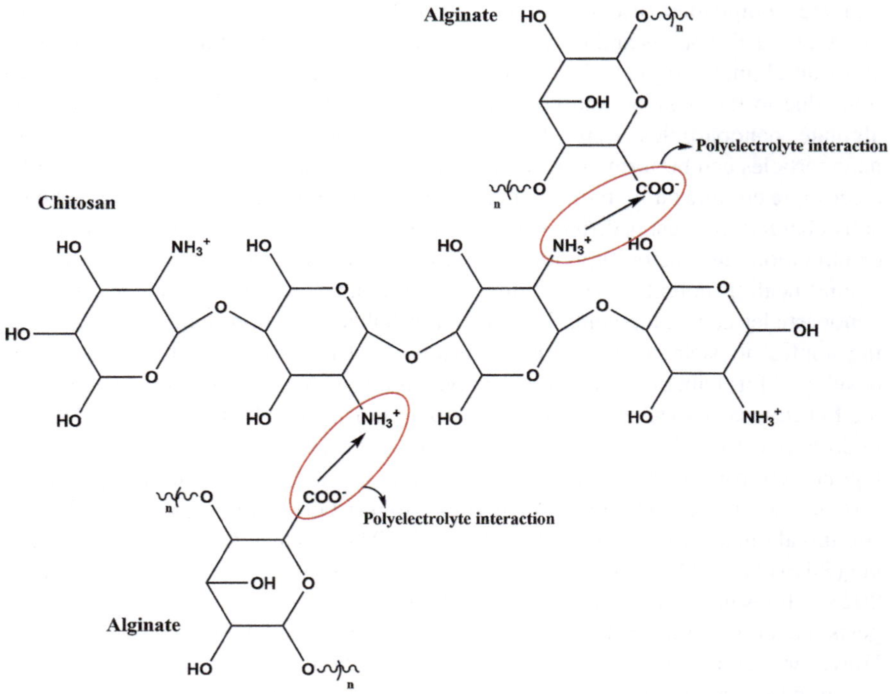

Fig. 14.2 Polyelectrolyte interaction of chitosan and alginate. (Krisanti et al. 2019)

hyperglycemic groups with plasticized chitosan-alginate membrane. On day 14, there was a slight concentration of thinner elastic fibers for the glycerol-plasticized chitosan-alginate membrane topical treatment group, with a more remarkable reorganization of papillary skin and improved red-orange birefringence collagen fibers, as well as net-shaped orientation, similar to intact skin (Azevedo et al. 2022). Amnion-conjugated chitosan-alginate membrane dressing was tested with a diabetic swine model. The bovine amniotic power growth factor contents. Twelve square-shaped wounds created on each pig's back were randomly divided into control ($n = 4$), chitosan (CS) ($n = 4$), and amnion-CS (AC; $n = 4$) groups and treated accordingly with different dressings. Results were evaluated with the contraction parameters over time, wound perfusion, indocyanine green angiography, and histological analysis of inflammatory markers. In vivo, results suggested that the amnion-chitosan group showed earlier wound contraction initiation, reepithelialization, and wound perfusion than the control group. The amnion-conjugated chitosan-alginate membrane showed a strong healing capacity and is favorable for diabetic wounds (Jeong et al. 2022). A carboxymethyl chitosan-sodium alginate composite film was developed, and S-nitroso glutathione was incorporated, which acts as an angiogenic and antibacterial agent. The developed composite films were porous, and the in vitro release study suggested that the NO was delivered continuously for 168 h. The in vivo results suggested that the developed films showed better wound-healing effects in diabetic rat models compared to the gauze and carboxymethyl chitosan-alginate composites (Razmjooee et al. 2022). Chitosan and alginate have been proven to heal wounds and have antibacterial properties. In addition, the combined chitosan-alginate polymers have been extensively studied for drug delivery applications due to the positive and negative charges on the system's surface. Chitosan-alginate nanoparticles were developed using the ionic gelation method; the nanoparticles can be formed between these two polymeric systems due to the polyelectrolyte complexation method. Further, the developed nanoparticles were chemically characterized using different analytical techniques. Wound closure, histological examination, and histomorphometry assessment show the nanoparticles' excellent wound-healing potential. Higher in collagen content, the results suggest that the nanoparticles can be a potential platform for diabetic and non-diabetic wound healing applications(Sheir et al. 2022). Various kind of chitosan with alginate were developed for diabetic wound healing application such as sodium alginate/carboxymethyl chitosan (SA/CMCS) (Qin et al. 2023), self-healing hydrogel (PBA-OSA-CS hydrogel) with carboxymethyl chitosan and gallic acid (Tan et al. 2023), sodium alginate-chitosan (Kannan et al. 2024), alginate/curdlan/agarose (Latiyan et al. 2024), sodium alginate-carboxymethyl chitosan (Li et al. 2024c; Ma et al. 2024b); sodium alginate and guar gum (Mondal et al. 2024), alginate/chitosan-based hydrogel (Silva et al. 2024), carboxymethyl chitosan and sodium alginate oxide (Wu et al. 2025), chitosan alginate nanoparticles (Sheir et al. 2021). A systematic review suggests that chitosan-alginate composite materials show better wound-healing ability. However, a clinical trial of chitosan-alginate-based composite materials needs more research to conclude its usage in real-life applications (Cai and Li 2020) (Fig. 14.3).

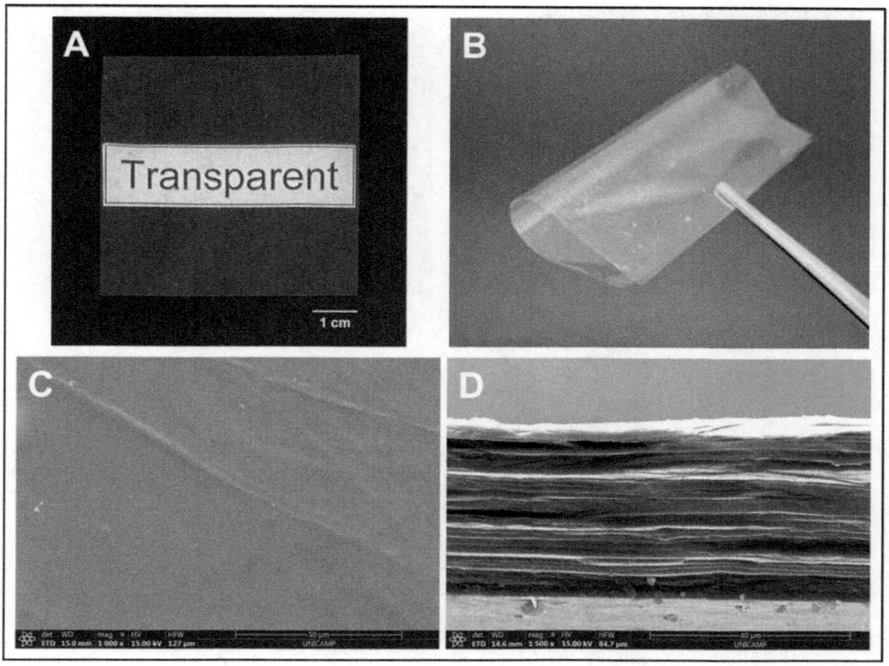

Fig. 14.3 Typical visual aspect of the membrane produced. Images showing (**a**) the transparency and (**b**) flexibility of the membrane analyzed with the naked eye (**c**) the microstructure of the surface, and (**d**) a cross-section of the membrane by SEM. (Azevedo et al. 2022)

6 Alginate Tricomponent Scaffolding System for Diabetic Wounds

Various tricomponent systems were developed with alginate along with chitosan and other bioactive ceramic materials. Chitosan-Alginate-fucoidan (Yanagibayashi et al. 2012), polyvinyl alcohol, polyvinyl pyrrolidone, and sodium alginate with neomycin (Choi et al. 2016), curcumin-collagen-alginate and chitosan (Karri et al. 2016), silk fibroin, chitosan, bioactive glass, and polyvinyl alcohol (Gao et al. 2017), chitosan-alginate-maltodextrin-pluronic (Akbar et al. 2018), calcium alginate hydrogel laden with protamine nanoparticles and hyaluronan oligosaccharide (Wang et al. 2019b), reduced graphene oxide in a poly(vinyl alcohol)-alginate (Wang et al. 2019a), collagen, carboxymethyl cellulose, alginate and ethylenediaminetetraacetic acid (Alavi et al. 2021), poly(N-isopropylacrylamide)/gelatin methacrylate (NG) layer and the upper highly stretchable alginate/polyacrylamide (AP) (Huang et al. 2023), poly (vinyl alcohol)/sodium alginate/carboxymethyl chitosan (Chai et al. 2024), Astragalus polysaccharide-carboxymethyl chitosan-sodium alginate (Tang et al. 2025), and calcium alginate nanoparticles (Choudhary et al. 2021). The advantages of tricomponent scaffolding along with alginate are to provide sufficient mechanical strength, cell proliferation, and cell adhesion (Fig. 14.4).

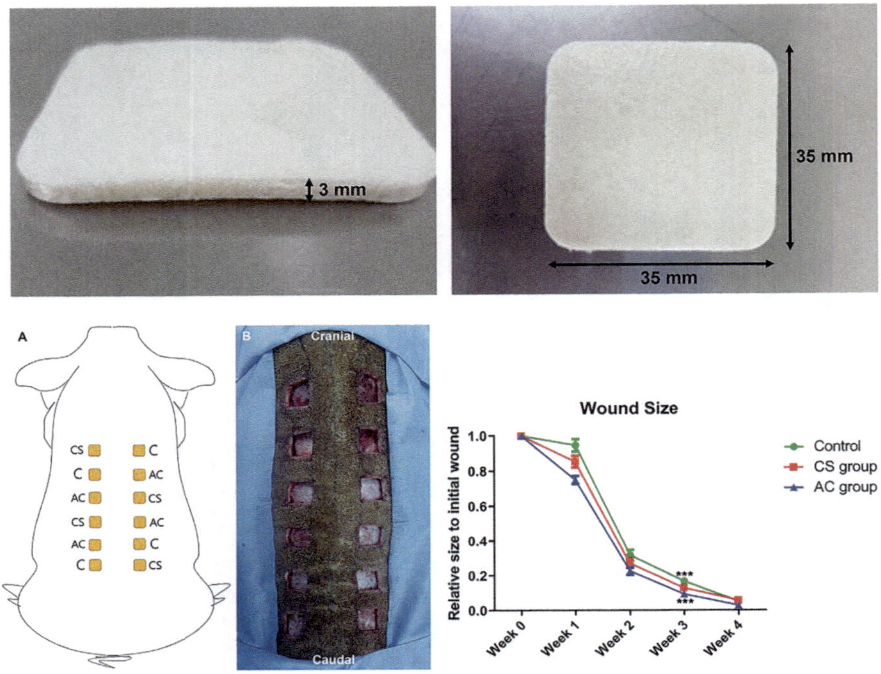

Fig. 14.4 Amnion-conjugated chitosan-alginate membrane. (Jeong et al. 2022)

7 Alginate for Drug Delivery

Alginate was used as a drug delivery system for diabetic treatment. Diosmin nanocrystals (Atia et al. 2019), Edaravone (Fan et al. 2019), Vicenin-2 (Tan et al. 2019), alginate-pectin hydrogel film loaded with Simvastatin (Rezvanian et al. 2021), dopamine-modified oxidized sodium alginate hydrogels (Chi et al. 2022), cross-linked sodium alginate hydrogels containing deferoxamin (Li et al. 2022), a paeoniflorin-sodium alginate-gelatin ski (Yu et al. 2022), Na-alginate with F-127 and poly(vinyl alcoholnaringenin-Loaded Hydrogel (Raina et al. 2023), sodium alginate-poly(vinyl alcohol)-taxifolin (Wang et al. 2023), sodium alginate-polyvinyl alcohol nanofibers loaded with Shikonin (Ding et al. 2024), poly(vinyl alcohol), sodium alginate, and tannic acid. (Hao et al. 2024), salvianolic acid B-sodium alginate-gelatin skin (Lihao et al. 2024), tannic acid-quaternized carboxymethyl chitosan-oxidized sodium alginate (Yang et al. 2024), gelatin and oxidized sodium alginate (Cao et al. 2023), hyaluronic acid-alginate (Khaliq et al. 2023).

8 Alginate with Growth Factor, Protein, and Peptide Delivery

Alginate is also used for growth factor, protein, and peptide-based delivery for diabetic wound healing applications. VEGF-loaded poly-Lactic-co-Glycolic-Acid--alginate microspheres (Gainza et al. 2013), polymeric alginate-poly-l-ornithine (A-PLO) microcapsules (MC) (Moore et al. 2014), alginate- and deoxyribonucleic acid (DNA)-based gels (Tellechea et al. 2015), 2-deoxy-d-ribose to clinically used alginate dressings (Azam et al. 2019), Triiodothyronine-impregnated alginate/gelatin/polyvinyl alcohol (Satish et al. 2019), exosome loaded alginate (Shafei et al. 2020), exosomes encapsulated nano hydrogel (Zhang et al. 2021), sprayable alginate hydrogel dressings with oxygen production and exosomes (Zhang et al. 2023), human growth hormone combined with alginate dressing (Zhou et al. 2021), platelet-rich plasma fibrin matrix combined with oxidized alginate (Garcia-Orue et al. 2022), chitosan-sodium alginate-velvet antler blood peptides (Hao et al. 2022), platelet-derived extracellular vesicles, and reduced graphene oxide (Hao et al. 2023), platelet Rich Plasma Loaded Multifunctional Hydrogel (Xu et al. 2023), alginate/polydeoxyribonucleotide (Dananjaya et al. 2024), liposome-embedded oxidized sodium alginate/carboxymethyl chitosan (Wu et al. 2024), and alginate/chondroitin sulfate (Liao et al. 2023).

9 Alginate with Metal Nanoparticles

Alginate, along with various metals, is used for diabetic wound applications. Metal nanoparticles have the ability to have antimicrobial activity (Dai et al. 2024). Alginate with silver (Ozaki et al. 2015), alginate/silver/nicotinamide (Montaser et al. 2016), rubidium-containing calcium alginate (He et al. 2019), copper and sodium alginate (Li et al. 2020), MoS_2nanoclusters with sodium alginate (SA) (Ma et al. 2020), chitosan-Mg-alginate (Wang et al. 2021), black phosphorus quantum dots encapsulated in hydrogel (Huang et al. 2022), ZnO (Loera-Valencia et al. 2022), zinc ions and sodium alginate (Yao et al. 2022), silver nanoparticles loaded in polysaccharide (Aldakheel et al. 2023), nanosilver-free alginate dressing (Tang et al. 2023), bioactive glass-metal oxides-alginate composites (Bargavi et al. 2024), lithium chloride-loaded silk fibroin-alginate (Bashiri et al. 2024), polyvinyl alcohol and sodium alginate-graphene oxide loaded with silver-platinum hybrid nanoparticles (Chen et al. 2024), conventional alginate-copper hydrogen phosphate (Feng et al. 2024), MSCs and MnO_2 (Hu et al. 2024b), polyvinyl alcohol, sodium alginate, and gelatin loaded with tea polyphenol (Hu et al. 2024a), bioactive glass and $MoO_3 - x$ in a biocompatible sodium alginate (Huang et al. 2024), alginate-derived hydrogels and metal-phenolic nanosphere (Li et al. 2024b), multi-enzyme-like activities nanocomposite (Mo,Fe/Cu,I-Ag@GOx) (Li et al. 2024a), hydrogels containing chitosan-modified gold nanoparticle(Meng et al. 2024), and insulin-loaded strontium-cross-linked alginate (Rajalekshmy et al. 2024).

10 Clinical Trial and Case Studies of Alginate Dressings for DFU

Various clinical trials were carried out to understand the usage and efficacy of alginate. Alginate clinical trial (Vassallo and Formosa 2015), silver with alginate (Tsang et al. 2017), clinical study of sodium alginate (Barbosa et al. 2018), collagen alginate dressing (Lullove et al. 2021), sodium alginate, and vitamins A (Barbosa et al. 2022), collagen alginate (Gould et al. 2022), alginate dressing (Hajimohammadi et al. 2023) and silver-incorporated calcium alginate shows better efficacy in treating DFU. (Jude et al. 2007).

11 Conclusion

Alginate-based materials were extensively studied and used for wound dressing application. However, alginate alone may not be valuable for solving several issues in DFU. Hence, adding other antimicrobial antibiotics and metal nanoparticles may solve the microbial issues in the DFU. Various growth factors, including VEGF and platelet-rich plasma, are incorporated into alginate materials and evaluated for efficiency. Overall, alginate-based composite biomaterials can help to treat DFU.

Acknowledgments The authors would like to thank Malla Reddy Vishwavidyapeeth—Deemed to be University, Hyderabad, Telangana, for their encouragement and seed grant support for writing this paper.

References

Hoversten KP, Kiemele LJ, Stolp AM, Takahashi PY, Verdoorn BP. Prevention, Diagnosis, and Management of Chronic Wounds in Older Adults. Mayo Clin Proc. 2020;95(9):2021-34. doi: https://doi.org/10.1016/j.mayocp.2019.10.014.

Martin CM, Hawkins L. Wound care basics for the pharmacist. Consult Pharm. 2013;28(6):344-52. doi: https://doi.org/10.4140/TCP.n.2013.344.

Bharambe SV, Darekar AB, Saudagar RB. Wound healing dressings and drug delivery systems: A review. Intl J Pharm Technol. 2013;5(3):2764-86.

Thomas DR. Clinical management of diabetic ulcers. Clin Geriatr Med. 2013;29(2):433-41. doi: https://doi.org/10.1016/j.cger.2013.01.007.

Narayana Reddy Karri VVS, Kuppusamy G, Mulukutla S, Sood S, Malayandi R. Understanding the implications of pharmaceutical excipients and additives in the treatment of diabetic foot ulcers. J Excipients Food Chem. 2015;6(1):7-22.

Borda LJ, Macquhae FE, Kirsner RS. Wound Dressings: A Comprehensive Review. Curr Dermatol Rep. 2016;5(4):287-97. doi: https://doi.org/10.1007/s13671-016-0162-5.

Macha IJ, Ben-Nissan B. Past and Future of Wound Dressing in Soft and Hard Tissue Surgery. Springer Series in Biomaterials Science and Engineering. Springer Series in Biomaterials

Science and Engineering: Springer Science and Business Media Deutschland GmbH; 2022. p. 1-14.

Armstrong DG, Tan T-W, Boulton AJ, Bus SA. Diabetic foot ulcers: a review. JAMA. 2023;330(1):62-75.

Wang X, Yuan C-X, Xu B, Yu Z. Diabetic foot ulcers: Classification, risk factors and management. World J Diabetes. 2022;13(12):1049.

Chang M, Nguyen TT. Strategy for treatment of infected diabetic foot ulcers. Acc Chem Res. 2021;54(5):1080-93.

Rao SS, Venkatesan J, Prabhu A, Rekha PD. Natural polymeric biomaterials in growth factor delivery for treating diabetic foot ulcers. J Drug Deliv Sci Technol. 2020;55. doi: https://doi.org/10.1016/j.jddst.2019.101385.

Rozman NAS, Tong WY, Leong CR, Tan WN, Ab Rashid S. Diabetic wound infection: A review on microbial population and infection control. Asia-Pacific Journal of Molecular Biology and Biotechnology. 2020;28(3):36-43. doi: https://doi.org/10.35118/apjmbb.2020.028.3.04.

Sahana TG, Rekha PD. Biopolymers: Applications in wound healing and skin tissue engineering. Mol Biol Rep. 2018;45(6):2857-67. doi: https://doi.org/10.1007/s11033-018-4296-3.

Lee KY, Mooney DJ. Alginate: properties and biomedical applications. Prog Polym Sci. 2012;37(1):106-26.

Mohapatra A, Nair A, Wagh NS, Lakkakula J. Alginate-based nanosystems for therapeutic use. Polymeric Nanosystems: Theranostic Nanosystems: Volume 1. Elsevier; 2023. p. 149-71.

Ahmed A, Getti G, Boateng J. Medicated multi-targeted alginate-based dressings for potential treatment of mixed bacterial-fungal infections in diabetic foot ulcers. Int J Pharm. 2021;606. doi: https://doi.org/10.1016/j.ijpharm.2021.120903.

Zainuddin SZ, Ramond NJMR, Anuar NK. Polysaccharide-Based Formulations for the Treatment of Diabetic Wounds: A Review. Res J Pharm Technol. 2023;16(6):2835-42. doi: https://doi.org/10.52711/0974-360X.2023.00467.

Zare-Gachi M, Sadeghi A, Choshali MA, Ghadimi T, Forghani SF, Pezeshki-Modaress M, et al. Degree of sulfation of freeze-dried calcium alginate sulfate scaffolds dramatically influence healing rate of full-thickness diabetic wounds. Int J Biol Macromol. 2024;283(Pt 2):137557. doi: https://doi.org/10.1016/j.ijbiomac.2024.137557.

Wang T, Gu Q, Zhao J, Mei J, Shao M, Pan Y, et al. Calcium alginate enhances wound healing by up-regulating the ratio of collagen types I/III in diabetic rats. Int J Clin Exp Pathol. 2015;8(6):6636-45.

Chin CY, Ng PY, Ng SF. Moringa oleifera standardised aqueous leaf extract-loaded hydrocolloid film dressing: in vivo dermal safety and wound healing evaluation in STZ/HFD diabetic rat model. Drug Deliv Transl Res. 2019;9(2):453-68. doi: https://doi.org/10.1007/s13346-018-0510-z.

Barros NR, Ahadian S, Tebon P, Rudge MVC, Barbosa AMP, Herculano RD. Highly absorptive dressing composed of natural latex loaded with alginate for exudate control and healing of diabetic wounds. Mater Sci Eng, C. 2021;119:111589. doi: https://doi.org/10.1016/j.msec.2020.111589.

Chin CY, Gan JE. Formulation and characterisation of alginate hydrocolloid film dressing loaded with gallic acid for potential chronic wound healing. F1000 Res. 2021;10:451. doi: https://doi.org/10.12688/f1000research.52528.1.

Shah SA, Sohail M, Khan SA, Kousar M. Improved drug delivery and accelerated diabetic wound healing by chondroitin sulfate grafted alginate-based thermoreversible hydrogels. Mater Sci Eng, C. 2021;126:112169. doi: https://doi.org/10.1016/j.msec.2021.112169.

Mahmood S, Khan NR, Razaque G, Shah SU, Shahid MG, Albarqi HA, et al. Microwave-Treated Physically Cross-Linked Sodium Alginate and Sodium Carboxymethyl Cellulose Blend Polymer Film for Open Incision Wound Healing in Diabetic Animals—A Novel Perspective for Skin Tissue Regeneration Application. Pharmaceutics. 2023;15(2). doi: https://doi.org/10.3390/pharmaceutics15020418.

Sheng W, Song Q, Su X, Lu Y, Bai Y, Ji F, et al. Sodium alginate/gelatin hydrogels loaded with adipose-derived mesenchymal stem cells promote wound healing in diabetic rats. J Cosmet Dermatol. 2023;22(5):1670-9. doi: https://doi.org/10.1111/jocd.15631.

Sadeghi A, Zare-Gachi M, Najjar-Asl M, Rajabi S, Fatemi MJ, Forghani SF, et al. Hybrid gelatin-sulfated alginate scaffolds as dermal substitutes can dramatically accelerate healing of full-thickness diabetic wounds. Carbohydr Polym. 2023;302:120404. doi: https://doi.org/10.1016/j.carbpol.2022.120404.

Contardi M, Summa M, Lenzuni M, Miracoli L, Bertora F, Mendez MD, et al. Combining Alginate/PVPI-Based Film with Frequency Rhythmic Electrical Modulation System (FREMS) Technology as an Advanced Strategy for Diabetic Wounds. Macromol Biosci. 2024;24(2):e2300349. doi: https://doi.org/10.1002/mabi.202300349.

Ilmi ZN, Wulandari PAC, Husen SA, Winarni D, Alamsjah MA, Awang K, et al. Characterization of alginate from sargassum duplicatum and the antioxidant effect of alginate-okra fruit extracts combination for wound healing on diabetic mice. Appl Sci. 2020;10(17). doi: https://doi.org/10.3390/app10176082.

Chen G, He L, Zhang P, Zhang J, Mei X, Wang D, et al. Encapsulation of green tea polyphenol nanospheres in PVA/alginate hydrogel for promoting wound healing of diabetic rats by regulating PI3K/AKT pathway. Mater Sci Eng, C. 2020;110:110686. doi: https://doi.org/10.1016/j.msec.2020.110686.

Kou SG, Peters LM, Mucalo MR. Chitosan: A review of sources and preparation methods. Int J Biol Macromol. 2021;169:85-94.

Ma S, Chen K, Ding Q, Zhang S, lu Y, Yu T, et al. Quaternized oxidized sodium alginate injectable hydrogel with high antimicrobial and hemostatic efficacy promotes diabetic wound healing. Int J Pharm. 2024a;661:124421. doi: https://doi.org/10.1016/j.ijpharm.2024.124421.

Astaneh ME, Hashemzadeh A, Fereydouni N. Recent advances in sodium alginate-based dressings for targeted drug delivery in the context of diabetic wound healing. Journal of Materials Chemistry B. 2024;12(40):10163-97. doi: https://doi.org/10.1039/d4tb01049c.

Breder JSC, Pires ALR, Azevedo FF, Apolinário PP, Cantaruti T, Jiwani SI, et al. Enhancement of cellular activity in hyperglycemic mice dermal wounds dressed with chitosan-alginate membranes. Brazilian Journal of Medical and Biological Research. 2020;53(1). doi: https://doi.org/10.1590/1414-431x20198621.

Ilhan E, Cesur S, Guler E, Topal F, Albayrak D, Guncu MM, et al. Development of Satureja cuneifolia-loaded sodium alginate/polyethylene glycol scaffolds produced by 3D-printing technology as a diabetic wound dressing material. Int J Biol Macromol. 2020;161:1040-54. doi: https://doi.org/10.1016/j.ijbiomac.2020.06.086.

Zhang P, He L, Zhang J, Mei X, Zhang Y, Tian H, et al. Preparation of novel berberine nano-colloids for improving wound healing of diabetic rats by acting Sirt1/NF-κB pathway. Colloids Surf B Biointerfaces. 2020;187:110647. doi: https://doi.org/10.1016/j.colsurfb.2019.110647.

Krisanti EA, Naziha GM, Amany NS, Mulia K, Handayani NA. Effect of biopolymers composition on release profile of iron (II) fumarate from chitosan-alginate microparticles. IOP Conference Series: Materials Science and Engineering: IOP Publishing; 2019. p. 012100.

Azevedo FF, Cantarutti TA, Remiro PDFR, Barbieri B, Azoubel RA, Nagahara MHT, et al. Histological and Molecular Evidence of the Positive Performance of Glycerol-Plasticized Chitosan-Alginate Membranes on Skin Lesions of Hyperglycemic Mice. Polymers. 2022;14(21). doi: https://doi.org/10.3390/polym14214754.

Jeong W, Hong J, Jung M, Jang M, An S, Jo T, et al. Therapeutic Effects of Amnion-Conjugated Chitosan-Alginate Membranes on Diabetic Wounds in an Induced Diabetic Swine Model: An In Vitro and In Vivo Study. Arch Plast Surg. 2022;49(2):258-65. doi: https://doi.org/10.1055/s-0042-1744429.

Razmjooee K, Oustadi F, Golaghaei A, Nassireslami E. Carboxymethyl chitosan-alginate hydrogel containing GSNO with the ability to nitric oxide release for diabetic wound healing. Biomedical Materials (Bristol). 2022;17(5). doi: https://doi.org/10.1088/1748-605X/ac877c.

Sheir MM, Nasra MMA, Abdallah OY. Phenytoin-loaded bioactive nanoparticles for the treatment of diabetic pressure ulcers: formulation and in vitro/in vivo evaluation. Drug Deliv Transl Res. 2022;12(12):2936-49. doi: https://doi.org/10.1007/s13346-022-01156-z.

Qin B, Wu S, Dong H, Deng S, Liu Y, Zhang W, et al. Accelerated Healing of Infected Diabetic Wounds by a Dual-Layered Adhesive Film Cored with Microsphere-Loaded Hydrogel Composite Dressing. ACS Appl Mater Interfaces. 2023;15(28):33207-22. doi: https://doi.org/10.1021/acsami.2c22650.

Tan WW, Lei SS, Long T, Xu ZL, Li DF, Mu CD, et al. Polysaccharide Based Injectable Self-healing Hydrogels with Glucose Responsive Drug Release Behavior. Acta Polym Sin. 2023;54(8):1155-65. doi: https://doi.org/10.11777/j.issn1000-3304.2023.23024.

Kannan M, Kalpana R, Kumar V. Sodium Alginate Chitosan Crosslinked Hydrogel Amoxycillin Absorption Properties. Journal of Pharmacy and Bioallied Sciences. 2024;16(Suppl 2):S1161-S3. doi: https://doi.org/10.4103/jpbs.jpbs_511_23.

Latiyan S, Kumar TSS, Doble M. Functionally multifaceted alginate/curdlan/agarose-based bilayer fibro-porous dressings for addressing full-thickness diabetic wounds. Biomater Adv. 2024;157:213757. doi: https://doi.org/10.1016/j.bioadv.2023.213757.

Li F, Liu T, Liu X, Han C, Li L, Zhang Q, et al. Ganoderma lucidum polysaccharide hydrogel accelerates diabetic wound healing by regulating macrophage polarization. Int J Biol Macromol. 2024c;260(Pt 2):129682. doi: https://doi.org/10.1016/j.ijbiomac.2024.129682.

Ma S, Ding Q, Xia G, Li A, Li J, Sun P, et al. Multifunctional biomaterial hydrogel loaded with antler blood peptide effectively promotes wound repair. Biomed Pharmacother. 2024b;170:116076. doi: https://doi.org/10.1016/j.biopha.2023.116076.

Mondal S, Hazra A, Paul P, Saha B, Roy S, Bhowmick P, et al. Formulation and evaluation of n-acetyl cysteine loaded bi-polymeric physically crosslinked hydrogel with antibacterial and antioxidant activity for diabetic wound dressing. Int J Biol Macromol. 2024;279(Pt 4):135418. doi: https://doi.org/10.1016/j.ijbiomac.2024.135418.

Silva LDS, Vila Nova BG, Sousa CEMD, Silva RG, Carvalho LRDS, Silva ISS, et al. Fabrication and characterization of physically crosslinked alginate/chitosan-based hydrogel loaded with neomycin for the treatment of skin infections caused by Staphylococcus aureus. Int J Biol Macromol. 2024;271(Pt 1):132577. doi: https://doi.org/10.1016/j.ijbiomac.2024.132577.

Wu X, Wang Y, Liu X, Ding Q, Zhang S, Wang Y, et al. Carboxymethyl chitosan and sodium alginate oxide pH-sensitive dual-release hydrogel for diabetes wound healing: The combination of astilbin liposomes and diclofenac sodium. Carbohydr Polym. 2025;349(Pt B):122960. doi: https://doi.org/10.1016/j.carbpol.2024.122960.

Sheir MM, Nasra MMA, Abdallah OY. Chitosan alginate nanoparticles as a platform for the treatment of diabetic and non-diabetic pressure ulcers: Formulation and in vitro/in vivo evaluation. Int J Pharm. 2021;607:120963. doi: https://doi.org/10.1016/j.ijpharm.2021.120963.

Cai H, Li G. Efficacy of alginate-and chitosan-based scaffolds on the healing of diabetic skin wounds in animal experimental models and cell studies: A systematic review. Wound Repair Regen. 2020;28(6):751-71. doi: https://doi.org/10.1111/wrr.12857.

Yanagibayashi S, Kishimoto S, Ishihara M, Murakami K, Aoki H, Takikawa M, et al. Novel hydrocolloid-sheet as wound dressing to stimulate healing-impaired wound healing in diabetic db/db mice. Bio-Medical Materials and Engineering. 2012;22(5):301-10. doi: https://doi.org/10.3233/BME-2012-0720.

Choi JS, Kim DW, Kim DS, Kim JO, Yong CS, Cho KH, et al. Novel neomycin sulfate-loaded hydrogel dressing with enhanced physical dressing properties and wound-curing effect. Drug Deliv. 2016;23(8):2806-12. doi: https://doi.org/10.3109/10717544.2015.1089958.

Karri VVSR, Kuppusamy G, Talluri SV, Mannemala SS, Kollipara R, Wadhwani AD, et al. Curcumin loaded chitosan nanoparticles impregnated into collagen-alginate scaffolds for diabetic wound healing. Int J Biol Macromol. 2016;93(Pt B):1519-29. doi: https://doi.org/10.1016/j.ijbiomac.2016.05.038.

Gao C, Wang F, Liang M, Meng X, Chen Z. Preparation of antibacterial dressing for diabetic wound and evaluation of its perfortmance of promoting healing. J Silk. 2017;54(3):13-9. doi: https://doi.org/10.3969/j.issn.1001-7003.2017.03.003.

Akbar MU, Zia KM, Akash MSH, Nazir A, Zuber M, Ibrahim M. In-vivo anti-diabetic and wound healing potential of chitosan/alginate/maltodextrin/pluronic-based mixed polymeric micelles: Curcumin therapeutic potential. Int J Biol Macromol. 2018;120(Pt B):2418-30. doi: https://doi.org/10.1016/j.ijbiomac.2018.09.010.

Wang T, Zheng Y, Shi Y, Zhao L. pH-responsive calcium alginate hydrogel laden with protamine nanoparticles and hyaluronan oligosaccharide promotes diabetic wound healing by enhancing angiogenesis and antibacterial activity. Drug Deliv Transl Res. 2019b;9(1):227-39. doi: https://doi.org/10.1007/s13346-018-00609-8.

Wang Y, Lu Y, Zhang J, Hu X, Yang Z, Guo Y, et al. A synergistic antibacterial effect between terbium ions and reduced graphene oxide in a poly(vinyl alcohol)-alginate hydrogel for treating infected chronic wounds. Journal of Materials Chemistry B. 2019a;7(4):538-47. doi: https://doi.org/10.1039/c8tb02679c.

Alavi A, Archer J, Coutts P. Use of an advanced collagen matrix dressing on patients with complex chronic lower extremity ulcers: A case series. SAGE Open Med Case Reports. 2021;9:2050313X211013684. doi: https://doi.org/10.1177/2050313X211013684.

Huang X, Xu L, Yu X, Li Y, Huang Z, Xu R, et al. Near-infrared light-responsive multifunctional hydrogel releasing peptide-functionalized gold nanorods sequentially for diabetic wound healing. Journal of Colloid and Interface Science. 2023;639:369-84. doi: https://doi.org/10.1016/j.jcis.2023.02.078.

Chai G, Wang N, Xu M, Ma L, Liu X, Ding Q, et al. Poly (vinyl alcohol)/sodium alginate/carboxymethyl chitosan multifunctional hydrogel loading HKUST-1 nanoenzymes for diabetic wound healing. Int J Biol Macromol. 2024;268(Pt 2):131670. doi: https://doi.org/10.1016/j.ijbiomac.2024.131670.

Tang L, Xie S, Wang D, Wei Y, Ji X, Wang Y, et al. Astragalus polysaccharide/carboxymethyl chitosan/sodium alginate based electroconductive hydrogels for diabetic wound healing and muscle function assessment. Carbohydr Polym. 2025;350:123058. doi: https://doi.org/10.1016/j.carbpol.2024.123058.

Choudhary M, Chhabra P, Tyagi A, Singh H. Scar free healing of full thickness diabetic wounds: A unique combination of silver nanoparticles as antimicrobial agent, calcium alginate nanoparticles as hemostatic agent, fresh blood as nutrient/growth factor supplier and chitosan as base matrix. Int J Biol Macromol. 2021;178:41-52. doi: https://doi.org/10.1016/j.ijbiomac.2021.02.133.

Atia NM, Hazzah HA, Gaafar PME, Abdallah OY. Diosmin Nanocrystal–Loaded Wafers for Treatment of Diabetic Ulcer: In Vitro and In Vivo Evaluation. J Pharm Sci. 2019;108(5):1857-71. doi: https://doi.org/10.1016/j.xphs.2018.12.019.

Fan Y, Wu W, Lei Y, Gaucher C, Pei S, Zhang J, et al. Edaravone-loaded alginate-based nanocomposite hydrogel accelerated chronic wound healing in diabetic mice. Mar Drugs. 2019;17(5). doi: https://doi.org/10.3390/md17050285.

Tan WS, Arulselvan P, Ng SF, Mat Taib CN, Sarian MN, Fakurazi S. Improvement of diabetic wound healing by topical application of Vicenin-2 hydrocolloid film on Sprague Dawley rats. BMC Complement Altern Med. 2019;19(1). doi: https://doi.org/10.1186/s12906-018-2427-y.

Rezvanian M, Ng SF, Alavi T, Ahmad W. In-vivo evaluation of Alginate-Pectin hydrogel film loaded with Simvastatin for diabetic wound healing in Streptozotocin-induced diabetic rats. Int J Biol Macromol. 2021;171:308-19. doi: https://doi.org/10.1016/j.ijbiomac.2020.12.221.

Chi J, Li A, Zou M, Wang S, Liu C, Hu R, et al. Novel dopamine-modified oxidized sodium alginate hydrogels promote angiogenesis and accelerate healing of chronic diabetic wounds. Int J Biol Macromol. 2022;203:492-504. doi: https://doi.org/10.1016/j.ijbiomac.2022.01.153.

Li S, Wang X, Chen J, Guo J, Yuan M, Wan G, et al. Calcium ion cross-linked sodium alginate hydrogels containing deferoxamine and copper nanoparticles for diabetic wound healing. Int J Biol Macromol. 2022;202:657-70. doi: https://doi.org/10.1016/j.ijbiomac.2022.01.080.

Yu H, Gong W, Mei J, Qin L, Piao Z, You D, et al. The efficacy of a paeoniflorin-sodium alginate-gelatin skin scaffold for the treatment of diabetic wound: An in vivo study in a rat model. Biomed Pharmacother. 2022;151:113165. doi: https://doi.org/10.1016/j.biopha.2022.113165.

Raina N, Haque S, Tuli HS, Jain A, Slama P, Gupta M. Optimization and Characterization of a Novel Antioxidant Naringenin-Loaded Hydrogel for Encouraging Re-Epithelization in Chronic Diabetic Wounds: A Preclinical Study. ACS Omega. 2023;8(38):34995-5011. doi: https://doi.org/10.1021/acsomega.3c04441.

Wang Y, Ding C, Zhao Y, Zhang J, Ding Q, Zhang S, et al. Sodium alginate/poly(vinyl alcohol)/taxifolin nanofiber mat promoting diabetic wound healing by modulating the inflammatory response, angiogenesis, and skin flora. Int J Biol Macromol. 2023;252:126530. doi: https://doi.org/10.1016/j.ijbiomac.2023.126530.

Ding C, Yang J, Wang N, Ding Q, Sun S, Gao Y, et al. Sodium alginate/polyvinyl alcohol nanofibers loaded with Shikonin for diabetic wound healing: In vivo and in vitro evaluation. Int J Biol Macromol. 2024;262(Pt 1):129937. doi: https://doi.org/10.1016/j.ijbiomac.2024.129937.

Hao J, Liu C, Zhou L, Wu N, Sun M, Kuang J, et al. Enhancing diabetic wound healing with a pH/glucose dual-responsive hydrogel for ROS clearance and antibacterial activity. Int J Biol Macromol. 2024;272(Pt 2):132935. doi: https://doi.org/10.1016/j.ijbiomac.2024.132935.

Lihao Q, Tingting L, Jiawei Z, Yifei B, Zheyu T, Jingyan L, et al. 3D bioprinting of Salvianolic acid B-sodium alginate-gelatin skin scaffolds promotes diabetic wound repair via antioxidant, anti-inflammatory, and proangiogenic effects. Biomed Pharmacother. 2024;171:116168. doi: https://doi.org/10.1016/j.biopha.2024.116168.

Yang Z, Wang C, Zhang Z, Yu F, Wang Y, Ding J, et al. A pH responsive tannic acid/quaternized carboxymethyl chitosan/oxidized sodium alginate hydrogels for accelerated diabetic wound healing and real-time monitoring. Int J Biol Macromol. 2024;264(Pt 2):130741. doi: https://doi.org/10.1016/j.ijbiomac.2024.130741.

Cao B, Wang C, Guo P, Zhang Q, Wang C, Sun H, et al. Photo-crosslinked enhanced double-network hydrogels based on modified gelatin and oxidized sodium alginate for diabetic wound healing. Int J Biol Macromol. 2023;245:125528. doi: https://doi.org/10.1016/j.ijbiomac.2023.125528.

Khaliq T, Sohail M, Minhas MU, Mahmood A, Munir A, Qalawlus AHM, et al. Hyaluronic acid/alginate-based biomimetic hydrogel membranes for accelerated diabetic wound repair. Int J Pharm. 2023;643:123244. doi: https://doi.org/10.1016/j.ijpharm.2023.123244.

Gainza G, Aguirre JJ, Pedraz JL, Hernández RM, Igartua M. RhEGF-loaded PLGA-Alginate microspheres enhance the healing of full-thickness excisional wounds in diabetised Wistar rats. Eur J Pharm Sci. 2013;50(3-4):243-52. doi: https://doi.org/10.1016/j.ejps.2013.07.003.

Moore K, Ghatnekar G, Gourdie RG, Potts JD. Impact of the controlled release of a connexin 43 peptide on corneal wound closure in an STZ model of type I diabetes. PLoS One. 2014;9(1). doi: https://doi.org/10.1371/journal.pone.0086570.

Tellechea A, Silva EA, Min J, Leal EC, Auster ME, Pradhan-Nabzdyk L, et al. Alginate and DNA gels are suitable delivery systems for diabetic wound healing. Int J Low Extrem Wounds. 2015;14(2):146-53. doi: https://doi.org/10.1177/1534734615580018.

Azam M, Dikici S, Roman S, Mehmood A, Chaudhry AA, U Rehman I, et al. Addition of 2-deoxy-d-ribose to clinically used alginate dressings stimulates angiogenesis and accelerates wound healing in diabetic rats. J Biomater Appl. 2019;34(4):463-75. doi: https://doi.org/10.1177/0885328219859991.

Satish A, Aswathi R, Caroline Maria J, Sai Korrapati P. Triiodothyronine impregnated alginate/gelatin/polyvinyl alcohol composite scaffold designed for exudate-intensive wound therapy. Eur Polym J. 2019;110:252-64. doi: https://doi.org/10.1016/j.eurpolymj.2018.11.032.

Shafei S, Khanmohammadi M, Heidari R, Ghanbari H, Taghdiri Nooshabadi V, Farzamfar S, et al. Exosome loaded alginate hydrogel promotes tissue regeneration in full-thickness skin wounds: An in vivo study. J Biomed Mater Res Part A. 2020;108(3):545-56. doi: https://doi.org/10.1002/jbm.a.36835.

Zhang Y, Zhang P, Gao X, Chang L, Chen Z, Mei X. Preparation of exosomes encapsulated nanohydrogel for accelerating wound healing of diabetic rats by promoting angiogenesis. Mater Sci Eng, C. 2021;120:111671. doi: https://doi.org/10.1016/j.msec.2020.111671.

Zhang Y, Fang M, Xie W, Zhang YA, Jiang C, Li N, et al. Sprayable alginate hydrogel dressings with oxygen production and exosome loading for the treatment of diabetic wounds. Int J Biol Macromol. 2023;242(Pt 3):125081. doi: https://doi.org/10.1016/j.ijbiomac.2023.125081.

Zhou DR, Deng HY, Pu LL, Lin SL, Gou R, Wang FL. The effectiveness and safety of recombinant human growth hormone combined with alginate dressing in the treatment of diabetic foot ulcer: A protocol for systematic review and meta-analysis. Medicine (United States). 2021;100(5):E23984. doi: https://doi.org/10.1097/MD.0000000000023984.

Garcia-Orue I, Santos-Vizcaino E, Sanchez P, Gutierrez FB, Aguirre JJ, Hernandez RM, et al. Bioactive and degradable hydrogel based on human platelet-rich plasma fibrin matrix combined with oxidized alginate in a diabetic mice wound healing model. Biomater Adv. 2022;135:112695. doi: https://doi.org/10.1016/j.msec.2022.112695.

Hao M, Ding C, Sun S, Peng X, Liu W. Chitosan/Sodium Alginate/Velvet Antler Blood Peptides Hydrogel Promotes Diabetic Wound Healing via Regulating Angiogenesis, Inflammatory Response and Skin Flora. J Inflamm Res. 2022;15:4921-38. doi: https://doi.org/10.2147/JIR.S376692.

Hao PC, Burnouf T, Chiang CW, Jheng PR, Szunerits S, Yang JC, et al. Enhanced diabetic wound healing using platelet-derived extracellular vesicles and reduced graphene oxide in polymer-coordinated hydrogels. J Nanobiotechnology. 2023;21(1). doi: https://doi.org/10.1186/s12951-023-02068-x.

Xu K, Deng S, Zhu Y, Yang W, Chen W, Huang L, et al. Platelet Rich Plasma Loaded Multifunctional Hydrogel Accelerates Diabetic Wound Healing via Regulating the Continuously Abnormal Microenvironments. Adv Healthc Mater. 2023;12(28):e2301370. doi: https://doi.org/10.1002/adhm.202301370.

Dananjaya SHS, Bandara N, Molagoda IMN, Sandamalika WMG, Kim D, Ganepola N, et al. Multifunctional alginate/polydeoxyribonucleotide hydrogels for promoting diabetic wound healing. Int J Biol Macromol. 2024;257(Pt 1):128367. doi: https://doi.org/10.1016/j.ijbiomac.2023.128367.

Wu X, Ding C, Liu X, Ding Q, Zhang S, Wang Y, et al. The role of phlorizin liposome-embedded oxidized sodium alginate/carboxymethyl chitosan in diabetic wound healing. Int J Biol Macromol. 2024;279(Pt 3):135324. doi: https://doi.org/10.1016/j.ijbiomac.2024.135324.

Liao W, Duan X, Xie F, Zheng D, Yang P, Wang X, et al. 3D-bioprinted double-crosslinked angiogenic alginate/chondroitin sulfate patch for diabetic wound healing. Int J Biol Macromol. 2023;236:123952. doi: https://doi.org/10.1016/j.ijbiomac.2023.123952.

Dai Y, Zhang Q, Gu R, Chen J, Ye P, Zhu H, et al. Metal ion formulations for diabetic wound healing: Mechanisms and therapeutic potential. Int J Pharm. 2024;667(Pt A):124889. doi: https://doi.org/10.1016/j.ijpharm.2024.124889.

Ozaki CK, Hamdan AD, Barshes NR, Wyers M, Hevelone ND, Belkin M, et al. Prospective, randomized, multi-institutional clinical trial of a silver alginate dressing to reduce lower extremity vascular surgery wound complications. J Vasc Surg. 2015;61(2):419-27.e1. doi: https://doi.org/10.1016/j.jvs.2014.07.034.

Montaser AS, Abdel-Mohsen AM, Ramadan MA, Sleem AA, Sahffie NM, Jancar J, et al. Preparation and characterization of alginate/silver/nicotinamide nanocomposites for treating diabetic wounds. Int J Biol Macromol. 2016;92:739-47. doi: https://doi.org/10.1016/j.ijbiomac.2016.07.050.

He X, Ding Y, Xie W, Sun R, Hunt NC, Song J, et al. Rubidium-Containing Calcium Alginate Hydrogel for Antibacterial and Diabetic Skin Wound Healing Applications. ACS Biomater Sci Eng. 2019;5(9):4726-38. doi: https://doi.org/10.1021/acsbiomaterials.9b00547.

Li Y, Xu T, Tu Z, Dai W, Xue Y, Tang C, et al. Bioactive antibacterial silica-based nanocomposites hydrogel scaffolds with high angiogenesis for promoting diabetic wound healing and skin repair. Theranostics. 2020;10(11):4929-43. doi: https://doi.org/10.7150/thno.41839.

Ma B, Dang W, Yang Z, Chang J, Wu C. MoS2 Nanoclusters-based biomaterials for disease- impaired wound therapy. Appl Mater Today. 2020;20. doi: https://doi.org/10.1016/j.apmt.2020.100735.

Wang M, Yang Y, Yuan K, Yang S, Tang T. Dual-functional hybrid quaternized chitosan/Mg/alginate dressing with antibacterial and angiogenic potential for diabetic wound healing. J Orthop Transl. 2021;30:6-15. doi: https://doi.org/10.1016/j.jot.2021.07.006.

Huang S, Xu S, Hu Y, Zhao X, Chang L, Chen Z, et al. Preparation of NIR-responsive, ROS-generating and antibacterial black phosphorus quantum dots for promoting the MRSA-infected wound healing in diabetic rats. Acta Biomater. 2022;137:199-217. doi: https://doi.org/10.1016/j.actbio.2021.10.008.

Loera-Valencia R, Neira RE, Urbina BP, Camacho A, Galindo RB. Evaluation of the therapeutic efficacy of dressings with ZnO nanoparticles in the treatment of diabetic foot ulcers. Biomed Pharmacother. 2022;155:113708. doi: https://doi.org/10.1016/j.biopha.2022.113708.

Yao S, Zhao Y, Xu Y, Jin B, Wang M, Yu C, et al. Injectable Dual-Dynamic-Bond Cross-Linked Hydrogel for Highly Efficient Infected Diabetic Wound Healing. Adv Healthc Mater. 2022;11(14):e2200516. doi: https://doi.org/10.1002/adhm.202200516.

Aldakheel FM, Mohsen D, El Sayed MM, Fagir MH, El Dein DK. Green Synthesized Silver Nanoparticles Loaded in Polysaccharide Hydrogel Applied to Chronic Wound Healing in Mice Models. Gels. 2023;9(8). doi: https://doi.org/10.3390/gels9080646.

Tang Y, Jia Z, Li X, Zhao X, Zhang S, Luo L, et al. Mechanism of wound repair in diabetic rats using nanosilver-free alginate dressing. J Wound Care. 2023;32(Sup8):CLI-CLX. doi: https://doi.org/10.12968/jowc.2023.32.Sup8.cli.

Bargavi P, Balakumar S, Raghunandhakumar S. Multi-functional bandage - bioactive glass/metal oxides/alginate composites based regenerative membrane facilitating re-epithelialization in diabetic wounds with sustained drug delivery and anti-bactericidal efficacy. Int J Biol Macromol. 2024;262(Pt 2):130054. doi: https://doi.org/10.1016/j.ijbiomac.2024.130054.

Bashiri Z, Sharifi AM, Ghafari M, Hosseini SJ, Shahmahmoodi Z, Moeinzadeh A, et al. In-vitro and in-vivo evaluation of angiogenic potential of a novel lithium chloride loaded silk fibroin / alginate 3D porous scaffold with antibacterial activity, for promoting diabetic wound healing. Int J Biol Macromol. 2024;277(Pt 4):134362. doi: https://doi.org/10.1016/j.ijbiomac.2024.134362.

Chen LP, Wang XY, Ren MJ, Wang Y, Zhao JM, Qiang TT, et al. Promoting the healing of infected diabetic wound by nanozyme-containing hydrogel with anti-bacterial inflammation suppressing, ROS-scavenging and oxygen-generating properties. J Biomed Mater Res Part B Appl Biomater. 2024;112(8):e35458. doi: https://doi.org/10.1002/jbm.b.35458.

Feng Y, Su L, Zhang Z, Chen Y, Younis MR, Chen D, et al. pH-Responsive Wound Dressing Based on Biodegradable CuP Nanozymes for Treating Infected and Diabetic Wounds. ACS Appl Mater Interfaces. 2024;16(1):95-110. doi: https://doi.org/10.1021/acsami.3c12997.

Hu M, Li Z, Liu Y, Feng Y, Wang Z, Huang R, et al. Multifunctional Hydrogel of Recombinant Humanized Collagen Loaded with MSCs and MnO2 Accelerates Chronic Diabetic Wound Healing. ACS Biomater Sci Eng. 2024b;10(5):3188-202. doi: https://doi.org/10.1021/acsbiomaterials.4c00019.

Hu X, He J, Qiao L, Wang C, Wang Y, Yu R, et al. Multifunctional Dual Network Hydrogel Loaded with Novel Tea Polyphenol Magnesium Nanoparticles Accelerates Wound Repair of MRSA Infected Diabetes. Adv Funct Mater. 2024a;34(22). doi: https://doi.org/10.1002/adfm.202312140.

Huang J, Huang J, Zhang X, Xie Q, Zheng Y, Shu C, et al. A bioactive multifunctional dressing with simultaneous visible monitoring of pH values and H2O2 concentrations for promoting diabetic wound healing. Mater horizons. 2024. doi: https://doi.org/10.1039/d4mh01142b.

Li D, Li M, Wang L, Zhang J, Wang X, Nie J, et al. The synergetic effect of alginate-derived hydrogels and metal-phenolic nanospheres for chronic wound therapy. Journal of Materials Chemistry B. 2024b;12(10):2571-86. doi: https://doi.org/10.1039/d3tb02685j.

Li Q, Dong M, Han Q, Zhang Y, Yang D, Wei D, et al. Enhancing diabetic wound healing with a pH-responsive nanozyme hydrogel featuring multi-enzyme-like activities and oxygen self-supply. Journal of Controlled Release. 2024a;365:905-18. doi: https://doi.org/10.1016/j.jconrel.2023.12.015.

Meng H, Zhao Y, Cai H, You D, Wang Y, Wu S, et al. Hydrogels Containing Chitosan-Modified Gold Nanoparticles Show Significant Efficacy in Healing Diabetic Wounds Infected with Antibiotic-Resistant Bacteria. International Journal of Nanomedicine. 2024;19:1539-56. doi: https://doi.org/10.2147/IJN.S448282.

Rajalekshmy GP, Ramesan RM, Geetha CS, Pratheesh KV, Shenoy SJ, Anilkumar TV. In Vivo Wound-Healing Efficacy of Insulin-Loaded Strontium-Cross-Linked Alginate-Based Hydrogels in Diabetic Rats. ACS Biomater Sci Eng. 2024;10(4):2552-66. doi: https://doi.org/10.1021/acsbiomaterials.3c01692.

Vassallo IM, Formosa C. Comparing Calcium Alginate Dressings to Vacuum-assisted Closure: A Clinical Trial. Wounds. 2015;27(7):180-90.

Tsang KK, Kwong EWY, To TSS, Chung JWY, Wong TKS. A Pilot Randomized, Controlled Study of Nanocrystalline Silver, Manuka Honey, and Conventional Dressing in Healing Diabetic Foot Ulcer. Evid Based Complement Alternat Med. 2017;2017:5294890. doi: https://doi.org/10.1155/2017/5294890.

Barbosa MAG, Paggiaro AO, De Carvalho VF, Isaac C, Gemperli R. Effects of hydrogel with enriched sodium alginate in wounds of diabetic patients. Plast Surg Nurs. 2018;38(3):133-8. doi: https://doi.org/10.1097/PSN.0000000000000228.

Lullove EJ, Liden B, Winters C, McEneaney P, Raphael A, Lantis JC. A Multicenter, Blinded, Randomized Controlled Clinical Trial Evaluating the Effect of Omega-3–Rich Fish Skin in the Treatment of Chronic, Nonresponsive Diabetic Foot Ulcers. Wounds. 2021;33(7):169-77. doi: https://doi.org/10.25270/wnds/2021.169177.

Barbosa MG, Carvalho VF, Paggiaro AO. Hydrogel Enriched with Sodium Alginate and Vitamins A and E for Diabetic Foot Ulcer: A Randomized Controlled Trial. Wounds. 2022;34(9):229-35. doi: https://doi.org/10.25270/wnds/20103.

Gould LJ, Orgill DP, Armstrong DG, Galiano RD, Glat PM, Zelen CM, et al. Improved healing of chronic diabetic foot wounds in a prospective randomised controlled multi-centre clinical trial with a microvascular tissue allograft. Int Wound J. 2022;19(4):811-25. doi: https://doi.org/10.1111/iwj.13679.

Hajimohammadi K, Parizad N, Bagheri M, Faraji N, Goli R. Maggot therapy, alginate dressing, and surgical sharp debridement: Unique path to save unresponsive diabetic foot ulcer. Int J Surg Case Rep. 2023;111:108907. doi: https://doi.org/10.1016/j.ijscr.2023.108907.

Jude EB, Apelqvist J, Spraul M, Martini J, Jones G, Harding K, et al. Prospective randomized controlled study of Hydrofiber® dressing containing ionic silver or calcium alginate dressings in non-ischaemic diabetic foot ulcers. Diabet Med. 2007;24(3):280-8. doi: https://doi.org/10.1111/j.1464-5491.2007.02079.x.

Chapter 15
Preclinical Animal Studies on the Antidiabetic Properties of Algae

Deepak Paramasivam, Prasannakumar Chinnamani, Manon Mani Vellingiri, and Nathiya Thiyagarajulu

Abstract Marinae macroalgae, with their rich diversity of bioactive compounds, have gained attention as potential sources of natural antidiabetic agents. This chapter provides a comprehensive analysis of preclinical studies exploring the antidiabetic properties of algae, focusing on their effectiveness across various animal models. Algal compounds, including polysaccharides, phenolics, and fatty acids, exhibit promising hypoglycemic effects by modulating glucose metabolism, enhancing insulin sensitivity, and reducing oxidative stress. Preclinical animal studies serve as crucial tools in understanding these mechanisms, using models such as diabetic rodents to replicate metabolic characteristics of human diabetes. This chapter examines the methodologies and approaches employed in these studies, including different dosages, modes of administration, and treatment durations. Key findings from the animal studies reported earlier highlight significant glucose-lowering effects of algal extracts, supporting the potential of algae-derived compounds as complementary antidiabetic agents. However, preclinical studies are not without limitations; challenges such as variability in algal composition, differences in animal metabolism, and issues in scaling up for clinical applications are addressed. The chapter concludes by discussing future directions in algal antidiabetic research, emphasizing the need for standardized extraction and testing protocols, and exploring the potential of algae as sustainable alternatives in diabetes management. The insights provided aim to bridge the gap between preclinical efficacy and clinical translation, fostering a better understanding of algae as a source of novel antidiabetic therapeutics.

Keywords Algae · Marine Bioactive Compounds · Natural Antidiabetic Agents · Polysaccharides · Preclinical animal studies

D. Paramasivam (✉) · P. Chinnamani · M. M. Vellingiri · N. Thiyagarajulu
Department of Life Sciences, School of Biological and Forensic Sciences, Kristu Jayanti (Deemed-to-be University), Bengaluru, Karnataka, India

R. Raja et al. (eds.), *Algae in Diabetes Management*,
https://doi.org/10.1007/978-981-95-0571-5_15

1 Introduction

Diabetes mellitus (DM) is a chronic metabolic disorder marked by high blood sugar levels due to inadequate insulin production by the pancreas or the body's inability to use insulin effectively. Insulin, a hormone essential for blood glucose regulation, is crucial in maintaining energy homeostasis (Baumeister et al. 2021). Persistent high blood glucose levels, or hyperglycemia, can cause severe long-term complications, affecting various organ systems, particularly nerves and blood vessels. DM is broadly classified into three main types: Type 1, Type 2, and gestational diabetes. Type 1 diabetes mellitus (T1DM), previously referred to as insulin-dependent or juvenile diabetes, results from insufficient insulin production and necessitates lifelong insulin administration. The exact cause of T1DM remains unknown, and prevention strategies are currently unavailable. In contrast, Type 2 diabetes mellitus (T2DM), formerly called non-insulin-dependent diabetes, typically arises from the body's inefficient use of insulin (Rindi et al. 2012; Baumeister et al. 2021). T2DM is prevalent globally, largely due to lifestyle factors such as obesity and physical inactivity, and affects adults and children alike. Often, it is diagnosed only after complications have emerged due to its subtler onset compared to T1DM. Gestational diabetes, occurring during pregnancy, results in elevated blood glucose levels that, while below those diagnostic of diabetes, increase the risk of complications for both mother and child. Additionally, individuals with impaired glucose tolerance (IGT) or impaired fasting glycemia (IFG) are at a higher risk of progressing to T2DM. The prevalence of diabetes has escalated dramatically, particularly in low- to middle-income countries, establishing DM as a leading cause of early mortality worldwide (Hostalek 2019). According to the International Diabetes Federation (IDF), around 537 million adults were living with diabetes as of recent data, with projections reaching 643 million by 2030 and potentially 783 million by 2045 in the absence of effective preventive strategies. Diabetes impacts various metabolic pathways-carbohydrate, protein, and lipid metabolism-leading to organ complications such as kidney failure, blindness, cardiovascular issues, and amputations, which considerably diminish both life quality and expectancy. While numerous pharmacological treatments exist, their high costs and adverse side effects underscore the need for affordable, safe, and effective alternatives (Hossain et al. 2024).

Algae encompass a vast and varied group of organisms that span multiple phylogenetic groups and taxonomic classifications. Generally, algae are recognized as plant-like, photosynthetic organisms that thrive in aquatic environments but lack true roots, stems, leaves, and vascular systems, and they possess simple reproductive structures. Algae are found in diverse habitats across the globe, including seawater, freshwater, soil, and wastewater. They are typically categorized based on their structure and size into two main groups: microalgae, which are either unicellular or form colonies, and macroalgae, which are larger, multicellular species (Rindi et al. 2012). Cyanobacteria, also known as blue-green algae, are photosynthetic prokaryotes that exhibit a range of forms and are adapted to various habitats, from freshwater to marine and terrestrial environments, often forming symbiotic relationships with other organisms such as plants and lichens. Algal cell sizes are highly diverse, ranging from tiny picoplanktonic algae measuring 0.2 to 2 μm to the impressive fronds of giant kelps that can reach lengths of up to 60 meters. "Seaweed" refers specifically to large, visible macroalgae that attach to rocks and other surfaces along shorelines. Seaweeds have numerous applications, including use as food for humans, animal feed, and fertilizers. They grow abundantly in shallow coastal waters and estuaries and are scientifically classified into three main groups: Rhodophyta (red algae), Phaeophyceae (brown algae), and Chlorophyta (green algae). This classification is based on various characteristics, including pigment composition, nutritional content, chemical properties, and structural features. Algae-derived natural products, including alkaloids, flavonoids, terpenoids, steroids, and phenols, have garnered significant attention due to their wide-ranging pharmacological benefits, such as antioxidant and antidiabetic properties (Thirunavukkarasu et al. 2013). Alkaloids, for example, possess cytotoxic properties that inhibit the formation of mitotic spindle fibers, essential for cell division. Terpenoids demonstrate broad antitumor effects, while steroids are valued for their antimicrobial and cardiotonic properties, contributing to their roles in nutrition, traditional medicine, and cosmetics. Tannins exhibit therapeutic effects, acting as antiviral, antibacterial, antiulcer, and antioxidant agents, and also function as anti-inflammatory and cytotoxic agents.

Flavonoids are known for their antimicrobial, antioxidant, and free radical-scavenging abilities, as well as spasmolytic properties. Other bioactive compounds like saponins are utilized for their antioxidant, anticancer, anti-inflammatory, and lipid-lowering effects, making them effective in conditions like hypercholesterolemia and hyperglycemia (Čižmárová et al. 2023). Phenolic compounds, a major bioactive class found in various algae extracts, are known for their potent antioxidant activities (Panche et al. 2016). These compounds play a significant role in algae's defense mechanisms against environmental stresses, such as oxidative damage, light exposure, temperature changes, and pathogenic attacks (Ullah et al. 2020).

Algae, particularly marine species like seaweeds, are gaining significant attention due to their diverse bioactive compounds, including polyunsaturated fatty acids, carotenoids, sulfated polysaccharides, and sterols. These compounds exhibit promising health benefits, notably antioxidative, antibacterial, and anti-inflammatory properties. Seaweed has been recognized as a nutritious food source, low in calories but rich in fiber and essential minerals, and it is now being explored as a potential source of functional ingredients for disease prevention and management (Ahmed et al. 2024). Seaweeds, classified as red (Rhodophyta), brown (Phaeophyta), or green algae (Chlorophyta), are non-vascular photosynthetic organisms commonly found in the littoral zone of oceans. Figure 15.1 represents the general overview of the extraction and purification of algal bioactive compound in the laboratory experiments.

Their pigment composition varies by type, with green algae containing carotenoids and chlorophylls, brown algae featuring fucoxanthin and other chlorophylls, and red algae harboring unique phycobilins (Menaa et al. 2021). These pigments and other bioactive constituents contribute to seaweeds' resilience in marine environments and confer their medicinal potential. Compounds derived from seaweeds

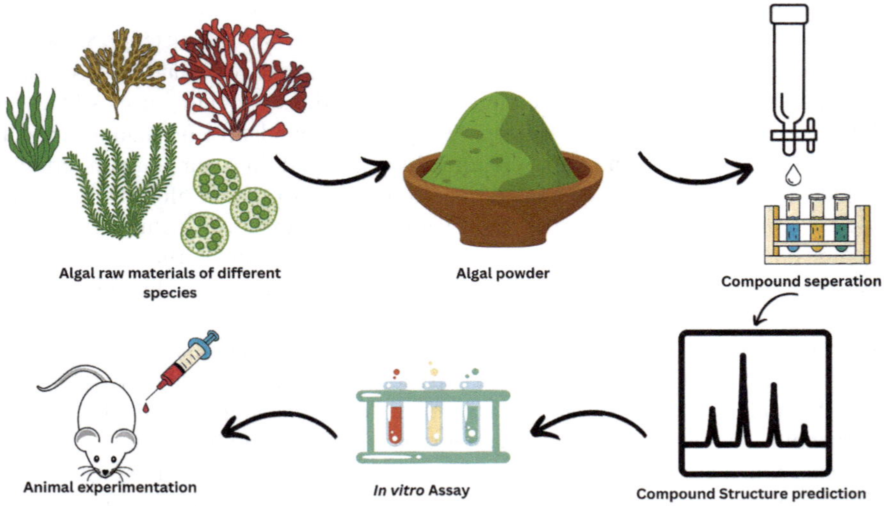

Fig. 15.1 Overview of the algal metabolite extraction and structural characterization of antidiabetic compounds

have shown substantial promise in biomedical applications, including anti-inflammatory, antiviral, antimicrobial, and anticancer activities (Salido et al. 2024a). Historically, extracts from seaweeds such as carrageenan, alginate, and agar have been used in pharmacology for their therapeutic effects. Moreover, seaweeds have proven effective in producing secondary metabolites with pharmacological potential, setting them apart from terrestrial plants due to the unique chemical environment of the ocean, which drives diverse biochemical adaptations. As a commercially renewable resource, seaweeds hold value beyond therapeutic applications. Species from genera such as *Sargassum*, *Padina*, and *Gracilaria* are utilized as fertilizers, food additives, and animal feed (Salido et al. 2024b). Their antioxidant and enzyme-inhibitory properties, especially among brown seaweeds along the Indian coast, are only beginning to be explored (El-Beltagi et al. 2022). Research into the bioactive compounds of algae continues to reveal their potential as sources of safe and effective alternative treatments, especially for conditions like diabetes, where metabolic regulation and antioxidative protection are crucial.

2 Mechanism of Antidiabetic Action of Algal Compounds

The marine macro-algae (seaweeds) are rich in bioactive compounds that could potentially be exploited as functional ingredients for both human and animal health applications. Seaweeds are considered to be a major source of bioactive compounds as they produce a great variety of secondary metabolites, characterized by a broad spectrum of biological activities. Many such bioactive compounds can be extracted from seaweeds. Bacterial species are responsible for mortality in the human population because their infection creates diseases like foodborne gastroenteritis, secondary infections, mastitis, and upper respiratory complications. The marine habitat of India has diverse seaweeds, spread through intertidal and deepwater regions of the Coast. It has already been documented that seaweeds from the Southeast coast of India have antibacterial property (Øverland et al. 2019).

The antidiabetic potential of algae-derived natural products is rooted in their diverse mechanisms of action. These bioactive compounds can influence critical pathways involved in glucose regulation, insulin sensitivity, and the mitigation of diabetes-related complications. Fully understanding these mechanisms is essential to leveraging the therapeutic benefits of algae-based products. While certain algae derivatives show strong potential for managing diabetes, a thorough evaluation of their relative effectiveness is still needed. Algae extracts contain a variety of bioactive substances that may target different aspects of diabetes-some may be particularly effective in regulating glucose levels, while others may enhance insulin sensitivity or prevent complications. Identifying the most promising extracts requires detailed study, including both in vitro and in vivo experiments, as well as comparative analyses of their therapeutic effects (Pereira and Valado 2023a).

Despite the intensive efforts of earlier researchers to isolate and identify new compounds with potential medicinal, health, or pharmaceutical activities, only very few compounds with real potency are available now. Bioactive compounds that have been most extensively researched include sulfated polysaccharides, phlorotannins, and diterpenes (Gupta et al. 2011). And such compounds have been reported to possess strong antiviral, antitumor, and anticancer properties. At the same time, the prebiotic health potential of the polysaccharides from seaweeds is also increasingly being studied either by direct feeding of whole seaweeds or purified seaweed-polysaccharides to laboratory and farm animals (Gupta et al. 2011).

Marine organic environment is considered to be a source of many functional materials, including polyunsaturated fatty acids, polysaccharides, essential minerals and vitamins, antioxidants, enzymes, and bioactive peptides. Among the marine organisms, the marine algae are rich sources of structurally diverse bioactive compounds with various biological activities. Of late, their importance as a source of novel bioactive substances is growing rapidly, and researchers have found that marine algal-originated compounds possess various biological properties (Thirunavukkarasu et al. 2013).

Seaweeds have been reported to possess various bio-efficacy effects like antimicrobial, antiviral, antiproliferative, anti-inflammatory, antioxidant, and antidiabetic properties (Liu et al. 2015). Many researchers have studied only the biological properties of their crude extractions instead of purified single compounds (Lim et al. 2014). The bioactive compounds isolation from seaweeds is a challenging task because seaweeds possess a complex set of phyco-molecules. However, some earlier studies have documented the antidiabetic activity of seaweed bioactive compounds. Some earlier researchers have reported the antidiabetic property of seaweed-bioactive compounds and their mechanism of enzyme inhibition (Lee et al. 2010). The bio-efficacy of the polyphenolic compounds isolated from seaweeds is currently focusing research area in which such compounds are considered to be a prominent antidiabetic agent. Polyphenolic compounds are known to form complexes upon their interaction with numerous proteins, and such compounds derived from vegetables and fruits exhibit various properties including the antidiabetic property (Sharifuddin et al. 2015).

Some earlier researchers have found that the different solvent extracts of seaweeds *viz*; *Caulerpa racemosa*, *Ecklonia stolonifera*, *Odonthalia corymbifera*, *Polyopes lancifolia*, *Symphylocladia latiuscula*, *Ecklonia cava*, *Sargassum binderi*, and *Turbinaria conoides* have shown antidiabetic property (Sharifuddin et al., 2015). In another study, researchers have isolated the antidiabetic compounds like Phlorofucofuroeckol-A, Dieckol, 7-Phloroeckol, Eckol, Dioxinodehydroeckol, and Phloroglucinol from the marine brown algae, *Ecklonia stolonifera* and *Eisenia bicyclis* (MOON et al. 2011). Similarly, the earlier report that the antidiabetic compound, dieckol, obtained from the brown alga, *Ecklonia cava* was found to have antioxidant, α-amylase, α-glucosidase, and reduction of post-prandial blood glucose properties(Lee et al. 2012). The present investigation focuses on the isolation of enzyme inhibitory and antioxidant compound from the seaweed, *Sargassum wightii* and its purification through the chromatographic techniques and spectral characterizations.

3 Methodologies in Preclinical Animal Testing for Antidiabetic Agents

Experimental diabetes mellitus in laboratory animals is commonly induced through chemical, surgical, and genetic approaches. These studies are primarily conducted on rodents, though larger animals are sometimes used. Animal models for screening antidiabetic agents fall into three main categories (Kumar et al. 2012):

3.1 Induced Diabetes Models

Involves methods like chemical induction with agents such as alloxan and streptozotocin, surgical techniques like pancreatectomy, and insulin deficiency through antibody interference. These approaches create insulin deficiency and hyperglycemia, mimicking human diabetes symptoms.

Animal models used to study are following:

- Wistar rats, neonatal STZ-induced rats, sprague dawley rats, alloxan-induced rabbits and high-fat diet-induced rats.

3.2 *Genetically Diabetic Animals*

Includes species with inherited diabetic characteristics, such as spontaneously diabetic rats and mice, transgenic animals, and specific diabetic strains like Chinese hamsters. These models are essential for studying genetic predispositions to diabetes.

Categories include:

- Spontaneously diabetic rats and mice, chinese hamsters, transgenic mice.

3.3 *Miscellaneous Models*

Comprises alternative approaches, including invertebrate models and diet-induced metabolic dysregulation, which offer unique perspectives on diabetes mechanisms and lifestyle-related metabolic issues.

Types include:

- Invertebrate models and diet-induces metabolic dysregulation.

4 Efficacy of Algal Compounds in Different Animal Models

Managing diabetes, a widespread chronic metabolic disorder, poses a significant challenge for healthcare systems worldwide. While traditional treatments can be effective, they frequently have drawbacks, including side effects and insufficient disease control. As a result, there is increasing interest in investigating alternative strategies that may provide safer and more effective options for diabetes care. Also it is a pervasive chronic metabolic disorder impacting millions globally, and remains a major challenge for healthcare systems. Although standard treatments offer some relief, they are often limited by side effects and an inability to fully control the disease. This has led to a growing focus on alternative approaches that could provide safer and more effective solutions for managing diabetes (Pereira and Valado 2023b).

The efficacy of algal compounds in various animal models for diabetes has been a key area of investigation, as different animal models provide unique insights into how these bioactive agents can influence diabetes progression and management. Algal compounds, such as polysaccharides, phenolics, and terpenoids, have shown promise in modulating glucose levels, improving insulin sensitivity, and reducing oxidative stress associated with diabetes. Here's an overview of their effects across common experimental models (Cotas et al. 2024):

1. **Rodent Models (Chemical Induction)**: Algal extracts have been tested extensively in rodent models, where diabetes is induced using agents like streptozotocin (STZ) or alloxan. These studies have demonstrated that algal compounds can lower blood glucose levels, reduce insulin resistance, and improve antioxidant

defenses, making them effective in chemically induced diabetic models (Truong et al. 2023).

2. **Genetically Diabetic Rodents**: In genetically diabetic mice or rats, algal bioactives compounds have shown the potential to mitigate insulin resistance and regulate metabolic pathways. These models help explore how chronic administration of algal compounds can influence long-term blood glucose control and delay complications.
3. **Obese Animal Models**: Given the link between obesity and Type 2 diabetes, obese rodent models (such as those fed a high-fat diet) are often used to study the effects of algae-derived compounds on insulin sensitivity and lipid profiles. Algal compounds, particularly those with antioxidant and anti-inflammatory properties, have shown efficacy in managing obesity-related metabolic dysregulation.
4. **Other Models (e.g., Invertebrates)**: While less common, invertebrate models have also been used to test algal extracts for antidiabetic effects. These models can offer a simpler system for studying specific pathways influenced by algal bioactives, especially those targeting glucose metabolism and oxidative stress (Kaushik et al. 2023).

Studies across these models emphasize that algae-derived compounds may offer multi-targeted benefits in diabetes management. The insulin signaling system, which controls blood glucose levels by improving insulin sensitivity or promoting insulin release to promote glucose metabolism in insulin-resistant conditions, is crucial for the prevention and treatment of diabetes. However, systematic and comparative studies are needed to identify the most effective algal species and compounds, as well as to clarify their specific mechanisms of action in each model (Tong et al. 2024).

5 Challenges and Limitations in Preclinical Animal Studies

Preclinical animal studies are foundational for exploring the efficacy and safety of potential antidiabetic therapies. However, they face several challenges and limitations that impact the translation of findings to human clinical outcomes.

1. **Physiological Differences**: One of the primary limitations is the physiological and metabolic disparity between animals and humans. While rodent models, such as mice and rats, are frequently used due to their cost-effectiveness and ease of genetic manipulation, their metabolic responses, especially in glucose and insulin regulation, often diverge significantly from human physiology. This limits the predictive validity of these models for human diabetes outcomes(Pandey et al. 2023).
2. **Variability in Diabetes Induction Methods**: Animal models of diabetes are commonly induced through chemical agents like streptozotocin (STZ) and alloxan, which selectively damage pancreatic beta cells to simulate diabetes.

However, the extent and permanence of diabetes induced by these agents vary across species, affecting the consistency and reliability of the disease model (Singh et al. 2024a).

3. **Lack of Complex Human Disease Progression**: Animal models often fail to fully replicate the complexity of human diabetes, which involves multiple genetic, environmental, and lifestyle factors. For instance, Type 2 diabetes in humans is typically a progressive condition associated with obesity and metabolic syndrome, which is difficult to reproduce in most animal models. This limitation restricts the exploration of long-term disease mechanisms and the testing of interventions over prolonged periods (Singh et al. 2024b).
4. **Ethical and Regulatory Constraints**: There is growing ethical concern and regulatory pressure to reduce animal use in research, especially for conditions like diabetes where humanized in vitro models may offer partial alternatives. However, despite advancements, these alternatives cannot yet fully replace animal studies in understanding systemic and chronic aspects of diabetes (Domínguez-Oliva et al. 2023).
5. **Technological and Economic Constraints in Advanced Models**: Recent advances, such as organ-on-chip (OoC) technology, offer promising avenues for studying diabetes mechanisms in a more human-representative manner. However, these models are costly and technically demanding, which limits their accessibility and scalability for large-scale antidiabetic drug screening. Additionally, OoC models may lack the capability to model whole-body interactions and complex metabolic feedback systems essential for understanding systemic diabetes (FarhangDoost and Srivastava 2024).
6. **Drug Translation Challenges**: Many antidiabetic drugs that show efficacy in animal models fail in human clinical trials. This challenge highlights the translational gap between preclinical and clinical research due to interspecies differences in drug absorption, metabolism, and action. Addressing this requires further refinement of models to better mimic human drug responses and an increased reliance on humanized models where possible (Jadalannagari and Ewart 2024).

6 Future Perspectives in Algal Antidiabetic Research

Efforts to improve preclinical antidiabetic studies could focus on developing more sophisticated models that incorporate human-derived cells and tissues, like humanized mouse models and patient-specific induced pluripotent stem cells (iPSCs) in organ-on-chip platforms (Ren et al. 2024). Additionally, integrating computational models to simulate human drug responses based on preclinical data may enhance translational accuracy. These advancements, combined with rigorous validation, could help bridge the gap between animal models and human diabetes research.

Future research on the antidiabetic properties of algae offers promising perspectives, given their rich biochemical diversity and potential for sustainable production.

Algae contain bioactive compounds such as polysaccharides, polyphenols, proteins, and pigments, which have been found to enhance glucose uptake, improve insulin sensitivity, and reduce oxidative stress associated with diabetes. These bioactive components can target several metabolic pathways relevant to diabetes management, providing alternatives to conventional antidiabetic drugs (Wu et al. 2023). One future direction in algal antidiabetic research is the isolation and identification of specific active compounds with strong hypoglycemic effects. Current studies have demonstrated the role of polysaccharides from various algae in regulating blood sugar levels by inhibiting α-glucosidase and α-amylase enzymes, which are involved in carbohydrate digestion. Additionally, phenolic compounds in certain seaweeds have shown antioxidant activities that reduce oxidative damage in pancreatic cells, potentially protecting insulin-producing cells from damage due to high glucose levels(Pereira and Valado 2023c). Another promising area is the development of nano-formulations that enhance the bioavailability and stability of these algal compounds. Many bioactive components in algae have low stability or poor absorption when taken orally, which limits their effectiveness. Nano-encapsulation techniques can improve the solubility and targeted delivery of these compounds, potentially maximizing their therapeutic effects. This approach could also enable controlled release formulations, allowing sustained action and reducing the frequency of dosage (Pateiro et al. 2021).

Integrating algal bioactive compounds into functional foods or dietary supplements is another promising direction. Algae-based food products with hypoglycemic benefits could serve as preventive measures or adjunct therapies for diabetes management, catering to the growing demand for natural and dietary-based approaches. Additionally, genetically engineered microalgae could be tailored to produce specific antidiabetic compounds in large quantities, ensuring a reliable and cost-effective source for these bioactives. This approach could significantly lower production costs, making algal compounds more accessible for clinical use (Tavares et al. 2023). The application of advanced technologies such as organ-on-chip (OoC) and in vitro models that simulate human physiological conditions can facilitate detailed studies of the mechanisms underlying the antidiabetic effects of algae. These models allow for precise testing of algal extracts or isolated compounds on pancreatic, hepatic, or adipose tissues, enabling researchers to study cellular and molecular responses without the limitations of animal models. Furthermore, exploring the effects of algae in combination with other antidiabetic therapies could reveal synergistic effects, enhancing overall therapeutic outcomes (Deng et al. 2023).

7 Conclusion

In conclusion, research on the antidiabetic properties of algae holds significant promise for developing new, sustainable, and effective diabetes treatments. Algae are rich in bioactive compounds, such as polysaccharides, peptides, and polyphenols, which show potential in lowering blood glucose levels, improving insulin sensitivity, and safeguarding pancreatic beta cells. These properties position algae as a valuable resource for addressing the increasing global prevalence of diabetes. To fully realize the potential of algae, future research should focus on improving isolation techniques to obtain pure and active compounds. Advanced extraction methods, such as supercritical fluid and enzyme-assisted extraction, can increase the yield and purity of these compounds, facilitating a better understanding of their mechanisms in glucose regulation. Enhancing bioavailability is also essential, as many algal compounds face limitations in absorption and stability. Strategies like nano-encapsulation or biocompatible carriers can improve their efficacy, making algal bioactives viable therapeutic agents. Additionally, exploring functional applications in food and nutraceuticals could help integrate algal extracts into everyday diabetes management. Developing algal-based supplements or functional foods offers a natural alternative or complement to traditional diabetes medications. These products could also meet the rising demand for plant-based and sustainable healthcare options. In summary, by advancing isolation techniques, improving bioavailability, and exploring practical applications, future research can unlock algae's potential for diabetes management, offering natural, effective, and sustainable solutions to support or complement conventional therapies.

Acknowledgments The authors are grateful to the authorities of Kristu Jayanti College, Bengaluru, Karnataka, India, for providing the lab facility and necessary tools needed to complete this work.

Funding: This research received no external funding.

Institutional Review Board Statement: Not applicable.

Informed Consent Statement: Not applicable.

Data availability statement: The authors confirm that the data findings that support this study are available upon request from the corresponding author.

Conflicts of interest: The authors declare that they have no conflicts of interest.

ORCID: Deepak P: https://orcid.org/0000-0003-0801-0441; Chinnamani Prasanna Kumar: https://orcid.org/0000-0002-1975-9258; Vellingiri Manon Mani: https://orcid.org/0000-0003-4248-9345; Thiyagarajulu Nathiya: https://orcid.org/0000-0001-9623-4538

References

Ahmed N, Sheikh MA, Ubaid M, et al (2024) Comprehensive exploration of marine algae diversity, bioactive compounds, health benefits, regulatory issues, and food and drug applications. Measurement: Food 14:100163. https://doi.org/10.1016/j.meafoo.2024.100163

Baumeister R, Thali MJ, Ampanozi G (2021) Postmortem radiological imaging of natural causes of death in adults—a review. Forensic Imaging 26:200473. https://doi.org/10.1016/j.fri.2021.200473

Čižmárová B, Hubková B, Tomečková V, Birková A (2023) Flavonoids as Promising Natural Compounds in the Prevention and Treatment of Selected Skin Diseases. Int J MolSci 24:6324. https://doi.org/10.3390/ijms24076324

Cotas J, Lomartire S, Pereira L, et al (2024) Seaweeds as Nutraceutical Elements and Drugs for Diabetes Mellitus: Future Perspectives. Mar Drugs 22:. https://doi.org/10.3390/md22040168
Deng S, Li C, Cao J, et al (2023) Organ-on-a-chip meets artificial intelligence in drug evaluation. Theranostics 13:4526–4558. https://doi.org/10.7150/thno.87266
Domínguez-Oliva A, Hernández-Ávalos I, Martínez-Burnes J, et al (2023) The Importance of Animal Models in Biomedical Research: Current Insights and Applications. Animals 13:1223. https://doi.org/10.3390/ani13071223
El-Beltagi HS, Mohamed AA, Mohamed HI, et al (2022) Phytochemical and Potential Properties of Seaweeds and Their Recent Applications: A Review. Mar Drugs 20:342. https://doi.org/10.3390/md20060342
FarhangDoost N, Srivastava SK (2024) A Comprehensive Review of Organ-on-a-Chip Technology and Its Applications. Biosensors (Basel) 14:225. https://doi.org/10.3390/bios14050225
Gupta AK, Garg SK, Pal SK, et al (2011) DEVELOPMENT AND CHARACTERIZATION OF SRM MICROSPHERES OF REPAGLINIDE. Journal of Drug Delivery and Therapeutics 1:. https://doi.org/10.22270/jddt.v1i1.4
Hossain MdJ, Al-Mamun Md, Islam MdR (2024) Diabetes mellitus, the fastest growing global public health concern: Early detection should be focused. Health Sci Rep 7:. https://doi.org/10.1002/hsr2.2004
Hostalek U (2019) Global epidemiology of prediabetes - present and future perspectives. Clin Diabetes Endocrinol 5:5. https://doi.org/10.1186/s40842-019-0080-0
Jadalannagari S, Ewart L (2024) Beyond the hype and toward application: liver complex in vitro models in preclinical drug safety. Expert Opin Drug MetabToxicol 20:607–619. https://doi.org/10.1080/17425255.2024.2328794
Kaushik A, Sangtani R, Parmar HS, Bala K (2023) Algal metabolites: Paving the way towards new generation antidiabetic therapeutics. Algal Res 69:102904. https://doi.org/10.1016/j.algal.2022.102904
Kumar S, Singh R, Vasudeva N, Sharma S (2012) Acute and chronic animal models for the evaluation of anti-diabetic agents. CardiovascDiabetol 11:9. https://doi.org/10.1186/1475-2840-11-9
Lee S-H, Min K-H, Han J-S, et al (2012) Effects of brown alga, Ecklonia cava on glucose and lipid metabolism in C57BL/KsJ-db/db mice, a model of type 2 diabetes mellitus. Food and Chemical Toxicology 50:575–582. https://doi.org/10.1016/j.fct.2011.12.032
Lee S-H, Park M-H, Heo S-J, et al (2010) Dieckol isolated from Ecklonia cava inhibits α-glucosidase and α-amylase in vitro and alleviates postprandial hyperglycemia in streptozotocin-induced diabetic mice. Food and Chemical Toxicology 48:2633–2637. https://doi.org/10.1016/j.fct.2010.06.032
Lim SJ, Wan Aida WM, Maskat MY, et al (2014) Isolation and antioxidant capacity of fucoidan from selected Malaysian seaweeds. Food Hydrocoll 42:280–288. https://doi.org/10.1016/j.foodhyd.2014.03.007
Liu Y-X, Si M-M, Lu W, et al (2015) Effects and molecular mechanisms of the antidiabetic fraction of Acoruscalamus L. on GLP-1 expression and secretion in vivo and in vitro. J Ethnopharmacol 166:168–175. https://doi.org/10.1016/j.jep.2015.03.014
Menaa F, Wijesinghe U, Thiripuranathar G, et al (2021) Marine Algae-Derived Bioactive Compounds: A New Wave of Nanodrugs? Mar Drugs 19:484. https://doi.org/10.3390/md19090484
Moon HE, Islam MdN, Ahn BR, et al (2011) Protein Tyrosine Phosphatase 1B and α-Glucosidase Inhibitory Phlorotannins from Edible Brown Algae, *Ecklonia stolonifera* and *Eiseniabicyclis*. BiosciBiotechnolBiochem 75:1472–1480. https://doi.org/10.1271/bbb.110137
Øverland M, Mydland LT, Skrede A (2019) Marine macroalgae as sources of protein and bioactive compounds in feed for monogastric animals. J Sci Food Agric 99:13–24. https://doi.org/10.1002/jsfa.9143
Panche AN, Diwan AD, Chandra SR (2016) Flavonoids: an overview. J NutrSci 5:e47. https://doi.org/10.1017/jns.2016.41

Pandey S, Chmelir T, ChottovaDvorakova M (2023) Animal Models in Diabetic Research—History, Presence, and Future Perspectives. Biomedicines 11:2852. https://doi.org/10.3390/biomedicines11102852
Pateiro M, Gómez B, Munekata PES, et al (2021) Nanoencapsulation of Promising Bioactive Compounds to Improve Their Absorption, Stability, Functionality and the Appearance of the Final Food Products. Molecules 26:. https://doi.org/10.3390/molecules26061547
Pereira L, Valado A (2023a) Algae-Derived Natural Products in Diabetes and Its Complications—Current Advances and Future Prospects. Life 13:1831. https://doi.org/10.3390/life13091831
Pereira L, Valado A (2023b) Algae-Derived Natural Products in Diabetes and Its Complications—Current Advances and Future Prospects. Life 13:1831. https://doi.org/10.3390/life13091831
Pereira L, Valado A (2023c) Unlocking Nature's Treasures: Algae-Derived Natural Products in Diabetes and Its Complications - Current Advances and Future Prospects
Ren R, Jiang J, Li X, Zhang G (2024) Research progress of autoimmune diseases based on induced pluripotent stem cells. Front Immunol 15:. https://doi.org/10.3389/fimmu.2024.1349138
Rindi F, Soler-Vila A, Guiry MD (2012) Taxonomy of Marine Macroalgae Used as Sources of Bioactive Compounds. In: Marine Bioactive Compounds. Springer US, Boston, MA, pp. 1–53
Salido M, Soto M, Seoane S (2024a) Seaweed: Nutritional and gastronomic perspective. A review. Algal Res 77:103357. https://doi.org/10.1016/j.algal.2023.103357
Salido M, Soto M, Seoane S (2024b) Seaweed: Nutritional and gastronomic perspective. A review. Algal Res 77:103357. https://doi.org/10.1016/j.algal.2023.103357
Sharifuddin Y, Chin Y-X, Lim P-E, Phang S-M (2015) Potential Bioactive Compounds from Seaweed for Diabetes Management. Mar Drugs 13:5447–5491. https://doi.org/10.3390/md13085447
Singh R, Gholipourmalekabadi M, Shafikhani SH (2024a) Animal models for type 1 and type 2 diabetes: advantages and limitations. Front Endocrinol (Lausanne) 15:. https://doi.org/10.3389/fendo.2024.1359685
Singh R, Gholipourmalekabadi M, Shafikhani SH (2024b) Animal models for type 1 and type 2 diabetes: advantages and limitations. Front Endocrinol (Lausanne) 15:. https://doi.org/10.3389/fendo.2024.1359685
Tavares JO, Cotas J, Valado A, Pereira L (2023) Algae Food Products as a Healthcare Solution. Mar Drugs 21:578. https://doi.org/10.3390/md21110578
Thirunavukkarasu R, Pandiyan P, Balaraman D, et al (2013) Isolation of bioactive compound from marine seaweeds against fish pathogenic bacteria Vibrio alginolyticus (VA09) and characterisation by FTIR. Journal of Coastal Life Medicinehttps://doi.org/10.12980/JCLM.1.20133D81
Tong A, Wang D, Jia N, et al (2024) Algal Active Ingredients and Their Involvement in Managing Diabetic Mellitus. Biology (Basel) 13:904. https://doi.org/10.3390/biology13110904
Truong TPT, Tran TM, Dai TXT, Tran CL (2023) Antihyperglycemic and anti-type 2 diabetic activity of marine hydroquinone isolated from brown algae (Dictyopterispolypodioides). J Tradit Complement Med 13:408–416. https://doi.org/10.1016/j.jtcme.2023.03.007
Ullah A, Munir S, Badshah SL, et al (2020) Important Flavonoids and Their Role as a Therapeutic Agent. Molecules 25:5243. https://doi.org/10.3390/molecules25225243
Wu JY, Tso R, Teo HS, Haldar S (2023) The utility of algae as sources of high value nutritional ingredients, particularly for alternative/complementary proteins to improve human health. Front Nutr 10:. https://doi.org/10.3389/fnut.2023.1277343

Chapter 16
Drug Delivery and Nanotechnology to Enhance Antidiabetic Properties of Algae

Neethu Asokan, Nalini Soni, Priya Gupta, Tamil Selvan Silambarasan, and Balasubramanian Velramar

Abstract Nanotechnology has revolutionized the drug delivery system offering novel solutions to enhance the therapeutic potential of algal bioactive compounds. Algae are rich in diverse bioactive compounds such as polysaccharides, polyphenols, and peptides that have promising antidiabetic properties. India has about 77 million people of above 18 years suffering from diabetes type 2 and almost 25 million are prediabetes. The potential of these compounds is usually limited due to challenges like poor bioavailability, rapid degradation, and non-specific targeting. This chapter delves into the application of nanotechnology to overcome these challenges, focusing on the development of advanced drug delivery systems specifically designed to enhance antidiabetic properties of algal bioactive compounds. The implementation of nanocarriers like nanofibers, nanoparticles, nanoliposomes, nanohydrogels, nanosomes, and nanoemulsions offers significant advantages, including improved solubility and targeted delivery. These carriers facilitate targeted drug release to specific tissues like pancreatic cells and thereby enhance therapeutic outcomes. The chapter focuses on key nanotechnological approaches in terms of their ability to enhance therapeutic potential. It also aims to provide an overview of the emerging field of nanotechnology-enhanced algal therapies for diabetes, highlighting the potential for innovative diabetic treatment strategies in conjunction with nanotechnological advancements.

N. Asokan
Department of Life Sciences, Sri Satya Sai university for Human Excellence, Kalaburagi, Karnataka, India

N. Soni · P. Gupta · B. Velramar (✉)
Amity Institute of Biotechnology, Amity University Chhattisgarh, Raipur, Chhattisgarh, India
e-mail: vbalasubramanian@rpr.amity.edu

T. S. Silambarasan
Helix Research Center, Department of General Medicine, Saveetha Medical College and Hospital, Saveetha Institute of Medical and Technical Sciences, Saveetha University, Chennai, Tamil Nadu, India

R. Raja et al. (eds.), *Algae in Diabetes Management*,
https://doi.org/10.1007/978-981-95-0571-5_16

Keywords Algae · Antidiabetic property · Drug delivery · Nanocarriers · Nanotechnology

1 Introduction

Nanotechnology has evolved as a transformative system for drug delivery, enhancing therapeutic potential of phytochemicals derived from algae. The unique properties of nanoscale materials help to improve the efficacy and bioavailability of therapeutics. The integration of nanotechnology into drug delivery has made it overcome several challenges like poor bioavailability, rapid degradation, and non-specific targeting. The utilization of nanocarriers like nanoparticles, nanofibers, nanoliposomes, nanohydrogels, nanosomes, and nanoemulsions has been reported to increase solubility, stability, and target delivery of algal bioactive compounds. Norio Taniguchi formally coined the word "nanotechnology" in 1974 however there are reports mentioning concept of nanotechnology in a 1959 lecture "There's Plenty of Room at the Bottom" by Richard Feynman. It is obvious that studying atoms, molecules, and molecular fragments is necessary to comprehend nanotechnology. In a broader sense, it can be stated that matter is composed of atoms, which are microscopic particles invisible to the human eye (Bayda et al. 2019). One well-known example is the administration of COVID-19 mRNA vaccines using lipid nanoparticles; this innovative method is the result of decades of basic research in nanotechnology and nanoscience. There are plenty of such instances in different spheres of life. A groundbreaking advancement in medicine, medication delivery by nanotechnology aims to increase therapy efficacy and precision. To offer controlled and targeted release, drugs are encased in nanoparticles like dendrimers, liposomes, nanocapsules, and polymer nanoparticles (Yetisgin et al. 2020). The encapsulation of drugs with nanoparticles was found to reduce the adverse effects of drugs while increasing their absorption rate simultaneously and minimizing off-target interactions. The modification of surface characteristics of nanoparticles was found by researchers to focus on drug targeting on specific cells or tissues like cancer cells to maximize their therapeutic effects (Tewabe et al. 2021). Additionally, nanotechnology in drug delivery systems could advance the treatment strategies by creating a stimulus-responsive delivery system that could promote targeted and controlled drug release under different pH or temperature (Rahim et al. 2021).

With an innumerous population suffering from diabetes, the potential of nanotechnology in this arena has become increasingly significant. The large surface area to volume ratio and capacity of nanoparticles to penetrate biological membranes/barriers make them an ideal candidate for targeted delivery. Various nanoparticles are involved in drug delivery like liposomes, hydrogels, solid-lipid nanoparticles, nanocrystals, or metal nanoparticles as discussed above (Yetisgin et al. 2020). They were found to have distinct characteristics and unique properties based on the need and application. Liposomes are typically preferred for their capacity to encapsulate both water-soluble and water-insoluble drugs (Guimarães et al. 2021). This enhances their solubility and stability for the treatment. Whereas polymer nanoparticles control releases the drugs allowing tailored responses to certain factors like pH or temperature. The mode of action of their carrier is of two types namely passive and active targeting. The former enables enhanced permeability and retention allowing the nanoparticle to attach to the tumor or target site. This lets the higher concentration of drug at the target site (Shah et al. 2022). In contrast, active targeting involves the attachment of specific ligands to the surface of nanoparticles. Another mechanism is controlled release. The nanoparticles are tailored to release drugs in a controlled manner ensuring good therapeutic efficacy enabling targeted release (Adepu and Ramakrishna 2021). This is attained by diffusion, degradation, or trigger-based release mechanisms. Nanotechnology promotes drug delivery systems with advancement in research and by addressing the research gap, challenges, and limitations. With rise in personalized medicine tailoring the design of nanoparticles can benefit the patients and increase the treatment outcomes (Yang et al. 2021). Nanotechnology and precision medicine have been wide research in the past few years. Nanotechnology has been promoting precision medicine thereby improving the delivery and efficacy of drugs with high number of patients being benefitted (Yang et al. 2021). A previous study on the optimization of nanoparticle delivery system with a one-size-fits-all solution focusing on the heterogenous barriers to delivery and tailoring the design for precision and personalized treatment.

Global prevalence of diabetics is escalating with an estimate of>537 million adults and this estimate is expected to surge to >643 million by 2030 and > 783 million by 2045. It has been estimated that about 77 million individuals are affected by type 2 diabetes with around 25 million classified as prediabetic in India (Xi and Xu 2021. The incidence of diabetes in India is high in population aged 20–79 years as per the International Diabetes Federation. As of 2024, around 101 million diabetics are estimated with 136 million being pre-diabetics. This has been attributed to populations with family history of diabetes and other metabolic disorders. This calls for new therapeutic approaches to treat and control diabetes effectively. Diabetes is primarily categorized into Type 1 and Type 2, where Type I is an autoimmune disorder accounting for 10% of the diabetic cases in India, while Type 2 is majority with 95% cases due to lifestyle factors like physical inactivity, unbalanced diet, stress, or obesity (Zaccardi et al. 2016). It is also associated with other health complications like cardiovascular diseases, neuropathy, foot ulcers, retinopathy, and nephropathy. It has been reported by the International Diabetes Federation that individuals with diabetes are likely to develop other organ failures four times more than normal individuals. Coming to the economic impact of diabetes in India, the healthcare costs are high and profound. It has been reported by ICMA that the healthcare expenditure related to diabetes care is estimated to reach $825 billion by 2030 which is a concern. The loss of productivity because of diabetes related complications like kidney failure, diabetic foot ulcers, and cardiovascular diseases has exacerbated the economic costs. Diabetes mellitus is a metabolic condition (DM) that is triggered by an absolute deficiency in insulin hormone, which is associated with persistent hyperglycemia. It is linked to serious consequences such as diabetic nephropathy, neuropathy, retinopathy resulting in blindness, numbness and tingling of nerves, sciatic pain, and amputations from diabetic foot ulcers (DFU), making it the third greatest cause of mortality globally (Mukhtar et al. 2020). It is categorized into Type I and Type II as discussed earlier. Despite the involvement of several pathogenic pathways, it is understood that type I is insulin-dependent, while type II diabetes is insulin-independent. Diabetes has a wide range of consequences that impact different organ systems. Diabetes-related chronic hyperglycemia causes both macrovascular and microvascular problems (Mukhtar et al. 2020): diabetic nephropathy causes damage to small arteries in kidneys, diabetic neuropathy affects nerves, and diabetic retinopathy causes blindness. Large blood arteries in the body are affected by macrovascular problems, which raise the risk of cardiovascular conditions such as peripheral arterial disease, coronary artery disease, and stroke. Therefore, for therapeutic approach study, the focus is on phytochemicals/naturals such as fucoxanthin, fucoidan, laminarins, alginic acid, and fucosterols present in algae and enhancing their bioactivity by nanotechnology. Studies and developments into diabetes prevention and management focusing on genetic, environmental, and lifestyle factors are to give a great contribution. Along with this the technological advances like glucose monitoring devices and insulin delivery devices have become more accessible than before (Renard 2023). This would enable controlling diabetes efficiently.

Algae, namely microalgae and macroalgae, are abundant in bioactive substances namely polysaccharides, phenolics, and fatty acids like DHA, Omega 3 fatty acids that may have antidiabetic properties. These compounds may enhance insulin, regulate glucose metabolism, and show antioxidant effects. The application of algae that derives bioactive compounds has acquired significance in several health domains like cancer, cardiovascular health, and diabetes management (Saadaoui et al. 2020). The exceptional phytochemicals with biological properties have made algae-derived compounds the most sought antidiabetic agents. Algae were found to be rich in certain compounds like ascorbic acids, DHA, EPA, gamma linolenic acids, fucoxanthin-like pigments, vitamins, proteins, adrenic acid, alginic acid, and several other biomolecules (Tripathi et al. 2024). Phenolic compounds, flavonoids, polysaccharides, fatty acids, vitamins, and minerals are the most researched natural compounds in algae. Polysaccharides like alginate, carrageenan, and fucoidan have been researched to have biological activities like enhancing insulin sensitivity, reducing hyperglycemia, and improving glucose metabolism (Tripathi et al. 2024). Phenolic compounds from algae were found to possess both antioxidant properties and insulin sensitivity. Similarly, omega-3 fatty acids are found to have association with insulin sensitivity and inflammation reduction. Studies on diabetic rats have demonstrated the administration of algal pigments and algal fucoidan to lower glucose levels in blood and improve insulin sensitivity (Mabate et al. 2021). There are also reports on spirulina improving glucose metabolism along with reduction in body weight. Algae have been used for green nanosynthesis like *Ulva* sp., *Xanthomonas* sp., *Chlorella vulgaris, Spirulina* sp., and *Lyngbya* sp., which are some potent algae species employed for nanoparticle synthesis like silver nanoparticles, gold nanoparticles, silica nanoparticles, etc. in comparison to other microorganisms, algae-mediated nanosynthesis has the potential and has beneficial applications in several areas of research (Mukherjee et al. 2021). Algae nano-based hybrid systems are innovative technologies that are in conjunction with algae and nanotechnology to address several issues in different domains. Algae drug delivery system with nanotechnology has been reported to have improved absorption rate and controlled release of drugs (Zhang et al. 2022). Several mechanisms are used to ensure improved drug delivery, and these include encapsulation techniques like nanoencapsulation or microencapsulation; nanoparticle conjugation where nanoparticles are conjugated with algal extracts; surface modification or combining algal compounds with nanomaterials or nanoparticles (Schwestka and Stoger 2021). Nano-algal systems are significant due to their remarkable potential in diabetes management and other healthcare benefits. Algal biocompounds have attained limelight due to their biological properties. The chapter includes comprehensive overview on the role of algae-derived bioactive compounds with nanotechnology for drug delivery for antidiabetic therapies. It can be understood that incorporation of multidisciplinary approaches will help overcome existing limitations and applications of algae-based nano-hybrid systems in diabetes management therapies. The chapter also describes how these innovations can contribute to more effective and sustainable diabetes therapies.

2 Algae-Derived Antidiabetic Compounds

It has been discovered that algae, with their amazing chemical diversity, consist of an assortment of bioactive chemicals that have antidiabetic effects. These substances provide an abundant supply of natural products that may be investigated for the creation of cutting-edge treatment approaches for the control of diabetes (Pereira and Valado 2023). Here, we describe some algae that derive bioactive substances with potential for modifying the major pathways governing glucose homeostasis and problems associated with diabetes (Abo-Shady et al. 2023).

2.1 Polysaccharide

Algae produce and synthesize several polysaccharides with possible antidiabetic properties (Bhuyan et al. 2023). Red algae possess sulfated polysaccharides like carrageenan and agar with antihyperglycemic properties by blocking the enzymes responsible for carbohydrate breakdown namely α-glucosidase and α-amylase, thereby dropping postprandial glucose levels (Lu and Chen 2022). Qian et al. (2021) have reported potent antioxidant activity of polysaccharide of brown algae. Similarly, Priatni et al. (2016) have recorded highest inhibition of glucosidase activity by exopolysaccharide isolated from cyanobacteria. There are also studies on Laminaria polysaccharides significantly reducing blood glucose levels at fasting and increasing the insulin levels in mice models (Li et al. 2012).

2.1.1 Polyphenols

Polyphenols are chief class of secondary metabolite abundant in several plants and algae. The possible antidiabetic effects of polyphenols from algae, such as flavonoids and phenolic acids, have been reported in many studies. These substances have anti-inflammatory and antioxidant qualities that may help lessen the inflammation that may be chronic and oxidative stress linked to diabetes (Rudrapal et al. 2022). Phlorotannin, among other polyphenols generated from brown algae, has demonstrated inhibitory effects on enzymes alpha-glucosidase and alpha-amylase, resulting in decreased glucose absorption and enhanced regulation of blood sugar levels (Rathod et al. 2023). There are reports asserting strong inhibition effect of alpha glucosidase activity both in vitro and in vivo with polyphenols isolated from algae namely phlorotannin (Iwai 2008). Lee and Han (2012) have also reported that polyphenols extracted from Sargassum possess alpha glycosidase and alpha amylase activity along with insulin stimulation.

2.1.2 Pigments

Pigments like chlorophylls, carotenoids, and phycobiliproteins are found in abundance in algae and were reported to have several biological activities (Figueiredo et al. 2016). Brown algae-derived pigment Fucoxanthin is a carotenoid with potent ability to stimulate the uptake of glucose into skeletal muscles and adipocytes thereby increasing the insulin sensitivity and absorption rate (Pereira and Valado 2023). Moreover, several pigments possess antioxidant properties and anti-inflammation properties, contributing to mitigating diabetes-related complications (Bocanegra et al. 2021). Li and Kim (2011) have reported that pigments from algae possess several biological properties like anticancer, antitumor, antidiabetic, and chronic disease therapies.

2.1.3 Lipids

The potential benefits of algae-derived lipids, especially Omega-3-fatty acids such as DHA and EPA, were thoroughly investigated in previous studies with respect to diabetes (Pereira and Valado 2023). These fats have been shown to have anti-inflammatory, antihyperglycemic, and insulin-sensitizing qualities. Furthermore, it has been demonstrated that microalgae-derived lipids, especially from *Spirulina* and *Chlorella*, were reported to improve lipid profiles and lower triglyceride levels in diabetics (Sinha et al. 2023).

2.1.4 Peptides and Proteins

Numerous peptides and proteins, including those having antidiabetic properties, are produced by algae. A few peptides produced from algae have been shown to block dipeptidyl peptidase-4 (DPP-4), an incretin hormone-degrading enzyme. Peptides from algae were found to increase the effects of incretins by inhibiting DPP-4, improve glucose regulation and insulin secretion (Conde et al. 2021). While Anderson et al. (2007) have reported in their study the potential activity of protein kinase on insulin activity especially for the treatment of diabetic neuropathy. Some important algal species were reported and tabulated as in Table 16.1 with their activity (Pereira et al. 2023). These are only a few of algae algae-derived bioactive compounds used in diabetes management while many are yet to be studied and researched. These algae-derived compounds could be breakthrough strategy to develop novel therapies for diabetes taking into consideration their diverse pharmacological activities like regulation of glucose metabolism, improve insulin sensitivity, and reduce the consequences of diabetes especially the secondary illness like foot ulcers (Pereira and Valado 2023). It can be said that further studies and quest for algae-derived compounds integrated with nanotechnology will help find new therapies and drugs to effectively treat diabetics and enhance patient outcomes simultaneously reducing healthcare costs (Zhao et al. 2018; Bocanegra et al. 2021).

3 Nanotechnology-Based Algal Bioactive Compound's Drug Delivery Systems

Fucosterols, fucoidan, alginic acids, phycoxanthin, and chlorotaninns are the most common algal compounds having beneficial properties like antioxidant property, antidiabetic property, anti-inflammatory, and cytotoxic activities (Nasab et al. 2020). Similarly, Silva et al. (2024) recently conferred advances in algae-derived compounds for nutraceutical applications for being antidiabetic, antioxidant, and anti-inflammatory in nature. The compounds derived from algae are advantageous in diabetes therapy, given the quest for effective antidiabetic medical therapies for the control of diabetes. While probing for potent antidiabetic drugs, marine algae were reported to be a potent source with great bioactivity (Bălaşa et al. 2020). It is expected that the isolation, categorization, and pharmacological examination of uncharted marine algae would prove beneficial in the quest for novel and valuable

Table 16.1 Glucose levels reduction by algae extracts via miscellaneous mechanisms (Periera et al. 2023)

Sl. no.	Species	Extract	Biological activities
1.	*Alaria marginata* (Phaeophyceae)	Ethyl acetate extract	α-Amylase inhibition (IC_{50} = 63.28 μg/mL) α-glucosidase inhibition (IC_{50} = 15.66 μg/mL)
2.	*Ascophyllum nodosum* (Phaeophyceae)	Aqueous ethanolic extract	α-Glucosidase inhibition (IC_{50} = 77 μg/mL)
3.	*A. nodosum* (Phaeophyceae)	Water extract	α-Amylase inhibition (IC_{50} = 1.34 μg/mL) α-glucosidase inhibition (IC_{50} = 0.24 μg/mL)
4.	*A. nodosum* (Phaeophyceae)	Phlorotannin-rich extract	α-Amylase inhibition (IC_{50} ~ 0.1 μg/mL) α-glucosidase inhibition (IC_{50} ~ 20 μg/mL)
5.	*A. nodosum* (Phaeophyceae)	Cold water and ethanol extracts	α-Amylase inhibition (water) (IC_{50} = 53.6 μg/mL) α-amylase inhibition (ethanol) (IC_{50} = 44.7 μg/mL)
6.	*A. nodosum* (Phaeophyceae)	Fucoidan	α-Amylase inhibition (IC_{50} = 0.12–4.64 mg/mL) α-glucosidase inhibition (IC_{50} = 0.013–0.047 mg/mL)
7.	*Ulva intestinalis* (Chlorophyta)	Methanolic extract	α-Amylase inhibition (500 μg/mL)
8.	*U. lactuca* (Chlorophyta)	Water extract	α-Amylase inhibition (IC_{50} = 67 μg/mL) α-glucosidase inhibition (IC_{50} = 53 μg/mL)
9.	*U. lactuca* (Chlorophyta)	Methanolic extract	α-Amylase inhibition (IC_{50} = 1.00 mg/mL) α-glucosidase inhibition (IC_{50} = 1.14 mg/mL)
10.	*Ecklonia bicyclis* (Phaeophyceae)	Phloroglucinol derivatives	α-Amylase inhibition at 1 mM by: Dieckol, inhibition = 97.5% 1-(3′,5′-dihydroxyphenoxy)-7-(2″,4″,6″-trihydroxyphenoxy)-2,4,9-trihydroxydibenzo-1,4-dioxin, inhibition = 89.5% Eckol, inhibition = 87.5%
11.	*E. bicyclis* (Phaeophyceae)	Phlorofucofuroeckol-A Dieckol 7-Phloroeckol Eckol Dioxinodehydroeckol Phloroglucinol	α-Glucosidase inhibition by: Phlorofucofuroeckol-A (IC_{50} = 1.37 μM) Dieckol (IC_{50} = 1.61 μM) 7-Phloroeckol (IC_{50} = 6.13 μM) Eckol (IC_{50} = 22.78 μM) Dioxinodehydroeckol (IC_{50} = 34.60 μM) Phloroglucinol (IC_{50} = 141.18 μM)

(continued)

Table 16.1 (continued)

Sl. no.	Species	Extract	Biological activities
12.	*E. bicyclis* (Phaeophyceae)	Fucofuroeckol A	α-Amylase inhibition (IC_{50} = 42.91 μM) α-glucosidase inhibition (IC_{50} = 131.34 nM)
13.		Dioxinodehydroeckol	α-Amylase inhibition (IC_{50} = 472.70 μM) α-glucosidase inhibition (IC_{50} = 93.33 nM)
14.	*E. cava* (Phaeophyceae)	Dieckol Fucodiphloroethol G Phlorofucofuroeckol A 6,6′-bieckol 7-Phloroeckol	α-Glucosidase inhibition by: Dieckol (IC_{50} = 10.8 μM) Fucodiphloroethol G (IC_{50} = 19.5 μM) Phlorofucofuroeckol A (IC_{50} = 19.7 μM) 6,6′-bieckol (IC_{50} = 22.2 μM) 7-Phloroeckol (IC_{50} = 49.5 μM)
15.		Dieckol 7-Phloroeckol 6,6′-bieckol Fucodiphloroethol G Phlorofucofuroeckol A	α-Amylase inhibition by: Dieckol (IC_{50} = 124.9 μM) 7-Phloroeckol (IC_{50} = 250.0 μM) 6,6′-bieckol (IC_{50} > 500 μM) Fucodiphloroethol G (IC_{50} > 500 μM) Phlorofucofuroeckol A (IC_{50} > 500 μM)
16.	*E. cava* (Phaeophyceae)	Dieckol	α-Amylase inhibition (IC_{50} = 0.66 mM) α-glucosidase inhibition (IC_{50} = 0.24 mM)
17.	*E. maxima* (Phaeophyceae)	Eckol Dibenzo [1,4] dioxine-2,4,7,9-tetraol Phloroglucinol	α-Glucosidase inhibition by: Eckol (IC_{50} = 11.16 μM) Dibenzo [1,4] dioxine-2,4,7,9-tetraol (IC_{50} = 33.69 μM) Phloroglucinol (IC_{50} = 1991 μM)
18.	*Ecklonia cava* subsp. *stolonifera* (formerly *E. stolonifera*) (Phaeophyceae)	Water extract	α-Glucosidase inhibition against: α-glucosidase (*Saccharomyces*—Yeast) (IC_{50} = 0.026 mg/mL) Rat intestinal maltase (IC_{50} = 4.213 mg/mL) Rat intestinal sucrose (IC_{50} = 10.10 mg/mL) Rat intestinal isomaltase (IC_{50} > 100 mg/mL) rat intestinal glucoamylase (IC_{50} > 100 mg/mL)
19.		Methanolic extract	α-Glucosidase inhibition against: α-glucosidase (Saccharomyces) (IC_{50} = 0.022 mg/mL) Rat intestinal maltase (IC_{50} = 0.772 mg/mL) Rat intestinal sucrose (IC_{50} = 4.056 mg/mL) Rat intestinal isomaltase (IC_{50} > 100 mg/mL) Rat intestinal glucoamylase (IC_{50} = 5.851 mg/mL)

(continued)

Table 16.1 (continued)

Sl. no.	Species	Extract	Biological activities
20.	*E. cava* subsp. *stolonifera* (Phaeophyceae)	Phlorofucofuroeckol-A Dieckol 7-Phloroeckol Eckol Dioxinodehydroeckol Phloroglucinol	α-Glucosidase inhibition by: Phlorofucofuroeckol-A (IC_{50} = 1.37 μM) Dieckol (IC_{50} = 1.61 μM) 7-Phloroeckol (IC_{50} = 6.13 μM) Eckol (IC_{50} = 22.78 μM) Dioxinodehydroeckol (IC_{50} = 34.60 μM) Phloroglucinol (IC_{50} = 141.18 μM)
21.	*Eucheuma denticulatum* (Rhodophyta)	Ethanolic extract	α-Amylase inhibition (10 mg/mL)
22.	*Fucus distichus* (Phaeophyceae)	Ethyl acetate extract	α-Amylase inhibition (IC_{50} = 13.98 μg/mL) α-glucosidase inhibition (IC_{50} = 0.89 μg/mL)
23.	*F. vesiculosus* (Phaeophyceae)	Cold water and ethanol extracts	α-Glucosidase inhibition (water) (IC_{50} = 0.32 μg/mL) α-glucosidase inhibition (ethanol) (IC_{50} = 0.49 μg/mL)
24.	*F. vesiculosus* (Phaeophyceae)	Fucoidan	α-Glucosidase inhibition (IC_{50} = 0.049 mg/mL)
25.	*Gracilaria corticata* (Rhodophyta)	Water extract	α-Amylase inhibition (IC_{50} = 82 μg/mL) α-glucosidase inhibition (IC_{50} = 87 μg/mL)
26.	*G. edulis* (Rhodophyta)	Water extract	α-Amylase inhibition (IC_{50} = 83 μg/mL) α-glucosidase inhibition (IC_{50} = 46 μg/mL)
27.	*G. edulis* (Rhodophyta)	Methanolic extract	α-Amylase inhibition (300 μg/mL)
28.		Acetone extract	α-Amylase inhibition (300 μg/mL)
29.		Acetate extract	α-Amylase inhibition (300 μg/mL)
30.	*G. gacilis* (Rhodophyta)	Methanolic extract	α-Amylase inhibition (300 μg/mL)
31.	*G. opuntia* (Rhodophyta)	Water extract (sulfated polygalactans)	α-Amylase inhibition (IC_{50} = 0.04 mg/mL) α-glucosidase inhibition (IC_{50} = 0.09 mg/mL)
32.	*Grateloupia elliptica* (Rhodophyta)	2,4,6-Tribromophenol	α-Glucosidase inhibition against: *B. stearothermophilus* (IC_{50} = 130.3 μM) *S. cerevisiae* (IC_{50} = 60.3 μM) Rat intestinal maltase (IC_{50} = 5.0 mM) Rat intestinal sucrose (IC_{50} = 4.2 mM)
33.		2,4-Dibromophenol	α-Glucosidase inhibition against: *Bacillus stearothermophilus* (IC_{50} = 230.3 μM) *S. cerevisiae* (IC_{50} = 110.4 μM) rat intestinal maltase (IC_{50} = 4.8 mM) rat intestinal sucrose (IC_{50} = 3.6 mM)

(continued)

Table 16.1 (continued)

Sl. no.	Species	Extract	Biological activities
34.	*Kappaphycus alvarezii* (Rhodophyta)	Water extract (sulfated polygalactans)	α-Amylase inhibition (IC_{50} = 0.15 mg/mL) α-glucosidase inhibition (IC_{50} = 0.09 mg/mL)
35.	*Odonthalia corymbifera* (Rhodophyta)	Bromophenols	α-Glucosidase (*S. cerevisiae*) inhibition by: 1,4-Butanediol diglycidyl ether (IC_{50} = 0.098 μM) 4-Bromo-2,3-dihydroxy-6-hydroxymethylphenyl 2,5- dibromo-6-hydroxy-3-hydroxymethylphenyl ether (IC_{50} = 25.0 μM) 4-Bromo-2,3-dihydroxy-6-methoxymethylphenyl 2,5-dibromo-6-hydroxy-3--methoxymethylphenyl ether (IC_{50} = 53.0 μM) 2,3-dibromo-4,5-dihydroxybenzyl alcohol (IC_{50} = 89.0 μM)
36.	*Polyopes lancifolius* (Rhodophyta)	1,4-Butanediol diglycidyl ether	α-Glucosidases inhibition against: *B. stearothermophilus* (IC_{50} = 0.12 μM) *S. cerevisiae* (IC_{50} = 0.098 μM) Rat Intestinal maltase (IC_{50} = 1.20 mM) Rat intestinal sucrose (IC_{50} = 1.00 mM)
37.	*Portieria hornemannii* (as *Chondrococcus hornemannii*) (Rhodophyta)	Methanolic extract	α-Amylase inhibition (300 μg/mL)
38.		Acetate extract	α-Amylase inhibition (300 μg/mL)
39.	*Sargassum patens*	2-(4-(3,5 (dihydroxyphenoxy)-3,5-dihydroxyphenoxy) benzene-1,3,5-triol	α-Amylase inhibition (IC_{50} = 3.2 μg/mL) α-glucosidase inhibition against: Rat intestinal maltase (IC_{50} = 114.0 μg/mL) Rat intestinal sucrose (IC_{50} = 25.4 μg/mL)
40.	*S. polycystum* (Phaeophyceae)	Water extract	α-Amylase inhibition (IC_{50} = 60 μg/mL) α-glucosidase inhibition (IC_{50} = 50 μg/mL)
41.	*S. polycystum* (Phaeophyceae)	Ethanolic extract	α-Amylase inhibition (10 mg/mL)
42.	*S. ringgoldianum* (Phaeophyceae)	Methanolic extract	α-Amylase inhibition (IC_{50} = 0.18 mg/mL) α-glucosidase inhibition (IC_{50} = 0.12 mg/mL)
43.	*S. swartzii* (as *S. wightii*) (Phaeophyceae)	Methanolic extract	α-Amylase inhibition (300 μg mL^{-1})

(continued)

Table 16.1 (continued)

Sl. no.	Species	Extract	Biological activities
44.	*S. swartzii* (as *S. wightii*) (Phaeophyceae)	Fucoidan	α-Glucosidase inhibition (IC_{50} = 132 μg/mL)
45.	*S. swartzii* (as *S. wightii*) (Phaeophyceae)	Ethanol extraction and $CaCl_2$ precipitation method (Fucoidan)	α-Amylase inhibition (IC_{50} = 103.83 μg/mL)
46.	*S. tenerrimum* (Phaeophyceae)	Methanolic extract	α-Amylase inhibition (300 μg/mL)
47.	*Spatoglossum asperum* (Phaeophyceae)	Methanolic extract	α-Amylase inhibition (IC_{50} = 55 μg/mL) α-glucosidase inhibition (IC_{50} = 61 μg/mL)
48.	*Spatoglossum schroederi* (Phaeophyceae)	Acetone crude extract	α-Amylase inhibition (IC_{50} = 0.58 mg/mL)
49	*Symphyocladia latiuscula* (Phaeophyceae)	Bromophenols	2,3,6-Tribromo-4,5-dihydroxybenzyl alcohol, (IC_{50} = 11.0 μM) Sucrose inhibition (IC_{50} = 2.4 μg/mL)
50..	*Turbinaria ornata* (Phaeophyceae)	Methanolic extract	α-Amylase inhibition (300 μg/mL)

antidiabetic compounds. Red and brown algae showed signs of preventing diabetes(Zhao et al. 2018; Motshakeri et al. 2013). Most research with phytochemicals derived from algae showed that these compounds might regulate blood glucose by inhibiting enzymes that hydrolyze carbohydrates and protein tyrosine-phosphatase 1B, as well as by improving insulin sensitivity, glucose absorption, and other effects that protect against diabetes (Zhao et al. 2018; Motshakeri et al. 2013). Sun et al. (2024) in a recent study have mentioned that protein tyrosine phosphatase plays a vital role in managing cardiovascular diseases, reversing obesity, improving insulin sensitivity, and controlling glucose level in blood.

Many people are interested in the biosynthesis of metal nanoparticles from medicinal plants as a safe alternative to dangerous chemical and physical synthesis processes. Due to their excellent production of AuNPs and unique metal tolerance, plants are utilized (Fawcett et al. 2017; Anand et al. 2017; Shah et al. 2015). A new study has reported the sustainable applications of algae-mediated synthesized metal nanoparticles in biomedicine (Yedoti and Supraja 2024). Algae-derived organic compounds/phytochemicals are non-toxic chemical constituents found in algae that make up a whole orchestra suitable for enhanced medical applications. It was reported that phytocomponents like terpenoids, polysaccharides, polyols, and flavones are involved in the bio-reduction, stabilization, and bio-capping processes that result in the generation of stable AuNPs and AgNPs. The mechanism and inhibitory activity of plant substances with diabetes are revealed by their inhibitory ability, the enzyme inhibitor kinetics, and ligand binding dynamics with the help of silico docking studies (Anand et al. 2017). Because marine algae contain bioactive components, a great deal of research in recent years has been conducted regarding their potential as antidiabetic agents. Odenthal et al. (2024) have mentioned the role of algae and cyanobacteria-derived compounds to possess enzymes and inhibitors for diabetes management.

One of the numerous bioactive compounds present in significant concentrations in brown algae are phenols which are therapeutic targets for a variety of human diseases, including diabetes. Phenol tannin subclasses, found in a wide range of brown algal species, show significant antidiabetic effects via a few different pathways. Overall, tests conducted *in vitro* and *in vivo* (animal models) have suggested that marine brown algae may have a hypoglycemic effect through different routes (Ogunwa et al. 2018). In rats treated with streptozotocin (STZ) to develop diabetes, fucosterol extracted from the brown alga *Pteris siliquosa* decreased serum glucose levels and prevented glycogen breakdown. A brown alga extract called *Pelvetia babingtonii* showed strong α-glucosidase inhibitory activity and proved to be efficacious in reducing postprandial hyperglycemia (Lee et al. 2004; Ohta et al. 2002). Previous studies have also reported similar results with bioactive compounds derived from brown algae (Abo-Shady et al. 2023) (Fig. 16.1).

Ultimately, polysaccharides produced from *U. lactuca,* known as sea lettuce, were reported to lower blood sugar levels by blocking enzymes that are directly involved in the breakdown and absorption of starch in the small intestine and plasma. In diabetic rats, ethanol extraction of *Ulva rigida* was found to reduce blood

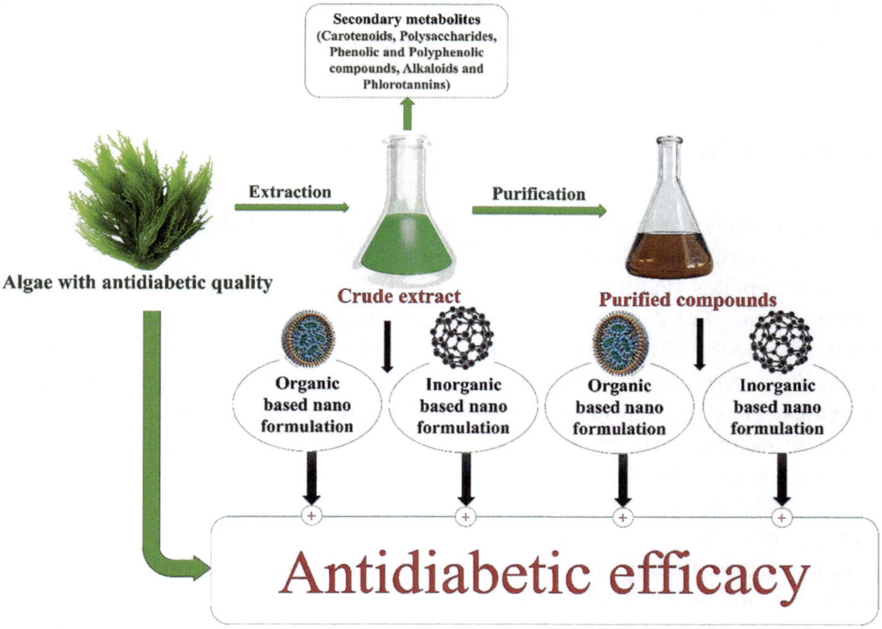

Fig. 16.1 Synthesis of algae-based nanoparticles for antidiabetic activity

glucose levels (Tian et al. 2015; BelHadj et al. 2013; Tas et al. 2011; Celikler et al. 2009). Several algae, including those in the Chlorophyceae, Phaeophyceae, Rhodophyceae, and Cyanophyceae families, have the potential to be used for AgNP biosynthesis due to their capacity for rapid growth, large levels of organic matter, and high metal accumulation. AgNPs from the marine red algae *Halymenia porphyroides* showed a dose-dependent *in vitro* antidiabetic activity by inhibiting the enzymes responsible for diabetes control in previous study reports (Khanna et al. 2019; Priya et al. 2021). Several nanotechnology approaches were developed to help patients with insulin-dependent diabetes live better lives. They aid blood glucose control by facilitating insulin delivery and non-invasive glucose monitoring, and customized formulation for diabetes therapy. At this crucial point, novel classes of tailored, specific nanoparticle drugs are created for various metabolic diseases (Ernst et al. 2019; Anand et al. 2017). The effect of age-related characteristics (e.g., child immaturity, physiological and metabolic changes in old age) is currently less explored in both nanomedicine and pharmacology since insufficient animal models are used (Ramos-Romero et al. 2021; Morran et al. 2015; Lin and Sun 2010). It is mentioned in studies that the bioavailability of insulin delivered by methods other than subcutaneous injection is only 60%. Additionally, the inability to translate animal model results into clinical human trial therapy may be due to change in gut permeability as in type 1 diabetes, insulin resistance in type 2 diabetes, which may impact the development of novel nanoparticulated drug preparations (Ramos-Romero et al. 2021; Morran et al. 2015; Lin and Sun 2010). While taken as a whole, the requirements for the medication and the metabolic alterations brought on by illness or aging must be considered while producing insulin in the future by NPs. In addition, appropriate animal models and disease prevention strategies are required.

4 Enhancing Antidiabetic Properties Using Nanotechnology

Nanotechnology has the capability to drastically change diabetes management by enabling improved medication administration, better monitoring, and maybe new treatment options (Patra et al. 2018). Algae-derived compounds show antidiabetic properties due to their diverse biochemical composition and activities. The high level of phytochemicals in algae makes it very effective against diabetes. Algae like *Chlorella, Spirulina* were found to possess strong antidiabetic and antioxidant activities. They are rich in polysaccharides, peptides that can reduce blood sugar and improve insulin secretion (Abo-Shady et al. 2023). They are known to increase glucose uptake and metabolism, improve pancreatic function, and reduce oxidative stress. Additionally, it was reported that some polysaccharides of algae exhibit hypoglycemic activity that may impact the digestion and absorption rate of carbohydrates. These findings may be a star finding to comprehend their efficacy and mode of action in diabetes care (Bocanegra et al. 2021). Nanoencapsulation of these algal compounds could prevent deterioration of these compounds and increase cell targeting and bioavailability. This may also help maintain stable blood glucose levels with controlled release of bioactive compounds/algae derived drugs to increase stability and absorption rate. This tailored transport of the algal drugs with functionalized nanoparticles will result in a long-lasting diabetic control.

4.1 Nanotechnology Techniques to Enhance Antidiabetic Property of Algae

Nanotechnology integrated with algae-derived compounds has enabled significant advancements in the field of medicine, pharmaceuticals especially in drug delivery and diabetes therapies (Anjum et al. 2021). The application of algal nanotechnology has brought about an improvement in diabetes care by increasing the stability, bioavailability, and distribution of antidiabetic compounds (Arshad et al. 2021). The key nanotechnology techniques employed for enhanced antidiabetic property of algae may include nanoparticle synthesis, encapsulation, nanocarriers, nanoemulsions, green nanosynthesis, and surface functionalization.

1. **Nanoparticle Encapsulation or Nanoencapsulation:** The absorption rate, stability, and distribution of algae-derived compounds like polysaccharides and phenols were enhanced by encapsulation of these into nanoparticles or nanomaterials. This technique improved the viability of compounds, controlling blood glucose levels. External stress like degradation, UV and sunlight exposure, heat and oxygen exposure was controlled with encapsulation technique. Previous reports have mentioned the nanoencapsulation of alpha glucosidase and alpha amylase enzymes in silver nanoparticles from marine algae namely *Colpomenia sinuosa* and other algae like *Argyreia nervosa, Punica granatum,* improved antidiabetic properties. In a study by using a simple microwave-assisted hydrothermal synthesis, brown macroalgae (BMG) were utilized as carriers for ZnO and brown algae and cobalt-doped ZnO (Co-ZnO/BMG). One potential commercial use for synthetic Co-ZnO/BMG composite is as an antioxidant and antidiabetic drug. Where, many essential physiologically active chemicals with strong antimicrobial, antifungal, antibacterial, anticancer, anti-inflammatory, and antioxidant qualities are derived from the brown macroalgae Sargassum species. Several types of nanoparticles are used for encapsulation like lipid-based nanoparticles like liposomes and polymeric nanoparticles like poly lactic co-glycolic acid and chitosan. Immobilization of enzymes with nanoscaffolds has been stated as a promising technique in biocatalysis, biomedicine, and biosensors. These nanoscaffolds encapsulated with enzymes were found to have potential drug delivery and stability (Raghavendra and Bhat 2022).
2. **Nanocarriers for Drug Delivery:** Nanocarriers are nanoscale delivery systems designed to transport drugs or compounds like algal-derived natural compounds. Algal extracts can be more effectively absorbed in the digestive tract and have

their antidiabetic benefits maximized by using nanocarriers (such as liposomes, nanosomes, or polymeric nanoparticles) to transport them (Alshawwa et al. 2022). These carriers improve the delivery by enhancing their solubility, facilitating targeted delivery, sustained release, and protecting degradation. Several research have highlighted the application and future of multifunctional nanocarriers for efficient drug delivery (Ahmad et al. 2022) have reported the advances of nanocarriers as drug delivery system for cerebral disorders offering a solution in managing the drug delivery through blood-brain barrier (BBB). Similarly, studies on progressive nanocarriers-based drug delivery systems are reported for antitumor metastasis (Peng et al. 2023); treatment of lung cancer (Kim et al. 2023), and infectious diseases (Saha et al. 2023). Apart from this, alginate-based nanocarriers are also an effective way to transport. Loaded bioactive compounds are used so that their techno-functional capabilities are enhanced by alginate-based nanocarriers at the target site. Gene delivery with silica nanoparticles (NPs) is an effective approach. Moreover, commonly used drug carriers include liposomes, micelles, and silicon oxide NPs; each has benefits and drawbacks. The utilization of marine resources in biomedicine is emphasized nowadays. Diatoms are brown algae that are inexpensive and simple to cultivate. They have an amorphous silica shell. The source of their fossils, known as diatomaceous earth (DE) or frustules, is porous silica (SiO_2) NPs. Silica nanoparticles (NPs) may be obtained from both live diatoms and DE from fossils.

3. **Surface Functionalization:** Surface functionalization involves the modification of the outer layer or surface of nanoparticles to enhance their interaction with biological system and therapeutic agents improving the targeting and specificity. By altering the surface of nanoparticles with certain ligands, one may promote the absorption of algae chemicals in target tissues or improve the interaction with insulin receptors, both of which improve glucose metabolism. The goal of NP surface functionalization is to enhance and/or add features that will make NPs more beneficial for usage in medicinal applications. Various kinds of nanomaterials can be employed in the initial stages of functionalization because of their exposed functional groups and distinctive chemical characteristics. The goal of the initial stage of surface modification is usually to add an organic functional group (R-NH_2, R-COOH, etc.) that will help bind biological molecules. This is achieved by using homo- or hetero-bifunctional cross-linkers. Amino silanes are the most often employed linkers for silica NPs because they provide an amino group to the NPs surface in preparation for the subsequent bioconjugation. Previous studies have reported that surface functionalization of nanoparticles can improve drug efficacy in cancer therapy and brain therapy (Hu et al. 2023). Similarly, surface functionalization of gold nanoparticles was done to target tumors (Tan et al. 2023a, 2023b) and other biomedical applications.
4. **Nanoemulsions:** Algal oils may be made into nanoemulsions to improve their solubility and absorption, which may improve the bioactive ingredients that help regulate blood sugar levels. A recent study has reported on nanoformulation of *Nigella sativa* seeds. A nanoformulation of bioactive compound thymoquinone

was reported to have improved absorption and high drug delivery efficiency (Raghavendra et al. 2024). have reported micro-nanoemulsion and nanoparticle-assisted drug delivery against tuberculosis infection which was found to have improved therapeutic efficacy and stability. Several bioactive compounds from algae were made into nanoemulsion to increase their bioavailability, potency, and stability. The nanoemulsion from bioactive lipids (Foroghi et al. 2025), algal oil (Oliyaei et al. 2023), carrageenan from red algae (James et al. 2023), DHA (Yan et al. 2024), *Spirulina* bioactive compounds (Shah et al. 2024), and *Sargassum* ethanolic extract (Samudra et al. 2023) were developed for improved therapeutic applications. In today's era, for benefiting the insulin drug delivery discovery of hydrogel-based system has been emerging as a new guide to diabetic-related problems. These hydrogels can be functionalized by the aid of nanotechnology, where immobilization of glucose oxidase enzyme-based nanoparticles can ease the delivery to target site and release insulin. This study has been depicted by Lemmerman et al.(2021) in mice.

5. **Metal Nanoparticle:** Gold or Silver nanoparticles with algal extracts help improve antioxidant qualities and lower oxidative stress which may be a reason for diabetes. Several metal nanoparticles such as silver and gold can be effective in treating oxidative stress levels and can indirectly aid in reducing the risk factors associated with diabetics where algae-based metal nanoparticle can be used as a great alternative, producing through green synthesis can be sustainable in biomedical era. Algal-based metal nanoparticles have been reported as a potential approach due to their attractive appl.;lication in several fields like remediation and biomedicine (Alprol et al. 2023). This composition of algae extract and nanoparticle can be an effective hybrid system for diabetics as well as different diseases. Metal nanoparticles were researched to be novel agents for lung cancer therapy and diagnosis (Zheng et al. 2023); to relieve tumor hypoxia (Shi et al. 2023); for cancer therapies (Villalobos Gutiérrez et al. 2023); theranostic applications periodontitis (Nasiri et al. 2023); biomedical imaging and healthcare applications (Truong et al. 2024), and COVID therapies.
6. **Gene Delivery Systems:** Certain genes or gene-editing instruments that enhance insulin synthesis or glucose absorption in cells can be delivered using nanotechnology. Nanotechnology can facilitate the delivery of genes or gene-editing tools (like CRISPR) to pancreatic cells. Algal extracts may be incorporated into nanostructured materials to create scaffolds that release bioactive molecules continuously, sustaining stable blood glucose levels over time (Mitchell et al. 2021). Li et al. (2021) have reported a novel design of non-viral gene delivery system with nanoparticle polymer that is supramolecular polymers with surface functionalization enhancing delivery of DNA into mammalian cells.
7. **Molecular approaches:** In another study by *in vitro* assessment of Caulerpa racemosa green algae by Glucosidase inhibition, anti-glycation, and alpha-amylase tests were used to evaluate the in vitro antidiabetic efficacy. Additionally, experiments using silico methods were carried out to examine the binding affinities of caulerpin with the alpha-glucosidase enzyme and the active sites of the

estrogen receptor (ER). In terms of both alpha-amylase inhibitory and anti-glycation actions, the crude polyphenolic extract showed a greater potential. Alpha-glucosidase inhibition was more effectively inhibited by the ethyl acetate fraction. Using H-bond interactions, molecular docking demonstrated a strong affinity for caulerpin and a strong binding affinity between the alpha-glucosidase enzyme and caulerpin exhibits antidiabetic properties in green algae.

Algal compounds' potential as antidiabetic medicines can be greatly increased by utilizing these nanotechnology techniques. These techniques open the door for new diabetic therapies by increasing the bioactive components' effectiveness while also providing creative options for targeted delivery and long-lasting therapeutic effects.

5 Algae-Nanoparticle Hybrid Systems

An innovative method for creating antidiabetic treatments is the use of algae-nanoparticle hybrid systems. These systems improve therapeutic effectiveness, stability, and targeted distribution by combining manufactured nanoparticles with naturally occurring bioactive chemicals from algae. Nanotechnology has improved drug delivery in medicine using nanoparticles. Drugs may be precisely delivered to specific cells using nanoparticles. The proper dosage and location of the medication significantly lowers drug use and adverse effects. This approach reduces expenses and adverse consequences. Gold is frequently recommended to improve memory. To enhance newborns' mental health, gold is added to several medicinal medicines. Here is a thorough explanation of these hybrid systems' operation and possible uses in the treatment of diabetes:

(i) **By Composition of Algae and Nanoparticles:** The two main parts of algal-nanoparticle hybrid systems are designed nanoparticles and bioactive substances extracted from algae, with their varied role in sustainable nanomaterials (Singh et al. 2024). Numerous advantageous bioactive compounds, like polysaccharides, polyphenol, carotenoid, and omega-3-fatty acid, are known for their anti-inflammatory, glucose-regulating, and antioxidant qualities which are abundant in the algal extracts. The algae extract has a wide application in varied fields when mixed with different nanoparticles such as metal-based nanoparticles, lipid, or with different polymers and can enhance the activity of algae compounds with further aid in better adsorption of bioactive compounds present in algae compounds.

(ii) **Synergistic Mechanism of Algae-NPs Hybrid System:** Various algal systems have been reported for their great biomedical application due to the bioactive compounds present in it as described earlier. But by the means of nanotechnology these algae bioactive compounds can be improved through a robust algae-NPs hybrid system that can as synergistically and can be proven as game changer approach for coming era. As a nanoparticle, it acts as a carrier which helps algal bioactive compounds to be absorbed based on its solubility and work better. Moreover, it also enhances the target site recognition by receptor-based functionalization. Once the target site has been attained it aids in reducing the inflammation and oxidative stress and therefore, can be effectively used in the treatment of clinical diseases (Kumari and Bharadvaja 2023).

(iii) **Site-directed drug delivery:** The algae-NPs hybrid systems are in pace to grow more due to their effective antidiabetic drug delivery. And for target-specific delivery nanoparticles can be engineered according to the dose-dependent delivery by adding ligands specific to receptor binding site. Moreover, with further improvements ligand can be used in the drug delivery system. This can boon the therapeutic efficacy and enhance traditional antidiabetic agents to work efficiently. And taking upon dose dose-dependent drug delivery system, nanoparticles can effectively control the release of bioactive algae compound which can mimic the side effects and will help to make a balanced glucose level in the blood (Behera et al. 2023).

(iv) **Management of Diabetics:** Algae being a prominent source of photosynthetic organism has an essential natural bioactive compound which has immense therapeutic properties to mimic the effect of diabetics. But in addition to nanoparticles, this algae-NPs hybrid system can do wonders to the biomedical fields. Where algae are known for its antioxidative effects, being rich in various antioxidants and anti-inflammatory effects due to the presence of metabolites can be a potent antidiabetic agent which can be further replace pre-existing diabetics treatment due to its highly efficient, anti-inflammatory, antioxidative, and less side effects. Moreover, with further research, this algae-NPs hybrid can be used in personalized medicines too to cure clinical diseases such as diabetics, cancer, etc.

5.1 Algae-Mediated Biosynthesis of Nanoparticles

Algae and nanoparticle hybrid systems can be obtained via algae-mediated biosynthesis of nanoparticles. Where, the two vital parts of algal-nanoparticle hybrid systems are designed by incorporating nanoparticles and bioactive substances extracted from algae. The algal biosynthesized NPs have also been investigated for their varied role in biomedical uses, which include anti-bacterial, antidiabetic, anti-fungal, anti-cancerous, anti-fouling, bioremediation, and biosensing activities, sustainable biomaterial, nanocarriers because of their biocompatibility, safety, and distinct physico-chemical characteristics (Singh et al. 2024). Numerous advantageous bioactive compounds, including polysaccharides, polyphenols, carotenoids, and omega-3 fatty acids which are well-known for their anti-inflammatory, glucose-regulating, and antioxidant qualities are abundant in algal extracts. These extracts are mixed with lipid-based, polymeric, or metallic (like gold or silver) nanoparticles. Algae chemicals solubility and stability are improved by the nanoparticles, which also help with targeted distribution to certain tissues or cells and the regulated release of bioactive substances. The hybrid systems seek to maximize the therapeutic potential of algae bioactives for better control of diabetes and associated metabolic diseases by combining these two elements in a synergistic manner. This hybrid system provides extracellular and intracellular production of algae-mediated nanoparticles. Gold and Silver are the most researched nanoparticle metals for intracellular and extracellular nanoparticle synthesis from algae. Metal nanoparticles (NPs) based on *Chlorella vulgaris, Chlorella pyrenoidosa, Nanochloropsis oculata,* and *Scenedesmus* sp. IMMTCC-25 has been reported to have antimicrobial and catalytic properties.

Synthesis of algae-derived nanoparticle can be assisted via different methods which contains physical, chemical as well as green synthesis of nanoparticles.

5.1.1 Green Nanosynthesis

Green nanosynthesis has emerged as a promising, ecofriendly, and sustainable method for the synthesis of nanoparticles and nanomaterials like metal oxide NP, nano-hybrids, and bio-inspired materials with potent applications. Metal nanoparticle synthesis is a significant area of nanotechnology for its unusual size (length ranging between 1 and 100 nm), shape, and appealing functions in biotechnology, bioenergy and fuel production, medicine, catalysis, and electronics. Therefore, green nanosynthesis is regarded as a crucial instrument to lessen the side effects of nanoparticles that are often utilized in industry or labs using traditional synthesis techniques. In most investigations, the following primary processes are seen in the production of metal nanoparticles utilizing algae:

(i) Heating or Boiling—in water or organic solution for stipulated time.
(ii) Preparation of Molar solutions of ionic metal compounds.
(iii) Incubation of the solutions with or without stirring for stipulated time.

Nanorods were initially produced in the lab using *Cyanobacteria* and *Nostoc ellipsosporum* for their intracellular biosynthesis. These were synthesized in cells at a temperature of 20 °C for 48 hours with gold metal solution as reported in earlier studies. Numerous methods may be used to easily synthesize gold nanoparticles (AuNPs) for commercial and medicinal applications (Mukherjee et al. 2021).

5.1.2 Physical and Chemical Methods

One of the most common methods for synthesis of metal nanoparticles is evaporation-reduction using a tube heater and air pressure. Physical methods provide the advantages of rapid radiation-based reduction and the absence of hazardous chemicals, but they also have drawbacks such as high energy costs, solvent pollution, and irregular patterns. Although nanoparticle size cannot be controlled exactly, it can be constrained by varying the system pressure, system geometry, or vaporization rate. Physical methods further consist of: Inert Gas Condensation method, Ion Sputtering method, Laser Ablation, and plasma jet method based on factors influencing nanoparticle formation include inert pressure of gas, rate of ablation, temperature, composition of gas, and sputtering technique.

Whereas, chemical methods include hydrothermal synthesis, solvothermal synthesis, cry synthesis, and aerosol method are cost-effective, easy, and high-producing approaches for controlling particle size. Hydrothermal synthesis involves direct crystallization, while solvothermal synthesis creates supersaturated solutions. Cry synthesis crystallizes solutions at low temperatures and eliminates dissolvents. Aerosol method consists of an atomization chamber, tube reactor, and particle collection.

5.2 *Characterization Technique*

There are various metal and non-metal, biological nanoparticles which are characterized using several analytical methods, which include Transmission or Scanning Electron Microscopy (TEM or SEM), Atomic Force Microscopy (AFM), Dynamic Light Scattering (DLS), X-Ray Photoelectron Spectroscopy (XPS), Powder X-Ray Diffractometric (XRD), Fourier infrared transform spectroscopy (FTIR), and UV-Vis spectroscopy (Mukherjee et al. 2021). The nanoparticles, size, shape, fractal dimensions, pore size, and surface area can all be characterized by these methods. Certain color-based characterization for nanoparticle is also possible here by exposing *Ulva intestinalis* and *Rhizoclonium fontinale* algae to chloroauric acid at 20 °C for 72 hours, AuNPs were produced intracellularly (Mukherjee et al. 2021). The production of AuNPs was verified by a discernible purple shift in thallus color from green. Whereas a report gave a surface plasmon peak at 530 nm, which indicated the involvement of proteins, enzymes, and biomolecules in algae-mediated NP synthesis, it documented the extracellular synthesis of AuNPs produced by *S. platensis* at different doses of chloroauric acid (Chaudhary et al. 2020). In certain cases, the synthesis of silver and gold non-metals has been visually shown by the change in color of a combination generated at a specified temperature. The development of silver nanoparticles is indicated by a change in the reaction mixture color to violet or brownish violet, whereas the formation of gold nanoparticles (AuNPs) is indicated by a change to purple or pink. Because nanometals have remarkable optical characteristics due to surface plasmon resonance (SPR), absorption spectroscopy in the UV-Vis spectral region (190–1100 nm) is used to monitor the biosynthesis of nanometals in the aqueous process (Mukherjee et al. 2021).

5.3 Applications of Algae-Based Nanoparticles

Algal-produced nanoparticles have been demonstrated to possess antibacterial, anticancer, and antifungal properties, and they can rival conventional medications. Metal nanoparticles offer a wide range of benefits in computers, optics, cosmetics, and other fields in addition to medicinal ones. By producing gold NP, silver NP, and other metallic NP (nanoparticles) both intracellularly and extracellularly, algae from the Cyanophyceae, Chlorophyceae, Phaeophyceae, and Rhodophyceae families have been employed as nano-machines. One of the areas of nanobiotechnology research that is most active is cancer therapy and targeted distribution of anticancer medications. The anticancer properties of algal-mediated nanoparticles have been published in recent years. During an experiment, Sargassum vulgare made silver NP with 10 nanometers (10 nm) showed significant anticancer properties against HeLa cells and HL60 cells (Fig. 16.2).

Gold nanoparticles from *Acanthophora spicifera* and *Chaetomorpha linum* have shown potent anticancer activity against colorectal adenocarcinoma and colon cancer cell lines. The inhibitory amount was 21.86 μg/mL, causing cell shrinkage, apoptosis, and morphological structure loss. Dose-dependent cytotoxic impacts have also been reported in previous studies.

The antibacterial efficacy of algae-derived nanoparticles against a wide range of bacterial types has prompted investigation. The biosynthetic Ag-nanoparticles from brown algae *Padina* sp. suppress the development of *Bacillus subtilis, Klebsiella planticola*, and *Pseudomonas aeuroginosa*. Acolloidal stable silver nanoparticles derived from marine green algae *Caulerpa serrulata* showed antibacterial ability at

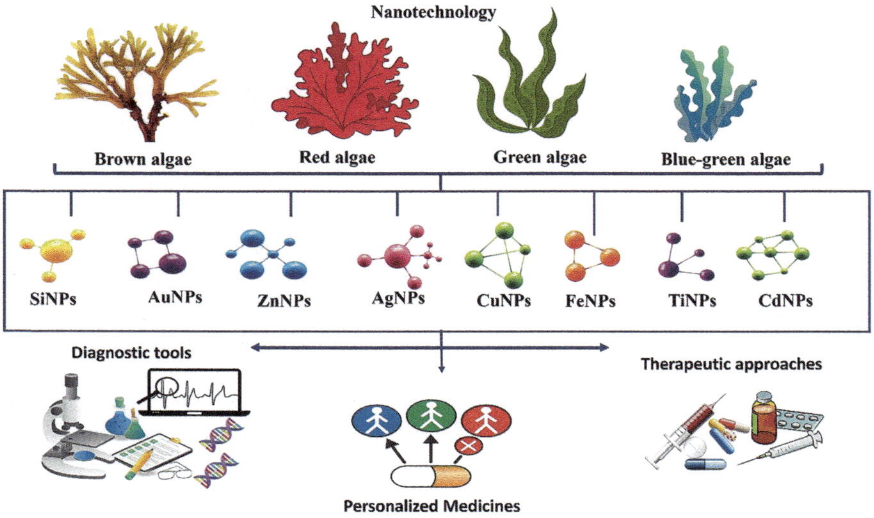

Fig. 16.2 Algae-Nanotechnology hybrid system applications

lower concentrations against *P. aeruginosa, S. aureus, Salmonella typhi, E. Coli,* and *Shigella* sp., according to another study.

Thanks to developments in nanotechnology and biotechnology, the use of algae-nanoparticle hybrid systems as antidiabetic treatments appears to have a bright future. It is anticipated that these hybrid systems will become more individualized, enabling customized therapies that fit the unique characteristics of each patient, boosting effectiveness and reducing adverse effects. In order to ensure prompt therapeutic effect, these systems may eventually incorporate intelligent delivery mechanisms that release bioactive substances in response to certain physiological cues, such as rising blood glucose levels, as research advances. Algae's sustainability as a renewable resource also fits with the increased focus on environmentally friendly medication development methods, which might result in more environmentally friendly pharmaceutical solutions. The safety and efficacy of these hybrids will be determined by ongoing clinical trials, opening the door for regulatory approvals and extensive clinical adaptation (Fig. 16.3).

6 Toxicity and Biocompatibility Assessment (Table-1; Figure-1)

The increased prevalence of diabetes has encouraged advancements in therapies. One such revolution has been attained by nanotechnology integrated with bioactive compounds derived from natural sources. With combined natural compounds with nanotechnology, it is expected to have an increased biocompatibility and minimum toxicity. Algal compounds have several advantageous properties like alginates that are widely used in biomedical applications or like phycocyanin that have high antioxidant properties. The bioactive compounds from algae that exhibit

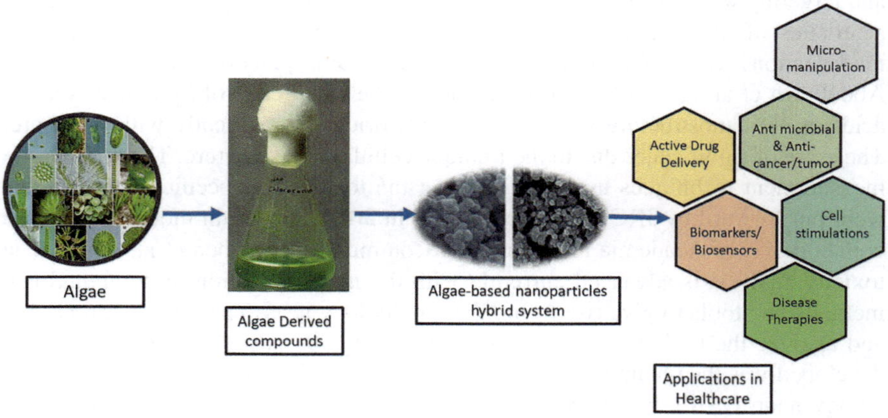

Fig. 16.3 Algae-based nanoparticle synthesis and healthcare applications

Table 16.2 Nanoparticle types and toxicity effects

Sl. no.	Types	Toxic effects
1.	Carbon NP (nanoparticle)	Metabolic activity, cell viability, mitochondria damages and mitochondria integrity affected, ROS (reactive oxygen species) produced
2.	Gold NP	
3.	Silver NP	
4.	Quantum dot	
5.	Fullerene	Oxidative stress, DNA damages, apoptosis, actin filament integrity, destruction of blood-brain barrier, gene expression alteration
6.	Aluminum NP	
7.	Zinc NP	
8.	Iron NP	
9.	Titanium NP	
10.	Silicon NP	

antidiabetic properties are effectively used in drug delivery system enhanced by nanotechnology; it is crucial to evaluate their toxicity and biocompatibility. The improved cellular intake and surface characteristics of nanoparticles can cause toxicity and influence biocompatibility. The toxicity assessment can be attained by several methods depending on the type of nanoparticles, experiment models, and toxic effects. Advanced techniques like flow cytometry, imaging techniques, microscopy, and biochemical assays are used to provide a better understanding of toxicity and its mechanism (Table 16.2).

Nanoparticles can have unique properties compared to their bulk counterparts, leading to different biological interactions. For instance, their small size allows for improved cellular uptake, but this may also enhance the likelihood of toxicity. A previous study has reported surface functionalization or modification of nanoparticles for effective drug delivery to increase cellular uptake and absorption. To make this possible, various strategies like polymer coating methods, targeting approaches, and targeting were focused (Abdelkawi et al. 2023). Furthermore, the surface characteristics of nanoparticles can significantly influence their biocompatibility, as modifications can affect how the immune system perceives these materials. Abdelkawi et al. (2023) have also mentioned the emergence of Spherical Nucleic acids, a 3D nanostructure with nucleic acids oriented spherically with nanocore. These are of significance due to their unique cellular uptake nature. The appropriate measurement techniques used for studying the toxicity of specific nanomaterials were put forward by JRC Reference Report in 2012 later again modified with the collaboration of Academia and European Community (Roebben et al. 2014). The toxicity analysis is advanced currently with the implementation of computational methods and tools to give rise to In-silico toxicology tests. These tools can predict and analyze the toxicity of materials used in drugs or as such. One such method developed was the Quantitative Structure Activity model (QSAR) and in nanotoxicology nano QSAR was implemented using physical and chemical properties as references (Fourches et al. 2011).

(i) ***In vitro* Toxicity Assessment Methods.**

To assess the toxicity in vitro methods is the most common as it is less costly, faster, and has minimum ethical concerns. Some of the most common assessment methods include apoptosis assay, necrosis assay, proliferation assay, oxidative stress assay, and DNA damage assays. Cell viability along with lethality are considered as the two important parameters used to measure the toxicity.

Proliferation assays are used to assess the cellular metabolism using MTT chemical. This essay produces quick results, are reproducible, and require least manipulation of cells. The MTT tetrazolium salt was reported to cross the plasma membrane of metabolically active cells and reduced to formazan to produce colored solution that can be spectrophotometrically measured. It is a standard assay that can measure cell viability. As compared to MTT, Alamar Blue was considered advantageous due to its simpler sample preparation, however it was found to react with certain compounds and nanomaterials used. Cologenic assay is another assay where proliferating cells are assessed and counted after nanoparticle exposure (Bahadar et al. 2016). Apoptosis is the next notable marker in in vitro assessment of nanoparticle toxicity. Excessive production of free radicals causes DNA damage and apoptosis, and these can be induced by nanoparticles. To assess apoptosis, different methods are employed namely Annexin V assay, Comet assay, TdT-mediated dUTP biotin nick end labelling assay or TUNEL assay and analysis of morphological changes. Apoptosis marks the irregular reduction of the cell size and DNA fragmentation. In Apoptosis assay, typical cell death marker, Annexin V is used that shows fluorescence when bound to phosphatidylserine. This can be measured to understand and study apoptosis in cells. The assay is usually followed by electrophoresis like Agarose gel electrophoresis. One such method is the single-cell gel electrophoresis assay also known as comet assay. It is known as a sensitive tool to detect mutation in the cells because of exposure to various parameters. This test can be used to study mutation and mutagenic potential both in vitro and in vivo. When electric current is passed through the sample, damaged DNA and broken fragments migrate out of cell while the undamaged remain intact inside the cell nucleus (Vandghanooni and Eskandani 2011). The analysis is usually done by corelating the shape and size of DNA and its distribution within the comet. The comet is presented as a head with whole genomic material and a tail with damaged or broken DNA fragments. The comet can be visualized using fluorescent stains or dyes. Comet assay can also be employed in vivo genotoxicity testing. TUNEL assay is another assay used by researchers to detect DNA damage on staining with dyes. The cells that have undergone death or apoptosis or fragmentation were stained for visualization. This assay is also used to study non-apoptosis necrotic cell death and DNA damages due to the exposure of toxic compounds (Huang et al. 2023). Necrosis assay is yet another assessment method usually used to study cell integrity and cell viability and membrane sturdiness. Dyes like trypan blue and neutral red are used to study the cell damages as the cells tend to uptake these dyes (Buzea and Pacheco 2019). The impact of nanoparticles, xenobiotics, or such toxic compounds on cells may cause cells to decrease the uptake of dyes and thus help to determine the dead and live cells. Alternative method for this is the Trypan Blue Exclusion Assay where the membrane stability is measured to understand the apoptosis and impact of nanoparticles on cells. Oxidative stress assay can measure ROS (reactive oxygen species) and RNS(reactive nitrogen species) on the exposure to nanoparticles and toxic compounds (Hellack et al. 2017). The method involves the reaction of chemical, TEMP with $O_2{}^-$ stable radical that can be detected using X-band electron paramagnetic resonance. The method is less used due to its high cost and expense. Oxidative stress, however, can also be assessed using C11-BIODIPY assay and TBA assay. There is also another similar assay that can be used to study oxidative stress (Patel et al. 2024). When toxic materials enter the cells and destroy the plasma membrane a cytosolic enzyme called lactose dehydrogenase is released into the extracellular plasma membrane space which is analyzed by LDH leakage assay (Salehipour et al. 2023). This enzyme brings the oxidation of lactate to pyruvate and NADH is produced. This reacts with the indicator chemical used in the assay to produce formazan (Joseph et al. 2023).

The toxicity of nanoparticles and bioactive compounds or drugs can be assessed using cell lines and its culturing. Several cell lines are available that may be used to study the effect of drugs or nanoparticles on them. There are several reports mentioning the use of human cell lines to study DNA damage and apoptosis. Another study has reported starch-coated silver nanoparticles toxicity assessment using glioblastoma cells and lung fibroblasts (human). The report indicated a dose-dependent rise in ATP and DNA damage due to the deposition of nanoparticles on the cells (Mahmud et al. 2023). Previous reports on gold nanoparticles exposed to human lung fibroblasts prompting autophagy accompanied by oxidative stress are notable. The study also assessed the cytotoxic effects of metal nanoparticles through cell motility assessing the cytotoxicity of gold quantum dots and nanorods in in-vitro. A study on CdTe quantum dots was of significance as they caused potential cytotoxicity with lysosomal damage and ROS production when studied with epidermal keratinocyte cell lines (Yao et al. 2023; Huang and Tang 2021). Furthermore, another study on aluminum nanoparticles impact on red blood cells was studied. It was found that there occurred a dose-dependent cytotoxicity causing DNA damage on exposure after 2 hours. Mechanistic studies on the effect of nanoparticles or drugs or bioactive compounds in vitro have become a significant method especially in medicine. It aims to study the biological interactions of compounds or drugs assessing its stability, safety, and efficacy. The key mechanisms used in mechanistic studies on biological pathways include cellular uptake, oxidative stress, inflammatory response, DNA damage, cell cycle arrest and apoptosis, and autophagy (Mohammadinejad et al. 2019). Various in vitro models can be used for mechanistic studies like cell lines, 3D cell lines, and co-culture systems. These studies can provide vital insights into the biological effects of the drugs or nanoparticles used thus paving a way for innovative therapies and treatment. These *in vitro* assessment methods can collect valuable data into the interactions between the drug or algal natural compounds with nanomaterials for drug delivery. These methods would give a primary understanding of the toxicity level and its impact when used in clinical setup.

(ii) **In Vivo Toxicity Assessment Methods.**

In vivo assessments for toxicity testing are done on animal models like mice, pig, and rats. The assessments used to study the impact of compounds in in-vivo models are biodistribution test, clearance, hematology, serum chemistry, and histopathology. Biodistribution studies examine the localization pathway of nanoparticles/bioactive compounds/drugs to the target organ or tissues. The compounds are usually detected using radiolabels in animal models. The common procedure to assess biodistribution involves the removal of main organs or tissues of the animal model to track the nanoparticle according to the characteristics. Several imaging techniques like MRI, CT, PET, or optimal imaging can be used for qualitative analysis (Yang et al. 2016a2016b). Clearance of nanoparticles is another test where the excretion and metabolism of nanoparticles in animals are examined at different time periods from exposure. Hepatic assessment is another assessment performed to analyze immunohistochemistry mainly to detect liver fibrosis and inflammation. The examination of serum chemistry is another important assessment in nanoparticle-exposed animal models. The common enzymes used to analyze the hepatic function include alanine aminotransferase, alkaline phosphatase, and aspartate aminotransferase, which may reflect functional changes (Yamagishi et al. 2013). Histopathology of cells, tissues, and organs after exposure to nanoparticles and bioactive compounds is determined to study the toxicity effects of the compounds. The pathological changes like glomerulosclerosis can be identified with the detection of fibrotic markers that can transform growth factors, interferons, or collagens. Electron microscopy techniques are usually used to observe pathological changes. There are reports mentioning several variables testing for renal function like total protein, albumin, glomerular filtration rate, and creatinine ratio to understand kidney function (Gandhi et al. 2013). Advanced technology to examine the toxicity in in vivo includes the use of micro-electrochemistry and microfluidics. There are reports indicating the toxicity of carbon nanotubes on several soil bacterial diversity (Kerfahi et al. 2015). Krause (2014) have mentioned the toxicity of the same CNTs causing mechanical damage to microorganisms especially bacteria. Another study reported the toxicity of nanoparticles on stem cells causing a reduced cell viability (Jeng and Swanson 2006) and silica nanoparticles causing a toxicity to human keratinocytes by decreasing the cell viability (Lu et al. 2011; (Fig. 16.4).

Animal model selection for nanotoxicity assessment or experimental study is the initial step in an in vivo toxicity analysis. Besides most common models like mice/rats, zebrafish, pigs, rabbits, and *Caenorhabditis elegans* are chosen with ethical clearance ((Yang et al. 2016a2016b). The assessments involve studying the tissue structure changes, apoptosis, inflammation infiltration, or target certain systems or cells for the drug delivery and its toxicity study. Another drug toxicity assessment is

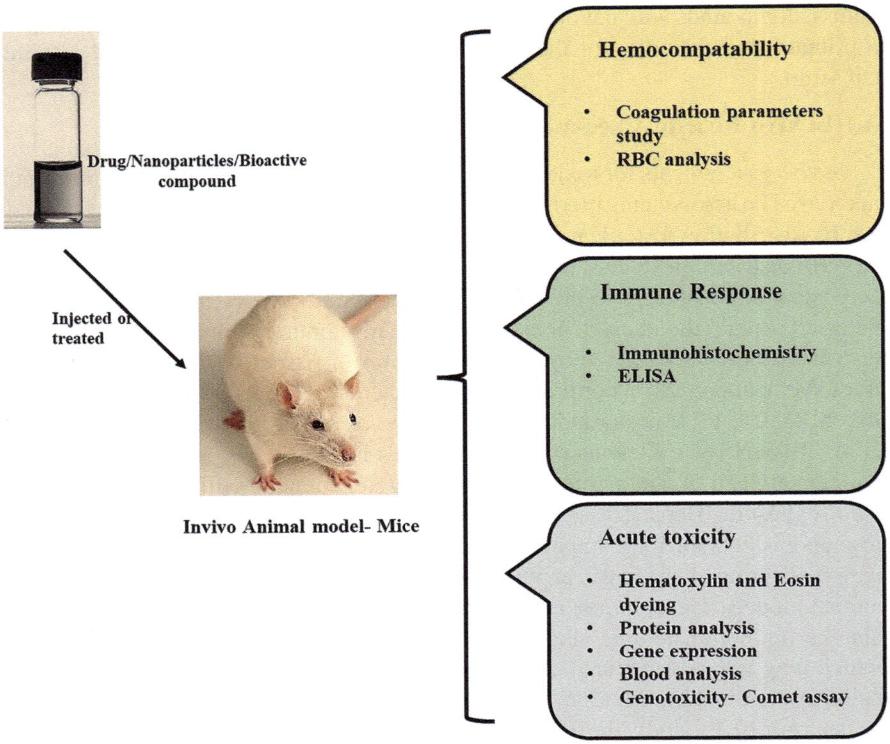

Fig. 16.4 In vivo toxicity assessment methods using animal models

through oral administration of drugs embedded with natural compounds and nanoparticles. Even though it is a limited assessment method, with improved nanotechnology several nanocapsules are developed for the treatment of gastrointestinal issues. These nanoparticles can cause several toxicities as they may interfere with nutrient absorption. This can be understood with the observation of mast cells quantity, microvilli, and epithelial atrophy (Yang et al. 2016a2016b). Xia et al. (2013) have reported the assessment of pulmonary toxicity in the case of inhaling particles. The accumulation of these compounds in pulmonary regions could be detected using MRI in animal models. Intratracheal instillation and bronchoalveolar lavage are some methods used for biochemical analysis to check pulmonary injury. The nanoparticles and some bioactive natural compounds used for drug delivery when exposed via intravenous injection can be impactful on the cardiovascular system. This can be detected in exposed animal models by assessing venous tolerance, symptoms of phlebitis, hemolysis, thrombosis, platelet aggregation, and analysis of complete blood count (Zhang et al. 2015; Sarhan and Hussein 2014). Reproductive system toxicity study is an extended experimental study. To understand the toxicity of nanoparticles or natural compounds in vivo models are developed where both

Table 16.3 Histopathologic assessment of systems

Sl. no.	Systems	Assessments (Histopathology)
1.	Hepatic	Hematoxylin and eosin staining, immunohistochemistry, TUNEL
2.	Renal	Kidney special dyeing, immunohistochemistry
3.	Gastrointestinal	GIT microvilli, epithelial atrophy
4.	Respiratory	Hematoxylin and eosin staining, TUNEL
5.	Cardiovascular	Hematoxylin and eosin staining
6.	Neuron	Brain histology, fluorescence imaging
7.	Immune	Hematoxylin and eosin staining
8.	Reproduction	Hematoxylin and eosin staining, TUNEL

male and female animals are used. Ahmad (2022) has stated the safety and toxicity implications of multifunctional drug delivery nanocarriers on reproductive systems in vivo. For male animals, testicle tissue structure, epididymal sperm parameters, the serous sexual hormones, and the concentration of nanoparticles or bioactive compounds or drugs used are tested (Li et al. 2016a, 2016b). Female sexual hormone levels in serum are analyzed for female animals. The function of major organs like ovary, uterus, and vaginal tract are evaluated to understand the effects of the drugs or nanoparticles used. Similarly, for male animals the major organs like testes or uteri that may absorb such compounds are taken for histopathological tests. (Qi et al. 2014). It has been noted that teratogenicity of nanoparticles on fetes is a critical issue that needs to be addressed. Therefore, the development, reproduction, survival, and growth of the fetus are assessed upon prenatal exposure (Di Bona et al. 2015) (Table 16.3).

In vivo assessment using animal models is significant in confirming the biocompatibility and toxicity of the compounds, drugs, or nanomaterials used for clinical therapies. These could give a wider focus and knowledge on algae-enhanced nanomaterials for drug delivery systems and contribute to significantly developing safe and effective nanomedicine strategies. There are several reports on the toxicity assessment assays for nanocarriers. One such report was by Tan et al. (2023a, 2023b) on the toxicity evaluation of silica nanoparticles as potent nanocarriers. Similarly, have also reviewed the toxicity of silver nanoparticles on human health. The review discussed the in vitro and in vivo toxicity studies to understand the effect of silver nanoparticles on cells. Noga et al. (2023) have also conversed in their review on the green's synthesis, application, and mainly the toxicity and risk assessment of nanoparticles in vivo and in vitro studies with special consideration to cells, organs, and tissues. The integration of algae bioactive compounds or natural compounds with nanotechnology will be a better method for enhancing biocompatibility while minimizing toxicity. Even though incorporation of bioactive or natural drugs with nanotechnology has potential advantages a comprehensive toxicity and biocompatibility assessments are essential for its application in clinical therapies and diagnostics. Several studies on elucidating the mechanism of toxicity at molecular level are being progressed. Advanced imaging techniques, multi-omics, in-silico, and artificial intelligence approaches are expected to provide a deeper insight into the biological interaction between nanomaterials, algae compounds, and biological systems.

(iii) **Future Directions.**

The need for innovative therapeutic strategies for diabetes has risen globally. This comes ahead as the traditional drug delivery system faces several challenges and limitations leading to suboptimal therapeutic outcomes. In this circumstance nanotechnology has emerged as a powerful approach for drug delivery thus increasing the efficacy of treatments. The incorporation of bioactive compounds from algae presents a unique opportunity to develop more effective and safer drug delivery systems. Behera et al. (2023) have stated the significance of algae-based nanoparticles to treat several infections. The potential bioactive compounds I algae were reported by them to have several clinical applications. Pereira and Valado (2023) have discussed algae-based natural products for diabetes management.

Future research focuses on designing multifunctional nanocarriers that can encapsulate bioactive compounds from algae. The nanocarriers may be nanoliposomes, dendrimers, or polymeric nanoparticles considering factors like the size, surface charge, and hydrophilicity that may impact the cellular uptake and distribution. Nanocarrier design is hence an important factor that may contribute to future direction in drug delivery systems. Another approach could be optimization of extraction processes using techniques like Supercritical Fluid Extraction, Ultrasound Assisted Extraction, and enzymatic extraction of drug or natural compound. Gayathri et al. (2023) have reported the approach of nanoformulation for drug delivery. It was mentioned that combining nanotechnology for the targeted delivery of curcumin enhanced its therapeutic activity, efficacy bioavailability, and solubility.

A significant challenge in the drug delivery system is the poor bioavailability of therapeutic agents. Nanotechnology however has become a boon by offering solutions to enhance the solubility and stability of poorly soluble drugs. One prospective study could explore the use of nanocarriers to improve oral bioavailability for diabetes management. Methods to increase the therapeutic index of antidiabetic compounds could be future research in targeted delivery systems. There are several deep learning techniques and artificial intelligence assimilated methods that could improve diabetes management. A research report by Vora et al. (2023) has stated that artificial intelligence will present a transformation in drug discovery, delivery, assessments, and formulation. Utilizing AI algorithms that could analyze the biological system, the dosages, interactions, genomics, and proteomics would enable researchers to develop better therapies as compared to traditional methods. Similarly, Bao et al. (2023) have also reported that machine learning directed drug formulation development would progress elf driven laboratories and help produce accurate drug delivery systems. The development of combinational therapies has been of great interest in nanotechnology and influenced the drug delivery system. Formulations amalgamating algal bioactive compounds, pharmaceuticals, and nanotechnology could be a more effective method compared to either alone. Assessments on the environmental impact of producing and disposing of nanomaterials are an issue to be addressed. Employing biodegradable nanocarriers made from polymers like alginates could minimize environmental concerns. Research focusing on developing such materials would be focused on nano-drug delivery research. One challenge worth mentioning in nano-drug delivery research is the transition of research from laboratory to clinical application. Further studies must focus on the production process, scalability, purification, in vivo and in vitro assessments to ensure an efficient product manufacture. Collaborative efforts of researchers, industrialists, pharmaceuticals, and regulatory bodies are essential to establish this transition. Development of clear guidelines for evaluation and approval is essential.

Advancements in nanotechnology, deep learning techniques, and artificial intelligence offer the potential for personalized medicine therapies in diabetes management. Future research should explore the development of person-centered therapies and treatment considering genetic, metabolic, and lifestyle factors. Personalized medical treatment should enhance the safety and efficacy of the treatment and drugs used (Alghamdi et al. 2022). A very recent study has discussed precision targeting with the help of nanotechnology along with tailored dosage forms, wearable drug delivery, and companion diagnostics as cutting-edge technologies in drug delivery systems (Puccetti et al. 2024). They have specified that nanomedicine could renovate several diseases therapy and its prevailing conditions with precise, personalized, and cell-specific drug delivery. Kannan et al. (2023) have also mentioned the significance of precision medicine that include clinical, genetic, lifestyle, and other biomarkers data for personalized treatment. Another report with discussion is the 3D printing technology in personalized medicine especially in nanomedicine to treat type 2 diabetes with high blood sugar. Sustained release 3D printed implants, stents, and other devices could be a breakthrough in diabetes management (Serrano et al. 2023).

The last and the most important measure is patient education and engagement to ensure successful implementation of novel drug delivery systems. Educating the patients and society about the benefits of nanotechnology-based therapies, their risks and concerns are essential for refining the drug delivery systems. With advancements in Artificial Intelligence (AI), several apps, deep learning software, and advanced digital technology are reported and surveyed to influence the knowledge improvement in patients with health issues. It has also been reported that ChatGPT could be used as a potential substitute to acquire more insights on diabetes management and several other healthcare issues (Moreira et al. 2023; Sharma et al. 2016).

The future of drug delivery using nanotechnology with algae-derived compounds is promising. With advanced technologies like formulation techniques, targeted delivery, personalized medicine, and safety assessments researchers can develop an innovative therapy for diabetes management. With the prevalence of increasing diabetes, innovative strategies would upsurge the efficacy of therapy while minimizing the risks. These holistic approaches are expected to transform clinical treatments especially for diabetes management and bring about sustainable and effective healthcare solutions in the future era.

7 Conclusion

Nanotechnology emerges as a promising solution to drug delivery systems in enhancing the therapeutic potential in the field of medicine. As mentioned in this chapter, diabetics are expected to increase further underscoring an urgent need for innovative therapies. The unique antidiabetic properties of algae bioactive compounds integrated with nanotechnology could be a solution to this. However, challenges such as bioavailability, rapid degradation, and non-specific targeting could be some limitations. However, integrating nanotechnology with natural compounds like encapsulating compounds in nanocarriers could help optimize biodistribution and pharmacokinetics of therapeutic agents. Controlled and targeted release of drugs or bioactive compounds from algae would be critical in managing diabetes while maintaining stable blood glucose and insulin. Tailored approaches could bring about improved patient outcomes with reduced risks and health concerns/side effects. The ability to enhance the stability of algal-derived compounds is another concern solved with the integration of nanotechnology. Furthermore, integrating these systems with digital health technology may make it easier to monitor and make modifications in real time, improving diabetes care even further. All things considered, the development of algae-nanoparticle hybrid systems has the potential to completely transform antidiabetic therapies by increasing their efficacy, sustainability, and individualization. The chapter discusses the importance of toxicity assessments and assays to be in line with the regulatory and safety guidelines. Ensuring early regulatory bodies could help facilitate smoother clinical translation. In conclusion, the integration of nanotechnology with algae-derived compounds heralds a novel epoch in diabetes care. By overcoming these challenges and limitations with the amalgamation of innovative disciplines like artificial intelligence, engineering, deep learning techniques, and nanotechnology, there could be an improvement in patient outcomes and a reduction in healthcare costs on a global scale.

References

Abdelkawi, A., Slim, A., Zinoune, Z., & Pathak, Y. (2023). Surface modification of metallic nanoparticles for targeting drugs. Coatings, 13(9), 1660.

Abo-Shady, A. M., Gheda, S. F., Ismail, G. A., Cotas, J., Pereira, L., & Abdel-Karim, O. H. (2023). Antioxidant and antidiabetic activity of algae. Life (Basel, Switzerland), 13(2), 460.

Adepu, S., & Ramakrishna, S. (2021). Controlled drug delivery systems: Current status and future directions. Molecules (Basel, Switzerland), 26(19), 5905.

Ahmad, A. (2022). Safety and toxicity implications of multifunctional drug delivery nanocarriers on reproductive systems in vitro and in vivo. Frontiers in Toxicology, 4, 895667.

Ahmad, A., Imran, M., & Sharma, N. (2022). Precision nanotoxicology in drug development: Current trends and challenges in safety and toxicity implications of customized multifunctional nanocarriers for drug-delivery applications. Pharmaceutics, 14(11), 2463.

Alghamdi, M. A., Fallica, A. N., Virzì, N., Kesharwani, P., Pittalà, V., & Greish, K. (2022). The promise of nanotechnology in personalized medicine. Journal of Personalized Medicine, 12(5), 673.

Alprol, A. E., Mansour, A. T., Abdelwahab, A. M., & Ashour, M. (2023). Advances in green synthesis of metal oxide nanoparticles by marine algae for wastewater treatment by adsorption and photocatalysis techniques. Catalysts (Basel, Switzerland), 13(5), 888.

Alshawwa, S. Z., Kassem, A. A., Farid, R. M., Mostafa, S. K., & Labib, G. S. (2022). Nanocarrier drug delivery systems: Characterization, limitations, future perspectives and implementation of artificial intelligence. Pharmaceutics, 14(4), 883.

Anand, K., Tiloke, C., Naidoo, P., & Chuturgoon, A. A. (2017). Phytonanotherapy for management of diabetes using green synthesis nanoparticles. Journal of Photochemistry and Photobiology. B, Biology, 173, 626–639.

Anderson, P. W., Mcgill, J. B., & Tuttle, K. R. (2007). Protein kinase C β inhibition: the promise for treatment of diabetic nephropathy. Current Opinion in Nephrology and Hypertension, 16, 397–402.

Anjum, S., Ishaque, S., Fatima, H., Farooq, W., Hano, C., Abbasi, B. H., & Anjum, I. (2021). Emerging applications of nanotechnology in healthcare systems: Grand challenges and perspectives. Pharmaceuticals (Basel, Switzerland), 14(8), 707.

Arshad, R., Gulshad, L., Haq, I.-U, Farooq, M. A., Al-Farga, A., Siddique, R., Manzoor, M. F., & Karrar, E. (2021). Nanotechnology: A novel tool to enhance the bioavailability of micronutrients. Food Science & Nutrition, 9(6), 3354–3361.

Bahadar, H., Maqbool, F., Niaz, K., & Abdollahi, M. (2016). Toxicity of nanoparticles and an overview of current experimental models. Iranian biomedical journal, 20(1), 1.

Bălaşa, A. F., Chircov, C., & Grumezescu, A. M. (2020). Marine biocompounds for neuroprotection—A review. Marine Drugs, 18(6), 290.

Bao, Z., Bufton, J., Hickman, R. J., Aspuru-Guzik, A., Bannigan, P., & Allen, C. (2023). Revolutionizing drug formulation development: The increasing impact of machine learning. Advanced Drug Delivery Reviews, 202(115108), 115108.

Bayda, S., Adeel, M., Tuccinardi, T., Cordani, M., & Rizzolio, F. (2019). The history of nanoscience and nanotechnology: From chemical-physical applications to nanomedicine. Molecules (Basel, Switzerland), 25(1), 112.

Behera, M., Behera, P. R., Bhuyan, P. P., Singh, L., & Pradhan, B. (2023). Algal nanoparticles and their antibacterial activity: current research status and future prospectives. Drugs and Drug Candidates, 2(3), 554–570.

BelHadj, S., Hentati, O., Elfeki, A., & Hamden, K. (2013). Inhibitory activities ofUlva lactucapolysaccharides on digestive enzymes related to diabetes and obesity. Archives of Physiology and Biochemistry, 119(2), 81–87.

Bhuyan, P. P., Nayak, R., Patra, S., Abdulabbas, H. S., Jena, M., & Pradhan, B. (2023). Seaweed-derived sulfated polysaccharides; The new age chemopreventives: A comprehensive review. Cancers, 15(3). https://doi.org/https://doi.org/10.3390/cancers15030715

Bocanegra, A., Macho-González, A., Garcimartín, A., Benedí, J., & Sánchez-Muniz, F. J. (2021). Whole alga, algal extracts, and compounds as ingredients of functional foods: Composition and action mechanism relationships in the prevention and treatment of type-2 diabetes mellitus. International Journal of Molecular Sciences, 22(8), 3816.

Buzea, C., & Pacheco, I. (2019). Toxicity of nanoparticles. In Nanotechnology in eco-efficient construction (pp. 705-754). Woodhead Publishing.

Celikler, S., Tas, S., Vatan, O., Ziyanok-Ayvalik, S., Yildiz, G., & Bilaloglu, R. (2009). Antihyperglycemic and antigenotoxic potential of Ulva rigida ethanolic extract in the experimental diabetes mellitus. Food and Chemical Toxicology: An International Journal Published for the British Industrial Biological Research Association, 47(8), 1837–1840.

Chaudhary, R., Nawaz, K., Khan, A. K., Hano, C., Abbasi, B. H., & Anjum, S. (2020). An overview of the algae-mediated biosynthesis of nanoparticles and their biomedical applications. Biomolecules, 10(11), 1498.

Conde, T. A., Zabetakis, I., Tsoupras, A., Medina, I., Costa, M., Silva, J., Neves, B., Domingues, P., & Domingues, M. R. (2021). Microalgal lipid extracts have potential to modulate the inflammatory response: A critical review. International Journal of Molecular Sciences, 22(18), 9825.

Di Bona, K. R., Xu, Y., Gray, M., Fair, D., Hayles, H., Milad, L., Montes, A., Sherwood, J., Bao, Y., & Rasco, J. F. (2015). Short- and long-term effects of prenatal exposure to iron oxide nanoparticles: Influence of surface charge and dose on developmental and reproductive toxicity. International Journal of Molecular Sciences, 16(12), 30251–30268.

Ernst, A. U., Bowers, D. T., Wang, L.-H., Shariati, K., Plesser, M. D., Brown, N. K., Mehrabyan, T., & Ma, M. (2019). Nanotechnology in cell replacement therapies for type 1 diabetes. Advanced Drug Delivery Reviews, 139, 116–138.

Fawcett, D., Verduin, J. J., Shah, M., Sharma, S. B., & Poinern, G. E. J. (2017). A review of current research into the biogenic synthesis of metal and metal oxide nanoparticles via marine algae and seagrasses. Journal of Nanoscience, 2017, 1–15.

Figueiredo, F., Encarnação, T., & G. Campos, M. (2016). Algae as functional foods for the elderly. Food and Nutrition Sciences, 07(12), 1122–1148.

Foroghi, S., Sarafbidabad, M., & Nasiri, R. (2025). Nanoemulsions of Bioactive Lipids Isolated from Cladophora Green Algae Declined NO and IL-6 Production in LPS-simulated RAW264. 7 Murine Macrophage Cells. Nano LIFE, 15(03).

Fourches, D., Pu, D., & Tropsha, A. (2011). Exploring quantitative nanostructure-activity relationships (QNAR) modeling as a tool for predicting biological effects of manufactured nanoparticles. Combinatorial Chemistry & High Throughput Screening, 14(3), 217–225.

Gandhi, S., Srinivasan, B. P., & Akarte, A. S. (2013). An experimental assessment of toxic potential of nanoparticle preparation of heavy metals in streptozotocin induced diabetes. Experimental and Toxicologic Pathology: Official Journal of the Gesellschaft Für Toxikologische Pathologie, 65(7–8), 1127–1135.

Gayathri, K., Bhaskaran, M., Selvam, C., & Thilagavathi, R. (2023). Nano formulation approaches for curcumin delivery-a review. Journal of Drug Delivery Science and Technology, 82.

Guimarães, D., Cavaco-Paulo, A., & Nogueira, E. (2021). Design of liposomes as drug delivery system for therapeutic applications. International Journal of Pharmaceutics, 601(120571), 120571.

Hellack, B., Nickel, C., Albrecht, C., Kuhlbusch, T. A., Boland, S., Baeza-Squiban, A., ... & Schins, R. P. (2017). Analytical methods to assess the oxidative potential of nanoparticles: a review. Environmental Science: Nano, 4(10), 1920-1934.

Hu, R., Wang, X., Han, L., & Lu, X. (2023). The developments of surface-functionalized selenium nanoparticles and their applications in brain diseases therapy. Biomimetics (Basel, Switzerland), 8(2), 259.

Huang, J., Zou, L., Bao, M., Feng, Q., Xia, W., & Zhu, C. (2023). Toxicity of polystyrene nanoparticles for mouse ovary and cultured human granulosa cells. Ecotoxicology and Environmental Safety, 249, 114371.

Huang, X., & Tang, M. (2021). Research advance on cell imaging and cytotoxicity of different types of quantum Dots. Journal of Applied Toxicology, 41(3), 342-361.

Iwai, K. (2008). Antidiabetic and antioxidant effects of polyphenols in brown alga Ecklonia stolonifera in genetically diabetic KK-Ay mice. Plant Foods for Human Nutrition, 63(4), 163–169.

James, J., Verma, M., & Sharma, N. (2023). Nanotechnology-driven improvisation of red algae-derived carrageenan for industrial and bio-medical applications. World Journal of Microbiology and Biotechnology, 40(1), 4.

Jeng, H. A., & Swanson, J. (2006). Toxicity of metal oxide nanoparticles in mammalian cells. Journal of Environmental Science and Health. Part A, Toxic/Hazardous Substances & Environmental Engineering, 41(12), 2699–2711.

Joseph, X., Akhil, Arathi, Megha, K. B., Vandana, U., & Mohanan, P. V. (2023). Toxicity Assessment of Nanoparticle. In Biomedical Applications and Toxicity of Nanomaterials (pp. 401-423). Singapore: Springer Nature Singapore.

Kannan, S., Chellappan, D. K., Kow, C. S., Ramachandram, D. S., Pandey, M., Mayuren, J., Dua, K., & Candasamy, M. (2023). Transform diabetes care with precision medicine. Health Science Reports, 6(11), e1642.

Kerfahi, D., Tripathi, B. M., Singh, D., Kim, H., Lee, S., Lee, J., & Adams, J. M. (2015). Effects of functionalized and raw multi-walled carbon nanotubes on soil bacterial community composition. PloS One, 10(3), e0123042.

Khanna, P., Kaur, A., & Goyal, D. (2019). Algae-based metallic nanoparticles: Synthesis, characterization and applications. Journal of Microbiological Methods, 163(105656), 105656.

Kim, S.-J., Puranik, N., Yadav, D., Jin, J.-O., & Lee, P. C. W. (2023). Lipid nanocarrier-based drug delivery systems: Therapeutic advances in the treatment of lung cancer. International Journal of Nanomedicine, 18, 2659–2676.

Krause, C. J. (2014). Evaluating the toxicity of nitrogen-functionalized carbon nanotubes MS (Master of Science) thesis.

Kumari, A., & Bharadvaja, N. (2023). A comprehensive review on algal nutraceuticals as prospective therapeutic agent for different diseases. 3 Biotech, 13.

Lee, C.-W., & Han, J.-S. (2012). Hypoglycemic effect of Sargassum ringgoldianum extract in STZ-induced diabetic mice. Preventive Nutrition and Food Science, 17(1), 8–13.

Lee, Y. S., Shin, K. H., Kim, B.-K., & Lee, S. (2004). Anti-Diabetic activities of fucosterol fromPelvetia siliquosa. Archives of Pharmacal Research, 27(11), 1120–1122.

Li, H., Peng, E., Zhao, F., Li, J., & Xue, J. (2021). Supramolecular surface functionalization of iron oxide nanoparticles with α-cyclodextrin-based cationic star polymer for magnetically-enhanced gene delivery. Pharmaceutics, 13(11), 1884.

Li, X., Yang, X., Yuwen, L., Yang, W., Weng, L., Teng, Z., & Wang, L. (2016a). Evaluation of toxic effects of CdTe quantum dots on the reproductive system in adult male mice. Biomaterials, 96, 24–32.

Li, Xiaodan, Yu, Z., Long, S., Guo, Y., & Duan, D. (2012). Hypoglycemic effect of Laminaria japonica polysaccharide in a type 2 diabetes mellitus mouse model. ISRN Endocrinology, 2012, 507462.

Li, Xiaohui, Yang, X., Yuwen, L., Yang, W., Weng, L., Teng, Z., & Wang, L. (2016b). Evaluation of toxic effects of CdTe quantum dots on the reproductive system in adult male mice. Biomaterials, 96, 24–32.

Li, Y.-X., & Kim, S.-K. (2011). Utilization of seaweed derived ingredients as potential antioxidants and functional ingredients in the food industry: An overview. Food Science and Biotechnology, 20(6), 1461–1466.

Lin, Y., & Sun, Z. (2010). Current views on type 2 diabetes. The Journal of Endocrinology, 204(1), 1–11.

Lu, L. W., & Chen, J.-H. (2022). Seaweeds as ingredients to lower glycemic potency of cereal foods synergistically—A perspective. Foods (Basel, Switzerland), 11(5), 714.

Lu, X., Qian, J., Zhou, H., Gan, Q., Tang, W., Lu, J., Yuan, Y., & Liu, C. (2011). In vitro cytotoxicity and induction of apoptosis by silica nanoparticles in human HepG2 hepatoma cells. International Journal of Nanomedicine, 6, 1889–1901.

Mabate, B., Daub, C. D., Malgas, S., Edkins, A. L., & Pletschke, B. I. (2021). Fucoidan structure and its impact on glucose metabolism: Implications for diabetes and cancer therapy. Marine Drugs, 19(1), 30.

Mahmud, F. B., Manisha, M., Kavvyasruthi, U. J., Krishnasamy, D., Chebwogen, S., Ramesh, V., ... & Senevirathne, H. (2023). A Review On Silver Nanoparticles And Its Role In Treating Diseases: Anticancer Activity Of AgNPs. International Research Journal of Modernisation in Engineering technology and Science. Vol 5. 1-16.

Mitchell, M. J., Billingsley, M. M., Haley, R. M., Wechsler, M. E., Peppas, N. A., & Langer, R. (2021). Engineering precision nanoparticles for drug delivery. Nature Reviews. Drug Discovery, 20(2), 101–124.

Mohammadinejad, R., Moosavi, M. A., Tavakol, S., Vardar, D. Ö., Hosseini, A., Rahmati, M., ... & Klionsky, D. J. (2019). Necrotic, apoptotic and autophagic cell fates triggered by nanoparticles. Autophagy, 15(1), 4-33.
Moreira, T. R., da Negreiros, F. D. S., Aquino, M. de J. N. de, Silva, L. M. S. da, Moreira, T. M. M., & Torres, R. A. M. (2023). Digital technology and its effects on knowledge improvement for diabetes management: An integrative review. International Journal of Nursing Practice, 29(1), e13029.
Morran, M. P., Vonberg, A., Khadra, A., & Pietropaolo, M. (2015). Immunogenetics of type 1 diabetes mellitus. Molecular Aspects of Medicine, 42, 42–60.
Motshakeri, M., Ebrahimi, M., Goh, Y. M., Matanjun, P., & Mohamed, S. (2013). Sargassum polycystum reduces hyperglycaemia, dyslipidaemia and oxidative stress via increasing insulin sensitivity in a rat model of type 2 diabetes. Journal of the Science of Food and Agriculture, 93(7), 1772–1778.
Mukherjee, A., Sarkar, D., & Sasmal, S. (2021). A review of green synthesis of metal nanoparticles using algae. Frontiers in Microbiology, 12, 693899.
Mukhtar, Y., Galalain, A., & Yunusa, U. (2020). A modern overview on diabetes mellitus: a chronic endocrine disorder. European Journal of Biology, 5(2), 1-14.
Nasab, S. B., Homaei, A., Pletschke, B. I., Salinas-Salazar, C., Castillo-Zacarias, C., & Parra-Saldívar, R. (2020). Marine resources effective in controlling and treating diabetes and its associated complications. Process Biochemistry (Barking, London, England), 92, 313–342.
Nasiri, K., Masoumi, S. M., Amini, S., Goudarzi, M., Tafreshi, S. M., Bagheri, A., Yasamineh, S., Alwan, M., Arellano, M. T. C., & Gholizadeh, O. (2023). Recent advances in metal nanoparticles to treat periodontitis. Journal of Nanobiotechnology, 21(1), 283.
Noga, M., Milan, J., Frydrych, A., & Jurowski, K. (2023). Toxicological aspects, safety assessment, and green toxicology of silver nanoparticles (AgNPs)-critical review: State of the art. International Journal of Molecular Sciences, 24(6).
Odenthal, K., Nunes, E., Nunes, N., Fernandes, T., Fernandes, I. A., & Pinheiro de Carvalho, M. A. A. (2024). Microalgae and cyanobacteria as natural sources of antioxidant enzymes and enzyme inhibitors for Alzheimer's and diabetes. Algal Research, 82(103610), 103610.
Ogunwa, T. H., Adeyelu, T. T., Fasimoye, R. Y., & Ayenitaju, F. C. (2018). In silico analysis of interaction between seaweed-derived bioactive compounds and selected diabetes-related targets. Biomedical Chemistry Research and Methods, 1(4), e00074.
Ohta, T., Sasaki, S., Oohori, T., Yoshikawa, S., & Kurihara, H. (2002). A-glucosidase inhibitory activity of a 70% methanol extract from ezoishige (Pelvetia babingtoniide Toni) and its effect on the elevation of blood glucose level in rats. Bioscience, Biotechnology, and Biochemistry, 66(7), 1552–1554.
Oliyaei, N., Tanideh, N., Moosavi-Nasab, M., Dehghanian, A. R., & Iraji, A. (2023). Development and characterization of a fucoidan-based nanoemulsion using Nigella sativa oil for improvement of anti-obesity activity of fucoxanthin in an obese rat model. International Journal of Biological Macromolecules, 235(123867), 123867.
Patel, K. D., Keskin-Erdogan, Z., Sawadkar, P., Sharifulden, N. S. A. N., Shannon, M. R., Patel, M., ... & Kim, H. W. (2024). Oxidative stress modulating nanomaterials and their biochemical roles in nanomedicine. Nanoscale horizons, 9(10), 1630-1682.
Patra, J. K., Das, G., Fraceto, L. F., Campos, E. V. R., Rodriguez-Torres, M. D. P., Acosta-Torres, L. S., Diaz-Torres, L. A., Grillo, R., Swamy, M. K., Sharma, S., Habtemariam, S., & Shin, H.-S. (2018). Nano based drug delivery systems: recent developments and future prospects. Journal of Nanobiotechnology, 16(1), 71.
Peng, S., Yuan, X., Li, H., Wei, Y., Zhou, B., Ding, G., & Bai, J. (2023). Recent progress in nanocarrier-based drug delivery systems for antitumour metastasis. European Journal of Medicinal Chemistry, 252(115259), 115259.
Pereira, L., & Valado, A. (2023). Algae-derived natural products in diabetes and its complications-current advances and future prospects. Life (Basel, Switzerland), 13(9), 1831.

Priatni, S., Budiwati, T. A., Ratnaningrum, D., Kosasih, W., Andryani, R., Susanti, H., & Susilaningsih, D. (2016). Antidiabetic screening of some Indonesian marine cyanobacteria collection. Biodiversitas: Journal of Biological Diversity, 17(2). https://doi.org/10.13057/biodiv/d170236

Priya, P., Sivamurugan, V., Murugesan, S., & Shanti, N. (2021). Biosynthesised Nano Materials As Nanomedicine. Biosynthesised Nano Materials As Nanomedicine. JPS Publications.

Puccetti, M., Pariano, M., Schoubben, A., Giovagnoli, S., & Ricci, M. (2024). Biologics, theranostics, and personalized medicine in drug delivery systems. Pharmacological Research: The Official Journal of the Italian Pharmacological Society, 201(107086), 107086.

Qi, W., Bi, J., Zhang, X., Wang, J., Wang, J., Liu, P., Li, Z., & Wu, W. (2014). Damaging effects of multi-walled carbon nanotubes on pregnant mice with different pregnancy times. Scientific Reports, 4(1), 4352.

Qian, W.-W., Yang, S.-Q., Hu, S.-M., Wang, X.-L., Zhu, Y., & Zhou, T. (2021). Enzymatic degradation, antioxidant and immunoregulatory activities of polysaccharides from brown algae Sargassum fusiforme. Journal of Food Measurement & Characterization, 15(2), 1960–1972.

Raghavendra, T., & Bhat, S. G. (2022). Enzyme immobilized nanomaterials. In Nanomaterials for Biocatalysis (pp. 17–65). Elsevier.

Raghavendra, V. B., Sagar, N., Kusha, L. M., Krishna, K. L., Sharma, M., Sridhar, K., & Puttasiddaiah, R. (2024). Recent advances on Nigella sativa-A Promising herb: Antihypertensive properties, thimoquinone nanoformulations, and health applications. Pharmacological Research-Natural Products.

Rahim, M. A., Jan, N., Khan, S., Shah, H., Madni, A., Khan, A., Jabar, A., Khan, S., Elhissi, A., Hussain, Z., Aziz, H. C., Sohail, M., Khan, M., & Thu, H. E. (2021). Recent advancements in stimuli responsive drug delivery platforms for active and passive cancer targeting. Cancers, 13(4), 670.

Ramos-Romero, S., Torrella, J. R., Pagès, T., Viscor, G., & Torres, J. L. (2021). Edible microalgae and their bioactive compounds in the prevention and treatment of metabolic alterations. Nutrients, 13(2), 563.

Rathod, N. B., Elabed, N., Punia, S., Ozogul, F., Kim, S.-K., & Rocha, J. M. (2023). Recent developments in polyphenol applications on human health: A review with current knowledge. Plants, 12(6), 1217.

Renard, E. (2023). Automated insulin delivery systems: from early research to routine care of type 1 diabetes. Acta Diabetologica, 60(2), 151-161.

Roebben, G., Rauscher, H., Amenta, V., Aschberger, K., Sanfeliu, A. B., Calzolai, L., & Stamm. (2014). Towards a review of the EC Recommendation for a definition of the term.

Rudrapal, M., Khairnar, S. J., Khan, J., Dukhyil, A. B., Ansari, M. A., Alomary, M. N., Alshabrmi, F. M., Palai, S., Deb, P. K., & Devi, R. (2022). Dietary polyphenols and their role in oxidative stress-induced human diseases: Insights into protective effects, antioxidant potentials and mechanism(s) of action. Frontiers in Pharmacology, 13, 806470.

Saadaoui, I., Rasheed, R., Abdulrahman, N., Bounnit, T., Cherif, M., Al Jabri, H., & Mraiche, F. (2020). Algae-derived bioactive compounds with anti-lung cancer potential. Marine Drugs, 18(4), 197.

Saha, S., Ali, M. R., Khaleque, M. A., Bacchu, M. S., Aly Saad Aly, M., & Khan, M. Z. H. (2023). Metal oxide nanocarrier for targeted drug delivery towards the treatment of global infectious diseases: A review. Journal of Drug Delivery Science and Technology, 86(104728), 104728.

Salehipour, M., Nikpour, S., Rezaei, S., Mohammadi, S., Rezaei, M., Ilbeygi, D., ... & Mogharabi-Manzari, M. (2023). Safety of metal–organic framework nanoparticles for biomedical applications: an in vitro toxicity assessment. Inorganic Chemistry Communications, 152, 110655.

Samudra, A. G., Ramadhani, N., Pertiwi, R., Fitriani, D., Sanik, F., & Burhan, A. (2023). Antihyperglycemic activity of nanoemulsion of brown algae (Sargassum sp.). Ethanol extract in glucose tolerance test in male mice. Annales Pharmaceutiques Francaises, 81(3), 484–491.

Sarhan, O. M. M., & Hussein, R. M. (2014). Effects of intraperitoneally injected silver nanoparticles on histological structures and blood parameters in the albino rat. International Journal of Nanomedicine, 9, 1505–1517.

Schwestka, J., & Stoger, E. (2021). Microparticles and nanoparticles from plants-the benefits of bioencapsulation. Vaccines, 9(4), 369.

Serrano, D. R., Kara, A., Yuste, I., Luciano, F. C., Ongoren, B., Anaya, B. J., Molina, G., Diez, L., Ramirez, B. I., Ramirez, I. O., Sánchez-Guirales, S. A., Fernández-García, R., Bautista, L., Ruiz, H. K., & Lalatsa, A. (2023). 3D printing technologies in personalized medicine, nanomedicines, and biopharmaceuticals. Pharmaceutics, 15(2), 313.

Shah, M. A. R., Zhang, Y., Cui, Y., Hu, X., Zhu, F., Kumar, S., Li, G., Kubar, A. A., Mehmood, S., & Huo, S. (2024). Ultrasonic-assisted green extraction and incorporation of Spirulina platensis bioactive components into turmeric essential oil-in-water nanoemulsion for enhanced antioxidant and antimicrobial activities. Food Chemistry, 452(139561), 139561.

Shah, M. R., Imran, M., & Ullah, S. (2022). Nanocarriers systems for brain targeted drug delivery and diagnosis. In Nanocarriers for Organ-Specific and Localized Drug Delivery (pp. 25–43). Elsevier.

Shah, M., Fawcett, D., Sharma, S., Tripathy, S., & Poinern, G. (2015). Green synthesis of metallic nanoparticles via biological entities. Materials, 8(11), 7278–7308.

Sharma, A., Sharma, S., Sharma, K., Chetri, S. P. K., Vashishtha, A., Singh, P., Kumar, R., Rathi, B., & Agrawal, V. (2016). Algae as crucial organisms in advancing nanotechnology: a systematic review. Journal of Applied Phycology, 28(3), 1759–1774.

Shi, P., Sun, X., Yuan, H., Chen, K., Bi, S., & Zhang, S. (2023). Nanoscale metal-organic frameworks combined with metal nanoparticles and metal oxide/peroxide to relieve tumor hypoxia for enhanced photodynamic therapy. ACS Biomaterials Science & Engineering, 9(10), 5441–5456.

Silva, A., Cassani, L., Grosso, C., Garcia-Oliveira, P., Morais, S. L., Echave, J., Carpena, M., Xiao, J., Barroso, M. F., Simal-Gandara, J., & Prieto, M. A. (2024). Recent advances in biological properties of brown algae-derived compounds for nutraceutical applications. Critical Reviews in Food Science and Nutrition, 64(5), 1283–1311.

Singh, S. K., Sharma, A., Sharma, L., & Sundaram, S. (2024). Green synthesis of algal nanoparticles: Harnessing nature's biofactories for sustainable nanomaterials. In Biogenic Wastes-Enabled Nanomaterial Synthesis (pp. 257–284). Springer Nature Switzerland.

Sinha, S., Haque, M., Lugova, H., & Kumar, S. (2023). The effect of omega-3 fatty acids on insulin resistance. Life (Basel, Switzerland), 13(6), 1322.

Sun, Y., Dinenno, F. A., Tang, P., & Kontaridis, M. I. (2024). Protein tyrosine phosphatase 1B in metabolic and cardiovascular diseases: from mechanisms to therapeutics. Frontiers in Cardiovascular Medicine, 11, 1445739.

Tan, K. F., In, L. L. A., & Vijayaraj Kumar, P. (2023a). Surface functionalization of gold nanoparticles for targeting the tumor microenvironment to improve antitumor efficiency. ACS Applied Bio Materials, 6(8), 2944–2981.

Tan, Y., Yu, D., Feng, J., You, H., Bai, Y., He, J., Cao, H., Che, Q., Guo, J., & Su, Z. (2023b). Toxicity evaluation of silica nanoparticles for delivery applications. Drug Delivery and Translational Research, 13(9), 2213–2238.

Tas, S., Celikler, S., Ziyanok-Ayvalik, S., Sarandol, E., & Dirican, M. (2011). Ulva rigida improves carbohydrate metabolism, hyperlipidemia and oxidative stress in streptozotocin-induced diabetic rats. Cell Biochemistry and Function, 29(2), 108–113.

Tewabe, A., Abate, A., Tamrie, M., Seyfu, A., & Abdela Siraj, E. (2021). Targeted drug delivery - from magic bullet to nanomedicine: Principles, challenges, and future perspectives. Journal of Multidisciplinary Healthcare, 14, 1711–1724.

Tian, H., Yin, X., Zeng, Q., Zhu, L., & Chen, J. (2015). Isolation, structure, and surfactant properties of polysaccharides from Ulva lactuca L. from South China Sea. International Journal of Biological Macromolecules, 79, 577–582.

Tripathi, G., Dubey, P., Ahmad, S., Farooqui, A., & Mishra, V. (2024). Role of algal-derived bioactive compounds in human health. Recent Patents on Biotechnology, 18(3), 190–209.

Truong, T. T., Mondal, S., Doan, V. H. M., Tak, S., Choi, J., Oh, H., Nguyen, T. D., Misra, M., Lee, B., & Oh, J. (2024). Precision-engineered metal and metal-oxide nanoparticles for biomedical imaging and healthcare applications. Advances in Colloid and Interface Science, 332(103263), 103263.

Vandghanooni, S., & Eskandani, M. (2011). Comet assay: a method to evaluate genotoxicity of nano-drug delivery system. Bioimpacts, 1(2), 87-97.

Villalobos Gutiérrez, P. T., Muñoz Carrillo, J. L., Sandoval Salazar, C., Viveros Paredes, J. M., & Gutiérrez Coronado, O. (2023). Functionalized metal nanoparticles in cancer therapy. Pharmaceutics, 15(7), 1932.

Vora, L. K., Gholap, A., Jetha, K., Thakur, R., Solanki, H. K., & Chavda, V. P. (2023). Artificial intelligence in pharmaceutical technology and drug delivery design. Pharmaceutics, 15. https://doi.org/https://doi.org/10.3390/pharmaceutics15071916

Xi, Y., & Xu, P.-F. (2021). Diabetes and gut microbiota. World Journal of Diabetes, 12(10), 1693–1703.

Xia, Z., Duan, X., Tao, S., Qiu, W., Liu, D., Wang, Y., Wei, S., Wang, B., Jiang, Q., Lu, B., Song, Y., & Hu, X. (2013). Pollution level, inhalation exposure and lung cancer risk of ambient atmospheric polycyclic aromatic hydrocarbons (PAHs) in Taiyuan, China. Environmental Pollution (Barking, Essex: 1987), 173, 150–156.

Yamagishi, Y., Watari, A., Hayata, Y., Li, X., Kondoh, M., Tsutsumi, Y., & Yagi, K. (2013). Hepatotoxicity of sub-nanosized platinum particles in mice. Die Pharmazie, 68(3), 178–182.

Yan, D.-D., Hu, B., Gao, P., Yin, J.-J., Wang, S., Yang, Y., Tan, L., Hu, C.-R., He, D.-P., & Zhong, W. (2024). Synthesis and characterization of emulsifiers based on the Maillard reaction and its application in stabilized DHA algal oil nanoemulsions. Foods (Basel, Switzerland), 13(11), 1667.

Yang, J., Jia, C., & Yang, J. (2021). Designing nanoparticle-based drug delivery systems for precision medicine. International Journal of Medical Sciences, 18(13), 2943–2949.

Yao, Y., Zhang, T., & Tang, M. (2023). The DNA damage potential of quantum dots: Toxicity, mechanism and challenge. Environmental Pollution, 317, 120676.

Yedoti, V., & Supraja, N. (2024). A review on algal mediated synthesized metallic nanoparticles: An Eco-friendly approach for sustainable nanotechnology. Current Journal of Applied Science and Technology, 43(6), 1–10.

Yetisgin, A. A., Cetinel, S., Zuvin, M., Kosar, A., & Kutlu, O. (2020). Therapeutic nanoparticles and their targeted delivery applications. Molecules (Basel, Switzerland), 25(9), 2193.

Zaccardi, F., Webb, D. R., Yates, T., & Davies, M. J. (2016). Pathophysiology of type 1 and type 2 diabetes mellitus: a 90-year perspective. Postgraduate medical journal, 92(1084), 63-69.

Zhang, S., Qamar, S. A., Junaid, M., Munir, B., Badar, Q., & Bilal, M. (2022). Algal polysaccharides-based nanoparticles for targeted drug delivery applications. Die Starke, 74(7–8), 2200014.

Zhang, W., Wang, G., See, E., Shaw, J. P., Baguley, B. C., Liu, J., Amirapu, S., & Wu, Z. (2015). Post-insertion of poloxamer 188 strengthened liposomal membrane and reduced drug irritancy and in vivo precipitation, superior to PEGylation. Journal of Controlled Release: Official Journal of the Controlled Release Society, 203, 161–169.

Zhao, C., Yang, C., Liu, B., Lin, L., Sarker, S. D., Nahar, L., Yu, H., Cao, H., & Xiao, J. (2018). Bioactive compounds from marine macroalgae and their hypoglycemic benefits. Trends in Food Science & Technology, 72, 1–12.

Zheng, X., Wu, Y., Zuo, H., Chen, W., & Wang, K. (2023). Metal nanoparticles as novel agents for lung cancer diagnosis and therapy. Small, 19(18), e2206624.

Chapter 17
Computational Insights into Algal Compounds for Diabetes Mellitus Management

Baby Anusha Satravada, Vairavanathan Jayavarshini, Subaranjani Ravichandiran, Sundararaj Rajamanikandan, and Dhamodharan Prabhu

Abstract The global health crisis of diabetes mellitus continues to worsen because of its increasing prevalence and insufficient treatment choices which lack effectiveness and affordability and individualized approaches. The diverse photosynthetic organism algae show promise as an underutilized source for developing new anti-diabetic bioactive compounds in our current situation. The chapter investigates how multi-omics approaches including genomics, transcriptomics, proteomics, metabolomics, lipidomics, and emerging omics can speed up the development and discovery of algal therapeutics. The diabetes management mechanisms of algal metabolites including sulfated polysaccharide, omega-3 fatty acid, carotenoid and bioactive peptides involved in blood glucose regulation, insulin sensitivity and anti-inflammatory properties. Research on Chlorella, Nannochloropsis, Spirulina, and *Haematococcus pluvialis* species demonstrates how combined omics approaches can identify, validate and enhance these compounds. Structural biology and the

B. Satravada
Computational Oncology, Memorial Sloan Kettering Cancer Centre, New York, NY, USA

V. Jayavarshini
Department of Biotechnology, Karpagam Academy of Higher Education, Coimbatore, India

Department of Biotechnology, Dr. N.G.P. Arts and Science College, Coimbatore, India

S. Ravichandiran
Center for Global Health Research, Saveetha Medical College & Hospital, Saveetha Institute of Medical and Technical Sciences, Chennai, Tamil Nadu, India

S. Rajamanikandan
School of Pharmaceutical Science & Technology, Tianjin University, Tianjin, People's Republic of China

Centre for Bioinformatics, Karpagam Academy of Higher Education, Coimbatore, India

D. Prabhu (✉)
Department of Biotechnology, Karpagam Academy of Higher Education, Coimbatore, India

Centre for Bioinformatics, Karpagam Academy of Higher Education, Coimbatore, India
e-mail: prabhu.dhamodharan@kahedu.edu.in

R. Raja et al. (eds.), *Algae in Diabetes Management*,
https://doi.org/10.1007/978-981-95-0571-5_17

System Model enable the function of the biosynthetic nerve pathway and the designation of the high-voltage target. Moreover, machine learning systems and machine learning transform facts integration and predictive models, allowing personalized and scalable creation of curative mixtures. Despite technical and infrastructure obstacles, the present part stresses the ability of multi-omics to revolutionize algal biotech and develop green, natural, and proficient solutions to diabetes management. Algal interventions could, in the long run, complement or even outperform traditional pharmaceutical approaches to treating metabolic disorders by convergent molecular biology, bioinformatics, and systems science.

Keywords Algae · Diabetes mellitus · OMICs · Metabolites · Bioinformatics

1 Introduction

Diabetes mellitus (DM) is a chronic metabolic disorder of global concern, affecting millions worldwide. It is characterized by persistent hyperglycemia resulting from inadequate insulin production, insulin resistance, or a combination of both. This condition poses serious health risks, including cardiovascular diseases, nephropathy, nephropathy, and retinopathy, which significantly reduce the quality of life. The World Health Organization (WHO) published statistics indicating diabetes affected 8.5% of adults worldwide in 2014 with a total of 1.5 million annual deaths occurring among people under 70. The increasing diabetic population forecast indicates a 592 million total by 2035 thus demonstrating the necessity for innovative treatment approaches (Antar et al. 2023; Gunathilaka et al. 2020).

Diabetes mellitus manifests primarily in two forms, such as Type 1 and Type 2. Type 1 diabetes mellitus (T1DM) emerges when pancreatic β-cell destruction leads to complete insulin deficiency that mostly occurs among younger patients. T1DM patients need insulin shots daily for survival and they face increased chances of developing diabetic ketoacidosis (Abo-Shady et al. 2023). Type 2 diabetes mellitus (T2DM) is the primary diabetes form and emerges mainly due to obesity combined with physical inactivity and unhealthy eating habits. The condition starts with insulin resistance and β-cell dysfunction, which affect mostly adults over 40 years old and results in cardiovascular diseases and kidney problems. The successful control of T2DM needs approaches that simultaneously prevent and treat the condition (Pereira and Valado 2023).

Microalgae to macroscopic seaweeds from marine algae group have demonstrated effectiveness in diabetes management because they contain numerous bioactive metabolites. The compounds found in marine algae include sulfated polysaccharides which boost insulin sensitivity and glucose metabolism alongside phlorotannins with antioxidant effects that fight inflammation and oxidative stress. While the fucoxanthin that improves lipid profiles and glucose metabolism, and omega-3 fatty acids were shown to reduce systemic inflammation while enhancing insulin sensitivity. The therapeutic mechanisms of algal compounds involve blocking α-glucosidase enzymes to delay carbohydrate digestion, activating insulin receptors and also protecting pancreatic β-cells through antioxidant effects while simultaneously managing lipid metabolism to enhance metabolic process.

The standard diabetes management practices include lifestyle changes, insulin therapy, and oral hypoglycemic drugs; however, these methods encounter challenges such as side effects and non-adherence issues in addition to expensive costs. Marine algae have emerged as a promising natural resource for diabetes management due to rising interest in their potential therapeutic value (T. Gunathilaka et al. 2022). New amalgamation of computational and experimental methods in drug discovery results on effective and efficient therapeutics. The computational methods incorporated in the discovery of new molecules from algae lead to faster development of therapeutics with higher precision and efficacy. This chapter deals with the computational approaches pertaining to the discovery and design of potential metabolites and associated bioactive compounds from algal sources toward the management of diabetes.

2 Computational Approaches for Screening Algal Compounds

In the current scenario, the advanced drug discovery pipelines were completely dependent on computational methods for the initial screening and preliminary drug-likeliness validations, which is very much cost-effective than the routine methods and in addition it aids in the identification of potential molecules rapidly with more efficacy. Moreover, these computational methods quickly evaluate and select potential algal compounds that possess promising therapeutic activity against various diseases like diabetes, cancer, and secondary inflammations (Ogunwa et al. 2018). The combination of network pharmacology and virtual screening including molecular docking methods empowers the researchers to predict potential algal metabolite and their molecular interactions with target proteins, which leads to the identification and further optimization of bioactive compounds for better efficacy and activity (Cakmak and Selvaraj 2017). Also, the computational methods such as molecular docking and molecular dynamics simulations provide detailed insights on the atomic level interactions between the selected metabolite and target protein, further these methods allow the researchers to understand the drug-like properties such as safety, efficacy, and specificity (Bhagat and Ghormade 2020). Computer-assisted drug discovery process has become highly effective due to the precise identification of compounds with high therapeutic potential.

Microalgae produce diverse bioactive compounds under stress conditions for its own survival, however those bioactive compounds are generally non-toxic and are therapeutically beneficial. The predominant bioactive compounds of microalgae are carbohydrates, carotenoids, sterols, and poly unsaturated fatty acids. The excessive fatty acid accumulation by Omega-3-enriched foods exhibited improved cellular adaptability and reduced insulin resistance. Moreover, the effective regulation of blood glucose levels was strongly associated with a lower risk of microvascular complications. Sixteen microalgae metabolites including astaxanthin, brassicasterol, canthaxanthin, β-carotene, β-stigma sterol, docosahexaenoic acid, and zeaxanthin were reported to show inhibitory activities against target proteins (Selvaraj et al. 2017). Enzymes such as glucokinase and fructose-1,6-bisphosphatase, essential for glucose regulation in the liver, were shown to increase glucose production from glycerol or gluconeogenic amino acids. Persistent glucose production was linked to severe non-insulin-dependent diabetes. Lutein and zeaxanthin analogs exhibited strong binding affinities to glucokinase and glycogen synthase. Similarly, lutein, fucoxanthin, and zeaxanthin displayed notable interactions with glucokinase, glycogen synthase, and fructose-1,6-bisphosphatase. Altered glycoprotein permeability was associated with genetic variations in transporter proteins, leading to decreased high-density lipoprotein levels, elevated blood glucose, cystic fibrosis, and retinal and kidney damage in diabetic patients. The potential of specific carotenoids, particularly lutein, fucoxanthin, and zeaxanthin, as promising candidates for novel antidiabetic drug development. Docking analyses revealed that the binding interaction of fucoxanthin is considerably high with fructose 1,6-bisphosphatase, human multidrug-resistant protein 1, and PPARγ. Lutein demonstrated high binding interactions with fructose-1,6-bisphosphatase, multidrug-resistant protein, glycogen synthase kinase, PPARγ, and cytochrome P450, while zeaxanthin showed strong energy interactions with glucokinase (Selvaraj et al. 2017).

The antidiabetic properties of fucosterol were determined by its capacity to block enzymes such as rat lens aldose reductase (RLAR), human recombinant aldose reductase (HRAR), protein tyrosine phosphatase 1B (PTP1B), and α-glucosidase. Fucosterol displayed moderate inhibitory effects on RLAR, HRAR, and PTP1B but showed no or minimal activity against advanced glycation end-products (AGE) formation and α-glucosidase. The kinetic studies revealed that fucosterol acts as a mixed-type inhibitor for RLAR and HRAR and a noncompetitive inhibitor for PTP1B. The inhibition of aldose reductase (AR) by fucosterol makes it a promising compound for diabetic complications treatment. The molecular docking program produced negative binding energies that demonstrated fucosterol's strong binding capability to the active sites of these enzymes. The results demonstrate that fucosterol has therapeutic potential for diabetes management and related complications (Jung et al. 2013).

The main pigment of *Spirulina platensis* known as phycocyanin (PC) demonstrates potential therapeutic effects on type-2 diabetes mellitus by blocking the activity of α-amylase and α-glucosidase enzymes. An in-silico analysis investigated the molecular interactions between PC and target enzyme by performing molecular docking simulations, which reports that PC showed significant interaction with the active site cavities in both enzyme systems, where PC functions as a promising natural therapeutic agent for antidiabetic treatment (Siti HalimatulMunawaroh et al. 2020).

The dipeptidyl peptidase-4 (DPP-4) enzyme has emerged as a crucial target for antidiabetic drug development because of its essential role in insulin signaling pathways. The computational methods provided insights on the inhibitory potential of DPP-4 from marine algae (*Caulerpa genus*). The molecular docking approach has identified seven compounds with a strong binding affinity range against the DPP-4 enzyme. Notably, four of these compounds demonstrated higher binding affinities compared to two FDA-approved DPP-4 inhibitors, sitagliptin and linagliptin (Balansa 2023).

3 Algal Compound Database and Bioinformatics Resources

The bioactive properties of secondary metabolites including phenolics, terpenoids, and alkaloids are substantial (Tripathi et al. 2024). The cellular activities require primary metabolites which include carbohydrates and lipids. The pharmaceutical and food industries use carrageenan and alginate polysaccharides as well as chlorophyll and fucoxanthin pigments for their antioxidant and anticancer effects and omega-3 fatty acids from microalgae serve as dietary supplements (Kachhwaha et al. 2022).

Academic researchers who study algal metabolites and their chemical properties and applications will find algal compound databases to be essential resources. The databases enhance bioprospecting and medication development and environmental research through their ability to gather complete information about algae species and their metabolites and biological activities (Kaushik et al. 2023). The materials serve as essential tools for biotechnology research and pharmacological studies and ecological investigations because they reveal information about algal compound taxonomy and bioactivity and metabolic capabilities.

4 Databases

(i) *Seaweed Metabolite Database (SWMD):* Bioactive constituents derived from seaweeds. The Seaweed Metabolite Database (SWMD) contains detailed records of 517 red algal-derived chemicals with complete chemical description and biological activity information. This information highlights the pharmacological value of these metabolites and helps scientists find compounds with antibacterial, antioxidant, and anticancer activities (Davis and Vasanthi 2011).

(ii) *Algal Bromophenol Database: Brominated Natural Compounds.* The database examines brominated natural chemicals which originate from algae and are classified as bromophenols. The database includes properties that enable the identification and separation of these compounds. The database proves especially useful for identifying bromophenols that exhibit antibacterial, antioxidant, and anti-inflammatory properties (Jacobtorweihen and Spiegler 2022).

(iii) *MarinLit:* MarinLit, a database that focuses on chemicals synthesized by marine organisms, especially by algae. It contains natural products with detailed information on chemical structures and biological activities as well as their origins, which gives vital information about the therapeutic properties and its potential applications (Lei and Zhou 2002).
(iv) *Algaebase:* The Algaebase serves as an international database that clearly provides the information related to the biodiversity, distribution, and taxonomy of algae species. It helps in ecological research, conservation projects, and the discovery of algae species with possible industrial and pharmacological applications (Guiry et al. 2014).
(v) *Comprehensive Marine Natural Product Database (CMNPD):* Bioactive marine natural products. The CMNPD gathers information about both the chemical structures and bioactivity of marine-derived chemicals including those from algae. This database acts as a crucial tool for researchers who seek to identify and screen marine-derived bioactives for pharmaceutical development (Lyu et al. 2021).
(vi) *PhycoBank:* Taxonomy and categorization of algae. Through PhycoBank users can access thorough taxonomic data which includes classification systems and nomenclature and type specimens of algae. The database helps scientists conduct systematic research and identify new algae species which could have biotechnological applications (Müller et al. 2022).
(vii) *Cyanobase:* Genomic analysis of cyanobacteria (blue-green algae). Cyanobase brings together genomic data from cyanobacteria which helps researchers understand their metabolic operations. The database enables genetic engineering and biotechnological applications including the production of biofuels and medicinal molecules (Teikari et al. 2022).
(viii) *Marine Algal Chemical Ecology Database (MACED):* A database focused on the chemical ecology of marine algae. MACED studies the ecological functions of chemical substances produced by marine algae through defense mechanisms and their roles in ecosystems. It serves as a vital tool for studying both environmental and ecological research (Lei and Zhou 2002).
(ix) *LipidMAPS (Lipid Metabolites and Pathways Strategy):* Algal lipidomics. The LipidMAPS database provides detailed information about algal lipids together with their metabolic pathways and their roles in biofuels and nutritional applications as well as pharmaceuticals. Researchers working in lipidomics and metabolic engineering find this platform particularly useful (Melo et al. 2015).
(x) *Database of Macromolecules from Macroalgae:* It contains the information about the three major macromolecules such as carbohydrates, lipids, and proteins of macroalgae (Fiset et al. 2019). The amalgamation of these data supports multiple investigations about algae biodiversity and ecology toward the basic understanding of algal biology and also for the development of novel bioactive compounds or new therapeutics from algae.

The medicinal development potential of algae-derived compounds is extremely high. Bioinformatics tools have revolutionized the field by enabling researchers to establish molecular targets while developing algal-derived compounds suitable for medical treatments. Tools for algal bioinformatics play an essential role in identifying and characterizing bioactive compounds produced by marine algae and cyanobacteria. These devices simplify complex data analysis which leads to the discovery of new substances with important uses in biotechnology and traditional medicine. The discovery of secondary metabolites necessary for medical research and agricultural advancement requires genomic data collection through platforms including anti-SMASH, MIBiG, and the Secondary Metabolite Bioinformatics Portal (Maurya et al. 2019). The integration of metabolomic and genomic information combined other analysis for about 2600 cyanobacterial compounds proves the diverse chemical nature and its vital medicinal properties (Ferrinho et al. 2024).

The recent advancements in omics technology specifically in marine microalgae research have expanded the opportunity to find species which produce metabolites. These omics methods are vital for identifying bioactive molecules which could serve medicine and fuel production and food technology (Lauritano et al. 2019). However, the major challenge remains the need for genomic and metabolomic data synchronization which hinders the complete comprehension of these compounds. The field of algae bioinformatics continues to face challenges that must be resolved before achieving its maximum potential. The standardization of data formats across individual databases stands as a critical requirement for achieving efficient integration and analysis. Advanced machine learning methods need development to improve prediction accuracy for chemical activities and interactions. The scientific community requires improved representation of underrepresented species to discover previously unknown bioactive potential (Vlachakis et al. 2013). The field of algae bioinformatics predicted that artificial intelligence applications in data mining and predictive modeling would trigger a revolutionary change. The integration of artificial intelligence and machine learning methods has largely improved the detection of bioactivity with maximum precision and aids to screen and optimize the drug discovery process very rapidly.

5 Machine Learning and Artificial Intelligence in Discovering Anti-diabetic Compounds in Algae

The identification process for anti-diabetic drugs from algae experiences a transformation through the automated optimization of bioactive molecule discovery using machine learning (ML) and artificial intelligence (AI). ML and AI algorithms examine algal metabolite databases to predict bioactivity and structure optimization for diabetes treatment applications (Singla et al. 2019).

6 Applications of ML/AI in Discovering Anti-diabetic Compounds from Algae

(a) *Supervised Learning Algorithms for Bioactivity Prediction*
Algal compound bioactivity predictions depend primarily on supervised learning algorithms which operate from known databases. The algorithms process labeled information about known antidiabetic compounds to forecast results from untested compounds.

- *Support Vector Machines (SVM):* SVM enables compound classification according to their bioactivity properties including inhibition of α-amylase or α-glucosidase. SVM performs well with extensive data sets thus making it appropriate for molecular descriptor-based predictions. The prediction model based on SVM successfully analyzed the inhibitory activity of marine-derived compounds against α-glucosidase which led to reliable antidiabetic compound screening from algae sources (Zernov et al. 2004).
- *Random Forest (RF):* RF blends multiple decision trees to classify or predict continuous values, making it suitable for predicting the bioactivity of algae compounds based on molecular fingerprints. Random Forest proved successful in α-amylase inhibitory activity prediction for compounds extracted from algae through its ability to achieve high bioactivity prediction precision (Grenier et al. 2010).
- *Neural Networks (NN):* Deep learning models including feedforward neural networks that demonstrate the ability to discover intricate molecular features and biological activity through the connections.
- *Deep neural networks (DNN):* DNN have been applied to predict the inhibitory potential of algal compounds on key diabetes-related enzymes, enhancing the prediction accuracy (Kaushik et al. 2023).

(b) *Unsupervised Learning Algorithms for Data Clustering and Pattern Discovery*
Unsupervised learning algorithms function to find patterns and clusters within unlabeled data which assists in discovering fresh bioactive compounds from algae.

- *K-means Clustering:* This algorithm groups algal compounds through molecular fingerprints to detect compounds which display similar bioactivity patterns. Sudarwati et al. (2022) employed K-means clustering to organize marine algae-derived compounds into clusters based on their chemical structures as well as their predicted activities thus enabling the identification of new antidiabetic potential.
- *Principal Component Analysis (PCA):* PCA reduces data dimensions which allows researchers to better understand large molecular datasets. PCA serves as a useful tool for determining which elements primarily lead to antidiabetic properties in algae. PCA served as an analytical tool to study the

chemical diversity of marine algae-derived compounds while revealing the essential characteristics that link to antidiabetic properties (Yang et al. 2022).

(c) *Reinforcement Learning for Compound Optimization*
Reinforcement learning (RL) within machine learning functions through incentives or penalties which provide feedback for agent decision-making. The reinforcement learning process allows researchers to enhance algal compound structures for improved antidiabetic properties.

- *Q-learning:* The optimization process of designing new algal compounds utilizes Q-learning to provide rewards for compounds which demonstrate strong antidiabetic target activity. The drug design application of Q-learning optimizes molecular structures of bioactive compounds to achieve enhanced binding affinity to key antidiabetic targets including α-amylase (Jaiswal and Gupta 2023).

(d) *Molecular Docking and AI-Driven Virtual Screening*
The combination of AI-driven virtual screening and molecular docking technologies enables scientists to determine which compounds effectively bind to diabetes-related biological targets such as enzymes. AI technology enables fast screening of extensive compound collections through virtual high-throughput screening.

- *Docking Algorithms:* These algorithms simulate molecular interactions to predict how well diabetes-related targets bind with algal compounds. The integration of AI tools with AutoDock and Glide screening systems led to the discovery of effective α-glucosidase inhibitors from algae-derived compounds for antidiabetic drug development (Ogunwa et al. 2018).
- *DeepDock (Deep Learning-based Docking)*: This AI model uses deep learning to predict protein-ligand interactions with higher accuracy than traditional docking methods. DeepDock has been used to predict the interaction between algae-derived metabolites and diabetes-related enzymes, significantly improving the speed and accuracy of virtual screening (Neelakandan and Rajanikant 2024).

(e) *Generative Models for Compound Discovery*
Generative models use AI to create novel molecular structures that can potentially exhibit antidiabetic properties. These models help discover compounds with unique chemical scaffolds that may not be found in existing compound libraries.

- *Variational Autoencoders (VAE):* VAE is a type of generative model that can be trained on molecular datasets to generate new, viable molecules with desirable properties, such as antidiabetic activity. VAEs have been used to generate novel molecular structures with predicted inhibitory activity against diabetes-related enzymes, accelerating the identification of new antidiabetic candidates (Tevosyan et al. 2022).

- *Generative Adversarial Networks (GAN):* The generator and discriminator neural networks that makeup GANs collaborate to produce realistic molecular structures. GANs have been applied to generate novel algal metabolites with improved binding affinity for diabetic targets, contributing to the development of new therapeutic agents (Jaiswal and Gupta 2023).

(f) *Predictive Toxicology Using ML*
Predicting the toxicity of algal compounds is critical in drug development. ML algorithms can predict the potential toxicity of compounds based on molecular structure, ensuring the safety of potential antidiabetic agents.

- *Gradient Boosting Machines (GBM):* GBM is used to predict the toxicity of algal compounds by analyzing the chemical structure and comparing it to known toxic compounds. ML models using GBM have been applied to assess the toxicity of marine algal compounds, ensuring that only safe compounds are prioritized for antidiabetic drug development (Gaur et al. 2022).

(g) *Integration with AI and Machine Learning:*

- *Enhancing Data Analysis and Pattern Recognition:* Artificial intelligence (AI) and machine learning (ML) offer transformative tools for managing and interpreting multi-omics data. By using machine learning algorithms, researchers can identify patterns and correlations across different omics layers that might not be evident through traditional methods. For example, ML models can analyze complex interactions between genes, proteins, and metabolites, providing insights into potential biosynthetic pathways responsible for anti-diabetic compound production.
- *Predictive Modeling for Compound Discovery:* AI can assist in predicting new bioactive compounds by analyzing multi-omics data to identify molecular signatures associated with therapeutic effects. Machine learning models can be trained to recognize features linked to glucose-regulating compounds, aiding in the discovery of novel molecules. This predictive capability enables researchers to prioritize candidate compounds for experimental testing, accelerating the identification of algal-derived therapeutics for diabetes.
- *Automation and Optimization of Experimental Design:* AI-driven tools can automate aspects of experimental design, such as selecting growth conditions that optimize compound production in algae. Through reinforcement learning, AI systems can test various parameters (e.g., light intensity, nutrient availability) and determine conditions that enhance the synthesis of anti-diabetic compounds. Additionally, AI can help refine experimental approaches by simulating the effects of genetic modifications, offering a valuable tool for metabolic engineering in algae (Fig. 17.1).

The integration of algal bioinformatics with advanced computational technologies marks a transformative step forward in natural product research and drug discovery. Algae, with their immense biodiversity and bioactive potential, have emerged as a vital resource for addressing global challenges in healthcare, biotechnology, and environmental sustainability. The implementation of ML and AI has revolutionized

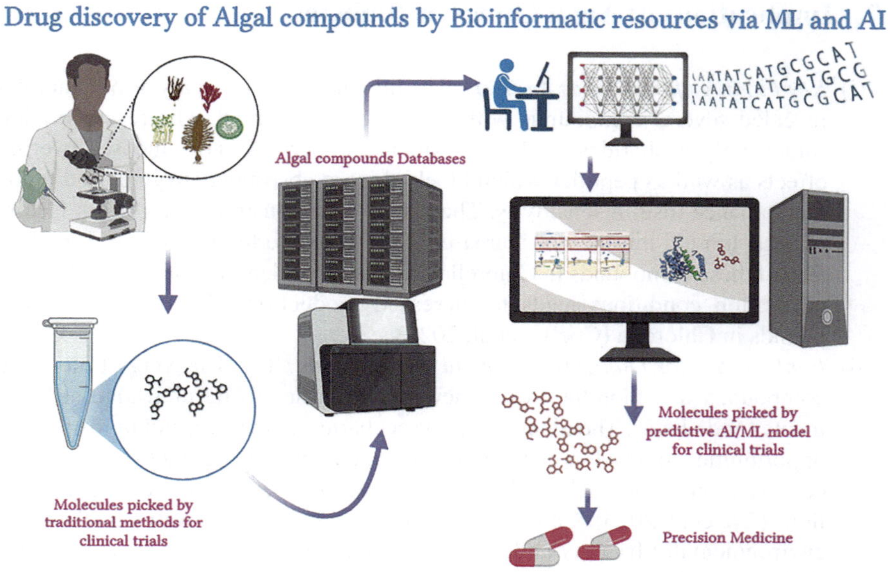

Fig. 17.1 AI/ML in anti-diabetic drug discovery in algal compounds

the discovery and development of algal metabolites, particularly for conditions like diabetes. Researchers can use advanced algorithms such as support vector machines, random forests, neural networks, molecular docking, and generative models to rapidly evaluate big datasets, find possible bioactive chemicals, and forecast therapeutic activity. These technologies not only streamline the drug discovery process but also open the door to exploring uncharted chemical spaces, facilitating the identification of new compounds with great therapeutic potential. Moreover, the convergence of algal bioinformatics with -omics technologies such as genomes, proteomics, and metabolomics offers a holistic view of algal biology, aiding in the identification of new biosynthetic pathways and metabolites. By integrating AI-driven approaches, researchers can optimize compound structures, predict interactions with biological targets, and analyze pharmacokinetics and toxicity profiles, drastically lowering the time and expense of traditional drug development. However, despite these advancements, challenges remain. Ensuring data quality, standardization, and accessibility across interdisciplinary teams is critical for harnessing the full potential of computational tools. Additionally, fostering global collaborations among bioinformaticians, marine biologists, chemists, and pharmaceutical scientists is essential to bridge knowledge gaps and translate computational findings into real-world applications.

Computational technologies and bioinformatics research has shown great promise to discover hidden algal compounds in the future. Algal research will produce major healthcare discoveries in diabetes treatment as well as sustainable biotechnology and environmental science applications by overcoming current obstacles through innovative inter-disciplinary approaches.

7 Implications of Algal Omics in Diabetes

(i) *Key Findings from Omics Analysis:* The integrative omics analysis of Chlorella revealed several compounds with anti-diabetic potential. The research team found polysaccharides with antioxidant properties and glucose-lowering effects as well as peptides which block glucose-absorbing enzymes and lipids that enhance insulin sensitivity. The research demonstrated that stress factors such as limited nitrogen or increased light exposure led to increased production of these compounds in Chlorella. The research demonstrates that adjusting cultivation conditions leads to increased production of anti-diabetic compounds in Chlorella (Coelho et al. 2022).

(ii) *Implications for Diabetes Treatment Development:* The discovery of bioactive compounds in Chlorella creates new opportunities to develop natural treatments for diabetes. The identified polysaccharides, peptides, and lipids present opportunities to develop new therapeutic agents that provide a safer and more comprehensive approach to blood glucose management than synthetic medications (Eze et al. 2023). Chlorella's ability to scale up its cultivation makes it an environmentally friendly and economically feasible source of these bioactive molecules. The validation of Chlorella-derived compounds through clinical studies could lead to their inclusion in complementary or alternative diabetes therapies which offer affordable solutions for glucose control and diabetes complication reduction (Mendes et al. 2024).

(iii) *Future Research Directions:* The omics-driven studies have shown Chlorella as a promising functional food ingredient as well as a nutraceutical source for diabetes management. Future focus on development of effective best cultivation methods to produce these active compounds and evaluation of the compounds in preclinical and clinical modes to validate its efficacy, effectiveness, and safety. The advancements in genetic engineering scientists could alter the expression of genes responsible for synthesis of anti-diabetic molecules resulting in forecasting Chlorella as an advanced tool for diabetes treatment.

8 Technical Challenges and Future Directions in Multi-Omics Algal Research for Diabetes

- *Data Complexity and Volume:* The complexity of data volume becomes a major challenge in multi-omics studies which collect extensive data across genomic and transcriptomic and proteomic and metabolomic and lipidomic levels. Scientists face major obstacles when trying to merge multiple omics data types because their unique data structures and formats, and noise levels differ between each data set. Genomic DNA sequence analysis differs from metabolomic small molecule analysis which demands specialized analytical approaches. The technical complexity of merging diverse data types into meaningful biological understanding presents a major challenge.

- *Pathway Interpretation and Validation:* The integration of multi-omics data into biological pathway interpretation continues to present significant challenges to researchers. Limited understanding exists about how algae produce bioactive compounds because numerous biosynthetic pathways have not received complete mapping. The production of bioactive molecules through multi-step pathways requires multiple regulatory elements which creates difficulties for scientists to identify targets to boost compound production. Research teams must use advanced tools to modify algal genes and enzymes while spending many resources on pathway validation procedures because these processes require extensive time.
- *Resource and Infrastructure Limitations:* The execution of multi-omics research needs extensive computational infrastructure for managing large amounts of data. Research institutions that have limited funding face difficulties obtaining advanced technologies such as high-throughput sequencing and mass spectrometry and specialized software platforms because these tools come with high costs. The successful handling of multi-omics data demands personnel who possess dual skills in omics technologies and bioinformatics since this field combines various scientific disciplines. Building infrastructure to manage and analyze multi-omics datasets is essential but remains a significant hurdle.

9 Algae Implications for Drug Development

Research on algae through multi-omics methods has established novel prospects for developing diabetes drugs. Scientists can identify blood glucose-regulating compounds in algae by using integrative analysis techniques alongside systems biology methods. Researchers have discovered plant-based diabetes treatment options through their findings which represent alternatives to conventional medications with their associated side effects. The large-scale renewable production of algae makes it a suitable source for creating sustainable healthcare solutions. The advancement of multi-omics methods will potentially enable researchers to develop individualized treatments that match the needs of each patient according to personalized medicine objectives.

10 Future Prospects

Development of Targeted Algal-Based Therapeutics: Multi-omics research provides promising opportunities to develop algal-derived compounds which can be used as diabetes treatments. Researchers who use systematic genetic and metabolic assessments of algae discover new compounds which enhance blood glucose regulation and improve insulin sensitivity and decrease oxidative stress. The results from multi-omics research will likely produce specific algal-derived treatments that can enhance current diabetes medications through natural and safer alternatives for patient care.

Scaling Sustainable Production of Bioactive Molecules: Large-scale bioactive molecule production becomes sustainable through algae cultivation because these organisms offer renewable and scalable features. The use of multi-omics approaches enables scientists to develop optimized algal cultivation methods that produce maximum amounts of anti-diabetic molecules for large-scale production. Future research directions include creating sustainable bioreactor systems alongside enhanced cultivation methods to support continuous production of therapeutic compounds at high volumes.

Personalized Therapeutics and Precision Medicine: The ability to scale up natural plant-based medicine production aligns with increasing market demand for these products. The implementation of multi-omics technology with personalized medicine will create individualized algal-based diabetes therapies. Scientists can identify the most suitable bioactive algal compounds for each patient through genetic and metabolic profile analysis to develop personalized treatments. The future use of multi-omics data could lead to personalized algal-derived medications that tailor treatment approaches to individual patient characteristics for better diabetes management.

Expanding Research into Other Therapeutic Areas: The multi-omics research conducted in algae offers multiple therapeutic opportunities which extend beyond diabetes treatment. The compounds that regulate glucose metabolism in the body could simultaneously benefit cardiovascular health while offering protection against inflammation and neural damage. Algae research through multi-omics approaches will likely develop into a versatile platform which produces treatments for different chronic diseases beyond diabetes thus expanding the therapeutic applications of this research field.

11 Closing Thoughts

The complete realization of algae's therapeutic value for diabetes management requires ongoing research and collaborative efforts between scientists who study multi-omics and algal biology. The combination of bioinformatics knowledge with molecular biology and pharmacology expertise will help researchers address present obstacles while improving research methods to advance algal compound clinical applications. The multidisciplinary collaboration between researchers and institutions will enable the multi-omics field of algae to deliver substantial contributions to diabetes care and healthcare in general. The adoption of this integrated method shows promise for developing sustainable natural algae-derived treatments that will enhance the quality of life for millions of diabetic patients.

12 Conclusion

The investigation of algal compounds for diabetes mellitus management presents a promising therapeutic opportunity because marine algae contain strong bioactive metabolites. The combination of Virtual Screening with Molecular Docking and Network Pharmacology computational tools has transformed the discovery and optimization process of these compounds which enables fast identification of promising diabetes treatment candidates. The algal compounds sulfated polysaccharides and phlorotannins together with fucoxanthin and omega-3 fatty acids show various mechanisms of action that affect different diabetes progression pathways through improved insulin sensitivity and glucose regulation and inflammation reduction. The advancement of computational methods will enhance their importance in finding the best algal metabolites for medical use. The incorporation of natural compounds into diabetes treatment plans could lead to safer and sustainable therapeutic options that provide better prospects for diabetes control and prevention.

References

Abo-Shady AM, Gheda SF, Ismail GA, Cotas J, Pereira L and Abdel-Karim OH 2023. Antioxidant and Antidiabetic Activity of Algae. *Life*, 13(2), 460. https://doi.org/10.3390/life13020460

Antar SA, Ashour NA, Sharaky M, Khattab M, Ashour NA, Zaid RT, Roh EJ, ElkamhawyAand Al-Karmalawy AA 2023. Diabetes mellitus: Classification, mediators, and complications; A gate to identify potential targets for the development of new effective treatments. *Biomedicine & Pharmacotherapy*, 168, 115734. https://doi.org/10.1016/j.biopha.2023.115734

Balansa W 2023. Exploring Metabolites Of Green Algae Caulerpa Spp. To Discover Putative Inhibitors Of Dpp-4, A New Antidiabetic Target Protein. *JurnalIlmiah PLATAX*, 11(2), 333–346. https://doi.org/10.35800/jip.v11i2.48061

Bhagat RP and Ghormade MB 2020. Docking Of Rigid Macromolecules Using Autodok. *International Journal of Emerging Technologies and Innovative Research*, 7 (2), 15–18.

Cakmak T and Selvaraj G 2017. Computational Screening of Anti-Diabetic Molecules from Microalgae Metabolites by Molecular Docking. *Journal of Bioinformatics and Proteomics Review*, 3(2), 1–7. https://doi.org/10.15436/2381-0793.17.1252

Coelho D, Ribeiro D, Osório H, de Almeida AM, Prates JAM 2022. Integrated Omics analysis of pig muscle metabolism under the effects of dietary *Chlorella vulgaris* and exogenous enzymes. *Scientific Reports,* 12, 16992. https://doi.org/10.1038/s41598-022-21466-z

Davis GDJ and Vasanthi AHR 2011. Seaweed metabolite database (SWMD): A database of natural compounds from marine algae. *Bioinformation*, 5(8), 361–364. https://doi.org/10.6026/97320630005361

Eze CN, Onyejiaka CK, Ihim SA, Ayoka TO, Aduba CC, Ndukwe JK, Nwaiwu O, Onyeaka H 2023. Bioactive compounds by microalgae and potentials for the management of some human disease conditions. *AIMS Microbiology*, 9(1):55–74. doi: https://doi.org/10.3934/microbiol.2023004.

Ferrinho S, Connaris H, Mouncey NJ and Goss RJM 2024. Compendium of Metabolomic and Genomic Datasets for Cyanobacteria: Mined the Gap. *Water Research*, 256, 121492. https://doi.org/10.1016/j.watres.2024.121492

Fiset C, Irwin AJ and Finkel ZV 2019. The macromolecular composition of noncalcified marine macroalgae. *Journal of Phycology*, 55(6), 1361–1369. https://doi.org/10.1111/jpy.12913

Gaur A, Pant G and Jalal AS 2022. Computer-aided cyanobacterial harmful algae blooms (CyanoHABs) studies based on fused artificial intelligence (AI) models. *Algal Research*, 67, 102842. https://doi.org/10.1016/j.algal.2022.102842

Grenier M, Lek S, Rodríguez MA, Rousseau AN and Campeau S 2010. Algae-based Biomonitoring: Predicting Diatom Reference Communities in Unpolluted Streams using Classification Trees, Random Forests, and Artificial Neural Networks. *Water Quality Research Journal*, 45(4), 413–425. https://doi.org/10.2166/wqrj.2010.041

Guiry MD, Guiry GM, MorrisonL, Rindi F, Miranda SV, Mathieson AC, Parker BC, Langangen A, John DM, BárbaraI, Carter CF, Kuipers P and Garbary DJ 2014. AlgaeBase: An On-line Resource for Algae. *CryptogamieAlgologie*, 35(2), 105–115. https://doi.org/10.7872/crya.v35.iss2.2014.105

Gunathilaka T, RangeeKeertihirathna L and Peiris D 2022. Advanced Pharmacological Uses of Marine Algae as an Anti-Diabetic Therapy. *In Natural Medicinal Plants*. IntechOpen. https://doi.org/10.5772/intechopen.96807

Gunathilaka TL, Samarakoon K, Ranasinghe P and Peiris LDC 2020. Antidiabetic Potential of Marine Brown Algae—a Mini Review. *Journal of Diabetes Research*, 2020, 1–13. https://doi.org/10.1155/2020/1230218

Jacobtorweihen J and Spiegler V 2022. The establishment of an analytical database of algal bromophenols and phytochemical characterization of the red alga Vertebrata lanosa. *Planta Medica*, 88(15): 1390. https://doi.org/10.1055/s-0042-1758883

Jaiswal S and Gupta P 2023. GLSTM: A novel approach for prediction of real and synthetic PID diabetes data using GANs and LSTM classification model. *International Journal of Experimental Research and Review*, 30, 32–45. https://doi.org/10.52756/ijerr.2023.v30.004

Jung H A, Islam MdN, Lee CM, Oh SH, Lee S, Jung JH and Choi JS 2013. Kinetics and molecular docking studies of an anti-diabetic complication inhibitor fucosterol from edible brown algae Eisenia bicyclis and Ecklonia stolonifera. *Chemico-Biological Interactions*, 206(1), 55–62. https://doi.org/10.1016/j.cbi.2013.08.013

Kachhwaha N, Kumar Sharma M, Khandelwal R and Kaushik P 2022. Marine Algal Bioactive Metabolites and their Pharmacological Applications, *Therapeutic Implications of Natural Bioactive Compounds Frontiers in Bioactive Compounds*, 3: 118—134. https://doi.org/10.2174/9789815080025122030009

Kaushik A, Sangtani R, Parmar HS and Bala K 2023. Algal metabolites: Paving the way towards new generation antidiabetic therapeutics. *Algal Research*, 69, 102904. https://doi.org/10.1016/j.algal.2022.102904

Lauritano C, Ferrante MI and Rogato A 2019. Marine Natural Products from Microalgae: An -Omics Overview. *Marine Drugs*, 17(5), 269. https://doi.org/10.3390/md17050269

Lei J and Zhou J 2002. A Marine Natural Product Database. *Journal of Chemical Information and Computer Sciences*, 42(3), 742–748. https://doi.org/10.1021/ci010111x

Lyu C, Chen T, Qiang B, Liu N, Wang H, Zhang L and Liu Z 2021. CMNPD: a comprehensive marine natural products database towards facilitating drug discovery from the ocean. *Nucleic Acids Research*, 49(D1), D509–D515. https://doi.org/10.1093/nar/gkaa763

Maurya SK, Niveshika and Mishra R 2019. Importance of Bioinformatics in Genome Mining of Cyanobacteria for Production of Bioactive Compounds. *In Cyanobacteria* (pp. 477–506). Elsevier. https://doi.org/10.1016/B978-0-12-814667-5.00024-6

Melo T, Alves E, Azevedo V, Martins AS, Neves B, Domingues P, Calado R, Abreu MH and Domingues MR 2015. Lipidomics as a new approach for the bioprospecting of marine macroalgae — Unraveling the polar lipid and fatty acid composition of Chondrus crispus. *Algal Research*, 8, 181–191. https://doi.org/10.1016/j.algal.2015.02.016

Mendes AR, Spínola MP, Lordelo M, Prates JAM 2024. Chemical Compounds, Bioactivities, and Applications of *Chlorella vulgaris* in Food, Feed and Medicine. *Applied Sciences*, 14(23),10810. https://doi.org/10.3390/app142310810

Müller A, Güntsch A, Jahn R, KohlbeckerA , Kusber WH and Zimmermann J 2022. PhycoBank: Repository for algal novelties. *Biodiversity Information Science and Standards*, 6. https://doi.org/10.3897/biss.6.90885

Neelakandan AR and Rajanikant GK 2024. A deep learning and docking simulation-based virtual screening strategy enables the rapid identification of HIF-1α pathway activators from a marine natural product database. *Journal of Biomolecular Structure and Dynamics*, 42(2), 629–651. https://doi.org/10.1080/07391102.2023.2194997

Ogunwa TH, Adeyelu TT, Fasimoye RY and Ayenitaju FC 2018. In Silico Analysis of Interaction between Seaweed-Derived Bioactive Compounds and Selected Diabetes-Related Targets. *Biomedical Chemistry: Research and Methods*, 1(4), e00074. https://doi.org/10.18097/BMCRM00074

Pereira L and Valado A 2023. Algae-Derived Natural Products in Diabetes and Its Complications—Current Advances and Future Prospects. *Life*, 13(9), 1831. https://doi.org/10.3390/life13091831

Selvaraj G, Kaliamurthi S, Çakmak ZE and Çakmak T 2017. In silico validation of microalgal metabolites against Diabetes mellitus. *Diabetes Mellitus*, 20(4), 301–307. https://doi.org/10.14341/DM8212

Singla R, Singla A, Gupta Y and Kalra S 2019. Artificial intelligence/machine learning in diabetes care. *Indian Journal of Endocrinology and Metabolism*, 23(4), 495. https://doi.org/10.4103/ijem.IJEM_228_19

Siti HalimatulMunawaroh H, Gumilar GG, Nurjanah F, Yuliani G, Aisyah S, Kurnia D, Wulandari AP, Kurniawan I, Ningrum A, Koyande AK and Show PL 2020. In-vitro molecular docking analysis of microalgae extracted phycocyanin as an anti-diabetic candidate. *Biochemical Engineering Journal*, 161, 107666. https://doi.org/10.1016/j.bej.2020.107666

Teikari J, Baunach M and Dittmann E 2022. Cyanobacterial Genome Sequencing, Annotation, and Bioinformatics.In: Skellam, E. (eds) *Engineering Natural Product Biosynthesis. Methods in Molecular Biology*, 2489, 269–287. https://doi.org/10.1007/978-1-0716-2273-5_14

Tevosyan A, Khondkaryan L, Khachatrian H, Tadevosyan G, Apresyan L, Babayan N, Stopper H and Navoyan Z 2022. Improving VAE based molecular representations for compound property prediction. *Journal of Cheminformatics*, 14(1), 69. https://doi.org/10.1186/s13321-022-00648-x

Tripathi G, Dubey P, Ahmad S, Farooqui A and Mishra V 2024. Role of Algal-derived Bioactive Compounds in Human Health. *Recent Patents on Biotechnology*, 18(3), 190–209. https://doi.org/10.2174/1872208317666230623141740

Vlachakis D, Pavlopoulou A, Kazazi D and Kossida S 2013. Unraveling microalgal molecular interactions using evolutionary and structural bioinformatics. *Gene*, 528(2), 109–119. https://doi.org/10.1016/j.gene.2013.07.039

Yang Z, Shi Y, Li P, Pan K, Li G, Li X, Yao S and Zhang D 2022. Application of Principal Component Analysis (PCA) to the Evaluation and Screening of Multiactivity Fungi. *Journal of Ocean University of China*, 21(3), 763–772. https://doi.org/10.1007/s11802-022-5096-x

Zernov VV, Balakin KV, Ivaschenko AA, Savchuk NP and Pletnev IV 2004. Drug Discovery Using Support Vector Machines. The Case Studies of Drug-Likeness, Agrochemical-Likeness, and Enzyme Inhibition Predictions. *Cheminform*, 35(5). https://doi.org/10.1002/chin.200405244

Chapter 18
Economic and Environmental Impact of Algae in Diabetic Management

Govindaraj A. Shivaranjani, and Gangatharan Muralitharan

Abstract Diabetes mellitus possesses serious health issues and it is increasing every year globally. For the past few years, research is going on in the treatment of diabetes where algae are being attempted as natural sources for the treatment. Microalgae and cyanobacteria are rich in antioxidants and antidiabetic properties which can reduce the hyperglycemic effect in blood. The secondary metabolites produced by micro and macroalgae can be used to treat diabetes. Cultivation and harvesting of algae are now being simplified with the usage of different mass cultivation strategies like open pond systems and photobioreactors. This chapter addresses the economic and environmental impacts of different cultivation systems and their environmental impact. Through life cycle assessment, different parameters possible for environmental impact were addressed.

Keywords Algae · Diabetes · Antidiabetic properties · Algal cultivation · Life cycle assessment

1 Introduction

Millions of individuals worldwide suffer from diabetes, a chronic metabolic disease that is connected to chronic hyperglycemia due to relative or total lack of the hormone insulin. It places a significant financial and medical burden on economies and healthcare systems. Type 2 diabetes is the most common type, and it is brought on

---

G. A. Shivaranjani
Molecular Evolution Laboratory, Department of Microbiology, School of Life Sciences, Bharathidasan University, Tiruchirappalli, Tamil Nadu, India
National Repository for Microalgae and Cyanobacteria (NRMC – F & M), Bharathidasan University, Tiruchirappalli, Tamil Nadu, Indiae-mail: drgm@bdu.ac.in

G. Muralitharan (✉)
National Repository for Microalgae and Cyanobacteria (NRMC – F & M), Bharathidasan University, Tiruchirappalli, Tamil Nadu, India

R. Raja et al. (eds.), *Algae in Diabetes Management*,
https://doi.org/10.1007/978-981-95-0571-5_18

by insufficient secretion and insulin resistance. An autoimmune disease called type 1 diabetes causes complete insulin insufficiency. Small blood vessels are affected during diabetes raising the risk of cardiovascular illnesses by causing microvascular and macrovascular problems. Targeted therapies, glucose-lowering drugs, and lifestyle changes are all part of management. Investigations are on in the disease modifications in all possible ways and improving the efficacy of therapeutic intervention (Gunathilaka et al. 2020).

Type 1, type 2, gestational diabetes mellitus, and other associated conditions are among all kinds of diabetes. Globally 31.6 crore people and 53.7 crore people are affected with glucose intolerance and diabetes mellitus, respectively. In type 1 diabetes, due to pancreatic β-cell degeneration, there is a lack of insulin secretion, causing the body to use lipids as an energy source instead of glucose. In type 2 diabetes, cells that secrete insulin are lost and or resistant to the hormones which is the root cause of type 2 diabetes. In order to deal with type 2 diabetes, neither the digestive enzymes α-amylase nor α-glucosidase is inhibited or endogenous production of insulin is induced and elevated. Dairy fibers, which are non-degraded and non-digestible, help in preventing and in the treatment of gastrointestinal issues, coronary heart disease, and type 2 diabetes. Reactive oxygen species (ROS) generation may need to be downregulated in order to manage issues related to diabetes. Propyl gallate (PG), butylated hydroxyanisole (BHA), and butylated hydroxytoluene (BHT) are examples of antioxidants that can aid in the neutralization and removal of ROS from the body's cells (Tavares et al. 2023). According to the International Diabetes Federation, 20–25% of the population globally suffers from metabolic syndrome. People with the syndrome are likely to have a heart attack or stroke three times and five times more likely to acquire type 2 diabetes than people without it. The growing use of algae-based components in nutraceutical applications will increase product demand due to their proven health advantages (https://www.insightaceanalytic.com/report/global-algae-based-ingredients-market/1461#:~:text=The%20Algae%2DBased%20Ingredients%20Market,forecast%20period%20for%202024%2D2031).

The existing therapies of type 2 diabetes mellitus that currently exist are either ineffective or have adverse reactions. The need for resilient, effective treatments has risen owing to the increasing prevalence of diabetes worldwide. In addition to their high nutritional value, antioxidant properties, and the ability to regulate blood sugar levels, algae, especially microalgae like *Spirulina* and *Chlorella*, have received an enormous amount of interest for their potential aid in managing diabetes. Hence, it is crucial to find novel plant-based chemicals with minimal negative consequences. Phlorotannin, one among the secondary metabolites derived from brown algal compounds, has been reported to have anti diabetic properties (Arunkumar et al. 2023).

The economic and environmental aspects of using algae for managing diabetes are investigated in this chapter. It explores the positive implications for local economies, production costs, and potential markets. It additionally evaluates the environmental impact of algae, highlighting carbon sequestration, resource efficiency, and potential ecological risks. To develop a sustainable model for using algae to treat diabetes, it seems essential to understand all of these elements.

1.1 Algae

Algae are basically classified into eleven major phyla: Charophyta, Chlorarachniophyta, Chlorophyta, Cryptophyta, Cyanophyta, Dinophyta, Euglenophyta, Glaucophyta, Haptophyta, Heterokontophyta, and Rhodophyta. Microalgae are unicellular algae that are between 1 and 50 m in diameter and comprise a highly diverse group of 200,000–800,000 species (Ramos-Romero et al. 2021). About 40, 000 different known species of microalgae exist, but less than 10% are currently cultivated and maintained in culture media for research or commercial purposes (Romagnoli et al. 2023).

1.2 Diversity and Distribution

The geographic distribution of algae reveals significant regional variations in species richness and endemism. Both temperate and tropical waters have the most diverse coastlines in terms of algae. Overall species richness appears to be unrelated to seaweeds' high levels of endemism. Although there is little data on microalgae, certain groups exhibit high endemism in areas with little diversity (Norton et al. 1996). Spectroscopic and chromatographic techniques have enabled us to discover a number of new classes in the picoplankton, another little-known ecosystem, and electron microscopy is crucial for many groups. Though its use has only just begun to be shown, molecular techniques are expected to become an indispensable tool for algae taxonomists and transform algal systematics, much like the electron microscope has (Norton et al. 1996).

Around 500 different kinds of seaweed can be found in some well-collected parts of the Pacific Ocean, which has the largest seaweed variety in the world's seas. With over 7000 islands comprising their vast coastline, the Philippines and Japan are at the heart of this wealth. Although it lacks a divided coastline with the different levels of shelter required to support the highest seaweed diversity, the eastern Pacific coast is nearly as rich. The subtropical to sub-Antarctic coastline of Chile would be home to many more species than the 550 that are currently known to exist (Norton et al. 1996). European countries like Northern Spain, France, and British Isles are rich in algal diversity. Over 80% of the tropical and subtropical western Atlantic's seaweed flora exists in the Caribbean. Some Atlantic islands are quite wealthy, whereas others located away from shores that are abundant in seaweed are poorer. Arctic seas are the places where diversity is said to be less at high latitudes.

Due to a reduced salinity and limited water exchange, enclosed seas can occasionally be species-sparse. Only 78 types of seaweed have been found in the Baltic Sea, compared to the Mediterranean's vast diversity and steady salinity. With a tropical flora on the east coast and a temperate flora on the west, seaweeds are abundant in Southern Africa. Japan also boasts a very diverse marine vegetation, with subtropical components in the south and cold water elements in the north (Norton et al. 1996). A survey in New South Wales, Australia, revealed over 530 taxa, including many new to science. Microalgae are cosmopolitan in diversity, with the members of coccolithophorids and dinoflagellates exhibiting huge diversity in the waters of tropics and polar seas. Prymnesiophytes and diatoms are more diverse in temperate and polar seas, with diatoms accounting for most species diversity. Dinoflagellate diversity is not significant between oceans (Norton et al. 1996). Both the microalgae and cyanobacteria were collected from different regions and conditions like sand, rain, and including snow by scientists (Wiśniewska et al. 2019).

2 Algae in Diabetic Management

Algae offer more nutritional value in foods like bread as they are high in proteins, minerals, and amino acids. Their beneficial health properties hold them as highly promising for utilization in nutritional products and health supplements. A red alga named Nori, rich in vitamins, protein, and minerals, can be used in gluten-free snacks and crackers for diabetic management and celiac disease (Arunkumar et al. 2023). *Palmaria palmata*, another red alga also known as Dulse, is rich in antioxidant properties, used to make gluten-free baking mixes for diabetes control and high level of proteins in bread production, potentially developing as a bioactive product (Derosa et al. 2019).

Recent studies showed that *Gracilaria* species reduce cholesterol. Incorporating extracts of *Gracilaria* into foods like gluten-free spreads or pasta could help individuals with diabetes and celiac disease while addressing health concerns related to cholesterol. Bioactive compounds produced by species of *Gelidium* (Rhodophyta) can act as antidiabetic agents. These extracts can be incorporated into functional beverages or health supplements targeting glycemic index control and diabetes management (Tavares et al. 2023). *Spirulina plantensis* is rich in natural antioxidants and antidiabetic properties that can reduce the hyperglycemic effect by reducing the levels of TNF-α, IL-1α, and IL-6 in blood (Wimmerova et al. 2022). The bioactive compounds like beta-carotene, algae oil, flavonoids, and fucoxanthin extracted from microalgae and cyanobacteria like *Dunaliella salina*, *Chlorella* sp., and *Nannochloropsis* are rich in antioxidants, anti-inflammatory activity, etc. Based on the secondary metabolites brown algae, red algae, and green algae are implemented in beverages, personal care, nutraceuticals, pharmaceuticals, and others (https://www.insightaceanalytic.com/report/global-algae-based-ingredients-market/1461#:~:text=The%20Algae%2DBased%20Ingredients%20Market,forecast%20period%20for%202024%2D2031).

3 Economic Importance of Algae

3.1 Mass Cultivation

Algal biomass is gaining attraction as it helps in reducing the carbon footprint, apart from this they also play a major role in pharmaceutical and other industries, as there is huge demand for algae-based natural product. The production of cyanobacteria and microalgae started booming across the world higher than before. The cost of algae production influences the pricing of algae-based products and the same applies to the algal-derived diabetes products. Companies producing algae on a large scale face the challenge of balancing production costs with market pricing to remain competitive with synthetic diabetes medications and other natural supplements. However, technological advancements in algae cultivation, such as automation and strain improvement, are gradually reducing the production costs, making algae-based products more accessible to a broader consumer base. The global nutraceutical and functional food market has shown considerable growth, with algae-based products being recognized nowadays for their health benefits, including glucose regulation, which is particularly relevant for diabetes management. Algae such as Spirulina, *Chlorella*, and certain types of seaweed contain bioactive compounds like phycocyanin, polyphenols, and carotenoids that have been shown to positively impact insulin sensitivity and glycemic control. The economic potential of algae-based diabetes products is significant, as they cater to a growing segment of health-conscious consumers seeking natural alternatives to synthetic medications with lesser side effects.

By 2032, the market for algae products is expected to increase by USD 9728.83 million due to increasing demand for nutrient-rich, sustainable products in sectors like food, beverages, biofuels, and pharmaceuticals. Reducing carbon footprints and increasing interest in natural and plant-based ingredients are important motivators. Omega-3 fatty acids, protein, and antioxidants make algae an ideal component for nutritional foods and dietary supplements. Because of growing disposable incomes and health consciousness, the Asia-Pacific area is predicted to develop at the quickest rate. Corbion N.V., Cyanotech Corporation, E.I.D.—Parry (India) Limited, and Koninklijile DSM N.V. are considered to be the important players that are making substantial investments in new products and increasing their product ranges (Peñalver et al. 2020).

As there is a huge demand for algae-based natural product, the production of cyanobacteria and microalgae started booming across the world higher than before. The cost of algae production influences the pricing of algae-based diabetes products. Companies producing algae on a large scale face the challenge of balancing production costs with market pricing to remain competitive with synthetic diabetes medications and other natural supplements. However, technological advancements in algae cultivation, such as automation and strain improvement, are gradually reducing production costs, making algae-based products more accessible to a broader consumer base. The most common method used for algal cultivation is open cultivation system which includes open ponds, round tanks, raceway ponds (RWPs), fiber-reinforced plastic (FRP) tanks, and closed cultivation system in which various types of photobioreactors (PBRs) are used.

3.1.1 Open Cultivation System

Open cultivation system includes FRPs, open tank, and raceway ponds in which raceway ponds are preferred, which is a primitive and broadly used system. Factors such as a proper land with appropriate pond depth of raceway pond which allows penetrating the light into the pond along with CO2 sparging system can be sufficient for a good cultivation practices. Microalgae like *Chlorella*, *Dunaliella salina,* and cyanobacteria like *Spirulina* are the organisms which are widely cultivated worldwide as they contain high amount of lipids, beta-carotene, and protein respectively. Where the protein content in *Spirulina* is 60–70 g/100 g is reported (Kumar et al. 2015; Costa et al. 2019).

Open ponds or raceway ponds are highly cost-effective and easy to maintain. Even though they are cost-effective the contamination rate is the main drawback of raceway pond systems, the contamination rate in open ponds are high than PBR. 100% axenic conditions cannot be assured in RWPs, still bacterial contamination is found in the harvested biomass which is due to the direct contact of medium with the atmosphere (Costa and de Morais 2014; Jerney and Spilling 2020) (Figs. 18.1, 18.2 and 18.3).

3.1.2 Closed Cultivation System

PBRs are referred as closed cultivation system and are fermentor kind of apparatus that is particularly used for the cultivation of algae in which all primary sources like pH, temperature, aeration, and light source are provided. There are different types of photobioreactors like tubular photobioreactor, stirred tank photobioreactors, flat-panel photobioreactor, bubble column photobioreactor, PBR bags, and plate

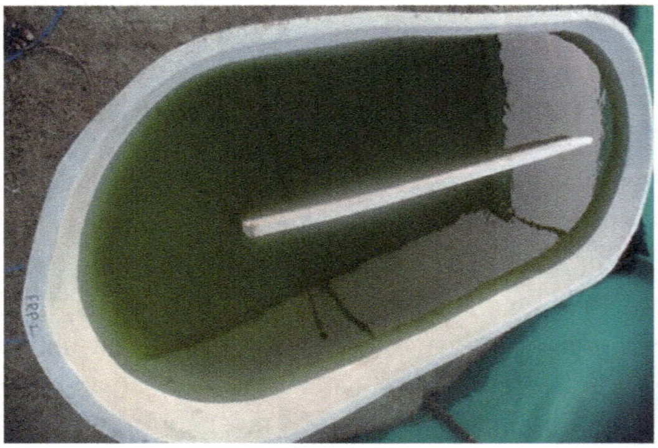

Fig. 18.1 FRP tank —500 L capacity

Fig. 18.2 Open raceway pond—5000 L capacity

Fig. 18.3 Open raceway—pond 35000L capacity

photobioreactors which are widely used in the cultivation of algal cultures (https://microbenotes.com/bioreactor/, https://pharmaguides.in/photobioreactor/, Chanquia et al. 2022).

Recently, PBRs are mostly used in companies for large-scale production as all the primary sources are provided so that the algal production can be done throughout the year in axenic conditions, as it reduces the risk of contamination (Chanquia et al. 2022) (Figs. 18.4 and 18.5).

3.2 Techno Economy

As both open pond system and PBRs are widely used for algal cultivations, the differences of both lie between operational cost and management (Table 18.1).

4 Environmental Impact Analysis

Algae production has major environmental benefits over other agricultural methods, even as its health benefits are becoming well acknowledged. Traditional agriculture can have detrimental effects on the environment, including deforestation, excessive

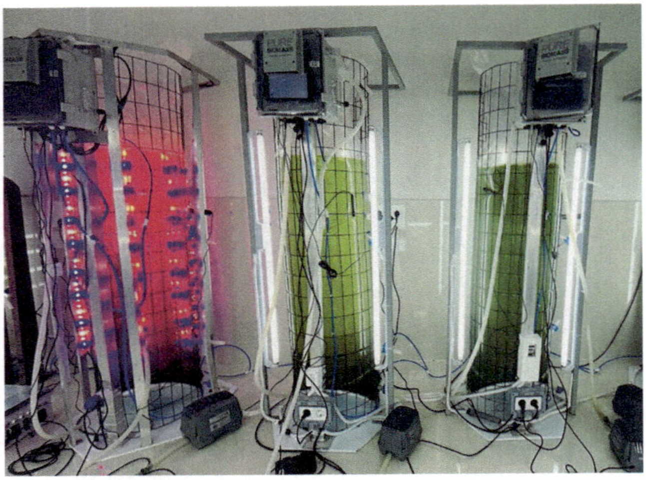

Fig. 18.4 Photobioreactor—250 L capacity

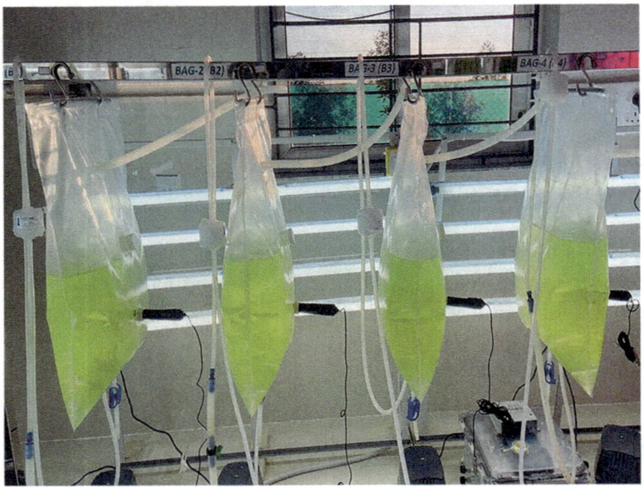

Fig. 18.5 PBR bags—25 L capacity

water consumption, and greenhouse gas emissions, particularly when it comes to crops used to make diabetes-related products. However, there are a number of important ways that algae farming offers an environmentally benign substitute.

Table 18.1 Comparison between open pond system and closed system

Open pond system	Closed system
(i) Raceway ponds can be constructed as per our requirements and it is cost-effective (ii) The operational cost and maintenance cost are less (iii) While handling algal culture the CO2 should be sparged by diffuser and a mechanical mixing of culture are recommended. The CO2 diffuser can be customized based the pore sizes (iv) Though they are cost-effective they possess few limitations; they are likely to get contaminated easily, consume large piece of land (v) Harvesting can be tough when compared to PBRs	(i) PBRs are made up of glass and plastic and require relatively smaller volume of space as they are compactly packed (ii) The cultures scaled up in PBRs remain pure and axenic as they are completely closed (iii) Culture yield will be high compared to open pond system (iv) The aerations, CO2 sparging is done automatically and no human handling is required (v) The limitation in PBRs is that they are costly compared, require skilled people for handling, and need to be maintained properly

4.1 Life Cycle Assessment

Life cycle assessment of algae in diabetes includes upstream and downstream process of algae. It involves the study of cultivation of algae in sustainable practices in which the parameters like utilization of water, nutrients along with the CO2 sequestration by reducing carbon footprints (Romagnoli et al. 2023). Nowadays due to global warming and usage of vehicles and industries the amount of CO2 is increasing rapidly. Report says that microalgae and cyanobacteria can capture 0.5 kg of C/kg of dry weight (Li and Yao 2024; de Souza et al. 2024).

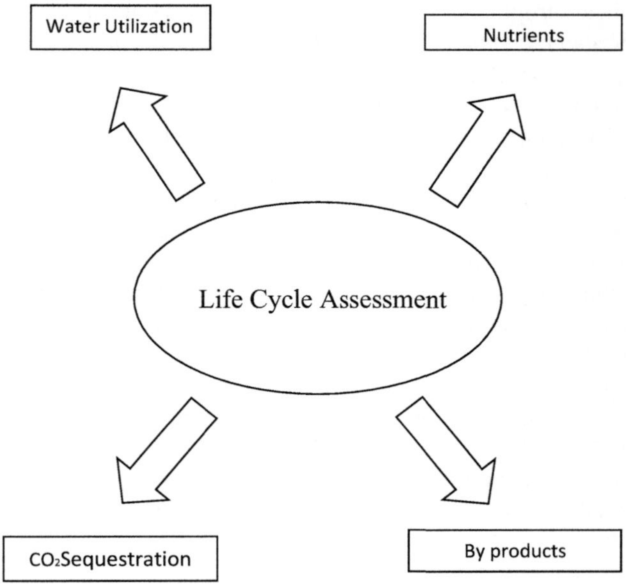

The Life Cycle Inventory (LCI) reveals a significant number of related processes and process links in microalgae cultivation systems. The study found that processes related to electricity and nutrients supply, along with waste disposal, have the highest number of impacts, including fossil fuel consumption, climate change, human toxicity, freshwater and marine ecotoxicity. This information applies to various software and data sources (Wimmerova et al. 2022). Bottlenecks were found in a study that compared microalgae cultivation techniques. The phototrophic cascade (48–100%), the flat-panel photobioreactor (1–12%), and the heterotrophic fermenter (93–100%) had the greatest environmental impact. This had to do with the energy source because the photobioreactor used energy from a facility that burned biomass. The study used sensitivity analysis to verify this. The main effects of the flat-panel photobioreactor were boosted by 30 times when the green energy source was switched to a conventional energy mix. However, as compared to the heterotrophic fermenter and the phototrophic cascade, the operation of this form of photobioreactor still produced half of the total consequences. According to the study, removing these obstacles may lessen the negative effects that microalgae growing methods have on the environment (Wimmerova et al. 2022).

4.1.1 CO_2 Sequestration

More than half of global CO_2 fixation is carried out by microalgae, the primary oxygen-evolving photosynthetic microorganisms on Earth. On the other hand, certain microalgae species can endure in reduced conditions due to their heterotrophic metabolism. Under certain circumstances, certain strains of algae can develop mixotrophically. Because it enables microalgae to store organic carbon in wastewater that would otherwise be released into the atmosphere, their ability to thrive under autotrophic, heterotrophic, or mixotrophic conditions is crucial (Ighalo et al. 2022). Dissolved inorganic carbon (DIC) from the aquatic environment may be absorbed by microalgae as H2CO3, $HCO3^{-}$, CO2, and $CO3^{2-}$.Conversely, the spectrum of DIC assimilation in terrestrial plants is much smaller. Numerous factors, including pH, mixing rates, microalgae concentrations, and others, affect the DIC forms (Ighalo et al. 2022).

The lifecycle analysis (LCA) technique provides quantitative and comprehensive evaluations of resource consumption and environmental implications, making it a useful tool for assessing the full life cycle of processes or products. It is especially helpful for comparing ecoperformance and assessing different setups in microalgae cultivation, a feasible solution for CO2 sequestration. The environmental effects of producing algal food, biodiesel, and biofuel have all been successfully analyzed using life cycle assessment (LCA), which has also been used to evaluate the effectiveness of pond systems, photobioreactors, and other cultivation characteristics. Goal and scope definition, inventory analysis, lifetime effect assessment, and findings interpretation are the four stages of life cycle assessment (LCA) as defined by the International Standard Organization (ISO) (Ighalo et al. 2022).

4.1.2 Other Impacts

The most efficient biofertilizer source for enhancing the physico-chemical properties of degraded land' soil, including its ability to retain water and its mineral nutrient status, is cyanobacterial biomass. Cyanobacteria's ubiquity, quick generation period, and capacity to repair atmospheric N2 are some of their distinctive features. Cyanobacteria, like other prokaryotic bacteria, are being used more and more as bio-inoculants to enhance environmental quality and soil fertility. The development of genetically modified cyanobacteria with unique genes for the synthesis of various biofuels, including biodiesel, bio-hydrogen, and bio-methane, has created new opportunities for the economically viable production of biofuels and related bioactive metabolites (Singh et al. 2016).

4.1.3 Future Perspective

Algae-based interventions in diabetes management have significant potential for future research and development. Despite progress, the full spectrum of bioactive compounds in algae and their effects on treating diabetes are yet to be fully explored. Further investigation into unexplored algae species and their unique metabolites may reveal novel bioactive compounds with potent antidiabetic properties. Advanced analytical techniques like metabolomics and proteomics can facilitate the identification and characterization of these compounds. Future studies should focus on elucidating molecular pathways and cellular targets influenced by algae-derived natural products (Pereira and Valado 2023).

Acknowledgement The authors acknowledge DBT (BT/PR29901/PBD/26/694/2018 dt 10/03/21), Govt. of India for the financial support.

References

Gunathilaka, T.L., Samarakoon, K., Ranasinghe, P. and Peiris, L.D.C., 2020. Antidiabetic potential of marine brown algae—a mini review. *Journal of diabetes research*, *2020*(1), p.1230218.

Tavares, J.O., Cotas, J., Valado, A. and Pereira, L., 2023. Algae food products as a healthcare solution. *Marine Drugs*, *21*(11), p.578.

Arunkumar, K., Nalluri, M., Anjana, K., Mohan, G. and Raja, R., 2023. Fucoxanthin as antioxidant, anti-hyaluronidase and cytotoxic agent: potential of brown seaweeds decoction for tea supplement. *Journal of Food Measurement and Characterization*, *17*(4), pp.3980–3989.

Romagnoli, F., Thedy, A., Ievina, B. and Feofilovs, M., 2023. Life cycle assessment of an innovative microalgae cultivation system in the Baltic region: Results from SMORP project. *Environmental and Climate Technologies*, *27*(1), pp.117–136.

Norton, T.A., Melkonian, M. and Andersen, R.A., 1996. Algal biodiversity. *Phycologia*, *35*(4), pp.308–326.

Wiśniewska, K., Lewandowska, A.U. and Śliwińska-Wilczewska, S., 2019. The importance of cyanobacteria and microalgae present in aerosols to human health and the environment–Review study. *Environment International, 131*, p.104964.

Ramos-Romero, S., Torrella, J.R., Pagès, T., Viscor, G. and Torres, J.L., 2021. Edible microalgae and their bioactive compounds in the prevention and treatment of metabolic alterations. *Nutrients, 13*(2), p.563.

Derosa, G., Pascuzzo, M.D., D'Angelo, A. and Maffioli, P., 2019. Ascophyllum nodosum, Fucus vesiculosus and chromium picolinate nutraceutical composition can help to treat type 2 diabetic patients. *Diabetes, Metabolic Syndrome and Obesity: Targets and Therapy*, pp.1861–1865.

Peñalver, R., Lorenzo, J.M., Ros, G., Amarowicz, R., Pateiro, M. and Nieto, G., 2020. Seaweeds as a functional ingredient for a healthy diet. *Marine Drugs, 18*(6), p.301.

https://www.credenceresearch.com/report/algae-products-market

Kumar, K., Mishra, S.K., Shrivastav, A., Park, M.S. and Yang, J.W., 2015. Recent trends in the mass cultivation of algae in raceway ponds. *Renewable and Sustainable Energy Reviews, 51*, pp.875–885.

Costa, J.A.V., Freitas, B.C.B., Santos, T.D., Mitchell, B.G. and Morais, M.G., 2019. Open pond systems for microalgal culture. In *Biofuels from algae* (pp. 199–223). Elsevier.

Costa, J.A.V. and de Morais, M.G., 2014. An open pond system for microalgal cultivation. In *Biofuels from algae* (pp. 1–22). Elsevier.

Jerney, J. and Spilling, K., 2020. Large scale cultivation of microalgae: open and closed systems. *Biofuels from algae: methods and protocols*, pp.1–8.

Chanquia, S.N., Vernet, G. and Kara, S., 2022. Photobioreactors for cultivation and synthesis: Specifications, challenges, and perspectives. *Engineering in Life Sciences, 22*(12), pp.712–724.

Li, G. and Yao, J., 2024. A Review of Algae-Based Carbon Capture, Utilization, and Storage (Algae-Based CCUS). *Gases, 4*(4), pp.468–503.

de Souza, M.F., Meers, E. and Mangini, S., 2024. The potential of microalgae for carbon capture and sequestration. *EFB Bioeconomy Journal*, p.100067.

Alrasheedi, A.A., Basnawi, A.I. and Althaiban, M.A., 2024. Effects of Spirulina platensis microalgae on renal function and antioxidant defence in rats with streptozotocin-induced diabetes. *Journal of Functional Foods, 122*, p.106485.

Wimmerova, L., Keken, Z., Solcova, O. and Vavrova, K., 2022. A Comparative Analysis of Environmental Impacts of Operational Phases of Three Selected Microalgal Cultivation Systems. *Sustainability, 15*(1), p.769.

Ighalo, J.O., Dulta, K., Kurniawan, S.B., Omoarukhe, F.O., Ewuzie, U., Eshiemogie, S.O., Ojo, A.U. and Abdullah, S.R.S., 2022. Progress in microalgae application for CO_2 sequestration. *Cleaner Chemical Engineering, 3*, p.100044.

Singh, J.S., Kumar, A., Rai, A.N. and Singh, D.P., 2016. Cyanobacteria: a precious bio-resource in agriculture, ecosystem, and environmental sustainability. *Frontiers in microbiology, 7*, p.529.

Pereira, L. and Valado, A., 2023. Algae-Derived Natural Products in Diabetes and Its Complications—Current Advances and Future Prospects. *Life, 13*(9), p.1831.

Chapter 19
Reactive Oxygen Species in Algae: Antidiabetic Benefits

G. Uma, T. Citarasu, J. R. Anusha, S. Jeraldin Nisha, D. Rameshbabu, and Grasian Immanuel

Abstract The need for new therapeutic techniques arises from the fact that diabetes is a major global health concern. Natural substances and their derivatives have gained popularity as appealing diabetic management solutions due to their diverse chemical compositions and pharmacological effects. Reactive oxygen species (ROS) are secondary messengers involved in a number of cellular functions, produced by algae. Enzymatic and non-enzymatic antioxidants are the two types of antioxidant defense mechanisms that provide protection against oxidative damage. With a focus on the antidiabetic properties of algae and the natural chemicals derived from algae for the management of diabetes, this chapter describes the various types of reactive oxygen species (ROS), their creation, and their antioxidant defense mechanisms.

Keywords ROS · Enzymatic and non-enzymatic antioxidants · Algae · Diabetes

1 Introduction

ROS are oxygen-containing molecules that have a short half-life, ranging from a few seconds to many days. They consist of carbonate radicals ($CO_3^{\bullet-}$), hydroxyl radicals ($HO^{\bullet}$), singlet oxygen (${}_1O^2$), hydrogen peroxide (H_2O_2), and superoxide ($O_2^{\bullet-}/HO_2$). The main ROS in marine and biotic systems arise from the successive monovalent reduction of molecular oxygen to water (Fridovich, 1998) (Fig. 19.1).

Superoxide is the conjugate base of the hydroperoxyl radical ($HOO^{\bullet}$) and the single-electron-reduced form of O_2. In marine waters, the hydroperoxyl radical is subordinated to the reactive anion superoxide ($O_2^{\bullet-}$), which has a pKa of 4.8. O_2 and hydrogen peroxide can be produced via the catalyzed or uncatalyzed dismutation of superoxide. Superoxide and hydrogen peroxide are broken down by a variety of

G. Uma (✉) · T. Citarasu · J. R. Anusha · S. J. Nisha · D. Rameshbabu · G. Immanuel
Centre for Marine Science & Technology, Manonmaniam Sundaranar University, Rajakkamangalam, Kanyakumari District, Tamil Nadu, India

R. Raja et al. (eds.), *Algae in Diabetes Management*,
https://doi.org/10.1007/978-981-95-0571-5_19

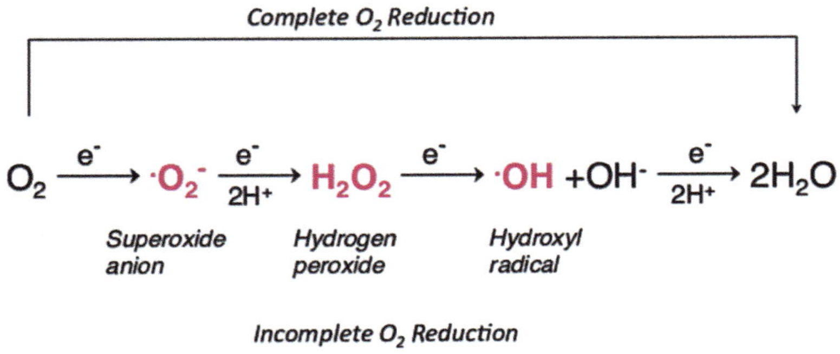

Fig. 19.1 Complete and incomplete reduction of molecular oxygen. (Source: Bartz and Piantadosi 2010).

processes, including enzyme activity, metal-organic matter reactions, and reactions with other substances. Superoxide typically has half-lives of seconds to minutes in marine environments, but hydrogen peroxide often has half-lives of hours to days (Hansel and Diaz 2021).

ROS can form both within (intracellular) and outside (extracellular) of cells in biological systems. The superoxide anion, which predominates under normal physiological pH levels, is not membrane permeable, but hydrogen peroxide can be transferred across biological membranes either actively or passively (Korshunov and Imlay 2002). Therefore, the creation of superoxide within cells usually contributes very little to the superoxide pools outside of cells (Hansel and Diaz 2021).

2 Extracellular Reactive Oxygen Species (ROS) Production in Algae

2.1 Extracellular ROS Production in Phytoplankton

Both freshwater and marine phytoplankton, including many that produce harmful algal blooms (HABs), frequently produce extracellular ROS. Since *Pleurochrysis carterae*'s early report of hydrogen peroxide production (Palenik et al. 1989), it has been identified in a variety of marine phytoplankton lineages, including raphidophytes, dinoflagellates, prymnesiophytes, diatoms, the cryptophyte *Geminigera cryophila* (Sutherland et al. 2019), and the pelagophyte *Aureococcus anophagefferens* (Diaz et al. 2018). Phytoplankton (about 10^3–10^6 cells mL^{-1}) can usually maintain steady-state quantities of superoxide between 0.1 and tens of nanomolar and extracellular hydrogen peroxide between 0.01 and several micromolar in culture (Schneider et al. 2016; Diaz and Plummer 2018; Diaz et al. 2018, 2019; Plummer et al. 2019; Sutherland et al. 2019). Within and between phytoplankton species, the net rates of superoxide production (10^{-5}–10^4amol $cell^{-1}$ h^{-1}) and extracellular hydrogen peroxide production (10^{-2}–10^5amol $cell^{-1}$ h^{-1}) differ by several orders of magnitude. Numerous physiological and environmental parameters, such as cell size, density, growth phase, and irradiance, regulate this variability. The highest rates of extracellular ROS production are produced by HAB-forming raphidophytes of the genus *Chattonella*; however, other species, including *Prorocentrum minimum* (Park et al. 2009) and *Karenia mikimotoi* (Kim et al., 2019), can achieve similar levels. Representative strains of the numerically dominant marine picocyanobacterium *Prochlorococcus* had the lowest documented rates of extracellular superoxide generation obtained from phytoplankton (Sutherland et al. 2019). Other phytoplankton taxa that have been shown to produce extracellular superoxide and/or hydrogen peroxide at intermediate levels include the bloom-forming marine coccolithophorid *Emiliania huxleyi*, a noncalcifying strain (Plummer et al. 2019), the HAB-forming *Pseudo-Nitzschia* sp. and *Kareniabrevis* (Diaz et al. 2018), and the phytoplankton species *Phaeocystis antarctica* from polar regions (Sutherland et al. 2019).

2.2 *Extracellular ROS Production in Seaweeds*

The formation of extracellular ROS is common in red, brown, and green macroalgae. The inducible oxidative burst, or seaweeds' capacity to quickly produce large amounts of ROS in response to environmental stimuli, has been the subject of a large portion of this field's research. However, as observed in the Fucales, constitutive generation of extracellular ROS without any external stimulus is also frequent (Kupper et al. 2002). The hydrogen peroxide explosion of the rhodophyte *Eucheuma platycladum* was the first documented instance of extracellular ROS generation among macroalgae (Collen et al. 1994). ROS are produced by several rhodophytes, including *Pyropia* spp. (Hou et al. 2015; Luo et al. 2015), *Chondrus crispus* (Collen and Davison 1999), and the Gracilariaceae (Weinberger et al. 2010). Among Chlorophytes, the ROS-producing species identified are *Ulva* spp. (Collen and Pedersen 1996), *Dasycladus vermicularis* (Ross et al. 2005), and *Cladophora glomerata* (Choo et al. 2004). ROS was generated in phaeophytes such as *Sargassum muticum*, the Desmarestiales, and the Laminariales, which include the giant kelp *Macrocystis pyrifera* (Kupper et al. 2002) and *Laminaria digitata* (Kupper et al. 2001). The most frequently detected ROS in the macroalgal oxidative burst is hydrogen peroxide (van Hees and Van Alstyne 2013). At the production site, concentrations of hydrogen peroxide can reach several millimolar, but in bulk seawater, they are diluted to micromolar levels (Bouarab et al., 1999; Kupper et al. 2001; Weinberger et al. 2002). Additionally, *Pyropiahaitanensis* (Luo et al. 2015), *L. digitata* sporophytes (Kupper et al. 2001), and the gametophytes of *L. digitata*, *Alaria esculenta*, and *Saccharina latissima* (Muller et al. 2012) have all been found to produce superoxide in oxidative bursts. As demonstrated in *L. digitata*, several seaweeds such as *Fucus serratus* (Kupper et al. 2002), *Saccharina* (McDowell et al. 2015), and *P. haitanensis* (Luo et al. 2014) also produce extracellular ROS beyond reproductive life stages.

3 Mechanism of Extracellular ROS Produced by Algae

3.1 Mechanism of ROS Produced by Phytoplankton

The NOX family of NADPH oxidases is one of the mechanisms by which phytoplankton produce extracellular superoxide (Fig. 19.2). It has been suggested that NOX enzymes are involved in a variety of phytoplankton species, such as *Prorocentrum minimum*, *Symbiodinium* spp., *Thalassiosira weissflogii*, *T. pseudonana*, *Phaeodactylum tricornutum*, *Dunaliella tertiolecta*, and *Chlamydomonas reinhardtii*, because the broad-spectrum flavoenzyme inhibitor diphenylene iodonium (DPI) eliminates the production of extracellular superoxide. However, DPI does not exclude the involvement of other oxidoreductases and peroxidases because it is not unique to the NOX family. However, two putative NOX enzymes belonging to the respiratory burst oxidase homolog (Rboh) family, known as Rbo1 and Rbo2, are present in the freshwater alga *C. reinhardtii*. According to Anderson et al. (2016), mutant *C. reinhardtii* lacking Rbo1 eliminates about 60% of extracellular superoxide production, indicating the possible existence of additional superoxide-generating mechanisms while also confirming the crucial role of NOX enzymes. A putative NOX homolog identified by antibodies produced against human NOX has been connected to extracellular superoxide generation in *Chattonella marina* and the closely related *C. ovata* (Kim et al. 2000). Additionally, the transcriptomes of *C. antiqua*, *C. subsalsa*, and a number of other red tide

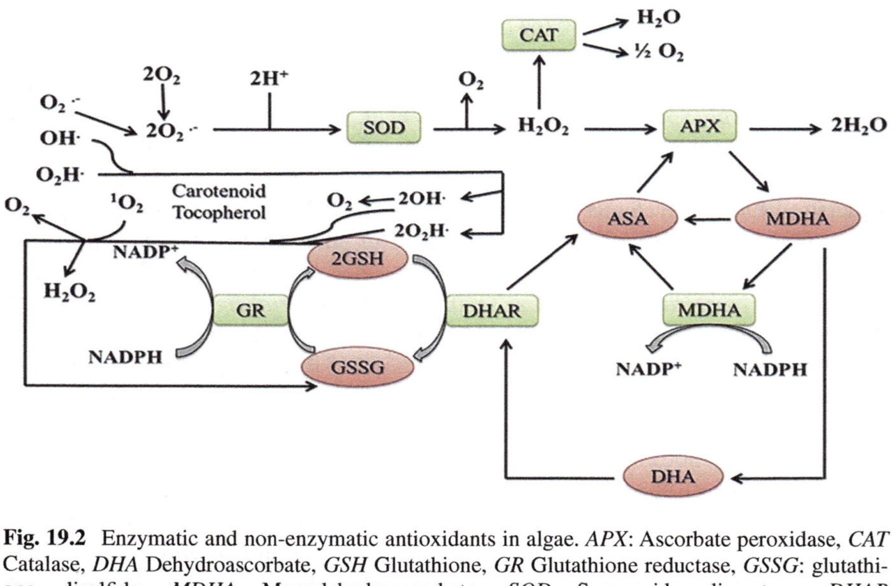

Fig. 19.2 Enzymatic and non-enzymatic antioxidants in algae. *APX*: Ascorbate peroxidase, *CAT* Catalase, *DHA* Dehydroascorbate, *GSH* Glutathione, *GR* Glutathione reductase, *GSSG*: glutathione disulfide, *MDHA* Monodehydroascorbate, *SOD* Superoxide dismutase, *DHAR* Dehydroascorbate reductase. (Source: Gill and Tuteja 2010)

flagellates have recently been found to include numerous putative Nox sequences (Shikata et al. 2019).

According to Diaz et al. (2019), bioinformatics analysis has also identified putative NOX homologs in the genomes of other phytoplankton, such as the dinoflagellate *Symbiodinium microadriaticum*, the marine picoeukaryotes *Micromonas pusilla* and *Ostreococcus tauri*, the diatoms *P. tricornutum*, *T. pseudonana*, *Thalassiosira oceanica*, and *Fragilariopsis cylindrus*. According to Magnani et al. (2017), NOX enzymes have also been found in a few cyanobacteria, including one marine strain that most likely obtained it through gene transfer from a eukaryotic organism (Zhang et al. 2013). However, nothing is known about these NOX enzymes' capacity to produce ROS in vivo. In addition to NOX, phytoplankton use a number of additional enzymes to produce extracellular ROS (Fig. 19.2). For instance, certain phytoplankton, such as *Pleurochrysis*, *Amphidinium*, and *Prymnesium parvum*, produce extracellular hydrogen peroxide via cell surface deaminases like amino acid oxidase and amine oxidase (Palenik et al. 1989). Furthermore, it was recently determined that a putative cell surface flavoenzyme with a significant sequence similarity to glutathione reductase (GR) was responsible for *T. oceanica*'s DPI-inhibitable, NADPH-dependent extracellular superoxide generation (Diaz et al. 2019). GR-mediated extracellular superoxide generation by phytoplankton may be more phylogenetically ubiquitous, as evidenced by the discovery of putative homologs of *T. oceanica* GR in model marine phytoplankton genomes and ocean metagenomes. Given that light can promote the generation of extracellular ROS in a variety of phytoplankton species, it is thought that photosynthesis provides NADPH for GR and NOX (Diaz and Plummer 2018; Diaz et al. 2019; Plummer et al. 2019). But even in the absence of light, phytoplankton continue to produce significant amounts of extracellular ROS, suggesting that there are other, as yet unidentified, processes at work (Diaz et al. 2019). For instance, cultures of various freshwater phytoplankton (Chaput et al. 2019) and marine HAB-forming species (Diaz et al. 2018) have been linked to unidentified cell-free superoxide-generating enzymes. Since NOX enzymes are extremely hydrophobic transmembrane enzymes, it is doubtful that these water-soluble proteins include them. As a result, they are excellent candidates for additional research.

3.2 Mechanism of ROS Produced by Seaweeds

At least three ROS-generating enzymes (Fig. 19.2) are implicated in the mechanisms of extracellular ROS production in macrophytes: amino acid oxidase, oligosaccharide oxidase, and NOX homologs (Rboh). Inhibition tests using DPI have generally suggested that NOX enzymes may play a role in the oxidative burst in red (Weinberger et al. 2010), green (Ross et al. 2005), and brown (Kupper et al. 2002) macroalgae. Several seaweeds, such as the red algae *Chondrus crispus* (Herve et al. 2006), *Pyropia* spp. (Luo et al. 2015), and *Gracilariopsis lemaneiformis* (Wang et al. 2018), have been shown to include Rboh gene sequences in their genomes.

Furthermore, a role for NOX homologs in the formation of extracellular ROS in seaweeds like *P. haitanensis* is supported by positive correlations between ROS production and Rboh gene copy number (Luo et al. 2014) and expression (Luo et al. 2015; Chen et al. 2016). According to McDowell et al. (2015), light is necessary for the kelp *S. latissima* to produce extracellular hydrogen peroxide through photosynthetic electron transport. This discovery may be related to the passive release of hydrogen peroxide generated during light reactions of photosynthesis, but it is also compatible with the light-dependent creation of photosynthetic NADPH as a substrate for NOX. Similarly, the green alga *Ulva rigida*, which increases the generation of extracellular hydrogen peroxide in response to rising irradiance (Collen and Pedersen 1996).

Well-known chemical elicitors of the ROS burst in seaweeds are oligosaccharides, which are the breakdown products of macroalgal cell walls (Hou et al. 2015; Chen et al. 2016). According to Hou et al. (2015), these oligosaccharides directly promote cell surface oligosaccharide oxidase's DPI-insensitive hydrogen peroxide synthesis. Purified hexose oxidase (HOX) from the red algae *C. crispus* and *Ptilophora subcostata*, is a member of the large family of oligosaccharide oxidases and is thought to contribute to the ROS burst (Ogasawara et al. 2016). The enzyme amino acid oxidase, which is triggered by the green algal parasite *Acrochaeteoperculata*'s release of L-asparagine, is another method of extracellular hydrogen peroxide production linked to the red alga *C. crispus* (Weinberger et al. 2002). In macroalgae, a variety of elicitors of extracellular ROS generation have been identified, in addition to oligosaccharides and amino acids. These include hormones such as abscisic acid (Wang et al. 2018), free fatty acids (Kupper et al. 2009), and bacterial flagellar peptides (Luo et al. 2015). These findings might help identify more signaling and enzymatic pathways that contribute to seaweeds' generation of extracellular ROS.

4 Generation of ROS in Algae

ROS are a byproducts of respiration and are produced in mitochondria, cytosol, peroxisomes, and chloroplasts. When light absorption exceeds the needs of the photosynthetic equipment, ROS are generated in thylakoid membranes. Because of the completion of the electron transfer in the stromal side of the thylakoid membrane, photosystem I (PSI) is the primary source of ROS production. When the electron transport chain between the photosystems and the excitation energy delivery to the PSII reaction center were constrained, PSII produced ROS. The type of ROS and the conditions of the reaction determine its activity. For instance, H_2O_2 has moderate activity while HO^* has very high chemical activity, which allows it to oxidize the majority of organic compounds. In biological contexts, ROS activity declines in the following order: $HO^\cdot > {}^1O_2 > H_2O_2 > O_2^{\cdot-}$ (Rezayian et al. 2019).

5 Antioxidative Defense System in Algae

Algae used two defense mechanisms to deal with ROS before they could harm various parts of the cell (Fig. 19.2). The first system consists of non-enzymatic (low molecular weight) components such as ascorbate, flavonoids, carotenoids, glutathione, tocopherols, and phenols, as well as antioxidant enzymes (high molecular weight) including superoxide dismutases, glutathione reductase, catalases, and ascorbate peroxidase. Repair enzymes, which fix and eliminate broken macromolecules, are the second method. While non-enzymatic antioxidants are more successful in preventing ROS production through excitation energy transfer ($^1O_2$), enzymatic antioxidants play a significant role in detoxifying the harmful effects of formed ROS by electron transport ($O_2^{\bullet-}$, H_2O_2, $HO^{\bullet}$) (Jahan et al. 2017; Pikula et al. 2019).

5.1 Enzymatic Components of Antioxidative Defense System

5.1.1 Superoxide Dismutase

Superoxide dismutase (SOD) is a metalloprotein that catalyzes the conversion of superoxide radicals into hydrogen peroxide and oxygen, making it the first line of defense against oxidative stress. Three isoforms of SOD are distinguished by their metal center cofactors: MnSOD, which is found in mitochondria; FeSOD, which is found in the chloroplast stroma; and Cu/ZnSOD, which is found in the cytosol and thylakoid membranes of higher plants, some dinoflagellates, and charophycean green algae. Abiotic stressors increase SOD activity, which serves as a defense mechanism (Davies 2000; Mallick and Mohn 2000).

5.1.2 Catalase

The primary function of the tetrameric enzyme catalase (CAT), which contains heme, is to convert H_2O_2 into H_2O and O_2. There are three groups in CAT. Prokaryotes have manganese catalases. Prokaryotes and certain eukaryotes include catalase peroxidases, which function as both catalases and peroxidases. Heme groups and the two-step conversion of H_2O_2 to H_2O and O_2 are features of classical catalases (cat). Initially, the catalase's Fe^{3+} is changed to cat(Fe[V]O) and one H_2O_2 molecule is converted to water. Second, another H_2O_2 molecule is reduced to H_2O and O_2, and the cat(Fe[V]O) is transformed back to Fe^{3+} (Mate et al. 1999).

5.1.3 Ascorbate Peroxidase

Heme enzyme ascorbate peroxidase (APX) uses ascorbate as an electron source to change H_2O_2 into H_2O. APX is found in algae and plants. The red algae *Galdieria partita* and *G. sulphuraria* have been found to contain two cytosolic APX (cAPX) isoenzymes (Oesterhelt et al. 2008).

5.1.4 Glutathione Reductase and Glutathione Peroxidase

Both prokaryotes and eukaryotes include the flavoprotein oxidoreductase glutathione reductase (GR). By keeping GSH in its reduced state, GR, a fantastic ascorbate-glutathione cycle enzyme, plays a vital role in the defense mechanism against ROS. Similar to ascorbate peroxidase, glutathione peroxidase uses GSH to break down H_2O_2 into H_2O (Romero-Puertas et al. 2006).

5.2 Nonenzymatic Components of Antioxidative Defense System

5.2.1 Glutathione

Glutamate, cysteine, and glycine combine to generate the tripeptide glutathione, which is the main low molecular weight thiol. Ascorbate regenerates when reduced glutathione oxidizes into oxidized glutathione through the ascorbate-glutathione cycle. Glutathione has a role in restoring protein disulfide links, scavenging oxidative damage, adjusting the redox potential of amino acids and proteins, acting as a non-specific reductant, scavenging oxidative damage, and suppressing H_2O_2 and organic peroxides (ROOH) (Halliwell and Gutteridge 2015).

5.2.2 Ascorbate

Ascorbic acid, often known as ascorbate or vitamin C, is a water-soluble antioxidant that serves as an electron donor for the OH radical, a good scavenger, a reducing antioxidant, and a substrate for ascorbate peroxidase. Because of the ascorbate-glutathione cycle, ascorbate inhibits H_2O_2 (Wheeler et al. 1998).

5.2.3 Carotenoids

There are 750 different types of carotenoids found in plants and microbes. Carotenoids are isoprenoid, lipophilic substances that can be yellow, orange, or red in color. Lycopene, α-carotene, β-carotene, and xanthophylls containing oxygen as hydroxyl groups (like lutein), oxy-groups (like canthaxanthin), or a combination of both (like astaxanthin) are examples of carotenoids. Microalgae generate some xanthophylls, such as violaxanthin, antheraxanthin, zeaxanthin, neoxanthin, lutein, loroxanthin, astaxanthin, and canthaxanthin, whereas brown algae also synthesize diatoxanthin, diadinoxanthin, and fucoxanthin. Carotenoids have antioxidant properties that help cells fight off free radicals, prevent lipid peroxidation, strengthen the photosynthetic system, and preserve membrane integrity (Di Pietro et al. 2016).

5.2.4 Tocopherols

Vitamin E or tocopherols are long hydrocarbon chains with an aromatic ring. Because of their aromatic ring, tocopherols participate in oxidation-reduction processes. Tocopherols shield unsaturated fatty acids from oxidation and eliminate reactive forms of oxygen. Because it may suppress $^1O_2$, reduce O_2, and complete the lipid peroxidation reaction, α-tocopherol is a crucial antioxidant among the four tocopherols (α-, β, ϒ, and ϐ-tocopherols) (Takenaka et al. 1991).

5.2.5 Phenolic Compounds

Based on their chemical structure, phenolic compounds which comprise over 8000 compounds are categorized into ten different classes. Through the transfer of hydrogen atoms and single electrons, phenolic chemicals are the primary natural antioxidants. Numerous investigations have demonstrated that microalgae have phenolic contents that are either less than or equivalent to the minimal levels reported in terrestrial plants. Additionally, microalgae create many forms of flavonoids, including dihydrochalcones, isoflavones, flavanones, and flavonols. Since complex phenolic compounds are generated by microalgae, it is necessary to identify these molecules. Numerous polyphenols, including phlorotannins, which have antioxidant properties, are found in marine algae. Defense strategies against environmental stressors require phenolics (Cornish and Garbary 2010).

6 ROS Signaling

Low concentrations of ROS play important functions in apoptosis, cell signaling, and infection defense. Due to their ability to transfer membranes through aquaporins, O_2^-, $OH^\cdot$, and H_2O_2 have greater signaling capacity (Reczek and Chandel 2015). The transduction of $H_2O_2^-$based signals is centered on sulfur chemistry, with the main player being the reversible oxidative modification of cellular sulfur-containing groups (e.g., cysteine residues and thioredoxin), which in turn results in disturbances of metabolism and signaling pathways in microorganisms. In mitogen-activated protein kinase (MAPK) pathways, which function as signaling responses to a variety of environmental stimuli, H_2O_2 oxidizes the cysteine thiol groups of phosphatases. According to recent research, ROS play a crucial role in resistance to oxidative stress through post-translational modifications and alteration of gene expression. JNK, p38, and extracellular signal-regulated kinases are members of the mitogen-activated protein kinase family that are involved in ROS-regulated cellular processes such as proliferation, differentiation, and apoptosis (Morano et al. 2012).

7 Diabetes and Oxidative Stress Related to Diabetes

Over the past few years, diabetes has become more common, mostly in low- to middle-income nations, and is now one of the leading causes of premature death globally. About 316 million people have impaired glucose tolerance and an increased risk of developing diabetes, whereas 537 million people have diabetes, according to the International Diabetes Federation (IDF) (2022). In the absence of prompt and precise preventative measures, these numbers are expected to rise to 643 million by 2030, and in less than 25 years, there will be 783 million individuals afflicted by this illness (IDF, 2022). Type 1 diabetes, type 2 diabetes, gestational diabetes mellitus, and other distinct types of diabetes with multiple causes, including drug-related or pancreatic diseases, monogenic diabetic syndrome, and chemical inducers, are among the different types of diabetes (American Diabetes Association 2021).

Reactive oxygen species (ROS) are intermediate products that can be produced by the body's biochemical activities. One or more unpaired electrons in dangerous free radicals (ROS) make them extremely reactive with other compounds. An imbalance in metabolism and the body's capacity to detoxify or offset the damaging oxidant effects of free radicals can result from an excess of ROS (Tiwari 2001). Oxidative stress is the term for this condition, which is dependent on the equilibrium between the generation of ROS and antioxidant defenses (Pirian et al. 2017). Accordingly, the imbalance between the production and neutralization of reactive oxygen and nitrogen species leads to oxidative stress (Poljsak and Fink 2014). Oxidase enzymes catalyze electron transfers to O_2 in order to oxidize substrates or generate chemical energy. In biological contexts, these enzymes may provide reduced Cu^{2+} derivatives; during catalysis, they also generate $O_2^{\bullet-}$ (Alfadda and Sallam 2012). O_2 is reduced to $O_2^{\bullet-}$ by the mitochondrial electron transport chain. $O_2^{\bullet-}$ radicals are reduced by dismutase enzymes to produce hydrogen peroxide (H_2O_2) and/or the hydroxyl secondary radical ($^{\bullet-}OH$), which is another kind of ROS (Styskal et al. 2012). Chronic metabolic illnesses, including obesity (Alfadda and Sallam 2012) and diabetes mellitus (Fig. 19.3), seem to be strongly associated with mitochondrial dysfunction; however, the cause-and-effect link is still unclear (Maiese 2015). Lipid peroxidation, protein denaturation, or DNA conjugation are examples of oxidative damage to biological components caused by this connection, which ultimately leads to cell death (Xin et al. 2015). For the reasons listed above, oxidative stress has been linked to a number of illnesses, including cancer,

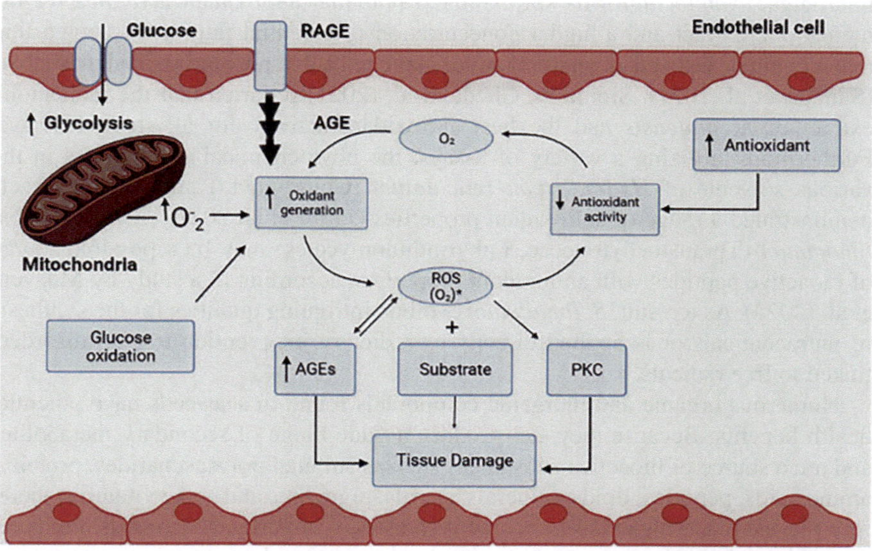

Fig. 19.3 The relationship between rates of antioxidant generation, antioxidant activity, oxidative stress, and oxidative damage in diabetes. (Source: Abo-Shady et al. 2023)

post-ischemic and brain degeneration, Parkinson's and Alzheimer's disease, AIDS, aging, and cardiovascular disorders (Maiese 2015).

8 Antioxidant and Antihyperglycemic Activity of Different Algal Extracts

Many studies have shown that the biological compounds of microalgae, including phenolic compounds, halogenated ketones, alkanes, cyclic polysulfides, phlorotannins, carotenoids, terpenoids, steroids, alkaloids, and phenols, as well as their abundance in proteins, vitamins, pigments, lipids, and vital minerals, are linked to their therapeutic potential (Coulombier et al. 2021). The butanol extract from the cyanobacterium *Arthrospira platensis* (previously *Spirulina platensis*) showed highest antioxidant activity demonstrated by DPPH, reducing power, hydroxyl radical, and nitric oxide scavenging activities (Gouda et al. 2015). Similarly, Jayshree et al. (2016) stated that highest antioxidant activities were observed in the methanolic extract of *Chlorella vulgaris* (Chlorophyta).

Scaglioni et al. (2018) observed that the aqueous extracts of *Arthrospira* sp. and *Nannochloropsis* sp. (Ochrophyta, Eustigmatophyceae) exhibited the DPPH scavenging capability. Furthermore, the presence of active secondary metabolites in the acetone extract of *C. vulgaris* demonstrated antioxidant activity by utilizing various standard assays (Abdel-Karim et al. 2019). According to recent studies, the microalgae *Nannochloropsis* sp. exhibited potential antioxidant activities for the methanolic extract and a higher concentration of essential fatty acids, suggesting that it could be utilized as a natural antioxidant and aid in preventing oxidative stress (Santhar et al. 2021). Similarly, Gheda et al. (2021) reported that the methanolic extract of *A. platensis* had the best antioxidant activity for all studied assays. Furthermore, utilizing a variety of assays, the phytochemical components in the various solvents of *Hydrodictyon reticulatum* (Chlorophyta) microalgal extracts demonstrated a range of antioxidant properties (Jaffer et al. 2021). *Schizochytrium limacinum* (Thraustochytriaceae, Labyrinthulomycetes) may be a possible source of bioactive peptides with antioxidant properties, according to a study by Moaveni et al. (2022). As a result, *S. limacinum* exhibits intriguing qualities for the synthesis of nutraceuticals or is applied directly as a dietary intervention to stop disorders linked to free radicals.

Numerous organic and inorganic compounds found in seaweeds have potential health benefits. Because they can produce a wide range of secondary metabolites and are a source of bioactive substances such as sulfated polysaccharides, proteins, amino acids, peptides, lipids, minerals, sterols, pigments, and some vitamins, therefore marine macroalgae have become increasingly important. Seaweeds are therefore acknowledged as renewable, high-quality antioxidant resources because of their ability to reduce the harmful effects of oxidative damage brought on by ROS (Zhao et al. 2018). Moreover, seaweeds are one of the best natural antioxidant sources due to their high phenolic molecule concentrations (Gunathilaka et al. 2022). According to Elangovan et al. (2019), phenolic compounds are defined as several hundred molecules that have at least one hydroxyl group connected to a benzene ring structure. By strengthening the body's natural antioxidant system and chelating metal ions like Cu and Fe that catalyze reactions which produce free radicals, phenolic substances can act as antioxidants (Al-Azzawie and Alhamdani 2006). Numerous research studies examined the antioxidant activity of various macroalgae extracts in light of this link. Using a variety of evaluated tests, the methanolic extract of *Centroceras clavulatum* (Rhodophyta) showed strong antioxidant activity (Elangovan et al. 2019). The methanolic extract from *Ulvalactuca* (Chlorophyta) showed exceptional ROS-scavenging potential in in vitro antioxidant tests (Ibrahim et al. 2021).

Additionally, the antioxidant activity of *Sargassum polycystum*'s ethyl acetate extract and *Sargassum wightii*'s acetone extract (Ochrophyta, Phaeophyceae) was determined by Unnikrishnan et al. (2015a, b). The protein hydrolysate-ultrafiltrate fractions obtained from *Fucusspiralis* (Ochrophyta, Phaeophyceae) also demonstrated a much greater FRAP according to the antioxidant analysis. The bioactive peptides and polyphenols liberated during enzymatic hydrolysis were the cause of this activity (Paiva et al. 2017). The findings of the study by Vijayan et al. (2018) unequivocally show that the *S. wightii* fraction that was extracted using *S. wightii* ethyl acetate exhibited strong antioxidant activity. Additionally, *Turbinaria decurrens*, a brown alga, had a significant content of polyphenols and a wide range of antioxidant activity in its methanolic extract (Arguelles and Sapin 2020).

A study by Ismail et al. (2019) revealed that *T. decurrens* acetone extract had the strongest antioxidant effect. Furthermore, based on different tests conducted by El-Sheekh et al. (2020), the ethanolic extract of *Taonia atomaria* brown algae demonstrated the strongest antioxidant capacity. The maximum DPPH radical scavenging activity was found in the methanolic extract of *Padina pavonica* (Ochrophyta, Phaeophyceae) (Shafay et al. 2021). Similar findings were found in the study by Abhishek et al. (2021), which showed that the polyphenols from *Padina boryana* (Ochrophyta, Phaeophyceae) could be successfully extracted using a methanolic solvent. These polyphenols were highly associated with the plant's antioxidant activity. The enzymatically broken-down polysaccharides from the brown algae *Sargassum fusiforme* had exceptional antioxidant activity (Qian et al. 2021). Also, the brown algae *P. pavonica* samples that were freeze-dried and extracted using ethanol exhibited higher levels of antioxidant activity (Cagalj et al. 2021). Nguyen et al. (2019) found that a maximum antioxidant activity was rendered by the ethyl acetate fraction from the red seaweed *Laurencia dendroidea*. Furthermore, *Jania rubens* (Rhodophyta) ethanolic crude extract showed the strongest antioxidant activity (El Nur et al. 2021). Similarly, the highest antioxidant capacity of the methanolic extract of *Pterocladiella capillacea* (red algae) was due to its high content of total polyphenols (Hmani et al. 2021). It was also discovered that the methanolic extract of *Chondrus crispus* (Rhodophyta) included a number of flavonoids, polyphenols, and tannins that had exceptional antioxidant activity (Alkhalaf 2021). Similar to this, Murugesan et al. (2021) noted that the red seaweed *Ahnfeltiopsispygmaea*'s methanolic extract was a useful source of antioxidants to treat illnesses brought on by oxidative stress.

According to Gouda et al. (2015), butanolic extract of *Spirulina* inhibits α-glucosidase activity. According to Priatni et al. (2016), the extract of exopolysaccharides from *Pseudanabaena* sp. (Cyanobacteria) exhibited the greatest inhibition of α-glucosidase activity. Because of their high phenolic content, the aqueous and methanolic extracts of *Euglenacantabrica* (Euglenozoa) microalgae showed the best antioxidant activity (Jerez-Martel et al. 2017). According to Ahmed et al. (2018), aqueous extracts of *Fischerella* BS1-EG (Cyanobacteria) showed possible α-glucosidase inhibitory activity, suggesting an antidiabetic benefit. Additionally, in a concentration-dependent way, fucoxanthin derived from *Phaeodactylum tricornutum* (Bacillariophyta) showed a substantial inhibitory effect towards α-amylase (Kawee-Ai et al. 2019). Similarly, the highest level of α-amylase inhibition was shown by the ethyl acetate extract of *Nannochloropsisoculata* (Deepa et al. 2020). A recent study by Gheda et al. (2021) showed the methanolic extract of *A. platensis* had the highest α-amylase enzyme inhibitory efficacy when compared to the pharmaceutical drug acarbose. In comparison to acarbose, the same extract exhibited a significant α-glucosidase inhibitory action. Priatni et al. (2021) demonstrated that the marine microalgae *Porphyridium* sp. (Rhodophyta) had the strongest methanolic extract among the microalgae they examined in terms of reducing the activity of the α-glucosidase enzyme.

Seaweeds are thought to be one of the best natural antioxidant sources because of their high content of phenolic compounds (Satpati et al. 2021). Numerous research studies have suggested that the seaweed extracts are potential antidiabetic medication candidates, because of their capacity to lessen the oxidative damage brought on by an excess of ROS and its detrimental effects. *Ulva reticulata* (Chlorophyta) ethanolic extract had the highest α-amylase and α-glucosidase inhibitory activity, according to Reka et al. (2017). By blocking the enzymes that break down carbohydrates, green (Chlorophyta) seaweed extracts from *Ulva intestinalis*, *Chaetomorpha aerea*, and *Cladophora rupestris* have demonstrated their ability to control diabetes (Unnikrishnan et al. 2015a, b). Mohapatra et al. (2016) reported that the ethyl acetate extract of *U. lactuca* has antidiabetic qualities by blocking the enzymes that hydrolyze carbohydrates, because of their bioactive components' activity. Furthermore, *S. polycystum*'s ethyl acetate extract and *S. wightii*'s acetone extract had demonstrated notable capacities to block α-amylase and α-glucosidase, hence averting postprandial hyperglycemia (Unnikrishnan et al. 2015a, b). The methanolic extract of *Spatoglossum asperum* (Ochrophyta) revealed a substantial α-glucosidase inhibitory action due to the presence of phytochemicals, flavonoids, tannins, and saponins (Pandithurai et al. 2015). The α-glucosidase enzyme was inhibited by the highest active methanolic extract of *Hormophysa cuneiformis* (Ochrophyta) (Osman et al. 2019). In comparison to the pharmaceutical medication, acarbose, a semi-purified phlorotannin fraction of *Fucus vesiculosus* (Ochrophyta, Phaeophyceae) showed strong inhibitory effects against the enzymes α-amylase and α-glucosidase, respectively (Catarino et al. 2019).

In contrast to employing acarbose and metformin as antidiabetic medications, Arguelles and Sapin (2020) reported a similar outcome, showing that the methanolic extract of *T. decurrens*(brown macroalgae) had a strong inhibitory action for α-glucosidase in vitro. Additionally, *T. decurrens* acetone extract had the strongest inhibitory effects for both α-amylase, andα-glucosidase enzymes that hydrolyze carbohydrates, which was explained by its capacity to have a total phenolic content (Ismail et al. 2019). The ethanolic extract of *T. atomaria* showed the highest α-amylase inhibitory activity among the suggested seaweed extracts (Admassu et al. 2018). Tessema (2018) documented that the protein fraction isolated from *Porphyra* sp., a red alga, inhibited the enzymes involved in carbohydrates. The red seaweed *L. dendroidea*'s ethyl acetate fraction showed potent in vitro α-glucosidase inhibitory activity (Nguyen et al. 2019). Similarly, a study by Sanger et al. (2019) demonstrated that a number of phytochemical components found in the aqueous extract of *Halymenia durvillei* (red algae) were in charge of the plant's antidiabetic properties through the suppression of α-glucosidase activity.

9 Antidiabetic Potential Bioactive Compounds in Algae

It has been shown that algae contain a wide variety of bioactive chemicals that may have antidiabetic effects due to their extraordinary chemical diversity. These substances provide a wealth of natural products that may be investigated for the creation of innovative treatment approaches for the control of diabetes. Polysaccharides, polyphenols, pigments, lipids, peptides, proteins, and other bioactive substances found in algae have demonstrated promise in modifying the major pathways implicated in glucose homeostasis and diabetes-related problems (Abo-Shady et al. 2023).

10 Regulation of Blood Glucose Level by Algae-Derived Products

Natural products made from algae have demonstrated potential in enhancing glycemic management and controlling blood glucose levels. The function of compounds derived from algae in glycemic control and their possible uses in the treatment of diabetes due to their anti-inflammatory and antioxidant properties, inhibition of enzymes that break down carbohydrates, stimulation of glucose absorption, modulation of insulin secretion, and control of hepatic glucose production. Insulin secretion, insulin sensitivity, adipocyte function, mitochondrial function, and the modulation of oxidative and inflammatory stress can all be used to assess how algae extracts affect insulin-related parameters and their implications for glycemic control (Pereira and Valado 2023). α-glucosidase and α-amylase inhibition were used to investigate the inhibitory effects of natural compounds derived from algae on enzymes that break down carbohydrates and their implications for glycemic management (Kashtoh and Baek 2022).

11 Algae-Derived Natural Products and Diabetic Complications

11.1 Preventing and Managing Diabetic Retinopathy

Natural medicines made from algae (including *Alaria*, *Palmaria*, and *Ulva*) have demonstrated promise in the treatment and prevention of diabetic retinopathy, a major eye consequence of diabetes (Lauritano and Ianora 2016). In addition to vascular alterations, diabetic retinopathy also causes retinal neuronal damage. By maintaining retinal ganglion cells, photoreceptor cells, and other retinal neurons, chemicals produced from algae (*Acetabularia acetabulum* Chlorophyta)) have demonstrated neuroprotective properties. These substances have the ability to alter neurotrophic factors, increase neuronal survival, and stimulate neurite outgrowth. Algal-derived chemicals help prevent and treat vision loss associated with diabetic retinopathy by safeguarding the retinal neurons (Fathalipour et al. 2020; Brillante et al. 2022).

11.2 Diabetic Nephropathy and Its Associated Complications

Natural compounds made from algae, such as *Taonia atomaria* and *Fucus spiralis*(Phaeophyceae), have potential as treatments for diabetic nephropathy and its aftereffects (Abo-Shady et al. 2023). Kidney damage and reduced renal function are hallmarks of diabetic nephropathy, a progressive kidney disease brought on by diabetes. By maintaining renal function and reducing kidney damage, extracts from the brown macroalga species *Fucus vesiculosus* have reno-protective benefits (Catarino et al. 2021). In models of diabetic nephropathy, extracts from the macroalga species *Gracilariopsis lemaneiformis* (Rhodophyta) have shown anti-fibrotic properties (Guo et al. 2018). It has been demonstrated that extracts from the cyanobacteria species *Arthrospira platensis* enhance insulin sensitivity and control blood glucose levels (Hannan et al. 2020). By improving glycemic management, *A. platensis* supplements lessen the strain on the kidneys and lower the chance of renal problems (Hatami et al. 2021).

11.3 Diabetic Neuropathy

Natural compounds made from algae have demonstrated promise as potential treatments for diabetic neuropathy, a common diabetic condition marked by malfunction and damage to the nerves. A wealth of bioactive chemicals with neuroprotective qualities can be found in the unique chemical elements of different algae species (Kumari et al. 2023). Extracts from *Ecklonia cava* have been demonstrated to improve nerve function in diabetic neuropathy, lower inflammation, and shield nerve cells from oxidative stress (Cho et al. 2023; Jo et al. 2023). Bioactive substances, including ulvan, which has anti-inflammatory and antioxidant properties, are present in *Ulvalactuca* extracts (Shibasaki and Ueda 2023). Sulfated polysaccharides, which have anti-inflammatory and antioxidant qualities, are present in the extracts from *Gracilaria edulis* (Mahendran et al. 2021). Extracts from *Sargassum* spp. have a variety of bioactive substances with anti-inflammatory, neuroprotective, and antioxidant properties, including phlorotannins, fucoidan, and fucoxanthin (Martins et al. 2022). Bioactive substances with anti-inflammatory and antioxidant qualities, such as agarobiose and agar-oligosaccharides, are found in *Gelidium elegans* extracts (Choi et al. 2018). The application of particular algae species, such as *Ulvalactuca*, *Gracilaria edulis*, *Ecklonia cava*, *Sargassum* spp., and *Gelidium elegans*, presents promising opportunities for the creation of natural treatments for diabetic neuropathy. Their anti-inflammatory, neuroprotective, and antioxidant qualities help to maintain the health and functionality of nerves (Pereira and Valado 2021).

12 Conclusion

Natural drugs made from algae have shown great promise in the treatment of diabetes and its consequences. Algae are appealing sources of bioactive chemicals with possible antidiabetic effects due to their varied chemical composition and complex pharmacological activity. We have emphasized the importance of ROS in the treatment of diabetes throughout this chapter, with a particular emphasis on antioxidant compounds derived from algae. We discussed the significance of natural products made from algae in the study of diabetes, highlighting their special capacity to alleviate the disease's worldwide burden. A wide range of bioactive substances, including polysaccharides, polyphenols, pigments, and peptides, have been discovered via the investigation of algae as a source of natural products. These substances may have antidiabetic properties. The capacity of these compounds produced from algae to control blood glucose levels, improve insulin production and sensitivity, and block enzymes that break down carbohydrates has been clarified by mechanistic insights into their therapeutic qualities. Interventions based on algae have also demonstrated potential in the management and prevention of diabetes complications, including retinopathy and nephropathy. Even while our knowledge of the potential of natural compounds derived from algae in the treatment of diabetes has advanced significantly, much more research remains. To fully realize the potential of algae and their bioactive components, more investigations and studies in this area are essential. Novel bioactive compounds with improved antidiabetic activities might be discovered as a result of more research on uncharted algae species and their metabolites.

References

Abdel-Karim OH, Gheda SF, Ismail GA and Abo-Shady AM 2019. Phytochemical Screening and antioxidant activity of *Chlorellavulgaris*. *Delta Journal of Science*, 41(1), 79–91.

Abhishek D, Sanjay S and Jadeja BA 2021. Cytotoxicity, antioxidant and antimicrobial activity of marine microalgae (*Iyengariastellata* and *Padinaboryana*) from the Gujarat coast. *Journal of the Maharaja Sayajirao University of Baroda*, 55, 1(14), 130–140.

Abo-Shady AM, Gheda SF, Ismail GA, Cotas J, Pereira L and Abdel-Karim OH 2023. Antioxidant and antidiabetic activity of algae. *Life*, 13(2), 460.

Admassu H, Gasmalla MAA, Yang R and Zhao W 2018. Evaluation of the *in vitro* α-amylase enzyme inhibition potential of commercial dried laver (*Porphyra* Species) seaweed protein hydrolysate. *Turkish Journal of Fisheries and Aquatic Sciences*, 18(4), 547–556.

Ahmed BE, Badawi MH, Mostafa SS and Higazy AM 2018. Human anticancers and antidiabetic activities of the cyanobacterium*Fischerella* sp. BS1-EG isolated from River Nile, Egypt. *International Journal of Current Microbiology and Applied Sciences*, 7(1), 3473–3485.

Al-Azzawie HF and Alhamdani MSS 2006.Hypoglycemic and antioxidant effect of oleuropein in alloxan induced diabetic rabbits. *Life Sciences*, 78(12), 1371–1377.

Alfadda AA and Sallam RM 2012.Reactive oxygen species in health and disease. *BioMed Research International*, 2012(1), 936486.

Alkhalaf MI 2021. Chemical composition, antioxidant, anti-inflammatory and cytotoxic effects of *Chondruscrispus* species of red algae collected from the Red Sea along the shores of Jeddah city. *Journal of King Saud University-Science*, 33(1), 101210.

American Diabetes Association 2021.Classification and diagnosis of diabetes: standards of medical care in diabetes. *Diabetes care*, 44(1), pp. S15-S33.

Anderson A, Laohavisit A, Blaby IK, Bombelli P, Howe CJ, Merchant SS, Davies JM and Smith AG 2016.Exploiting algal NADPH oxidase for biophotovoltaic energy. *Plant Biotechnology Journal*, 14(1), 22–28.

Arguelles ED and Sapin AB 2020.*In vitro* antioxidant, alpha-glucosidase inhibition and antibacterial properties of *Turbinariadecurrens*Bory (Sargassaceae, Ochrophyta). *Asia-Pacific Journal of Science and Technology*, 25(3).

Bartz RR and Piantadosi CA 2010. Clinical review: oxygen as a signaling molecule. *Critical Care*, 14, 1–9.

Brillante S, Galasso C, Lauritano C and Carrella S 2022. From the Sea for the sight: Marine derived products for human vision. *Frontiers in Aging Neuroscience*, 14, 892764.

Cagalj M, Skroza D, Tabanelli G, Ozogul F and Simat V 2021. Maximizing the antioxidant capacity of *Padinapavonica* by choosing the right drying and extraction methods. *Processes*, 9(4), 587.

Catarino MD, Marcal C, Bonifacio-Lopes T, Campos D, Mateus N, Silva AM, Pintado MM and Cardoso SM 2021. Impact of phlorotannin extracts from *Fucusvesiculosus* on human gut microbiota. *Marine Drugs*, 19(7), 375.

Catarino MD, Silva AM, Mateus N and Cardoso SM, 2019. Optimization of phlorotanninsextraction from *Fucusvesiculosus* and evaluation of their potential to prevent metabolic disorders. *Marine Drugs*, 17(3), 162.

Chaput DL, Fowler AJ, Seo O, Duhn K, Hansel CM and Santelli CM 2019. Mn oxide formation by phototrophs: Spatial and temporal patterns, with evidence of an enzymatic superoxide-mediated pathway. *Scientific Reports*, 9, 18244

Chen H, Jian Q, Luo Q, Zhu Z, Yang R and Yan X 2016. Application of oligoagars as elicitors for field aquaculture of *Pyropiahaitanensis*. *Journal of Applied Phycology*, 28, 1783–1791

Cho CH, Yoo G, Kim M, Kurniawati UD, Choi IW and Lee SH 2023.Dieckol, derived from the edible brown algae *Eckloniacava*, attenuates methylglyoxal-associated diabetic nephropathy by suppressing AGE–RAGE interaction. *Antioxidants*, 12(3), 593.

Choi J, Kim KJ, Koh EJ and Lee BY 2018.*Gelidiumelegans* extract ameliorates type 2 diabetes via regulation of MAPK and PI3K/Aktsignaling. *Nutrients*, 10(1), 51, 1–14.

Choo KS, Snoeijs P and Pedersen M 2004. Oxidative stress tolerance in the filamentous green algae *Cladophoraglomerata*, and *Enteromorphaahlneriana*. *Journal of Experimental Marine Biology and Ecology*, 298, 111–123.

Collen J and Davison IR 1999. Stress tolerance and reactive oxygen metabolism in the intertidal red seaweeds *Mastocarpusstellatus*, and *Chondruscrispus*. *Plant, Cell and Environment*, 22,1143–1151.

Collen J and Pedersen M 1996. Production, scavenging and toxicity of hydrogen peroxide in the green seaweed *Ulvarigida*. *European Journal of Phycology*, 31, 265–271.

Collen J, Pedersen M and Bornman C 1994. A stress-induced oxidative burst in *Eucheumaplatycladum* (Rhodophyta). *Plant Physiology*, 92, 417–422.

Cornish ML and Garbary DJ 2010. Antioxidants from macroalgae: potential applications in human health and nutrition. *Algae*, 25(4), 155–171.

Coulombier N, Jauffrais T and Lebouvier N 2021. Antioxidant compounds from microalgae: A review. *Marine Drugs*, 19(10), 549.

Davies KJ 2000. Oxidative stress, antioxidant defenses, and damage removal, repair, and replacement systems. *IUBMB Life*, 50(4–5), 279–289.

Deepa PK, Subramanian A and Manjusha WA 2020. Phytochemical screening and evaluation o fantidiabetic activity of the marine microalgae: *Nannochloropsis* sp., *International Journal of Life Science and Pharma Research*, 10(4), L36–40.

Di Pietro N, Di Tomo P and Pandolfi A 2016. Carotenoids in cardiovascular disease prevention. *JSM Atheroscler*, 1(1), 1002.

Diaz JM and Plummer S 2018. Production of extracellular reactive oxygen species by phytoplankton: Past and future directions. *Journal of Plankton Research*, 40, 655–666.

Diaz JM, Plummer S, Hansel CM, Andeer PF, Saito MA and McIlvin MR 2019. NADPH-dependent extracellular superoxide production is vital to photophysiology in the marine diatom *Thalassiosiraoceanica*. *Proceedings of the National Academy of Sciences*, 116, 16448–16453.

Diaz JM, Plummer S, Tomas C and Alves-de-Souza C 2018. Production of extracellular superoxide and hydrogen peroxide by five marine species of harmful bloom-forming algae. *Journal of Plankton Research*, 40, 667–677.

El Nur EE, Ali LI, Fadul E and Mohamed E 2021. Antioxidant, antibacterial and cytotoxic potential of selected macroalgae from the Red Sea, Sudan Coast. *International Research Journal of Biological Sciences*, 10, 1–11.

Elangovan M, Noorjahan A and Anantharaman P 2019. Extraction of metabolites and screening their antioxidant potential from marine macro algae. *International Journal of Scientific and Technology Research*, 8, 1059–1064.

El-Sheekh MM, El-Shenody RAEK, Bases EA and El Shafay SM 2020. Comparative assessment of antioxidant activity and biochemical composition of four seaweeds, Rocky Bay of Abu Qir in Alexandria, Egypt. *Food Science and Technology*, 41(1), 29–40.

Fathalipour M, Fathalipour H, Safa O, Nowrouzi-Sohrabi P, Mirkhani H and Hassanipour S 2020. The therapeutic role of carotenoids in diabetic retinopathy: A systematic review. *Diabetes, Metabolic Syndrome and Obesity*, 2347–2358.

Gheda SF, Abo-Shady AM, Abdel-Karim OH and Ismail GA 2021. Antioxidant and antihyperglycemic activity of *Arthrospiraplatensis* (*Spirulinaplatensis*) methanolic extract: *in vitro* and *in vivo* study. *Egyptian Journal of Botany*, 61(1), 71–93.

Gill SS and Tuteja N 2010. Reactive oxygen species and antioxidant machinery in abiotic stress tolerance in crop plants. *Plant Physiology and Biochemistry*, 48(12), 909–930.

Gouda MKG, Kavitha MD and Sarada R 2015. Antihyperglycemic, antioxidant and antimicrobial activities of the butanol extract from *Spirulinaplatensis*. *Journal of Food Biochemistry*, 39(5), 594–602.

Gunathilaka T, Keertihirathna LR and Peiris D 2022. Advanced pharmacological uses of marine algae as an anti-diabetic therapy. *Natural Medicinal Plants*, *11*, 79.

Guo D, Yu K, Sun XY and Ouyang JM 2018. Structural characterization and repair mechanism of *Gracilarialemaneiformis*sulfated polysaccharides of different molecular weights on damaged renal epithelial cells. *Oxidative Medicine and Cellular Longevity*, 2018(1), 7410389.

Halliwell B and Gutteridge JM 2015. *Free Radicals in Biology and Medicine*. Oxford University Press, USA.

Hannan JMA, Ansari P, Azam S, Flatt PR and Wahab YHA 2020. Effects of *Spirulinaplatensis* on insulin secretion, dipeptidyl peptidase IV activity and both carbohydrate digestion and absorption indicate potential as an adjunctive therapy for diabetes. *British Journal of Nutrition*, 124(10), 1021–1034.

Hansel CM, and Diaz JM 2021. Production of extracellular reactive oxygen species by marine biota. *Annual Review of Marine Science*, 13(1), 177–200.

Hatami E, Ghalishourani SS, Najafgholizadeh A, Pourmasoumi M, Hadi A, Clark CC, Assaroudi M, Salehi-Sahlabadi A, Joukar F and Mansour-Ghanaei F 2021. The effect of spirulina on type 2 diabetes: A systematic review and meta-analysis. *Journal of Diabetes & Metabolic Disorders*, 20, 883–892.

Herve, C, Tonon T, Collen, J, Corre E and Boyen C 2006. NADPH oxidases in eukaryotes: Red algae provide new hints!. *Current genetics*, 49(3), 190–204.

Hmani I, Ktari L, Ismail A, M'dallel C and El Bour M 2021. Assessment of the antioxidant and antibacterial properties of red algae (Rhodophyta) from the north coast of Tunisia. *Euro-Mediterranean Journal for Environmental Integration*, 6, 1–9.

Hou Y, Wang J, Simerly T, Jin W, Zhang H and Zhang Q 2015. Hydrogen peroxide released from *Pyropiayezoensis* induced by oligo-porphyrans: Mechanisms and effect. *Journal of Applied Phycology*, 27, 1639–1649.

Ibrahim RYM, Saber AA and Hammad HBI 2021. The possible role of the seaweed *Ulvafasciata* on ameliorating hyperthyroidism-associated heart inflammations in a rat model. *Environmental Science and Pollution Research*, 28, 6830–6842.

International Diabetes Federation (IDF) IDF Diabetes Atlas. 2022. [(accessed on 21 November 2024)]. Available online: https://www.diabetesatlas.org/

Ismail GA, Gheda SF, Abo-Shady AM and Abdel-Karim OH 2019.*In vitro* potential activity of some seaweeds as antioxidants and inhibitors of diabetic enzymes. *Food Science and Technology*, 40, 681–691.

Jaffer M, Ashraf H and Shaheen S 2021. Phytochemical, antioxidant and antimicrobial activity of biological important algae *Hydrodictyonreticulatum* L.: Algae act as a potential source as phytochemical, antioxidant and antimicrobial agent. *Biological Sciences-PJSIR*, 64(3), 244–250.

Jahan S, Yusoff IB, Alias YB and Bakar AFBA 2017. Reviews of the toxicity behavior of five potential engineered nanomaterials (ENMs) into the aquatic ecosystem. *Toxicology Reports*, 4, 211–220.

Jayshree A, Jayashree S and Thangaraju N 2016.*Chlorellavulgaris* and *Chlamydomonasreinhardtii*: Effective antioxidant, antibacterial and anticancer mediators. *Indian Journal of Pharmaceutical Sciences*, 78(5), 575–581.

Jerez-Martel I, Garcia-Poza S, Rodriguez-Martel G, Rico M, Afonso-Olivares C and Gomez-Pinchetti JL 2017. Phenolic profile and antioxidant activity of crude extracts from microalgae and cyanobacteria strains. *Journal of Food Quality*, 2017(1), 2924508.

Jo SL, Yang H, Jeong KJ, Lee HW and Hong EJ 2023.Neuroprotective effects of *Eckloniacava* in a chronic neuroinflammatory disease model. *Nutrients*, 15(8), 2007.

Kashtoh H and Baek KH 2022. Recent updates on phytoconstituent alpha-glucosidase inhibitors: An approach towards the treatment of type two diabetes. *Plants*, 11(20), 2722.

Kawee-Ai A, Kim AT and Kim SM 2019. Inhibitory activities of microalgal fucoxanthin against α-amylase, α-glucosidase, and glucose oxidase in 3T3-L1 cells linked to type 2 diabetes. *Journal of Oceanology and Limnology*, 37, 928–937.

Kim D, Nakamura A, Okamoto T, Komatsu N, Oda T, Iida T, Ishimatsu A and Muramatsu T 2000. Mechanism of superoxide anion generation in the toxic red tide phytoplankton *Chattonellamarina*: Possible involvement of NAD (P) H oxidase. *BiochimicaetBiophysicaActa (BBA)-General Subjects*, 1524(2–3), 220–227.

Korshunov SS and Imlay JA 2002. A potential role for periplasmic superoxide dismutase in blocking the penetration of external superoxide into the cytosol of Gram-negative bacteria. *Molecular Microbiology*, 43(1), 95–106.

Kumari A, Garima and Bharadvaja N 2023. A comprehensive review on algal nutraceuticals as prospective therapeutic agent for different diseases. *3 Biotech*, 13(2), 44.

Kupper FC, Gaquerel E, Cosse A, Adas F, Peters AF, Müller DG, Kloareg B, Salaun JP and Potin P 2009. Free fatty acids and methyl jasmonate trigger defense reactions in *Laminariadigitata*. *Plant and Cell Physiology*, 50(4), 789–800.

Kupper FC, Kloareg B, Guern J and Potin P 2001.Oligoguluronates elicit an oxidative burst in the brown algal kelp *Laminariadigitata*. *Plant physiology*, 125(1), 278–291.

Kupper FC, Muller DG, Peters AF, Kloareg B and Potin P 2002.Oligoalginate recognition and oxidative burst play a key role in natural and induced resistance of sporophytes of Laminariales. *Journal of Chemical Ecology*, 28, 2057–2081.

Lauritano C and Ianora A 2016.Marine organisms with anti-diabetes properties. *Marine Drugs*, 14(12), 220.
Luo Q, Zhu Z, Yang R, Qian F, Yan X and Chen H 2015. Characterization of a respiratory burst oxidase homologue from *Pyropiahaitanensis* with unique molecular phylogeny and rapid stress response. *Journal of Applied Phycology*, 27, 945–955.
Luo Q, Zhu Z, Zhu Z, Yang R, Qian F, Chen H and Yan X 2014. Different responses to heat shock stress revealed heteromorphic adaptation strategy of *Pyropiahaitanensis* (Bangiales, Rhodophyta). *PloS ONE*, 9(4), e94354.
Mahendran S, Maheswari P, Sasikala V, Jaya Rubika J and Pandiarajan J 2021.*In vitro*antioxidant study of polyphenol from red seaweeds dichotomously branched gracilaria*Gracilariaedulis* and robust sea moss *Hypneavalentiae*. *Toxicology Reports*, 8, 1404–1411.
Maiese K 2015.mTOR: Driving apoptosis and autophagy for neurocardiac complications of diabetes mellitus. *World Journal of Diabetes*, 6(2), 217.
Mallick N and Mohn FH 2000. Reactive oxygen species: Response of algal cells. *Journal of Plant Physiology*, 157(2), 183–193.
Martins B, Vieira M, Delerue-Matos C, Grosso C, and Soares C 2022. Biological potential, gastrointestinal digestion, absorption, and bioavailability of algae-derived compounds with neuroprotective activity: A comprehensive review. *Marine Drugs*, 20(6), 362.
Mate MJ, Zamocky M, Nykyri LM, Herzog C, Alzari PM, Betzel C, Koller F and Fita I 1999. Structure of catalase-A from *Saccharomycescerevisiae*. *Journal of Molecular Biology*, 286(1), 135–149.
McDowell RE, Amsler MO, Li Q, Lancaster Jr JR and Amsler CD 2015. The immediate wound-induced oxidative burst of *Saccharinalatissima* depends on light via photosynthetic electron transport. *Journal of Phycology*, 51(3), 431–441.
Moaveni S, Salami M, Khodadadi M, McDougall M and Emam-Djomeh Z 2022.Investigation of *S. limacinum*microalgae digestibility and production of antioxidant bioactive peptides. *Lebensmittel-Wissenschaft&Technologie,* 154, 112468.
Mohapatra L, Bhattamisra SK, Panigrahy RC and Parida SK 2016. Evaluation of the antioxidant, hypoglycaemic and anti-diabetic activities of some seaweed collected from the East Coast of India. *Biomedical and Pharmacology Journal*, 9(1), 365–375.
Morano KA, Grant CM and Moye-Rowley WS 2012. The response to heat shock and oxidative stress in *Saccharomycescerevisiae*. *Genetics*, 190(4), 1157–1195.
Muller R, Desel C, Steinhoff FS, Wiencke C and Bischof K 2012. UV-radiation and elevated temperatures induce formation of reactive oxygen species in gametophytes of cold-temperate/ Arctic kelps (Laminariales, Phaeophyceae). *Phycological Research*, 60(1), 27–36.
Murugesan S, Vinoth Kumar R, Kotteswari M and Shanthi N 2021.*In vitro* antioxidant activity of marine red alga *Gymnogongruspygmaeus* J. Agardh. *International Journal of Pharmaceutical Research*, 13, 892–899.
Nguyen TH, Nguyen TLP, Tran TVA, Do AD and Kim SM 2019. Antidiabetic and antioxidant activities of red seaweed *Laurenciadendroidea*. *Asian Pacific Journal of Tropical Biomedicine*, 9(12), 501–509.
Oesterhelt C, Vogelbein S, Shrestha RP, Stanke M and Weber APM 2008. The genome of the thermoacidophilic red microalga *Galdieriasulphuraria* encodes a small family of secreted class III peroxidases that might be involved in cell wall modification. *Planta*, 227, 353–362.
Ogasawara K, Yamada K, Hatsugai N, Imada C and Nishimura M 2016. Hexose oxidase-mediated hydrogen peroxide as a mechanism for the antibacterial activity in the red seaweed *Ptilophorasubcostata*. *PloS ONE*, 11, e0149084
Osman NA, Siam AA, El-Manawy I and Jeon YJ 2019. Anti-microbial and anti-diabetic activity of six seaweeds collected from the Red Sea, Egypt. *Catrina: The International Journal of Environmental Sciences*, 19(1), 55–60.

Paiva L, Lima E, Neto AI and Baptista J 2017. Angiotensin I-converting enzyme (ACE) inhibitory activity, antioxidant properties, phenolic content and amino acid profiles of *Fucusspiralis* L. protein hydrolysate fractions. *Marine Drugs*, 15(10), 311.

Palenik B, Kieber DJ and Morel FMM 1989. Dissolved organic nitrogen use by phytoplankton: The role of cell-surface enzymes. *Biological Oceanography*, 6(3–4), 347–354.

Pandithurai M, Murugesan S, Bhuvaneswari S and Thennarasan S 2015.*In vitro* α-amylase and α-glucosidase inhibition activity of methanolic extract of marine brown alga *Spatoglossumasperum*. *International Journal of Advanced Pharmaceutical Sciences*, 4(5), 83–87.

Park SY, Choi ES, Hwang J, Kim D, Ryu TK and Lee TK 2009. Physiological and biochemical responses of *Prorocentrumminimum* to high light stress. *Ocean Science Journal*, 44, 199–204.

Pereira L and Valado A 2021. The seaweed diet in prevention and treatment of the neurodegenerative diseases. *Marine Drugs*, 19(3), 128.

Pereira L and Valado A 2023. Algae-derived natural products in diabetes and its complications - Current advances and future prospects. *Life*, 13(9), 1831.

Pikula KS, ZakharenkoAM, Aruoja V, Golokhvast KS and Tsatsakis AM 2019. Oxidative stress and its biomarkers in microalgal ecotoxicology. *Current Opinion in Toxicology*, 13, 8–15.

Pirian K, Moein S, Sohrabipour J, Rabiei R and Blomster J 2017. Antidiabetic and antioxidant activities of brown and red macroalgae from the Persian Gulf. *Journal of Applied Phycology*, 29, 3151–3159.

Plummer S, Taylor AE, Harvey EL, Hansel CM and Diaz JM 2019. Dynamic regulation of extracellular superoxide production by the coccolithophore*Emilianiahuxleyi* (CCMP 374). *Frontiers in Microbiology*, 10, 1546.

Poljsak B and Fink R 2014. The protective role of antioxidants in the defence against ROS/RNS-mediated environmental pollution. *Oxidative Medicine and Cellular Longevity*, 2014(1), 671539.

Priatni S, Budiwati TA, Ratnaningrum D, Kosasih W, Andryani R, Susanti H and Susilaningsih D 2016.Antidiabetic screening of some Indonesian marine cyanobacteria collection. *Biodiversitas Journal of Biological Diversity*, 17(2), 642–646.

Priatni S, Ratnaningrum D and Kosasih W 2021. The screening of antidiabetic activity and the cultivation study of local marine microalgae. In: *IOP Conference Series: Materials Science and Engineering*, 1011(1), 012066.

Qian WW, Yang SQ, Hu SM, Wang XL, Zhu Y and Zhou T 2021. Enzymatic degradation, antioxidant and immunoregulatory activities of polysaccharides from brown algae *Sargassumfusiforme*. *Journal of Food Measurement and Characterization*, 15, 1960–1972.

Reczek CR and Chandel NS 2015. ROS-dependent signal transduction. *Current Opinion in Cell Biology*, 33, 8–13.

Reka P, Banu T and Seethalakshmi M 2017. Alpha amylase and alpha glucosidase inhibition activity of selected edible seaweeds from south coast area of India. *International Journal of Pharmacy and Pharmaceutical Sciences*, 9(6), 64–68.

Rezayian M, Niknam V and Ebrahimzadeh H 2019. Oxidative damage and antioxidative system in algae. *Toxicology Reports*, 6, 1309–1313.

Romero-Puertas MC, Corpas FJ, Sandalio LM, Leterrier M, Rodriguez-Serrano M, Del Rio LA and Palma JM 2006. Glutathione reductase from pea leaves: Response to abiotic stress and characterization of the peroxisomal isozyme. *New Phytologist*, 170(1), 43–52.

Ross C, Kupper FC, Vreeland V, Waite JH and Jacobs RS 2005. Evidence of a latent oxidative burst in relation to wound repair in the giant unicellular chlorophyte*Dasycladusvermicularis*. *Journal of Phycology*, 41, 531–541.

Sanger G, Rarung LK, Damongilala LJ, Kaseger BE and Montolalu LADY 2019. Phytochemical constituents and antidiabetic activity of edible marine red seaweed (*Halymeniadurvilae*). In: *IOP Conference Series: Earth and Environmental Science*, 278(1), 012069.

Santhar DT, Haq MB, Marudhupandi T, Vaseeharan B, Rajan DK and Moovendhan M 2021. Evaluation of chemical compositions and antioxidant potential of marine microalgae of the genus *Nannochloropsis*. *Biomass Conversion and Biorefinery*, 1–7.

Satpati GG, Mal N and Pal R 2021. Seaweed-based interventions for diabetic complications: An analytical discourse. *Systematic Bioscience and Engineering*, 1(1),53–66.

Scaglioni PT, Quadros L, de Paula M, Furlong VB, Abreu PC and Badiale-Furlong E 2018. Inhibition of enzymatic and oxidative processes by phenolic extracts from *Spirulina* sp. and *Nannochloropsis* sp. *Food Technology and Biotechnology*, 56(3), 344.

Schneider RJ, Roe KL, Hansel CM and Voelker BM 2016. Species-level variability in extracellular production rates of reactive oxygen species by diatoms. *Frontiers in Chemistry*, 4(5), 1–13.

Shafay SE, El-Sheekh M, Bases E and El-Shenody R 2021. Antioxidant, antidiabetic, anti-inflammatory and anticancer potential of some seaweed extracts. *Food Science and Technology*, 42, e20521.

Shibasaki S and Ueda M 2023. Utilization of macroalgae for the production of bioactive compounds and bioprocesses using microbial biotechnology. *Microorganisms*, 11(6), 1499, 1–19.

Shikata T, Takahashi F, Nishide H, Shigenobu S, Kamei Y, Sakamoto S, Yuasa K, Nishiyama Y, Yamasaki Y and Uchiyama I 2019. RNA-seq analysis reveals genes related to photoreception, nutrient uptake, and toxicity in a noxious red-tide raphidophyte*Chattonellaantiqua*. *Frontiers in Microbiology*, 10, 1764.

Styskal J, Van Remmen H, Richardson A and Salmon AB 2012. Oxidative stress and diabetes: what can we learn about insulin resistance from antioxidant mutant mouse models?. *Free Radical Biology and Medicine*, 52(1), 46–58.

Sutherland KM, Coe A, Gast RJ, Plummer S, Suffridge CP, Diaz JM, Bowman JS, Wankel SD and Hansel CM 2019. Extracellular superoxide production by key microbes in the global ocean. *Limnology and Oceanography*, 64(6), 2679–2693.

Takenaka Y, Miki M, Yasuda H and Mino M 1991. The effect of α-tocopherol as an antioxidant on the oxidation of membrane protein thiols induced by free radicals generated in different sites. *Archives of Biochemistry and Biophysics*, 285(2), 344–350.

Tiwari AK 2001. Imbalance in antioxidant defence and human diseases: Multiple approach of natural antioxidants therapy. *Current Science*, 1179–1187.

Unnikrishnan PS, Suthindhiran K and Jayasri MA 2015a. Alpha-amylase inhibition and antioxidant activity of marine green algae and its possible role in diabetes management. *Pharmacognosy Magazine*, 11(4), S511.

Unnikrishnan PS, Suthindhiran K and Jayasri MA 2015b. Antidiabetic potential of marine algae by inhibiting key metabolic enzymes. *Frontiers in Life Science*, 8(2), 148–159.

van Hees DH and Van Alstyne KL 2013. Effects of emersion, temperature, dopamine, and hypoxia on the accumulation of extracellular oxidants surrounding the bloom-forming seaweeds *Ulvalactuca* and *Ulvariaobscura*. *Journal of Experimental Marine Biology and Ecology*, 448, 207–213.

Vijayan R, Chitra L, Penislusshiyan S and Palvannan T 2018. Exploring bioactive fraction of *Sargassumwightii*: *In vitro* elucidation of angiotensin-I-converting enzyme inhibition and antioxidant potential. *International Journal of Food Properties*, 21(1), 674–684.

Wang F, Lv Y, Lin L, Xu N, Lu K and Sun X 2018. Characterization of a respiratory burst oxidase homolog from *Gracilariopsislemaneiformis* (Rhodophyta) during stress and phytohormone treatments. *Botanica Marina*, 61(5), 511–519.

Weinberger F, Guillemin ML, Destombe C, Valero M, Faugeron S, Correa JA, Pohnert G, Pehlke C, Kloareg B and Potin P 2010.Defense evolution in the gracilariaceae (rhodophyta): Substrate-regulated oxidation of agar oligosaccharides is more ancient than the oligoagar-activated oxidative burst. *Journal of Phycology*, 46(5), 958–968.

Weinberger F, Pohnert G, Kloareg B and Potin P 2002. A signal released by an endophytic attacker acts as a substrate for a rapid defensive reaction of the red alga *Chondruscrispus*. *ChemBioChem*, 3(12), 1260–1263.

Wheeler GL, Jones MA and Smirnoff N 1998. The biosynthetic pathway of vitamin C in higher plants. *Nature*, 393(6683), 365–369.

Xin YJ, Yuan B, Yu B, Wang YQ, Wu JJ, Zhou WH and Qiu Z 2015. Tet1-mediated DNA demethylation regulates neuronal cell death induced by oxidative stress. *Scientific Reports*, 5(1), p.7645.

Zhang X, Krause KH, Xenarios I, Soldati T and Boeckmann B 2013. Evolution of the ferric reductase domain (FRD) superfamily: modularity, functional diversification, and signature motifs. *PloSONE*, 8, e58126

Zhao C, Yang C, Liu B, Lin L, Sarker SD, Nahar L, Yu H, Cao H and Xiao J 2018. Bioactive compounds from marine macroalgae and their hypoglycemic benefits. *Trends in Food Science and Technology*, 72, 1–12.

Chapter 20
Future Prospects and Potential Challenges of Understanding Antidiabetic Potential of Algae

Kulvinder Bajwa and Tamil Selvan Silambarasan

Abstract Diabetes requires innovative therapeutic strategies. Natural products, especially algae-derived compounds, show promise. This chapter examines algae-based natural products for diabetes management. Recent developments, mechanisms, and future prospects are discussed. Algae's diverse bioactive constituents and antidiabetic properties are highlighted, including polysaccharides, polyphenols, and fatty acids known for their potential to improve insulin sensitivity and regulate glucose metabolism. Additionally, the chapter explores how specific algae species, such as *Spirulina* and *Chlorella*, exhibit unique mechanisms of action that could complement existing treatments.

Challenges and limitations are addressed, such as variability in bioactive compound concentrations, the need for standardized extraction methods, and regulatory hurdles for clinical applications. Recommendations for optimization include advancing biotechnological methods for compound extraction and enhancing bioavailability through novel delivery systems. By fostering collaboration between researchers, clinicians, and industry, we can accelerate the development of effective algae-based therapies and pave the way for new, sustainable approaches to diabetes management.

Keywords Diabetes · Natural products · Algae · Prospects · Challenges

K. Bajwa (✉)
University School for Graduates Studies, Chaudhary Devi Lal University, Sirsa, India

T. S. Silambarasan
Helix Research Center, Department of General Medicine, Saveetha Medical College and Hospital, Saveetha Institute of Medical and Technical Sciences, Saveetha University, Chennai, Tamil Nadu, India

R. Raja et al. (eds.), *Algae in Diabetes Management*,
https://doi.org/10.1007/978-981-95-0571-5_20

1 Introduction

Diabetes mellitus, characterized by elevated blood sugar and insufficient/ineffective insulin production or utilization, is exacerbated by oxidative stress and chronic inflammation is a metabolic disorder with escalating global prevalence, threatening to reach pandemic levels soon. The global diabetes epidemic affects approximately 500 million people, with healthcare costs expected to skyrocket to $845 billion by 2045 (Kaushik et al. 2023; Ciecierska et al. 2019). Unfortunately, an effective cure for diabetes and its associated complications remains an unmet medical need. Microalgae have shown promise in diabetes treatment due to their rich nutrient profile and bioactive. To avoid the side effects of conventional medications, natural remedies like microalgae biomass, extracts, and bioactive compounds are gaining attention for their potential therapeutic benefits. Research has shown that microalgae-derived nutraceuticals and supplements exhibit anti-inflammatory, antioxidant, and pancreas-protective properties, enhancing insulin sensitivity and offering a promising complementary approach to managing diabetes (Pereira and Valado 2023). This chapter underscores the prospects of microalgae-derived supplements in diabetes management, while also addressing current limitations and future research directions, emphasizing the need for rigorous clinical trials to translate their potential into practical applications.

2 Importance of Algal-Based Natural Products in Diabetes Treatment

A class of metabolic disorders known as diabetes mellitus is typified by insulin resistance and hyperglycemia. According to Saeedi (Saeedi et al. 2019), diabetes mellitus is one of the main chronic diseases that cause disability in humans and has been increasing in incidence and prevalence worldwide due to changes in diet and lifestyle. The most recent (9th edition) Global Diabetes Map from the International Diabetes Federation (IDF) indicates that there are currently over 463 million verified cases of diabetes worldwide, with an average growth rate of 51%. According to estimates, there will be 700.2 million diabetics worldwide by 2045 (Lin et al. 2021; Yuen et al. 2019). According to Saeedi et al. (2019)), the expected number of deaths attributable to diabetes in 2019 was 4.2 million. This translates to one person dying from diabetes mellitus every 8 s. Unfortunately, some of the medications now in use for the treatment of diabetes mellitus have negative side effects in addition to being ineffective and short-lived. In recent years, there has been growing interest in exploring natural products as potential therapeutic agents for managing diabetes. Chronic treatment with acarbose and all thiazolidinediones, therapeutic drugs for diabetes, has been proven to cause liver damage (Spiller and Sawyer 2006).

Sulfonylurea users may also have agitation, perspiration, fatigue, disorientation, tachycardia, and confusion (Lorenzati et al. 2010). According to Lorenzati et al. (2010)), patients using thiazolidinediones run the risk of fracture as well as macular oedema. Thus, research into novel hypoglycemic medications that are safer, more targeted, and more efficient is required. Recently, natural products and their derivatives have attracted considerable attention as potential therapies for diabetes and its complications. These compounds, derived from plants, fungi, animals, and microorganisms, offer a variety of bioactive components with diverse pharmacological application.

Using natural products in diabetes research provides access to a wide range of new compounds while also tapping into traditional medicinal knowledge passed down through generations (Wen et al. 2021; Lin et al. 2021). A key benefit of natural products is their rich chemical diversity. For example, algae and plants generate a remarkable variety of secondary metabolites, such as flavonoids, alkaloids, phenolic compounds, and terpenoids, which display numerous biological activities. This diverse array of chemical structures serves as a valuable resource for developing targeted therapies that can influence different molecular targets related to glucose regulation, insulin secretion, and insulin sensitivity (Pereira and Valado 2023). Moreover, natural products from algae could serve as valuable supplements to current diabetes treatments, addressing therapeutic needs. They may enhance the effectiveness of traditional antidiabetic medications or provide synergistic benefits. Additionally, these algae-derived compounds could present alternative treatment options for individuals who are intolerant to or resistant to existing drugs (Subermaniam et al. 2021). Algae are photosynthetic organisms that can be found in a wide variety of aquatic environments. They are remarkably adapting at synthesizing a wide range of distinct and biologically active chemicals. These natural compounds made from algae provide fresh perspectives on the etiology of diabetes and the creation of creative therapies for its control (Pereira and Valado 2023).

3 Bioactive Compounds Found in Algae with Potential Antidiabetic Effects

Algae are rich sources of bioactive compounds that have garnered attention for their potential antidiabetic properties. These compounds, including polysaccharides, lipids, peptides, and phenolic compounds, have shown promise in regulating blood sugar levels, improving insulin sensitivity, and reducing inflammation. Their diverse chemical profiles make algae a valuable resource in the search for natural alternatives to manage diabetes and its associated complications. Here are some key types of algae and their bioactive components that show promise in managing diabetes as shown in Fig. 20.1.

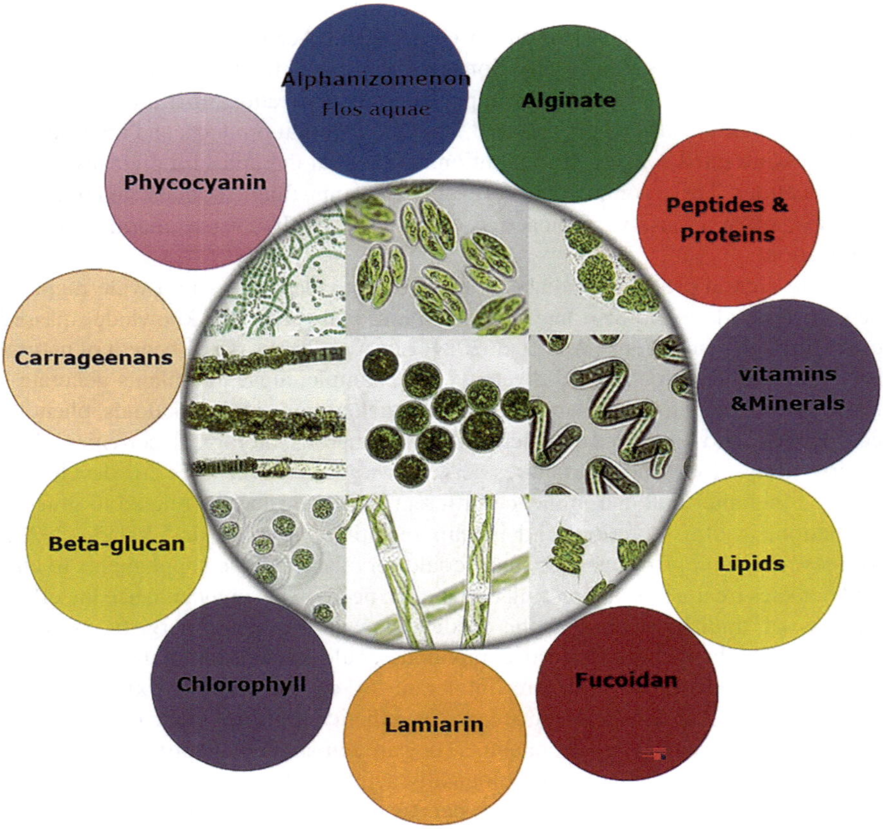

Fig. 20.1 Bioactive compounds in algae for diabetic management

3.1 Fucoidan

This polysaccharide has been studied for its ability to enhance insulin sensitivity and reduce blood glucose levels. Algae are recognized for producing various polysaccharides that may have antidiabetic benefits. For instance, sulfated polysaccharides from red algae, like agar and carrageenan, have shown anti-hyperglycemic effects by hindering the enzymes α-glucosidase and α-amylase, which play a role in carbohydrate digestion, leading to lower postprandial glucose levels. In addition, these polysaccharides possess antioxidant effects and can enhance insulin sensitivity (Lu and Chen 2022; Bhuyan et al. 2023).

3.2 *Alginate*

Alginate is known for its gel-forming properties, alginate can help in slowing down carbohydrate absorption and modulating blood glucose levels. By making specific chemical and biological modifications to change the structure and properties of alginates, there is potential for innovative applications in the food industry, especially since these modifications can yield high bioactivities at low concentrations. As a result, collaboration among food manufacturers, alginate producers, and food and nutrition scientists will be essential to explore and develop new uses of alginate in food and beverages (Qin et al. 2018).

3.3 *Laminarin*

Marine algal species contain a significant number of polysaccharides that exhibit multifunctional properties and health benefits for the treatment and prevention of various human diseases. Laminarin, also known as β-glucan, is a storage polysaccharide found in brown algae that has been identified to possess potential pharmacological properties, including antioxidant, anti-tumor, anti-coagulant, anticancer, immunomodulatory, anti-obesity, antidiabetic, anti-inflammatory, wound healing, and neuroprotective effects (Karuppusamy et al. 2022).

3.4 *Chlorophyll*

Some studies suggest that chlorophyll may improve insulin sensitivity. Various pigments found in marine algae, including fucoxanthin, zeaxanthin, β-carotene, lutein, anthocyanins, phlorotannin, and phycobiliproteins, have demonstrated multiple health benefits, such as antioxidant properties and antidiabetic effects (Gateau et al. 2017; Imchen and Singh 2023).

3.5 *Beta-Glucan*

Beta-glucan is a type of dietary fiber that offers several health benefits, including the prevention and treatment of specific digestive disorders and support for the immune system (Bobade et al. 2022). Its biological activity is influenced by factors such as size, molecular weight, structure, bond frequency, solubility, and structural changes. Additionally, beta-glucan helps lower cholesterol and glucose levels in the blood, thereby reducing the risk of cardiovascular disease and diabetes (Singla et al. 2024).

3.6 Carrageenan

Algae-based hydrocolloids like carrageenan, agar, and alginate are commonly utilized in the food industry for their thickening, gelling, and emulsifying properties. Carrageenan are sulfated polysaccharides that have various structural characteristics. Recently, the safety of using carrageenan in food has come under scrutiny due to concerns about its potential pro-inflammatory effects and the possibility that gut microbiota may digest carrageenan, leading to the production of pro-inflammatory oligosaccharides (Liao et al. 2021; Pereira 2016; Tavares et al. 2023).

3.7 Phycocyanin

Certain microalgae species, like spirulina, are notable for their remarkable nutritional value. *Spirulina* exhibits a variety of pharmacological benefits, primarily due to phycocyanin. Phycocyanin is a protein compound recognized for its antidiabetic effects and is considered a nutraceutical (Zheng et al. 2013; Ziyaei et al. 2023).

3.8 Aphanizomenonflos-Aquae

Blue-green algae (BGA) are some of the earliest life forms on Earth and have been used by humans as food and medicine for centuries. BGA is rich in various bioactive compounds, including phycocyanin, carotenoids, γ-linolenic acid, fiber, and plant sterols, all of which can support human health (Kumar et al. 2019). Aphanizomenonflos-aquae (AFA) is a freshwater unicellular blue-green alga from the Cyanophyta group, known for being a nutrient-rich food source and for its health-boosting benefits. Blue-green algae (BGA), specifically Aphanizomenonflos-aquae (AFA), is one of the dietary plants recognized for its ability to help lower hyperlipidemia (Zahkouk et al. 2019).

3.9 Lipid

Lipids derived from algae, particularly omega-3 fatty acids such as docosahexaenoic acid (DHA) and eicosapentaenoic acid (EPA), have been thoroughly researched for their positive effects on diabetes management (Sinha et al. 2023). These fatty acids exhibit properties that lower blood sugar levels, reduce inflammation, and enhance insulin sensitivity. Moreover, lipids sourced from microalgae, including *Chlorella* and *Spirulina*, have demonstrated the ability to improve lipid profiles and decrease triglyceride levels in people with diabetes (Conde et al. 2021).

3.10 Peptides and Proteins

Algae are known to generate a diverse array of peptides and proteins that exhibit various biological functions, including properties beneficial for diabetes management. Certain peptides derived from algae have shown the ability to inhibit dipeptidyl peptidase-4 (DPP-4), an enzyme responsible for breaking down incretin hormones. By blocking DPP-4, these peptides can extend the effects of incretins, resulting in increased insulin secretion and better regulation of blood glucose levels (Echave et al. 2022).

3.11 Vitamins and Minerals

Microalgae are fascinating and valuable natural sources of important bioactive compounds, including vitamins, essential amino acids, polyunsaturated fatty acids, minerals, carotenoids, enzymes, and fibers. Their potential has made microalgae some of the most promising and innovative sources for new food and functional products. Additionally, microalgae can serve as functional ingredients to boost the nutritional value of foods, positively impacting health and overall life satisfaction, while also lowering the risk of diseases and health conditions (Matos et al. 2017; Sathasivam et al. 2019).

These examples represent only a small portion of the bioactive compounds in algae that demonstrate potential for managing diabetes (Zhao et al. 2018). The wide-ranging pharmacological effects of these compounds, such as their ability to regulate glucose metabolism, enhance insulin sensitivity, and reduce diabetes-related complications, make them promising candidates for further research and development. Continued exploration of algae-derived bioactive compounds could uncover new therapeutic options, leading to improved diabetes management and better patient outcomes (Bocanegra et al. 2021).

4 Action Mechanisms of Algae for Diabetes Management

Algae are increasingly attracting attention in the pharmaceutical and cosmetic sectors because they contain a wide variety of bioactive compounds that have potential uses in pharmacology and nutraceuticals. Additionally, due to lifestyle changes driven by rapid urbanization, diabetes mellitus has become the third leading cause of death worldwide. The search for effective natural antidiabetic treatments is essential for addressing diabetes and its complications, especially given the negative side effects of existing antidiabetic medications. Exploring the potential benefits of algae in managing diabetes is vital for developing natural therapeutics.

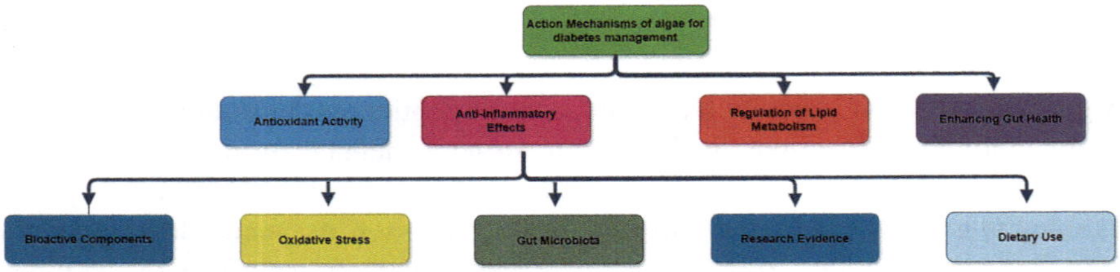

Fig. 20.2 The bioactive compounds in algae showing their antidiabetic effects through various mechanisms

The bioactive compounds in these algae may exert their antidiabetic effects through various mechanisms as shown in Fig. 20.2, including:

(A) ***Antioxidant Activity***
Biochemical processes in the body can generate intermediate products known as reactive oxygen species (ROS). Among these, harmful free radicals possess one or more unpaired electrons, making them highly reactive with other molecules. An excess of ROS can disrupt metabolic balance and hinder the body's ability to detoxify or mitigate the damaging effects of these free radicals (Abo-Shady et al. 2023). This state is referred to as oxidative stress, which arises from the imbalance between ROS production and the body's antioxidant defenses (Finkel and Holbrook 2000). Therefore, oxidative stress results from the disparity between the creation and neutralization of reactive oxygen and nitrogen species. Reducing oxidative stress, which is often elevated in diabetic conditions (Abo-Shady et al. 2023; Tiwari and Rao 2002).

(B) ***Anti-inflammatory Effects***
Algae have gained attention for their potential anti-inflammatory properties, which may be beneficial in managing diabetes. Here are some key points regarding the anti-inflammatory activity of algae in the context of diabetes:

(i) ***Bioactive Components***: Algae are rich in bioactive substances like polyphenols, carotenoids, and omega-3 fatty acids, which can help reduce chronic inflammation linked to diabetes (Nacer et al. 2020).
(ii) ***Oxidative Stress***: Algae are abundant in antioxidants, which can combat oxidative stress and lower inflammation levels, addressing one of the key issues in diabetes (Zang et al. 2023).
(iii) ***Gut Microbiota***: Metabolic syndrome is a cluster of conditions that heightens the risk of issues like abdominal obesity, diabetes, atherosclerosis, cardiovascular diseases, and cancer. The gut microbiota plays a significant role in the development of metabolic syndrome, and both its diversity and function are heavily influenced by diet (Kaushik et al. 2023). Recent epidemiological studies suggest that consuming seaweed may help prevent metabolic syndrome by modulating the gut microbiota. Certain algae, like spirulina, seaweeds can improve gut health, promoting a balanced microbiome that supports reduced inflammation and better metabolic outcomes (Zang et al. 2023).

(iv) ***Research Evidence***: Some studies indicate that algae supplements can improve blood glucose levels and inflammatory markers in diabetes patients, though further research is needed for conclusive evidence (Ahda et al. 2024; Kim et al. 2008).

(v) ***Dietary Use***: Adding algae (such as *spirulina, chlorella*, and *seaweeds*) to the diet is a straightforward way to gain their health benefits while providing essential nutrients (Ibañez and Cifuentes 2013; Wells et al. 2017).

(C) ***Regulation of Lipid Metabolism*:** Enhancing lipid profiles can contribute positively to overall metabolic health. By promoting healthy cholesterol levels and reducing triglycerides, effective lipid metabolism helps lower the risk of cardiovascular diseases and supports better insulin sensitivity. Additionally, maintaining a balanced lipid profile is crucial for energy regulation and can mitigate inflammation, further benefiting metabolic processes in the body (Wan et al. 2018; Wells et al. 2017).

(D) ***Enhancing Gut Health***: Prebiotic effects can improve the composition of gut microbiota, which is essential for various metabolic processes. A diverse and balanced microbiome aids in the fermentation of dietary fibers, producing short-chain fatty acids that support gut integrity and reduce inflammation. This, in turn, can enhance nutrient absorption and influence energy metabolism, potentially lowering the risk of metabolic disorders. Furthermore, a healthy gut microbiome is linked to improved immune function and may play a role in regulating appetite and weight management (Furbeyre et al. 2017; Patel et al. 2021; Wan et al. 2018).

5 Challenges and Future Directions

Although algae-derived natural products hold significant promise for diabetes management, several limitations and challenges must be tackled for their effective use (Fig. 20.3). Recognizing these issues and identifying appropriate solutions is essential for the future advancement and implementation of algae-based interventions. In this discussion, we will outline the existing limitations and suggest potential strategies to address these challenges.

The bioavailability and pharmacokinetic characteristics of compounds derived from algae play a crucial role in determining their therapeutic effectiveness. Certain bioactive compounds may suffer from issues like poor solubility, inadequate absorption, or swift metabolism, which can result in lower bioavailability and reduced efficacy. To improve these properties, strategies such as nanoencapsulation, formulation optimization, and prodrug design could be investigated.

1. ***Complex Composition***: Algae possess a wide range of bioactive compounds, including polysaccharides, lipids, and proteins, each of which can influence glucose metabolism in different ways. This complexity complicates the task of isolating specific compounds that exhibit antidiabetic properties. Moreover, the

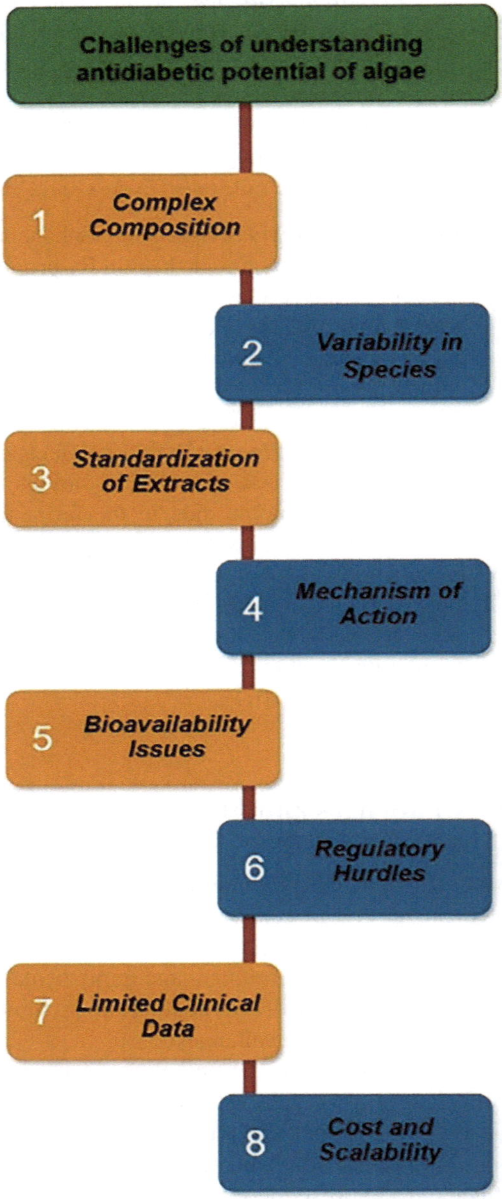

Fig. 20.3 Potential challenges of understanding antidiabetic treatment using algae

interactions between these various components can further obscure their individual effects, making it challenging to pinpoint which compounds contribute most significantly to their potential benefits in diabetes management. As a result, comprehensive profiling and characterization of algal extracts are essential for understanding their therapeutic potential (Pereira and Valado 2023).

2. ***Variability in Species:*** Different algae species can exhibit a wide range of effects on glucose metabolism, influenced by their unique chemical compositions. Additionally, environmental factors such as nutrient availability, light conditions, and water quality can significantly alter the biochemical profiles of these species. This variability makes it challenging to draw consistent conclusions across studies, as findings from one species or environmental context may not apply to another. Therefore, thorough investigation of specific algae under controlled conditions is necessary to better understand their antidiabetic potential and to ensure reliable outcomes in research and application (Rayapu et al. 2021; Unnikrishnan and Jayasri 2018; Wen et al. 2021).
3. ***Standardization of Extracts***: The absence of standardized methods for bio-extraction and analysis can result in significant inconsistencies across studies, complicating the comparison of results. Variations in bio-extraction techniques, solvent types, and processing conditions can lead to different yields and compositions of bioactive compounds, which may skew findings (Fabrowska et al. 2015). This lack of uniformity hinders the establishment of reliable benchmarks for evaluating the efficacy of algae-derived products. To address this issue, developing standardized protocols for extraction and analysis is essential, as it would enhance reproducibility and allow for more meaningful comparisons between studies, ultimately advancing the field (Wen et al. 2021; Thiyagarasaiyar et al. 2020).
4. ***Mechanism of Action***: Deciphering the specific biological mechanisms by which algae exert their antidiabetic effects is a complex challenge that often necessitates sophisticated research methodologies. The interactions between algal compounds and metabolic pathways can be intricate, involving various signaling cascades and cellular responses (Thiyagarasaiyar et al. 2020). Moreover, the multifaceted nature of these compounds means they may act through multiple mechanisms simultaneously, complicating the ability to pinpoint a singular mode of action. Advanced techniques such as molecular profiling, metabolomics, and in vivo studies are essential for gaining deeper insights into how these bioactive compounds influence glucose metabolism and insulin sensitivity, paving the way for targeted therapeutic applications (Pereira and Valado 2023).
5. ***Bioavailability Issues***: The bioavailability of active compounds derived from algae can be relatively low within the human digestive system, presenting a significant hurdle for their therapeutic use. Factors such as poor solubility, complex molecular structures, and interactions with other dietary components can limit absorption (Thiyagarasaiyar et al. 2020). To enhance understanding, comprehensive studies are needed to investigate how these compounds are absorbed, metabolized, and utilized in the body. Additionally, exploring innovative delivery systems, such as encapsulation or formulation modifications, could improve bioavailability and maximize the therapeutic potential of algal bioactive compounds. Addressing these bioavailability challenges is crucial for translating research findings into effective dietary interventions for diabetes management (Pereira and Valado 2023).

6. ***Regulatory Hurdles***: Any potential therapeutic applications of algae must comply with stringent regulatory frameworks, which can pose significant barriers to research and development. These processes often require extensive safety and efficacy data, demanding rigorous clinical trials and comprehensive documentation (Barros de Medeiros et al. 2022). The complexity of demonstrating the benefits of algae-derived products, alongside ensuring their safety for human consumption, can slow down progress and increase costs. As a result, fostering collaboration between researchers, regulatory bodies, and industry stakeholders is essential to streamline these processes and facilitate the translation of algal research into viable therapeutic options. Addressing these regulatory challenges will be key to unlocking the full potential of algae in healthcare applications (Fabris et al. 2020).
7. ***Limited Clinical Data***: A significant portion of research on algae's antidiabetic properties remains preliminary, often relying on in vitro or animal models rather than human clinical trials (Wen et al. 2021). While these studies can provide valuable insights into the mechanisms and potential benefits of algal compounds, they may not accurately reflect human physiology or responses (Udayan et al. 2021). The scarcity of robust clinical data raises questions about the efficacy and safety of these products for human use, hindering their acceptance in the medical community. To move forward, there is a pressing need for well-designed clinical trials that assess the effects of algae-derived interventions in diverse human populations. These studies should focus on various factors, including dosage, long-term effects, and potential interactions with other medications, to establish a comprehensive understanding of their therapeutic potential (Khan et al. 2018; Udayan et al. 2021).
8. ***Cost and Scalability***: Cultivating and processing algae for commercial purposes can be both resource-intensive and costly, presenting significant economic challenges to widespread application. Factors such as land and water use, energy requirements for cultivation, and the technology needed for harvesting and processing contribute to high production costs (Ahda et al. 2024). Additionally, scaling up production to meet commercial demand can lead to further financial constraints, particularly for smaller enterprises. To overcome these challenges, it is essential to explore innovative cultivation techniques, such as closed-loop systems or offshore farming, which can enhance efficiency and reduce costs. Investing in research and development to optimize processing methods and increase yields could also help make algae-based products more economically viable, paving the way for their broader adoption in the health and wellness sectors (Wen et al. 2021).
9. ***Public Perception and Acceptance***: Skepticism and a lack of awareness about the health benefits of algae can significantly impact consumer acceptance and marketability. Many people may have preconceived notions about algae being unappealing or associated primarily with environmental issues rather than health benefits (Ahmed et al. 2024). This perception can deter potential users from exploring algae-derived products, even if they offer valuable therapeutic properties. To enhance acceptance, it is crucial to engage in educational campaigns that

highlight the nutritional benefits, safety, and efficacy of algae. Collaborating with influencers, healthcare professionals, and nutritionists to promote these products can also help shift public attitudes. Furthermore, developing appealing product formulations and transparent labeling can improve consumer trust and interest, ultimately driving market demand for algae-based interventions (Mellor et al. 2022).

6 Future Directions

Future directions for diabetes treatment using algae hold exciting potential, particularly as research expands into their bioactive compounds and mechanisms of action (Fig. 20.4). Here are several key areas to explore:

1. **Bioactive Compound Identification**: Continued research is needed to isolate and characterize specific bioactive compounds in various algae species that demonstrate antidiabetic properties. This includes polysaccharides, peptides, and polyphenols that could enhance insulin sensitivity and glucose metabolism (Lourenço-Lopes et al. 2020; Popović-Djordjević et al. 2021; Zhao et al. 2018).
2. **Clinical Trials**: More rigorous clinical trials are essential to establish the efficacy and safety of algae-derived products for diabetes management. These studies should assess various forms of algae, including whole foods, extracts, and

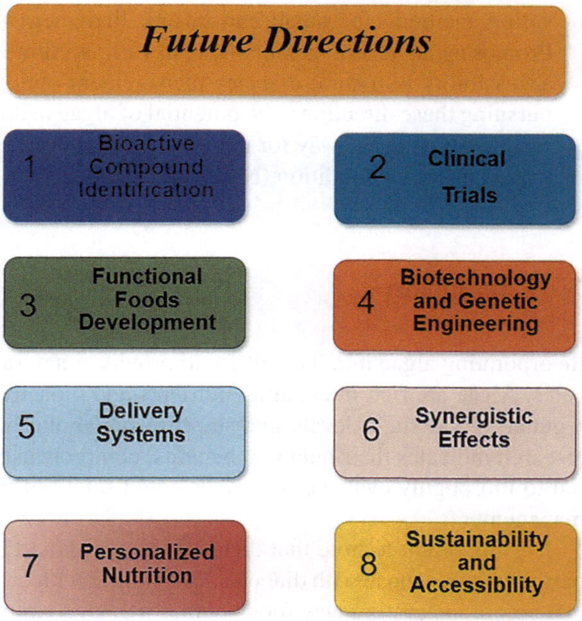

Fig. 20.4 Future directions for diabetes treatment using algae

supplements, to determine optimal dosages and formulations (Menaa et al. 2021; Zhao et al. 2018).

3. **Functional Foods Development**: Integrating algae into functional foods could offer a practical approach for consumers (Wells et al. 2017). Developing products that combine algae with other health-promoting ingredients may enhance their benefits and appeal to a broader audience (Admassu et al. 2015; Plaza et al. 2008).
4. **Biotechnology and Genetic Engineering**: Advances in biotechnology could enable the genetic modification of algae to boost their nutritional profiles or increase the concentration of beneficial compounds. This could lead to more effective treatments and sustainable production methods (Mohibbullah et al. 2022; Unnikrishnan and Jayasri 2018).
5. **Delivery Systems**: Exploring innovative delivery systems to enhance the bioavailability of algae-derived compounds can improve their therapeutic effects. Techniques such as nanoencapsulation or combining algae with other carriers may enhance absorption in the digestive tract (Thiyagarasaiyar et al. 2020).
6. **Synergistic Effects**: Investigating the synergistic effects of algae with other natural products or conventional diabetes medications could lead to more effective treatment strategies. This approach may help to reduce side effects and improve overall outcomes (Gheda et al. 2023; Mabate et al. 2021).
7. **Personalized Nutrition**: The integration of algae into personalized nutrition plans for diabetes management could optimize health benefits (Naik et al. 2024). Tailoring dietary recommendations based on individual metabolic responses may enhance the effectiveness of algae-based interventions (Ibañez and Cifuentes 2013; Jyoti et al. 2022).
8. **Sustainability and Accessibility**: Focusing on sustainable harvesting and cultivation methods for algae can ensure that these resources remain available. Promoting algae as a dietary staple in various cultures could also enhance global accessibility to natural diabetes management options (Kapoor et al. 2024). By pursuing these directions, the potential of algae in diabetes treatment can be fully realized, paving the way for innovative and holistic approaches to managing this prevalent health condition (Naik et al. 2024).

7 Conclusion

Incorporating algae into the diet could provide a natural approach to managing diabetes. Algae are rich in essential nutrients and bioactive compounds that may help regulate blood sugar levels and improve metabolic health. Although preliminary research indicates their potential benefits, comprehensive clinical studies are essential to thoroughly evaluate the efficacy and safety of these compounds in diabetes management.

It is important to note that dietary changes should be approached with caution, particularly for those with diabetes. Consulting with healthcare professionals is crucial before integrating new food sources like algae into one's diet, as they can offer personalized advice and monitor any potential interactions with existing treatments. As research progresses, algae may become a valuable component in holistic diabetes care, contributing to better health outcomes and improved quality of life.

References

Bhuyan, P. P., Nayak, R., Patra, S., Abdulabbas, H. S., Jena, M., & Pradhan, B. (2023). Seaweed-derived sulfated polysaccharides; the new age chemopreventives: a comprehensive review. *Cancers, 15*(3), 715.

Ciecierska, A., Drywien, M., Hamulka, J., & Sadkowski, T. (2019). Nutraceutical functions of beta-glucans in human nutrition. *RocznikiPaństwowegoZakładuHigieny, 70*(4).

Liao, Y.-C., Chang, C.-C., Nagarajan, D., Chen, C.-Y., & Chang, J.-S. (2021). Algae-derived hydrocolloids in foods: applications and health-related issues. *Bioengineered, 12*(1), 3787–3801.

Abo-Shady, A. M., Gheda, S. F., Ismail, G. A., Cotas, J., Pereira, L., & Abdel-Karim, O. H. (2023). Antioxidant and Antidiabetic Activity of Algae. *Life, 13*(2), 460. https://www.mdpi.com/2075-1729/13/2/460

Admassu, H., Zhao, W., Yang, R., Gasmalla, M., & Alsir, E. (2015). Development of functional foods: seaweeds (algae) untouched potential and alternative resource—a review. *Int J Sci Technol Res, 4*(1).

Ahda, M., Suhendra, & Permadi, A. (2024). Spirulina platensis microalgae as high protein-based products for diabetes treatment. *Food Reviews International, 40*(6), 1796–1804.

Ahmed, N., Sheikh, M. A., Ubaid, M., Chauhan, P., Kumar, K., & Choudhary, S. (2024). Comprehensive exploration of marine algae diversity, bioactive compounds, health benefits, regulatory issues, and food and drug applications. *Measurement: Food*, 100163.

Barros de Medeiros, V. P., da Costa, W. K. A., da Silva, R. T., Pimentel, T. C., & Magnani, M. (2022). Microalgae as source of functional ingredients in new-generation foods: Challenges, technological effects, biological activity, and regulatory issues. *Critical reviews in food science and nutrition, 62*(18), 4929–4950.

Bobade, H., Gupta, A., & Sharma, S. (2022). Beta-glucan. In *Nutraceuticals and health care* (pp. 343–358). Elsevier.

Bocanegra, A., Macho-González, A., Garcimartín, A., Benedí, J., & Sánchez-Muniz, F. J. (2021). Whole alga, algal extracts, and compounds as ingredients of functional foods: composition and action mechanism relationships in the prevention and treatment of type-2 diabetes mellitus. *International Journal of Molecular Sciences, 22*(8), 3816.

Conde, T. A., Zabetakis, I., Tsoupras, A., Medina, I., Costa, M., Silva, J., Neves, B., Domingues, P., & Domingues, M. R. (2021). Microalgal lipid extracts have potential to modulate the inflammatory response: A critical review. *International Journal of Molecular Sciences, 22*(18), 9825.

Echave, J., Otero, P., Garcia-Oliveira, P., Munekata, P. E., Pateiro, M., Lorenzo, J. M., Simal-Gandara, J., & Prieto, M. A. (2022). Seaweed-derived proteins and peptides: promising marine bioactives. *Antioxidants, 11*(1), 176.

Fabris, M., Abbriano, R. M., Pernice, M., Sutherland, D. L., Commault, A. S., Hall, C. C., Labeeuw, L., McCauley, J. I., Kuzhiuparambil, U., & Ray, P. (2020). Emerging technologies in algal biotechnology: toward the establishment of a sustainable, algae-based bioeconomy. *Frontiers in plant science, 11*, 279.

Fabrowska, J., Łęska, B., Schroeder, G., Messyasz, B., & Pikosz, M. (2015). Biomass and extracts of algae as material for cosmetics. *Marine Algae Extracts: Processes, Products, and Applications*, 681–706.

Finkel, T., & Holbrook, N. J. (2000). Oxidants, oxidative stress and the biology of ageing. *nature, 408*(6809), 239–247.

Furbeyre, H., Van Milgen, J., Mener, T., Gloaguen, M., & Labussière, E. (2017). Effects of dietary supplementation with freshwater microalgae on growth performance, nutrient digestibility and gut health in weaned piglets. *Animal, 11*(2), 183–192.

Gateau, H., Solymosi, K., Marchand, J., & Schoefs, B. (2017). Carotenoids of microalgae used in food industry and medicine. *Mini reviews in medicinal chemistry, 17*(13), 1140–1172.

Gheda, S., Hamouda, R. A., Naby, M. A., Mohamed, T. M., Al-Shaikh, T. M., & Khamis, A. (2023). Potent effect of phlorotannins derived from Sargassum linifolium as antioxidant and antidiabetic in a streptozotocin-induced diabetic rats model. *Applied Sciences, 13*(8), 4711.

Ibañez, E., & Cifuentes, A. (2013). Benefits of using algae as natural sources of functional ingredients. *Journal of the Science of Food and Agriculture, 93*(4), 703–709.

Imchen, T., & Singh, K. S. (2023). Marine algae colorants: Antioxidant, anti-diabetic properties and applications in food industry. *Algal Research, 69*, 102898.

Jyoti, A., Nehra, M., & Khan, M. (2022). Algae as a nutritional and functional food source. *Madridge J Food Technol, 7*(1), 189–199.

Kapoor, D. U., Kukkar, M. R., Gaur, M., Prajapati, B. G., Suttiruengwong, S., & Sriamornsak, P. (2024). Algae as Third-Generation Materials: Exploring the Emerging Role in Pharmaceutical Applications. *Materials Today Sustainability*, 100935.

Karuppusamy, S., Rajauria, G., Fitzpatrick, S., Lyons, H., McMahon, H., Curtin, J., Tiwari, B. K., & O'Donnell, C. (2022). Biological properties and health-promoting functions of laminarin: A comprehensive review of preclinical and clinical studies. *Marine Drugs, 20*(12), 772.

Kaushik, A., Sangtani, R., Parmar, H. S., & Bala, K. (2023). Algal metabolites: Paving the way towards new generation antidiabetic therapeutics. *Algal Research, 69*, 102904.

Khan, M. I., Shin, J. H., & Kim, J. D. (2018). The promising future of microalgae: current status, challenges, and optimization of a sustainable and renewable industry for biofuels, feed, and other products. *Microbial Cell Factories, 17*, 1–21.

Kim, M. S., Kim, J. Y., Choi, W. H., & Lee, S. S. (2008). Effects of seaweed supplementation on blood glucose concentration, lipid profile, and antioxidant enzyme activities in patients with type 2 diabetes mellitus. *Nutrition research and practice, 2*(2), 62–67.

Kumar, J., Singh, D., Tyagi, M. B., & Kumar, A. (2019). Cyanobacteria: applications in biotechnology. In *Cyanobacteria* (pp. 327–346). Elsevier.

Lin, G., Wan, X., Liu, D., Wen, Y., Yang, C., & Zhao, C. (2021). COL1A1 as a potential new biomarker and therapeutic target for type 2 diabetes. *Pharmacological research, 165*, 105436.

Lorenzati, B., Zucco, C., Miglietta, S., Lamberti, F., & Bruno, G. (2010). Oral Hypoglycemic Drugs: Pathophysiological Basis of Their Mechanism of ActionOral Hypoglycemic Drugs: Pathophysiological Basis of Their Mechanism of Action. *Pharmaceuticals, 3*(9), 3005–3020. https://www.mdpi.com/1424-8247/3/9/3005

Lourenço-Lopes, C., Fraga-Corral, M., Jimenez-Lopez, C., Pereira, A. G., Garcia-Oliveira, P., Carpena, M., Prieto, M. A., & Simal-Gandara, J. (2020). Metabolites from macroalgae and its applications in the cosmetic industry: A circular economy approach. *Resources, 9*(9), 101.

Lu, L. W., & Chen, J.-H. (2022). Seaweeds as ingredients to lower glycemic potency of cereal foods synergistically—A perspective. *Foods, 11*(5), 714.

Mabate, B., Daub, C. D., Malgas, S., Edkins, A. L., & Pletschke, B. I. (2021). A Combination Approach in Inhibiting Type 2 Diabetes-Related Enzymes Using Ecklonia radiata Fucoidan and Acarbose. *Pharmaceutics, 13*(11), 1979.

Matos, J., Cardoso, C., Bandarra, N. M., & Afonso, C. (2017). Microalgae as healthy ingredients for functional food: a review. *Food & function, 8*(8), 2672–2685.

Mellor, C., Embling, R., Neilson, L., Randall, T., Wakeham, C., Lee, M. D., & Wilkinson, L. L. (2022). Consumer Knowledge and Acceptance of "Algae" as a Protein Alternative: A UK-Based Qualitative Study. *Foods, 11*(12), 1703. https://www.mdpi.com/2304-8158/11/12/1703

Menaa, F., Wijesinghe, U., Thiripuranathar, G., Althobaiti, N. A., Albalawi, A. E., Khan, B. A., & Menaa, B. (2021). Marine algae-derived bioactive compounds: a new wave of nanodrugs? *Marine Drugs, 19*(9), 484.

Mohibbullah, M., Haque, M. N., Sohag, A. A. M., Hossain, M. T., Zahan, M. S., Uddin, M. J., Hannan, M. A., Moon, I. S., & Choi, J.-S. (2022). A systematic review on marine algae-derived fucoxanthin: An update of pharmacological insights. *Marine Drugs, 20*(5), 279.

Nacer, W., Baba Ahmed, F. Z., Merzouk, H., Benyagoub, O., & Bouanane, S. (2020). Evaluation of the anti-inflammatory and antioxidant effects of the microalgae Nannochloropsis gaditana in streptozotocin-induced diabetic rats. *Journal of Diabetes & Metabolic Disorders, 19*, 1483–1490.

Naik, B., Mishra, R., Kumar, V., Mishra, S., Gupta, U., Rustagi, S., Gupta, A. K., Preet, M. S., Bhatt, S. C., & Rizwanuddin, S. (2024). Micro-algae: Revolutionizing food production for a healthy and sustainable future. *Journal of Agriculture and Food Research, 15*, 100939.

Patel, A. K., Singhania, R. R., Awasthi, M. K., Varjani, S., Bhatia, S. K., Tsai, M.-L., Hsieh, S.-L., Chen, C.-W., & Dong, C.-D. (2021). Emerging prospects of macro-and microalgae as prebiotic. *Microbial Cell Factories, 20*(1), 112.

Pereira, L. (2016). *Carrageenans: sources and extraction methods, molecular structure, bioactive properties and health effects*. Nova Science Publishers, Incorporated.

Pereira, L., & Valado, A. (2023). Algae-Derived Natural Products in Diabetes and Its Complications—Current Advances and Future Prospects. *Life, 13*(9), 1831.

Plaza, M., Cifuentes, A., & Ibáñez, E. (2008). In the search of new functional food ingredients from algae. *Trends in Food Science & Technology, 19*(1), 31–39.

Popović-Djordjević, J. B., Katanić Stanković, J. S., Mihailović, V., Pereira, A. G., Garcia-Oliveira, P., Prieto, M. A., & Simal-Gandara, J. (2021). Algae as a source of bioactive compounds to prevent the development of type 2 diabetes mellitus. *Current Medicinal Chemistry, 28*(23), 4592–4615.

Qin, Y., Jiang, J., Zhao, L., Zhang, J., & Wang, F. (2018). Applications of alginate as a functional food ingredient. In *Biopolymers for food design* (pp. 409–429). Elsevier.

Rayapu, L., Chakraborty, K., & Valluru, L. (2021). Marine algae as a potential source for antidiabetic compounds-A brief review. *Current Pharmaceutical Design, 27*(6), 789–801.

Saeedi, P., Petersohn, I., Salpea, P., Malanda, B., Karuranga, S., Unwin, N., Colagiuri, S., Guariguata, L., Motala, A. A., & Ogurtsova, K. (2019). Global and regional diabetes prevalence estimates for 2019 and projections for 2030 and 2045: Results from the International Diabetes Federation Diabetes Atlas. *Diabetes research and clinical practice, 157*, 107843.

Sathasivam, R., Radhakrishnan, R., Hashem, A., & Abd_Allah, E. F. (2019). Microalgae metabolites: A rich source for food and medicine. *Saudi journal of biological sciences, 26*(4), 709–722.

Singla, A., Gupta, O. P., Sagwal, V., Kumar, A., Patwa, N., Mohan, N., Ankush, Kumar, D., Vir, O., & Singh, J. (2024). Beta-Glucan as a Soluble Dietary Fiber Source: Origins, Biosynthesis, Extraction, Purification, Structural Characteristics, Bioavailability, Biofunctional Attributes, Industrial Utilization, and Global Trade. *Nutrients, 16*(6), 900.

Sinha, S., Haque, M., Lugova, H., & Kumar, S. (2023). The effect of omega-3 fatty acids on insulin resistance. *Life, 13*(6), 1322.

Spiller, H. A., & Sawyer, T. S. (2006). Toxicology of oral antidiabetic medications. *American Journal of Health-System Pharmacy, 63*(10), 929–938.

Subermaniam, K., Teoh, S. L., Yow, Y.-Y., Tang, Y. Q., Lim, L. W., & Wong, K. H. (2021). Marine algae as emerging therapeutic alternatives for depression: A review. *Iranian Journal of Basic Medical Sciences, 24*(8), 997.

Tavares, J. O., Cotas, J., Valado, A., & Pereira, L. (2023). Algae food products as a healthcare solution. *Marine Drugs, 21*(11), 578.

Thiyagarasaiyar, K., Goh, B.-H., Jeon, Y.-J., & Yow, Y.-Y. (2020). Algae Metabolites in Cosmeceutical: An Overview of Current Applications and Challenges. *Marine Drugs, 18*(6), 323. https://www.mdpi.com/1660-3397/18/6/323

Tiwari, A. K., & Rao, J. M. (2002). Diabetes mellitus and multiple therapeutic approaches of phytochemicals: Present status and future prospects. *Current science*, 30–38.

Udayan, A., Pandey, A. K., Sharma, P., Sreekumar, N., & Kumar, S. (2021). Emerging industrial applications of microalgae: challenges and future perspectives. *Systems Microbiology and Biomanufacturing*, *1*(4), 411–431.

Unnikrishnan, P. S., & Jayasri, M. A. (2018). Marine algae as a prospective source for antidiabetic compounds–a brief review. *Current Diabetes Reviews*, *14*(3), 237–245.

Wan, X., Li, T., Liu, D., Chen, Y., Liu, Y., Liu, B., Zhang, H., & Zhao, C. (2018). Effect of marine microalga Chlorella pyrenoidosa ethanol extract on lipid metabolism and gut microbiota composition in high-fat diet-fed rats. *Marine Drugs*, *16*(12), 498.

Wells, M. L., Potin, P., Craigie, J. S., Raven, J. A., Merchant, S. S., Helliwell, K. E., Smith, A. G., Camire, M. E., & Brawley, S. H. (2017). Algae as nutritional and functional food sources: revisiting our understanding. *Journal of applied phycology*, *29*, 949–982.

Wen, Y., Gao, L., Zhou, H., Ai, C., Huang, X., Wang, M., Zhang, Y., & Zhao, C. (2021). Opportunities and challenges of algal fucoidan for diabetes management. *Trends in Food Science & Technology*, *111*, 628–641.

Yuen, L., Saeedi, P., Riaz, M., Karuranga, S., Divakar, H., Levitt, N., Yang, X., & Simmons, D. (2019). Projections of the prevalence of hyperglycaemia in pregnancy in 2019 and beyond: Results from the International Diabetes Federation Diabetes Atlas. *Diabetes research and clinical practice*, *157*, 107841.

Zahkouk, S. A., Eid, F. A., Abdelhafez, H. M., & El-Wahsh, A. M. (2019). Effect of Aphanizomenon flos-aquae (AFA) or Orlistat on hyperlipidemic adult male Albino rats. *Journal of Medical and Life Science*, *1*(4), 8–103.

Zang, L., Baharlooeian, M., Terasawa, M., Shimada, Y., & Nishimura, N. (2023). Beneficial effects of seaweed-derived components on metabolic syndrome via gut microbiota modulation. *Frontiers in Nutrition*, *10*, 1173225.

Zhao, C., Yang, C., Liu, B., Lin, L., Sarker, S. D., Nahar, L., Yu, H., Cao, H., & Xiao, J. (2018). Bioactive compounds from marine macroalgae and their hypoglycemic benefits. *Trends in Food Science & Technology*, *72*, 1–12.

Zheng, J., Inoguchi, T., Sasaki, S., Maeda, Y., McCarty, M. F., Fujii, M., Ikeda, N., Kobayashi, K., Sonoda, N., & Takayanagi, R. (2013). Phycocyanin and phycocyanobilin from Spirulina platensis protect against diabetic nephropathy by inhibiting oxidative stress. *American Journal of Physiology-Regulatory, Integrative and Comparative Physiology*, *304*(2), R110-R120.

Ziyaei, K., Abdi, F., Mokhtari, M., Daneshmehr, M. A., & Ataie, Z. (2023). Phycocyanin as a nature-inspired antidiabetic agent: a systematic review. *Phytomedicine*, 154964.

GPSR Compliance

The European Union's (EU) General Product Safety Regulation (GPSR) is a set of rules that requires consumer products to be safe and our obligations to ensure this.

If you have any concerns about our products, you can contact us on ProductSafety@springernature.com

In case Publisher is established outside the EU, the EU authorized representative is:

Springer Nature Customer Service Center GmbH
Europaplatz 3
69115 Heidelberg, Germany

Batch number: 10755329

Printed by Printforce, the Netherlands